c8/83 8·60

CW01662363

Pure mathematics for advanced lev

Butterworths Publications on Related Subjects

Applied mathematics for advanced level
H. Mulholland and J. H. G. Phillips

Fundamentals of statistics
H. Mulholland and C. R. Jones

Also by H. Mulholland

Calculus made simple
(W. H. Allen, 1976)

Guide to Eire's 3000-foot mountains
(Mulholland Wirral, 1980)

Guide to Wales' 3000-foot mountains
(Mulholland Wirral, 1982)

Guide to Lakeland's 3000-foot mountains
(Mulholland Wirral, 1983)

Pure mathematics for advanced level

Second edition

B. D. Bunday, BSc, PhD

Bradford University

H. Mulholland, MSc

formerly at Liverpool Polytechnic

BUTTERWORTHS

London Boston Durban Singapore Sydney Toronto Wellington

First published 1967
Reprinted (with revisions) 1970
Reprinted 1972
Reprinted 1973
Reprinted 1975
Reprinted 1976
Reprinted 1977
Reprinted 1978
Reprinted 1979
Reprinted 1980
Second edition 1983

British Library Cataloguing in Publication Data

Bunday, B. D.
Pure mathematics for advanced level.
1. Mathematics—1961–
I. Title II. Mulholland, H.
510 QA39.2

ISBN 0–408–70958–8

Filmset in Monophoto Times by
Northumberland Press Ltd, Gateshead
Printed and Bound in Great Britain by
Robert Hartnoll Ltd,
Bodmin, Cornwall

Preface to the second edition

Since the first edition was published there have been changes in the content of the pure mathematics syllabuses of the various examining boards. The most important was the recent agreement by an inter-board committee to draft a national 'common core' in mathematics. While the final form has not yet (1982) been approved formally by all boards, it is intended that it should eventually be featured in the syllabuses of all GCE boards who will however be free to add extra topics as they wish. This common core will give a basis for comparison between the various syllabuses but it is *not* intended that the boards will set common questions. At the same time, the Joint Matriculation Board will have (from 1984) rationalised their mathematical syllabuses into: Pure Mathematics I and II, Mechanics I and II, Applied Mathematics I and II and Statistics. Candidates can offer selected combinations of two of the papers as a single mathematical subject. The Associated Examining Board already have, effectively, a pure mathematics paper I which is common to their subjects Mathematics: 636 (Pure and Applied), 646 (Pure with Statistics), 632 (Pure) and 647 (Pure with Computations).

This edition covers the contents of the common core and virtually all of the extra pure mathematics included by each of the various boards in their syllabuses which are equivalent to that of the JMB's Pure Mathematics I.

Two new chapters have been added: Numerical Methods and Vectors. The remainder and factor theorems, graphical solution of $y = a\cos\theta + b\sin\theta$ and angles between lines and planes have been introduced in the appropriate places, the work on polar co-ordinates has been extended and the introduction to the calculus has been rewritten. To allow space for these the sections on co-ordinate geometry have been revised and condensed and one or two minor topics, such as linear equations in one unknown, have been deleted.

In addition, the general lay-out has been improved and many of the original questions from the examination papers of the Joint Matriculation Board and London University have been replaced with recent ones. Also questions set by the Oxford, Associated Examining and Cambridge Boards have now been included. A large number of worked examples remains a feature of the book and there are plenty of exercises for the student to develop and test his or her skills. A candidate studying on his or her own is recommended to obtain a copy of the appropriate syllabus and copies of immediate past examination papers to check that all topics are covered.

Finally, we are grateful to the Joint Matriculation Board (JMB), the London

University Entrance and Schools Examinations Council (LU), the Associated Examining Board (AEB), the Oxford Delegacy of Local Examinations (O) and the Cambridge Local Examinations Syndicate (C) for granting us permission to use questions from their more recent examinations. The abbreviations shown have been used to indicate the source of such questions.

BDB
HM

Preface to the first edition

This is a textbook of pure mathematics written to meet the needs of the student studying for the General Certificate of Education at Advanced Level. The book assumes a knowledge of mathematics up to Ordinary Level and covers all the pure mathematics necessary for the Advanced Level examination in mathematics (A26), of the Northern Universities Joint Matriculation Board, together with the great majority of the work required for the Advanced Level examinations of the Southern Universities Joint Board, the Welsh Joint Committee and London University.

The teaching method adopted is for the most part that suggested by the various reports of the Mathematical Association. The emphasis throughout has been on technique, although we have tried to indicate where a particular result needs more rigorous justification than is given in this book. In this way, we hope that all students can progress quickly in the understanding and application of these techniques without the hindrance of having to justify everything they do. This latter step comes at a later stage in their mathematical development.

For convenience, the book has been prepared in the order algebra (Chapters 1–5), trigonometry (Chapters 6–8), calculus (Chapters 9–16), and co-ordinate geometry (Chapters 17–20), but this is not to imply that the chapters should be read in this order. For the student at school, this will be decided by the teacher; for the student working alone, we would recommend an advance on a broad front through Chapters 1, 3, 6, 9, 10, 12 (the first two sections), 13, 17, 18. This lays the foundations for all the main topics and this broad advance can then be maintained. We would suggest that each of Chapters 7, 11, 14, 15, 16 and 20 be read in at least two stages. Not only will this make for easier digestion of the many ideas and techniques discussed in these chapters, but will also provide for constant revision and extension of this material.

The book includes over 350 worked examples and about 1800 examples for the student to solve. The worked examples indicate the main applications of the ideas and techniques discussed. The exercises set at the ends of the sections within the chapters are for the most part fairly straightforward. All our readers should attempt these exercises. The exercises at the ends of the chapters are a 'mixed bag'. Some are of a routine type, others are more testing; many are from past papers set by the various examining boards. Finally, there are some (indicated with an asterisk) which are of a more difficult nature. The student should not be too dismayed if he is unable to solve all of these.

For convenience, the results and formulae obtained have been labelled, the first number of the label identifying the chapter in which the result is derived. Thus formula 10.7 is the seventh result obtained in Chapter 10. It is not suggested that all these formulae be memorised.

We should like to express our thanks to the Joint Matriculation Board (JMB), the Southern Universities Joint Board (SUJB), the Welsh Joint Committee (WJC) and London University (LU) for granting us permission to use questions from their examinations in this book. The abbreviations above have been used to indicate the source of such questions.

Finally, we should like to thank our publishers for the care and trouble they have taken over the general presentation of the text.

BDB
HM

Contents

1
Operations with real numbers

1.1 The real numbers

Algebra is concerned with operations with numbers and we shall begin with a brief review of these operations and the numbers involved.

The first set of numbers usually encountered is the set of positive integers including zero: 0, 1, 2, 3,.... These by themselves are insufficient for the solution of many actual problems and need to be supplemented by fractions which can all be expressed in the form a/b, where a and b are positive integers (b non-zero). This set of numbers includes the positive integers which arise when $b = 1$.

The solution of a particular problem might require the solution of the equation $x + a = b$. If a is greater than b, in order to interpret the result $x = b - a$, we need to extend our number system to include negative numbers. The integers are then the set

$$\ldots -3,\ -2,\ -1,\ 0,\ 1,\ 2,\ 3,\ 4,\ldots$$

and the rational numbers (fractions), which include the integers, are of the form a/b, where a and b are integers (b non-zero).

There are still quantities which cannot be expressed in terms of the rational numbers. For example, the length of a diagonal of a square of side 1 unit is $\sqrt{2}$ units, and $\sqrt{2}$ cannot be expressed in the form a/b, where a and b are integers. Tables of square roots show $\sqrt{2} \approx 1{\cdot}414 = \frac{1414}{1000}$ but this is only an approximation to the value of $\sqrt{2}$, as the squaring of 1·414 will soon show. This property is not unique to $\sqrt{2}$; $\sqrt{3}$, $\sqrt{5}$, $\sqrt[3]{1{\cdot}6}$, $\sqrt[5]{11{\cdot}61}$, etc. all have the same property. These numbers are examples of algebraic numbers. They are all of them solutions of algebraic equations which involve only rational numbers. $\sqrt{3}$ is a solution of $x^2 = 3$, $\sqrt[3]{1{\cdot}6}$ is a solution of $x^3 = 1{\cdot}6$, $\sqrt[5]{11{\cdot}61}$ is a solution of $x^5 = 11{\cdot}61$, etc.

There are still other numbers which do not fall into any of the categories mentioned so far. Such numbers, of which $\pi(\approx 3{\cdot}142)$, $\log_{10} 2(\approx 0{\cdot}301)$, $\sin 74°(\approx 0{\cdot}9613)$ are but three examples, are called transcendental numbers. Our system of real numbers with which we shall be mainly concerned, will consist of the rational, algebraic and transcendental (irrational) numbers.

It is often convenient to represent these numbers by points on a line (*Figure 1.1*), letting 0 be an origin on the line $x'x$. Conventionally, we let points to the right of 0 represent positive numbers and points to the left of

0 represent negative numbers. Points on the line distant 1 unit, 2 units, ... to the right of 0 will represent the numbers 1, 2, 3,.... Points on the line distant 1 unit, 2 units ... to the left of 0 will represent the numbers -1, -2, -3,.... The rational numbers will be represented by intermediate points.

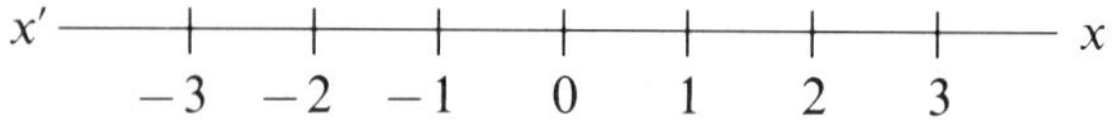

Figure 1.1

The set of integers (positive, negative and zero) is often referred to collectively by the symbol $\mathbb{Z}$, while the set of all real numbers is referred to by the symbol $\mathbb{R}$.

The fundamental operations of algebra are addition and multiplication. Subtraction can be regarded as the addition of the corresponding negative number, and division as multiplication by the reciprocal. We are all familiar with these operations, although it is perhaps worth reminding ourselves of the fundamental laws governing these operations.

If a, b, c are any three real numbers:

I $a + b = b + a$, the commutative law of addition
II $(a + b) + c = a + (b + c)$, the associative law of addition
III $ab = ba$, the commutative law of multiplication
IV $(ab)c = a(bc)$, the associative law of multiplication
V $a(b + c) = ab + ac$, the distributive law of multiplication and addition.

1.2 Equations

Readers will already be familiar with the solution of simple equations and quadratic equations involving one unknown, as well as with pairs of simple simultaneous equations.

For the equation $ax + b = 0$, where a and b are real numbers

$$x = -\frac{b}{a} \tag{1.1}$$

For the equation $ax^2 + bx + c = 0$

$$x = \frac{-b \pm \sqrt{(b^2 - 4ac)}}{2a} \tag{1.2}$$

In more complicated situations, one has to use one's native wit. The following examples illustrate some useful techniques.

Example 1 Solve the equation $x^2 + 2x - 4 + \dfrac{3}{x^2 + 2x} = 0$.

With $z = x^2 + 2x$

$$z - 4 + \frac{3}{z} = 0$$

$$z^2 - 4z + 3 = 0$$

$$(z - 3)(z - 1) = 0$$

$$z = 1 \quad \text{or} \quad 3$$

With $z = 3$

$$x^2 + 2x = 3$$

$$x^2 + 2x - 3 = 0$$

$$(x+3)(x-1) = 0$$

$$x = -3 \quad \text{or} \quad x = 1$$

With $z = 1$

$$x^2 + 2x - 1 = 0$$

$$x = \frac{-2 \pm \sqrt{[4 - 4 \times (1) \times (-1)]}}{2}$$

$$= \frac{-2 \pm \sqrt{8}}{2} = \frac{-2 \pm 2\sqrt{2}}{2}$$

$$= -1 \pm \sqrt{2}$$

Therefore, the solutions are 1, -3, $-1 + \sqrt{2}$, $-1 - \sqrt{2}$.

Example 2 Solve the equation

$$\sqrt{(4-x)} - \sqrt{(6+x)} = \sqrt{(14+2x)}.$$

Squaring both sides, we have

$$4 - x + 6 + x - 2\sqrt{[(4-x)(6+x)]} = 14 + 2x$$

$$-2\sqrt{[(4-x)(6+x)]} = 4 + 2x$$

$$-\sqrt{[(4-x)(6+x)]} = 2 + x$$

Again squaring both sides, we now have

$$(4-x)(6+x) = 4 + 4x + x^2$$

$$24 - 2x - x^2 = 4 + 4x + x^2$$

$$2x^2 + 6x - 20 = 0$$

$$2(x+5)(x-2) = 0$$

$$x = 2 \quad \text{or} \quad x = -5$$

It is easy to see that it is only the value $x = -5$ which satisfies the original equation. $x = 2$ is a solution of the equation $\sqrt{(4-x)} + \sqrt{(6+x)} = \sqrt{(14+2x)}$. If we square both sides of this equation, we obtain $\sqrt{[(4-x)(6+x)]} = 2 + x$, which in turn leads to $2x^2 + 6x - 20 = 0$. The original equation gave $-\sqrt{[(4-x)(6+x)]} = 2 + x$ but when we square, the distinction between the two cases is lost. Thus we must always verify the correctness of our solutions after we have carried out such operations. As a trivial example, consider the equation $2x = 2$ which has solution $x = 1$. If we square both sides, we obtain the equation $4x^2 = 4$, i.e. $x^2 = 1$ which has solutions $x = 1$ or $x = -1$!

We shall assume that readers are familiar with the procedure for the solution of a pair of linear simultaneous equations in two unknowns. The solution of two equations in two unknowns when one or both of the equations contain quadratic terms is a more interesting problem. We first consider two cases where a systematic method of solution exists.

Example 3 Solve the equations $x + y = 3$, $x^2 + xy + 2y^2 + x + 2y = 12$, in which one equation is linear and the other quadratic.

We use the linear equation to express one unknown in terms of the other. Thus we have

$$x = 3 - y$$

We now substitute this expression for x into the second equation to obtain a quadratic equation for y. Thus

$$\begin{aligned}
(3 - y)^2 + (3 - y)y + 2y^2 + (3 - y) + 2y &= 12 \\
9 - 6y + y^2 + 3y - y^2 + 2y^2 + 3 - y + 2y &= 12 \\
2y^2 - 2y &= 0 \\
y(y - 1) &= 0
\end{aligned}$$

$$y = 0 \quad \text{or} \quad y = 1$$

When $y = 0$, $x = 3$; when $y = 1$, $x = 2$ (since $x = 3 - y$). Thus the solutions are $x = 2$, $y = 1$; $x = 3$, $y = 0$.

We could, of course, have used $y = 3 - x$ and obtained an equation for x on substituting this into the second equation.

Example 4 Solve the equations $x^2 - y^2 = 3$, $2x^2 + xy - 2y^2 = 4$, in which the terms involving the unknowns are all quadratic in both equations.

The solution can generally be obtained by writing $y = mx$ and proceeding as follows.

The equations can be written

$$\begin{aligned}
x^2(1 - m^2) &= 3 \\
x^2(2 + m - 2m^2) &= 4
\end{aligned}$$

By division

$$\begin{aligned}
\frac{1 - m^2}{2 + m - 2m^2} &= \frac{3}{4} \\
4 - 4m^2 &= 6 + 3m - 6m^2 \\
2m^2 - 3m - 2 &= 0 \\
(2m + 1)(m - 2) &= 0
\end{aligned}$$

$$m = 2 \quad \text{or} \quad m = -\tfrac{1}{2}$$

With $m = -\frac{1}{2}$, we have $\frac{3}{4}x^2 = 3$. Therefore

$$x^2 = 4 \quad \text{i.e.} \quad x = \pm 2$$

The corresponding values for y are ∓ 1 (since $y = mx$). With $m = 2$, we have $x^2(-3) = 3$; therefore, $x^2 = -1$, and this equation has no solution in the domain of real numbers. Thus, the solutions are $x = 2$, $y = -1$; $x = -2$, $y = 1$.

It is not usually possible to give general procedures for the solution of simultaneous equations which do not fall within the categories just mentioned.

Rather, each problem must be considered on its merits and the solver must use his or her own ingenuity.

Example 5 Solve the equations $x + \dfrac{1}{y} = 1$ and $y + \dfrac{1}{x} = 4$.

The equations can be rewritten in the form

$$xy + 1 = y$$
$$xy + 1 = 4x$$

Therefore, $y = 4x$, which on substitution gives

$$4x^2 + 1 = 4x$$
$$4x^2 - 4x + 1 = 0$$
$$(2x - 1)^2 = 0$$

Therefore, $x = \frac{1}{2}$ and, since $y = 4x$, $y = 2$. The solution is thus $x = \frac{1}{2}$, $y = 2$.

Exercises 1a

1 Solve the equation $\dfrac{x+1}{3} - \dfrac{x-2}{4} = \dfrac{2x+3}{6}$.
2 Solve the equation $x^2 - 5x - 11 = 0$.
3 Solve the equation $\dfrac{x+1}{2x+3} = \dfrac{5x-1}{7x+3}$.
4 Solve the equation $\sqrt{x} - \dfrac{6}{\sqrt{x}} = 1$.
5 Solve the equation $y^2 + 5y - \dfrac{36}{y^2 + 5y} = 0$.
6 Solve the equation $x^4 - 25x^2 + 144 = 0$.
7 Find the values of x which satisfy the equation $2\sqrt{(x+5)} - \sqrt{(2x+8)} = 2$.
8 Solve the equation $\sqrt{(x+1)} + \sqrt{(5x+1)} = 2\sqrt{(x+6)}$.
9 Solve the equation $x^4 - 2x^3 - 6x^2 - 2x + 1 = 0$.
$\left[\text{Hint: let } v = x + \dfrac{1}{x}\right]$
10 Solve the equation $y^4 - 2y^3 - 2y^2 + 2y + 1 = 0$.
$\left[\text{Hint: let } z = y - \dfrac{1}{y}\right]$

Solve the simultaneous equations 11–20:

11 $x + 2y = 3$, $x^2 - xy + 5y^2 + 2y = 7$
12 $2x + y = 1$, $x^2 + xy + 3x - y = 4$
13 $2x - 3y = 1$, $x^2 + xy - 4y^2 = 2$
14 $x^2 + 2xy = 3$, $3x^2 - y^2 = 26$
15 $x^2 + y^2 = 13$, $x^2 - 3xy + 2y^2 = 35$
16 $x^2 - xy + 7y^2 = 27$, $x^2 - y^2 = 15$
17 $x^2 + y^2 = 5$, $\dfrac{1}{x^2} + \dfrac{1}{y^2} = \dfrac{5}{4}$

18 $x^2 + y^2 = 10, \quad \dfrac{1}{x} + \dfrac{1}{y} = \dfrac{4}{3}$

19 $x^2 - y^2 = 24, \quad \dfrac{1}{x+y} + \dfrac{3}{x-y} = \dfrac{11}{12}$

20 $\dfrac{x}{y} + \dfrac{y}{x} = \dfrac{17}{4}, \quad x^2 - 4xy + y^2 = 1$

1.3 Elimination

In Section 1.2, we considered methods for the solution of two equations in two unknown quantities. If we have more equations than unknowns, two equations in one unknown, or three equations in two unknowns, then in order to obtain a consistent solution to the equations the coefficients must satisfy some relationship. This relationship is known as the eliminant of the system. It is obtained by forming from the given equations an equation which does not involve the unknowns. This process is known as the elimination of the unknowns. It is a technique which is of great value in co-ordinate geometry.

Example 1 Eliminate t from the equations $x = at^2$, $y = 2at$.

From the second equation, we can solve for t in terms of y, i.e. $t = y/2a$. Substitution into the first equation gives

$$x = a\left(\frac{y}{2a}\right)^2 = \frac{y^2a}{4a^2}$$

Therefore, $y^2 = 4ax$, which is the required result.

Example 2 . Eliminate t from the equations

$$x = \frac{t}{1+t^2} \quad \text{and} \quad y = \frac{t^2}{1+t^2}.$$

We have $y/x = t$. Substitution in the first equation gives

$$x = \frac{y/x}{1 + y^2/x^2} = \frac{y/x}{(x^2+y^2)/x^2}$$

$$x = \frac{xy}{x^2+y^2}$$

$$x(x^2+y^2) = xy$$

Example 3 Eliminate l and m from the equations $lx + my = a$, $mx - ly = b$, $l^2 + m^2 = 1$.

The straightforward procedure would be to solve the first two equations for l and m in terms of a, b, x and y. Substitution of these expressions into the last equation would then provide the eliminant. However, in some cases it is possible to use more subtle methods. In the present example, if we square the first two equations and add the results, we obtain

$$l^2x^2 + m^2x^2 + m^2y^2 + l^2y^2 = a^2 + b^2$$

$$(x^2+y^2)(l^2+m^2) = a^2 + b^2$$

and since $l^2 + m^2 = 1$

$$x^2 + y^2 = a^2 + b^2$$

which is the required eliminant.

Exercises 1b

1 Eliminate t from the equations $x = 1 + t$, $y = 1 + \frac{1}{t}$.

2 Eliminate t from the equations $x = 3 + t^3$, $y = 2 + \frac{1}{t}$.

3 Eliminate t from the equations $x = \frac{1}{t} - t$, $y = \frac{1}{t^2} - 1$.

4 If $x = \frac{2at}{1+t^2}$, $y = \frac{b(1-t^2)}{1+t^2}$, show that $\frac{x^2}{a^2} + \frac{y^2}{b^2} = 1$.

5 Eliminate θ from the equations $x - a\cos\theta = 0$, $y - b\sin\theta = 0$.

6 If $x = 1 + t^2$, $y = 2t$, show that $y^2 = 4(x-1)$.

7 If $x = 1 - t^2$ and $y = 1 + 5t - t^2$, show that $(x-y)^2 = 25(1-x)$.

8 Eliminate x and y from the equations $x - y = a$, $x^2 + y^2 = b^2$, $xy = 1$.

9 Eliminate x and y from the equations $x - y = a$, $x + y = b$, $xy = c$.

10 If $x + 2y^2 = a$, $x - 2y^2 = b$, $xy = 2$, show that $(a+b)(a^2 - b^2) = 64$.

1.4 Inequalities

In this section, we shall consider the rules governing the relationships between numbers which are not equal. For any two real numbers a and b, we say that a is greater than b ($a > b$) if $a - b$ is positive. We say that a is less than b ($a < b$) if $a - b$ is negative. In terms of the representations of numbers on a line (*Figure 1.1*), $a > b$ if a is to the right of b; $a < b$ if a is to the left of b. Thus we have by definition

$$a > b \quad \text{if} \quad a - b > 0 \quad \text{and} \quad a < b \quad \text{if} \quad a - b < 0 \tag{1.3}$$

For example, $5 > -3$ since $5 - (-3) = 8$ is positive, i.e. >0. Also $-3 < -1$ since $-3 - (-1) = -2$ is negative, i.e. <0.

RULE I We first show that if $a > b$, then

$$a + x > b + x \tag{1.4}$$

where x is any real number.

If $a > b$, $a - b = c > 0$

$$a + x - (b + x) = a + x - b - x = a - b = c$$
$$a + x - (b + x) = c > 0$$
$$a + x > b + x \quad \text{by definition}$$

In the same way, if $a < b$

$$a + x < b + x \tag{1.5}$$

Thus, as with equations, we may add the same number to both sides of an inequality and still preserve the inequality. For example,

$$5 > -2$$

and after adding 6 to both sides

$$11 > 4$$

Also

$$6 < 9$$

and after adding -3 (i.e. subtracting 3) on both sides

$$3 < 6$$

We cannot, however, treat inequalities in the same way as equations if we multiply both sides of the inequality by the same number. Rather, we have

RULE II If $a > b$ $ax > bx$ if x is *positive*
$ax < bx$ if x is *negative* (1.6a)
If $a < b$ $ax < bx$ if x is *positive*
$ax > bx$ if x is *negative* (1.6b)

We shall prove this for the case $a > b$.

If, $a - b = c$, where c is positive, then

$$ax - bx = cx$$

which is positive if x is positive, but negative if x is negative. Therefore,

$$ax - bx > 0 \quad \text{if} \quad x > 0$$
$$ax - bx < 0 \quad \text{if} \quad x < 0$$

which is the required result.

Thus if we multiply both sides of an inequality by a negative number, the inequality sign must be reversed.

For example, $7 > 3$, which becomes $21 > 9$ after multiplication by 3 but $-14 < -6$ after multiplication by -2.

RULE III If $a > b$ and $c > d$, then

$$a + c > b + d \tag{1.7}$$

For example, $7 > 3$ and $-4 > -7$, and $3 > -4$.

Note: it does *not* follow that $a - c > b - d$. For example, $11 > 10$ and $9 > 2$ but $11 - 9 < 10 - 2$.

RULE IV If $a > b$ and $b > c$, then

$$a > c \tag{1.8}$$

For example, $8 > 7$ and $7 > 2$, and $8 > 2$.

Note: if $a > b$ and $b < c$, we can say nothing about the relative magnitudes of a and c. For example, $9 > 2$ and $2 < 8$ and of course $9 > 8$, but we could equally well have had $9 > 2$ and $2 < 11$ with, of course, $9 < 11$.

RULE V If $a > b$ and $c > d$ and a, b, c, d *are all positive*, then

$$ac > bd \quad \text{and} \quad \frac{a}{d} > \frac{b}{c} \tag{1.9}$$

For example, $9 > 2$ and $6 > 3$ and of course $9 \times 6 > 2 \times 3$, i.e. $54 > 6$.

RULE VI *If $a > b$ and a and b are both positive*, it follows that

$$a^2 > b^2, \quad a^3 > b^3, \quad \ldots \frac{1}{a} < \frac{1}{b}, \quad \frac{1}{a^2} < \frac{1}{b^2} \ldots$$

Indeed

$$\begin{aligned} a^n > b^n \quad &\text{if} \quad n > 0 \\ a^n < b^n \quad &\text{if} \quad n < 0 \end{aligned} \tag{1.10}$$

For example, $3 > 2$ and $3^3 > 2^3$, i.e. $27 > 8$, but $3^{-2} < 2^{-2}$, i.e. $\frac{1}{9} < \frac{1}{4}$.

Example 1 For what values of x are both the inequalities $9 + 2x > 0$ and $7 - 3x > 0$ true?

If $9 + 2x > 0$, $2x > -9$, i.e. $x > -\frac{9}{2}$.

If $7 - 3x > 0$, $-3x > -7$, i.e. $x < \frac{7}{3}$. (Note the reversal of the sign.)

From *Figure 1.2* we see at once that both inequalities are true for $-\frac{9}{2} < x < \frac{7}{3}$.

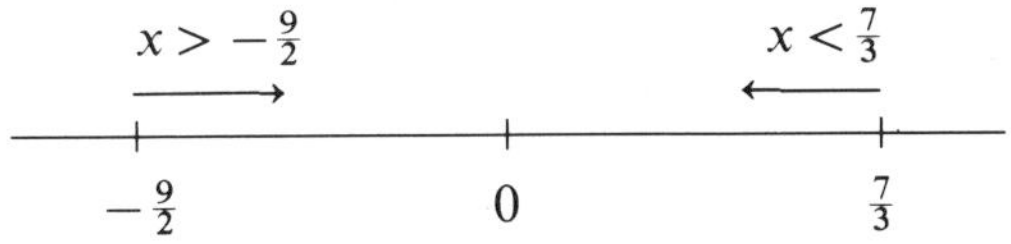

Figure 1.2

Example 2 Find the range of values of x for which $\dfrac{2x+1}{x+2} > \dfrac{1}{2}$.

We multiply both sides of the inequality by $(x + 2)^2$ which is positive. Thus we can be sure that the inequality sign is preserved correctly. Thus we have

$$\begin{aligned} (2x+1)(x+2) &> \tfrac{1}{2}(x+2)^2 \\ 2(2x+1)(x+2) &> (x+2)^2 \\ (x+2)(4x+2) - (x+2)^2 &> 0 \\ (x+2)(3x) &> 0 \end{aligned}$$

This will be true if $x > 0$ *and* $x + 2 > 0$, or if $x < 0$ *and* $x + 2 < 0$, i.e. if both factors are positive or both factors are negative.

The first two inequalities are true if $x > 0$ and the latter two inequalities are true if $x < -2$. This can be clearly seen if the following table showing the signs of the factors is drawn up. The individual factors change sign at 0 and -2.

	$x < -2$	$-2 < x < 0$	$x > 0$
$3x$	$-$ve	$-$ve	$+$ve
$x + 2$	$-$ve	$+$ve	$+$ve
$3x(x + 2)$	$+$ve	$-$ve	$+$ve

Thus the original inequality is true if $x > 0$ *or* $x < -2$.

Example 3 Determine the range of values of x for which

$$\frac{x^2 + x - 2}{x^2 + 4} > \frac{1}{2}$$

We notice that $x^2 + 4$, being the sum of two squares, is always positive. Thus we can multiply both sides of the inequality by $x^2 + 4$ and still preserve the sign. Thus

$$\begin{aligned} x^2 + x - 2 &> \tfrac{1}{2}(x^2 + 4) \\ 2x^2 + 2x - 4 &> x^2 + 4 \\ x^2 + 2x - 8 &> 0 \\ (x + 4)(x - 2) &> 0 \end{aligned}$$

We draw up our table showing the signs of the individual factors. The individual factors change sign at -4 and 2.

	$x < -4$	$-4 < x < 2$	$x > 2$
$x + 4$	$-$ve	$+$ve	$+$ve
$x - 2$	$-$ve	$-$ve	$+$ve
$(x + 4)(x - 2)$	$+$ve	$-$ve	$+$ve

Thus the inequality is true if $x < -4$ *or* $x > 2$.

Example 4 Solve the inequality $\dfrac{x + 3}{x - 2} > \dfrac{x + 1}{x - 3}$.

Here, we must multiply by the positive factor $(x - 2)^2(x - 3)^2$ to obtain

$$(x + 3)(x - 2)(x - 3)^2 > (x + 1)(x - 3)(x - 2)^2$$

which yields

$$\begin{aligned} (x - 3)(x - 2)[(x + 3)(x - 3) - (x + 1)(x - 2)] &> 0 \\ (x - 3)(x - 2)[x^2 - 9 - (x^2 - x - 2)] &> 0 \\ (x - 3)(x - 2)(x - 7) &> 0 \end{aligned}$$

Again, we draw up our table showing the signs of the individual factors, which change sign at 2, 3 and 7.

	$x < 2$	$2 < x < 3$	$3 < x < 7$	$x > 7$
$x - 2$	$-$ve	$+$ve	$+$ve	$+$ve
$x - 3$	$-$ve	$-$ve	$+$ve	$+$ve
$x - 7$	$-$ve	$-$ve	$-$ve	$+$ve
$(x - 2)(x - 3)(x - 7)$	$-$ve	$+$ve	$-$ve	$+$ve

Thus the original inequality is true if $2 < x < 3$ *or* $x > 7$.

We shall in later chapters have cause to use the notion of the modulus of a number x:

Modulus of x is the positive number having the same magnitude as x.

It is written $|x|$. Thus $|3| = 3$, $|-6| = 6$, $|-2| = 2$, $|-1| = 1$.

In general if x is positive $|x| = x$, but if x is negative $|x| = -x$.

With this notation the range of values of x specified by the inequality $-1 < x < 1$ can be specified more concisely by $|x| < 1$.

Example 5 Find x if $|x + 1| = 5$.

We have $x + 1 = 5$ or $x + 1 = -5$. Therefore

$$x = 4 \quad \text{or} \quad x = -6$$

Example 6 Find x if $|2x + 1| > 7$.

$$|2x + 1| > 7 \quad \text{means} \quad 2x + 1 > 7 \quad \textit{or} \quad 2x + 1 < -7$$

Thus, we have $2x > 6$ or $2x < -8$. Therefore

$$x > 3 \quad \text{or} \quad x < -4$$

Example 7 For what values of x is $|x - 1| > 2|x + 3|$?

Since $|x + 3|$ is positive, we can write the inequality as

$$\frac{|x-1|}{|x+3|} > 2 \quad \text{i.e.} \quad \left|\frac{x-1}{x+3}\right| > 2$$

This is true if

$$\frac{x-1}{x+3} > 2 \quad \text{or} \quad \frac{x-1}{x+3} < -2$$

In the first case, multiplication by the positive quantity $(x + 3)^2$ gives

$$(x - 1)(x + 3) > 2(x + 3)^2$$
$$(x + 3)(x - 1 - 2x - 6) > 0$$
$$(x + 3)(-x - 7) > 0$$

Multiplying by -1 gives

$$(x + 7)(x + 3) < 0$$

If we draw up a table as in Examples 2, 3 and 4, we find that $-7 < x < -3$.

In the second case, multiplication by the same positive quantity $(x + 3)^2$ gives

$$(x - 1)(x + 3) < -2(x + 3)^2$$
$$(x + 3)(x - 1 + 2x + 6) < 0$$
$$(x + 3)(3x + 5) < 0$$

which is satisfied if $-3 < x < -\frac{5}{3}$.

We can combine these two regions and observe that the original inequality is satisfied if $-7 < x < -\frac{5}{3}$ but *excluding* the value $x = -3$.

The inequality of the means

The arithmetic mean $\dfrac{a+b}{2}$ of two positive numbers a and b is greater than or equal to their geometric mean $\sqrt{ab}$. For we have, if a and b are positive,

$$(\sqrt{a}-\sqrt{b})^2 \geqslant 0$$
$$a+b-2\sqrt{ab} \geqslant 0$$
$$\frac{a+b}{2} \geqslant \sqrt{ab} \qquad (1.11)$$

which proves the result.

Example 8 If a, b, c, d are any real numbers, prove that (i) $a^4+b^4 \geqslant 2a^2b^2$ and (ii) $a^4+b^4+c^4+d^4 \geqslant 4abcd$.

(i) By (1.11), we have

$$\frac{a^4+b^4}{2} \geqslant \sqrt{a^4b^4} = a^2b^2$$
$$a^4+b^4 \geqslant 2a^2b^2$$

(ii) By the previous result, we have

$$a^4+b^4+c^4+d^4 \geqslant 2a^2b^2+2c^2d^2$$

But $2a^2b^2$ and $2c^2d^2$ are two positive numbers and so by (1.11)

$$\frac{2a^2b^2+2c^2d^2}{2} \geqslant \sqrt{(4a^2b^2c^2d^2)}$$
$$2a^2b^2+2c^2d^2 \geqslant 4abcd$$
$$a^4+b^4+c^4+d^4 \geqslant 4abcd \quad \text{by (1.8)}$$

Note: the result will certainly be true if some of a, b, c, d are negative, so that $abcd$ is negative, since the left-hand side is certainly positive.

Example 9 Show that if a, b, c are real numbers, $a^2+b^2+c^2-bc-ca-ab$ cannot be negative.

We have

$$a^2+b^2 \geqslant 2ab$$
$$b^2+c^2 \geqslant 2bc$$
$$c^2+a^2 \geqslant 2ac \quad \text{by (1.11)}$$

On adding these results, we obtain by (1.7)

$$2(a^2+b^2+c^2) \geqslant 2(ab+bc+ca)$$
$$a^2+b^2+c^2 \geqslant ab+bc+ca$$

which is the required result.

Exercises 1c

1 Solve the inequalities $3x+11>0$ and $8-7x>0$.

2 Find the values of x which satisfy $2x^2-7x+9<x^2-2x+3$.

3 For what values of x is $\dfrac{1}{x-3}<-1$?

4 For what values of x is $\dfrac{2x-1}{x+3}<\dfrac{2}{3}$?

5 Solve the inequality $\dfrac{x-1}{x-2} > \dfrac{x-2}{x-3}$.

6 Solve the inequality $\dfrac{2x^2+5x+7}{3x+5} \geqslant 2$.

7 Solve the inequality $\dfrac{2x^2-3x-5}{x^2+2x+6} < \dfrac{1}{2}$.

8 Find x if $|x+3| = 2$.

9 Find x if $\left|\dfrac{1}{x+1}\right| = 1$.

10 Find x if $|x+3| > 5$.

11 Find x if $|x-1| > 3|x-2|$.

12 If a and b are positive numbers, show that (i) $a + \dfrac{1}{a} \geqslant 2$ and (ii) $(a+b)\left(\dfrac{1}{a}+\dfrac{1}{b}\right) \geqslant 4$.

13 If a, b and c are three positive numbers, show that $(a+b)(b+c)(c+a) \geqslant 8abc$.

14 Show that $x^3 + y^3 > x^2y + xy^2$ if $(x+y) > 0$.

15 Verify that $a^3 + b^3 + c^3 - 3abc = (a+b+c)(a^2+b^2+c^2-ab-bc-ca)$. Hence show that, if a, b, c are all positive, then $a^3+b^3+c^3 > 3abc$.

1.5 The remainder and factor theorems

We assume that readers are familiar with the division process in algebra as applied to simple polynomial expressions. It is analogous to the division process in arithmetic and is illustrated by the following example.

Example 1 Find the quotient and remainder when $2x^4 - x^3 + 4x^2 + 1$ is divided by $x^2 + 3x + 1$.

$$
\begin{array}{r|l}
 & \quad + 2x^2 - 7x + 23 \\
x^2+3x+1 & 2x^4 - x^3 + 4x^2 + 0x + 1 \\
 & 2x^4 + 6x^3 + 2x^2 \\
 & \qquad - 7x^3 + 2x^2 + 0x \\
 & \qquad - 7x^3 - 21x^2 - 7x \\
 & \qquad\qquad 23x^2 + 7x + 1 \\
 & \qquad\qquad 23x^2 + 69x + 23 \\
 & \qquad\qquad\qquad - 62x - 22
\end{array}
$$

Thus the quotient is $2x^2 - 7x + 23$ and the remainder is $-62x - 22$.

The result above can be put in the form

$$2x^4 - x^3 + 4x^2 + 1 \equiv (x^2+3x+1)(2x^2-7x+23) - 62x - 22$$

The symbol $\equiv$ means that the two expressions on each side of it are identical even though they are expressed in different forms. They will be equal to each other for every value of x.

The idea can be extended. If the polynomial $P(x) = a_n x^n + a_{n-1}x^{n-1} + \ldots + a_1 x + a_0$ is divided by the polynomial $D(x) = d_m x^m + d_{m-1}x^{m-1} + \ldots + d_1 x + d_0$, where $m < n$, the quotient will be another polynomial $Q(x)$ (of degree $n - m$) and a remainder $R(x)$ which will also be a polynomial. From the nature of the process, the degree of $R(x)$ will be less than m, the degree of the divisor.

We will further have

$$P(x) \equiv D(x)Q(x) + R(x) \tag{1.12}$$

In the case where $D(x)$ is a first-degree polynomial of the form $(x - \alpha)$, $R(x)$ will simply be a constant c, say:

$$P(x) \equiv (x - \alpha)Q(x) + c \tag{1.13}$$

We can find c without working through the division process. If we substitute $x = \alpha$ into both sides of (1.13), we obtain

$$P(\alpha) = c$$

the other term on the right-hand side being zero when $x = \alpha$, whatever the form of $Q(x)$.

$P(\alpha)$ is the value of $P(x)$ when $x = \alpha$ and is given by

$$a_n\alpha^n + a_{n-1}\alpha^{n-1} + \ldots + a_1\alpha + a_0$$

This result is often known as the *remainder theorem*, which states that when the polynomial $P(x)$ is divided by $x - \alpha$ the remainder is $P(\alpha)$.

Example 2 Find the remainder when $x^3 + x^2 + x - 2$ is divided by $x + 3$.

Now $x + 3 = x - (-3)$, so the remainder is

$$\begin{aligned} R &= (-3)^3 + (-3)^2 + (-3) - 2 \\ &= -27 + 9 - 3 - 2 \\ &= -23 \end{aligned}$$

Alternatively,

$$x^3 + x^2 + x - 2 \equiv (x + 3)Q(x) + R$$

where $Q(x)$ is some quadratic polynomial which we need not calculate. If we substitute $x = -3$ on both sides, the result follows. We could, of course, carry out the division.

Example 3 The remainder when $2x^3 + ax^2 + bx + 1$ is divided by $x - 1$ is 7. When it is divided by $x - 2$, the remainder is 39. Find a and b. Also find the remainder when the expression is divided by $(x - 1)(x - 2)$.

If we denote $2x^3 + ax^2 + bx + 1$ by $P(x)$, we have by the remainder theorem

$$\begin{aligned} P(1) &= 7 = 2 + a + b + 1 \\ P(2) &= 39 = 16 + 4a + 2b + 1 \end{aligned}$$

Thus we have

$$\begin{aligned} a + b &= 4 \\ 4a + 2b &= 22 \end{aligned}$$

Whence

$$\begin{aligned} 2a + 2b &= 8 \\ 4a + 2b &= 22 \\ 2a \qquad &= 14 \end{aligned}$$

Therefore

$$a = 7 \quad \text{and} \quad b = -3$$

Thus $P(x) \equiv 2x^3 + 7x^2 - 3x + 1$.

When $P(x)$ is divided by $(x-1)(x-2)$, the remainder will be a polynomial of degree one (the divisor is a quadratic), which we can denote by $mx + n$. Thus

$$P(x) \equiv (x-1)(x-2)Q(x) + mx + n$$

$$\text{When } x = 1 \qquad P(1) = m + n = 7$$

$$\text{When } x = 2 \qquad P(2) = 2m + n = 39$$

On solving these equations, we have $m = 32$ and $n = -25$. The remainder is thus $32x - 25$.

Example 4 Find the remainder when $x^3 - x^2 - x + 1$ is divided by $x - 1$. Hence factorise this polynomial.

$$P(x) = x^3 - x^2 - x + 1$$
$$P(1) = 1 - 1 - 1 + 1 = 0$$

Thus, the remainder is zero and we deduce that $x - 1$ is a factor of the polynomial. The division process gives

$$x^3 - x^2 - x + 1 \equiv (x-1)(x^2-1)$$

so that

$$x^3 - x^2 - x + 1 \equiv (x-1)(x-1)(x+1)$$

Example 4 illustrates the factor theorem, which is immediately clear from (1.13). If $P(\alpha) = 0$, then $x - \alpha$ is a factor of $P(x)$.

Exercises 1d

1 Find the remainder when $3x^3 + 6x - 1$ is divided by $x - 3$.
2 Find a if $x^3 + ax^2 - x - 8$ is divisible by $x - 1$.
3 For what value of m is $x^3 + 3x^2 - x + m$ divisible by $x + 8$?
4 Show that the remainder when the polynomial $P(x)$ is divided by $ax - b$ is $P\left(\frac{b}{a}\right)$.
5 Find m and n if the remainders when $8x^3 + mx^2 - 6x + n$ is divided by $x - 1$ and $2x - 3$ are 2 and 8 respectively.
6 Factorise $x^3 - x^2 + 4x - 4$.
7 Factorise $x^3 - 6x^2 + 11x - 6$.
8 Factorise $12x^3 - 31x^2 + 2x + 24$.
9 Find a and b if $x^4 + ax^3 - 2x^2 + bx - 8$ is divisible by $x^2 - 4$.

10 Find the remainder when $2x^4 + x^3 - x^2 + x + 4$ is divided by $x^2 - 1$. Verify the result by doing the division.

1.6 Partial fractions

Readers will already be familiar with the technique of forming the sum or difference of two or more algebraic fractions. For example,

$$\frac{1}{x-1} + \frac{2}{x+1} + \frac{x+1}{x^2+1} = \frac{x+1+2(x-1)}{x^2-1} + \frac{x+1}{x^2+1}$$

$$= \frac{3x-1}{x^2-1} + \frac{x+1}{x^2+1}$$

$$= \frac{(x^2+1)(3x-1)+(x+1)(x^2-1)}{x^4-1}$$

$$= \frac{4x^3+2x-2}{x^4-1}$$

For the purposes of expanding such a complicated algebraic fraction in powers of x, or for integrating such a fraction with respect to the variable x, it is often necessary to carry out the reverse procedure, i.e. to resolve such a fraction into the sum of two or more partial fractions. The denominators of these partial fractions are the factors of the denominator of the original fraction. The technique is governed by five simple rules:

RULE I If the degree of the numerator is greater than or equal to the degree of the denominator, it is possible to carry out a division to obtain a quotient together with a fraction whose numerator is of lower degree than its denominator. This latter fraction is then resolved into partial fractions.

RULE II To each linear factor of the form $x - a$ in the denominator, there corresponds a partial fraction of the form $\frac{A}{x-a}$, where A is constant.

Example 1 Resolve into partial fractions $\frac{x^3+x^2+4x}{x^2+x-2}$.

The numerator is of degree 3, the denominator is of degree 2, so we divide

$$\begin{array}{r} x \\ x^2+x-2\overline{)x^3+x^2+4x} \\ \underline{x^3+x^2-2x} \\ 6x \end{array}$$

Therefore

$$\frac{x^3+x^2+4x}{x^2+x-2} = x + \frac{6x}{x^2+x-2} = x + \frac{6x}{(x+2)(x-1)}$$

We set

$$\frac{6x}{(x+2)(x-1)} \equiv \frac{A}{x-1} + \frac{B}{x+2}$$

To determine A and B, we multiply throughout by $(x+2)(x-1)$ to obtain

$$6x \equiv A(x+2) + B(x-1)$$

By putting $x = 1$, we obtain $6(1) = A(1+2)$, i.e. $A = 2$. By putting $x = -2$, we obtain $6(-2) = B(-2-1)$, i.e. $B = 4$. After substitution, we get

$$\frac{x^3 + x^2 + 4x}{x^2 + x - 2} = x + \frac{2}{x-1} + \frac{4}{x+2}$$

An alternative procedure for determining A and B is as follows.

We can make $A(x+2) + B(x-1) = (A+B)x + 2A - B$ identically equal to $6x$ by choosing A and B so that the coefficient of x, namely $(A+B)$, is equal to 6, and the term independent of x, namely $(2A - B)$, is equal to zero. Thus, we would have $A + B = 6$, $2A - B = 0$, whence $A = 2$, $B = 4$ as before. We shall find that both these techniques for determining the unknown quantities are valuable.

Example 2 Resolve into partial fractions

$$\frac{3x^2 - 4x + 5}{(x+1)(x-3)(2x-1)}$$

The degree of the numerator is less than the degree of the denominator. Thus we set

$$\frac{3x^2 - 4x + 5}{(x+1)(x-3)(2x-1)} \equiv \frac{A}{x+1} + \frac{B}{x-3} + \frac{C}{2x-1}$$

Multiplication by $(x+1)(x-3)(2x-1)$ gives

$$3x^2 - 4x + 5 \equiv A(x-3)(2x-1) + B(x+1)(2x-1) + C(x+1)(x-3)$$

With $x = -1$

$$3(-1)^2 - 4(-1) + 5 = A(-4)(-3) \qquad \text{therefore } A = 1$$

With $x = 3$

$$3(3)^2 - 4(3) + 5 = B(4)(5) \qquad \text{therefore } B = 1$$

With $x = \frac{1}{2}$

$$3(\tfrac{1}{2})^2 - 4(\tfrac{1}{2}) + 5 = C(\tfrac{3}{2})(-\tfrac{5}{2}) \qquad \text{therefore } C = -1$$

Therefore

$$\frac{3x^2 - 4x + 5}{(x+1)(x-3)(2x-1)} = \frac{1}{x+1} + \frac{1}{x-3} - \frac{1}{2x-1}$$

If we use the second technique (not so convenient in this case) to determine A, B, and C, we obtain the equations

$$\begin{aligned} 2A + 2B + C &= 3 \\ -7A + B - 2C &= -4 \\ 3A - B - 3C &= 5 \end{aligned}$$

which have solutions $A = 1$, $B = 1$, $C = -1$.

RULE III To each quadratic factor in the denominator of the form $ax^2 + bx + c$ which does *not* have linear factors, there corresponds a partial fraction of the form $\dfrac{Ax + B}{ax^2 + bx + c}$, where A and B are constants.

Example 3 Resolve into partial fractions $\dfrac{3x^2 + 8x + 13}{(x-1)(x^2 + 2x + 5)}$.

We set

$$\frac{3x^2 + 8x + 13}{(x-1)(x^2 + 2x + 5)} \equiv \frac{A}{x-1} + \frac{Bx + C}{x^2 + 2x + 5}$$

therefore

$$\begin{aligned} 3x^2 + 8x + 13 &\equiv A(x^2 + 2x + 5) + (x-1)(Bx + C) \\ &\equiv x^2(A + B) + x(2A - B + C) + 5A - C \end{aligned}$$

With $x = 1$, we obtain $24 = 8A$, therefore $A = 3$. If we make the coefficients of x^2 equal, $A + B = 3$, so $B = 0$. If we make the terms independent of x equal, $5A - C = 13$, and so $C = 2$.

It is easy to see that with these values for A, B, C, the coefficients of x are also equal. Therefore

$$\frac{3x^2 + 8x + 13}{(x-1)(x^2 + 2x + 5)} = \frac{3}{x-1} + \frac{2}{x^2 + 2x + 5}$$

Example 4 Resolve into partial fractions $\dfrac{2x^2 + 2x + 10}{(x+1)(x^2 + 9)}$.

We set

$$\frac{2x^2 + 2x + 10}{(x+1)(x^2 + 9)} \equiv \frac{A}{x+1} + \frac{Bx + C}{x^2 + 9}$$

therefore

$$2x^2 + 2x + 10 = A(x^2 + 9) + (x+1)(Bx + C)$$

With $x = -1$, $10A = 10$, therefore $A = 1$. If we make the coefficients of x^2 equal, $A + B = 2$, so $B = 1$. If we make the terms independent of x equal, $10 = 9A + C$, and so $C = 1$. Therefore

$$\frac{2x^2 + 2x + 10}{(x+1)(x^2 + 9)} = \frac{1}{x+1} + \frac{x+1}{x^2 + 9}$$

RULE IV To each repeated linear factor in the denominator of the form $(x - a)^2$, there correspond partial fractions of the form

$$\frac{A}{x-a} + \frac{B}{(x-a)^2}$$

For repeated linear factors of the form $(x - a)^3$, there are partial fractions of the form $\dfrac{A}{x-a} + \dfrac{B}{(x-a)^2} + \dfrac{C}{(x-a)^3}$ etc.

RULE V To each repeated quadratic factor in the denominator of the form $(ax^2 + bx + c)^2$, there correspond partial fractions of the form

$$\frac{Ax + B}{ax^2 + bx + c} + \frac{Cx + D}{(ax^2 + bx + c)^2}$$

Example 5 Resolve into partial fractions $\dfrac{x^3 - x^2 - 3x + 5}{(x - 1)(x^2 - 1)}$.

The denominator $(x - 1)(x^2 - 1) = x^3 - x^2 - x + 1$ is of the same degree as the numerator. We divide

$$\begin{array}{r@{}l} & \phantom{x^3 - x^2 - 3x + {}}1 \\ x^3 - x^2 - x + 1 \,\big) & x^3 - x^2 - 3x + 5 \\ & x^3 - x^2 - x + 1 \\ \hline & - 2x + 4 \end{array}$$

therefore

$$\frac{x^3 - x^2 - 3x + 5}{(x - 1)(x^2 - 1)} = 1 + \frac{4 - 2x}{(x - 1)(x^2 - 1)}$$

$$= 1 + \frac{4 - 2x}{(x - 1)^2(x + 1)}$$

We set

$$\frac{4 - 2x}{(x - 1)^2(x + 1)} \equiv \frac{A}{x + 1} + \frac{B}{x - 1} + \frac{C}{(x - 1)^2}$$

therefore

$$4 - 2x \equiv A(x - 1)^2 + B(x - 1)(x + 1) + C(x + 1)$$

With $x = 1$, $2C = 2$, therefore $C = 1$. With $x = -1$, $4A = 6$, so $A = \frac{3}{2}$. If we make the coefficients of x^2 equal, $A + B = 0$, and so $B = -\frac{3}{2}$. Therefore

$$\frac{x^3 - x^2 - 3x + 5}{(x - 1)^2(x + 1)} = 1 + \frac{3}{2(x + 1)} - \frac{3}{2(x - 1)} + \frac{1}{(x - 1)^2}$$

Example 6 Resolve into partial fractions $\dfrac{7 - 2x}{(x + 1)(x - 2)^2}$.

We set

$$\frac{7 - 2x}{(x + 1)(x - 2)^2} \equiv \frac{A}{x + 1} + \frac{B}{x - 2} + \frac{C}{(x - 2)^2}$$

therefore

$$7 - 2x \equiv A(x - 2)^2 + B(x + 1)(x - 2) + C(x + 1)$$

With $x = 2$, $3C = 3$, therefore $C = 1$. With $x = -1$, $9A = 9$, so $A = 1$. If we make the coefficients of x^2 equal, $A + B = 0$, and so $B = -1$. Therefore

$$\frac{7-2x}{(x+1)(x-2)^2} = \frac{1}{x+1} - \frac{1}{x-2} + \frac{1}{(x-2)^2}$$

Exercises 1e

Resolve 1–10 into partial fractions and verify the results.

1 $\dfrac{6x-10}{x^2-2x-3}$

2 $\dfrac{x^3-x^2-4}{x^2-1}$

3 $\dfrac{4x+11}{(x^2+4x-5)}$

4 $\dfrac{13-5x^2}{(x^2-1)(x+3)}$

5 $\dfrac{x^2+4x-7}{(x+1)(x^2+4)}$

6 $\dfrac{x^2+2}{(x^2+2x+3)(2x+1)}$

7 $\dfrac{2x^4-2x^3+4x^2-2x}{(x-1)(x^2+1)}$

8 $\dfrac{7x+2}{(2x-3)(x+1)^2}$

9 $\dfrac{2x^3+2x^2+2}{(x+1)^2(x^2+1)}$

10 $\dfrac{x^2}{(x+1)^3}$

1.7 Indices

The product of a number with itself, $a \times a$, is called the second power of a and is written a^2. The product $a \times a \times a$, written a^3, is called the third power of a. The product $a \times a \times a \times \ldots$ to m factors, written a^m, is called the mth power of a. The number which expresses the power is called the *index* or the *exponent*. Thus, the index of a^2 is 2, the index of a^3 is 3, the index of a^m is m.

When the algebraic processes of multiplication and division are carried out with different powers of the *same* number, the indices combine according to certain fundamental laws. In the proofs of these laws which follow, we assume m and n are positive integers with $m > n$.

RULE I
$$a^m \times a^n = a^{m+n} \tag{1.14}$$

For $a^m \times a^n = (a \times a \times a \times \ldots$ to m factors$) \times (a \times a \times \ldots$ to n factors), which is clearly $a \times a \times a \times a \times \ldots$ to $(m+n)$ factors $= a^{m+n}$ by definition.

RULE II
$$a^m \div a^n = a^{m-n} \tag{1.15}$$

For

$$\begin{aligned}\frac{a^m}{a^n} &= \frac{a \times a \times a \times \ldots \times a \times \ldots \times a(m \text{ factors})}{a \times a \times a \times \ldots \times a(n \text{ factors})} \\ &= a \times a \times a \times \ldots \text{ to } (m-n) \text{ factors} \\ &= a^{m-n}\end{aligned}$$

RULE III
$$(a^m)^n = a^{mn} \tag{1.16}$$

For

$$\begin{aligned}(a^m)^n &= a^m \times a^m \times a^m \times \dots \text{ to } n \text{ factors}\\ &= (a \times a \times a \times \dots \text{ to } m \text{ factors})\\ &\qquad \times (a \times a \times a \times \dots \text{ to } m \text{ factors}) \dots n \text{ times}\\ &= a \times a \times a \times a \times \dots \text{ to } mn \text{ factors}\\ &= a^{mn}\end{aligned}$$

The rules (1.14), (1.15), (1.16) have been proved for m and n positive integers. Indeed, we have no meaning for a^m unless m is a positive integer, the definition of a^m as the product of m factors each equal to a being meaningless unless m is a positive integer. We shall generalise our concept of power to include indices which are fractional and negative. This generalisation is carried out in such a way that the rules (1.14), (1.15), (1.16) remain valid. We do not want one set of rules for positive integer indices and another set of rules for fractional indices. The rules stated in Section 1.1 are true whether a, b, c are integers, rationals, etc; indeed, when they are any real numbers. In the same way, we require the rules (1.14), (1.15), (1.16) to be universally true.

The meaning of $a^{1/n}$, where n is a positive integer Since (1.14) is to remain valid,

$$a^{1/n}.a^{1/n}.a^{1/n} \text{ to } n \text{ factors} = a^{1/n+1/n+1/n+\dots} = a$$

$$(a^{1/n})^n = a$$

Therefore

$$a^{1/n} = \sqrt[n]{a} \tag{1.17}$$

The meaning of $a^{m/n}$, where m and n are positive integers Since (1.14) is to remain valid,

$$a^{m/n}.a^{m/n} \dots \text{ to } n \text{ factors} = a^{m/n+m/n+\dots} = a^m$$

$$(a^{m/n})^n = a^m$$

Therefore

$$a^{m/n} = \sqrt[n]{a^m} \tag{1.18}$$

An alternative and equally valid interpretation is

$$a^{m/n} = (\sqrt[n]{a})^m \tag{1.19}$$

The meaning of a^0 Since we require (1.14) to remain valid for all m and n, we have with $m = 0$

$$\begin{aligned}a^0.a^n &= a^{n+0} = a^n\\ a^0 &= a^n/a^n = 1\end{aligned}$$

Therefore

$$a^0 = 1 \quad \text{for all values of } a \text{ except } a = 0 \tag{1.20}$$

The meaning of a^{-n} Since we require (1.14) to remain valid for all m and n, we have with $m = -n$

$$a^n . a^{-n} = a^{n-n} = a^0 = 1$$

Therefore

$$a^{-n} = \frac{1}{a^n} \tag{1.21}$$

With these definitions it is easy to see that the rules (1.14), (1.15), (1.16) remain valid for all values of m and n.

Example 1 Evaluate (i) $(81)^{3/4}$, (ii) $(16)^{-5/4}$.

(i) $(81)^{3/4} = (\sqrt[4]{81})^3 = 3^3 = 27$

Note: it is more convenient to use (1.19) than (1.18):

$$(81)^{3/4} = \sqrt[4]{81^3} = \sqrt[4]{531\,441}$$

which we are unlikely to recognise as 27.

(ii) $16^{-5/4} = 1/16^{5/4} = 1/(\sqrt[4]{16})^5 = 1/2^5 = 1/32$

Example 2 Show from the definition that $(5^{1/4})^{1/2} = 5^{1/4 . 1/2} = 5^{1/8}$.

$$5^{1/4} = \sqrt[4]{5} \quad \text{where} \quad (\sqrt[4]{5})^4 = 5$$
$$(5^{1/4})^{1/2} = \sqrt[2]{5^{1/4}} = \sqrt[2]{(\sqrt[4]{5})} = \sqrt[8]{5} \quad \text{where} \quad (\sqrt[8]{5})^8 = 5$$
$$(5^{1/4})^{1/2} = 5^{1/8}$$

Example 3 Show that $(a^m)^n = a^{mn}$ for all m and n.

We show this by allowing m to be any value and considering in turn the cases where n is (i) a positive integer, (ii) a positive fraction, and (iii) any negative value.

(i) If n is a positive integer

$$(a^m)^n = a^m . a^m \text{ to } n \text{ factors}$$
$$= a^{m+m+m+\dots} = a^{mn}$$

(ii) If n is a positive fraction, say p/q, where p, q are positive integers, then

$$(a^m)^n = (a^m)^{p/q}$$

Now by (i)

$$[(a^m)^{p/q}]^q = (a^m)^{p/q . q} = (a^m)^p = a^{mp}$$

therefore

$$(a^m)^{p/q} = \sqrt[q]{a^{mp}} = a^{mp/q} = a^{mn}$$

(iii) If n is negative, we replace it by $-k$, where k is positive. Therefore

$$(a^m)^n = (a^m)^{-k}$$

and

$$(a^m)^{-k} = \frac{1}{(a^m)^k} = \frac{1}{a^{mk}} = a^{-mk} = a^{mn}$$

as required.

Exercises 1f

1 Evaluate (i) $27^{5/3}$, (ii) $(36)^{-3/2}$, (iii) $(8)^{7/3}$, (iv) $16^{-1/4}$.

2 Express with positive indices (i) $\dfrac{x^{-2}y^3z^{-4}}{6} \times \dfrac{9}{x^3y^{-3}z^4}$, (ii) $\dfrac{\sqrt[3]{abc^{-4}}}{\sqrt[4]{a^3b^{-3}c}}$.

3 Show that $\sqrt{x} - \sqrt{a} = \dfrac{x-a}{\sqrt{x}+\sqrt{a}}$. $\left[\text{Hint: multiply by } \dfrac{\sqrt{x}+\sqrt{a}}{\sqrt{x}+\sqrt{a}}\right]$

Deduce that $\dfrac{1}{\sqrt{x}-\sqrt{a}} = \dfrac{\sqrt{x}+\sqrt{a}}{x-a}$.

4 Simplify $(4{\cdot}2^{n+1} - 2^{n+2})/(2^{n+1} - 2^n)$.

5 Show that $(xy)^n = x^ny^n$. Treat separately the cases n is (i) a positive integer, (ii) a positive fraction, (iii) a negative quantity.

1.8 Logarithms

The logarithm of a positive number N to the base a is defined as the power of a which is equal to N. Thus, if

$$a^x = N \tag{1.22}$$

then x is the logarithm of N to the base a, written

$$x = \log_a N \tag{1.23}$$

(1.22) and (1.23) are by definition equivalent, so we have

$$a^{\log_a N} = N \tag{1.24}$$

Since $a^1 = a$ and $a^0 = 1$, it follows that

$$\log_a a = 1 \tag{1.25}$$
$$\log_a 1 = 0 \tag{1.26}$$

for all $a(\neq 0)$.

Example 1 Evaluate (i) $\log_3 9$, (ii) $\log_9 3$, (iii) $\log_4 64$.

(i) Since $3^2 = 9$, $\log_3 9 = 2$
(ii) Since $9^{1/2} = 3$, $\log_9 3 = \frac{1}{2}$
(iii) Since $4^3 = 64 \log_4 64 = 3$

The laws for the manipulation of logarithms are derived directly from the laws of indices:

RULE I $$\log_a (bc) = \log_a b + \log_a c \tag{1.27}$$

For, if $\log_a b = x$ and $\log_a c = y$,

$$b = a^x \quad \text{and} \quad c = a^y$$
$$bc = a^x.a^y = a^{x+y} \quad \text{by (1.14)}$$
$$\log_a (bc) = x + y = \log_a b + \log_a c$$

RULE II $$\log_a \left(\frac{b}{c}\right) = \log_a b - \log_a c \tag{1.28}$$

For, with the notation above,

$$\frac{b}{c} = \frac{a^x}{a^y} = a^{x-y} \quad \text{by (1.15)}$$

therefore

$$\log_a\left(\frac{b}{c}\right) = x - y = \log_a b - \log_a c$$

RULE III $$\log_a(b^p) = p\log_a b \tag{1.29}$$

For, with the notation above,

$$b^p = (a^x)^p = a^{px}$$
$$\log_a b^p = px = p\log_a b$$

We have just seen in Example 1 that the logarithm of a number may be calculated to any base. Tables of logarithms to the base 10 (common logarithms) are in existence and are very useful for arithmetical calculations. It is not difficult to use these tables to calculate the logarithm of any number to any specified base. We need the following transformation rule:

$$\log_a N = \log_a b . \log_b N \tag{1.30}$$

For if $y = \log_b N$, $N = b^y$

$$\log_a N = \log_a(b^y) = y\log_a b$$
$$\log_a N = \log_a b . \log_b N$$

If we put $N = a$ in (1.30), we obtain

$$\log_a a = \log_a b . \log_b a = 1$$
$$\log_a b = \frac{1}{\log_b a} \tag{1.31}$$

Another useful form for (1.30) is then

$$\log_a N = \frac{\log_b N}{\log_b a} \tag{1.32}$$

Example 2 Use the table of common logarithms to evaluate (i) $\log_2 9$, (ii) $\log_3 16$.

(i) $\log_2 9 = \dfrac{\log_{10} 9}{\log_{10} 2} = \dfrac{0{\cdot}9542}{0{\cdot}301} = 3{\cdot}17$

(ii) $\log_3 16 = \dfrac{\log_{10} 16}{\log_{10} 3} = \dfrac{1{\cdot}2041}{0{\cdot}4771} = 2{\cdot}524$

number	log
0·9542	$\bar{1}{\cdot}9796$
0·301	$\bar{1}{\cdot}4786$
	0·5010
1·2041	0·0806
0·4771	$\bar{1}{\cdot}6786$
	0·4020

by calculator or by tables as shown.

Example 3 Show that $\log_a(a^2 - x^2) = 2 + \log_a\left(1 - \frac{x^2}{a^2}\right)$.

$$\begin{aligned}\log_a(a^2 - x^2) &= \log_a\left[a^2\left(1 - \frac{x^2}{a^2}\right)\right]\\ &= \log_a a^2 + \log_a\left(1 - \frac{x^2}{a^2}\right)\\ &= 2 + \log_a\left(1 - \frac{x^2}{a^2}\right)\end{aligned}$$

Example 4 Show that $\log_a b . \log_b c . \log_c a = 1$.

$$\log_a b . \log_b c = \log_a c \quad \text{by (1.30)}$$

therefore, multiplying by $\log_c a$

$$\begin{aligned}\log_a b . \log_b c . \log_c a &= \log_a c . \log_c a\\ &= \log_a a = 1\end{aligned}$$

Exercises 1g

1 Evaluate (i) $\log_3 27$, (ii) $\log_{1/3} 27$, (iii) $\log_2 16$, (iv) $\log_{1/2} 32$, (v) $\log_x x^3$, (vi) $\log_{y^3} y$, (vii) $\log_{1/x} x^6$, (viii) $\log_{1/x} x^n$.

2 Show that $\log_a(a + b)^2 = 2 + \log_a\left(1 + \frac{2b}{a} + \frac{b^2}{a^2}\right)$.

3 Evaluate (i) $\log_6 12$, (ii) $\log_3 24$.

4 If $\log_a b = \log_b c = \log_c a$, show that $a = b = c$.

5 If u, v, s, t are all positive, show that

$$\log\left(\frac{u}{v}\right).\log\left(\frac{s}{t}\right) = \log\left(\frac{u}{s}\right).\log\left(\frac{v}{t}\right) + \log\left(\frac{u}{t}\right).\log\left(\frac{s}{v}\right)$$

the logarithms all being to the same base.

1.9 Equations in which the unknown is an index

Some of the techniques used to solve this type of equation are illustrated by the following examples.

Example 1 Solve the equation $3^{x^2} = 9^{x+4}$.

$$3^{x^2} = 9^{x+4} = (3^2)^{x+4} = 3^{2(x+4)}$$

Taking logarithms to the base 3, we obtain

$$\begin{aligned}x^2 &= 2(x + 4)\\ x^2 - 2x - 8 &= 0\\ (x - 4)(x + 2) &= 0\\ x = 4 \quad &\text{or} \quad x = -2\end{aligned}$$

Example 2 Solve the equation $2^{3x+1} = 5^{x+1}$.

Taking logarithms on both sides, we obtain

$$(3x+1)\log_{10}2 = (x+1)\log_{10}5$$
$$(3\log_{10}2 - \log_{10}5)x = \log_{10}5 - \log_{10}2$$
$$(\log_{10}8 - \log_{10}5)x = \log_{10}5 - \log_{10}2$$
$$x = \frac{\log_{10}5 - \log_{10}2}{\log_{10}8 - \log_{10}5} = \frac{\log_{10}\frac{5}{2}}{\log_{10}\frac{8}{5}}$$

$$x = \frac{0{\cdot}3979}{0{\cdot}2041} = 1{\cdot}95$$

number	log
0·3979	$\bar{1}$·5998
0·2041	$\bar{1}$·3098
	0·2900

Example 3 Solve for x: $9^x - 4 \times 3^x + 3 = 0$.

$$(3^2)^x - 4 \times 3^x + 3 = 0$$
$$3^{2x} - 4 \times 3^x + 3 = 0$$
$$(3^x)^2 - 4 \times 3^x + 3 = 0$$
$$(3^x - 1)(3^x - 3) = 0$$
$$3^x = 1 \quad \text{or} \quad 3^x = 3$$
$$x = 0 \quad \text{or} \quad x = 1$$

Example 4 Solve for x: $\log_x 9 + \log_{x^2} 3 = 2{\cdot}5$.

We first try to express all the logarithms to the same base.

Now by (1.30) $\log_{x^2} 3 = \log_{x^2} x . \log_x 3$ and $\log_{x^2} x = \frac{1}{2}$ since $(x^2)^{1/2} = x$. Therefore

$$\log_{x^2} 3 = \tfrac{1}{2}\log_x 3 = \log_x 3^{\frac{1}{2}} = \log_x \sqrt{3}$$

Thus, the equation becomes

$$\log_x 9 + \log_x \sqrt{3} = 2{\cdot}5$$
$$\log_x 9\sqrt{3} = 2{\cdot}5$$
$$9\sqrt{3} = x^{2{\cdot}5}$$

Whence, by inspection, we see that $x = 3$.

Otherwise

$$\log_{10} 9\sqrt{3} = 2{\cdot}5\log_{10} x$$
$$\log_{10} x = \frac{\log_{10} 9\sqrt{3}}{2{\cdot}5} = \frac{\log_{10} 3^{2{\cdot}5}}{2{\cdot}5} = \log_{10} 3$$
$$x = 3$$

Example 5 Solve the equations $2^{x+y} = 8$, $3^{2x-y} = 27$.

From the first equation, we have, after taking logarithms to the base 2,

$$x + y = 3$$

From the second equation, after taking logarithms to the base 3,

$$2x - y = 3$$

Thus our equations are equivalent to

$$x + y = 3 \qquad 2x - y = 3$$

(These equations would have resulted, after some simplification, whatever base had been chosen for the logarithms.)

Adding the last two equations gives $3x = 6$, therefore

$$x = 2 \quad \text{and} \quad y = 1$$

Exercises 1h

1 Find x if $3^x = 7{\cdot}83$.
2 Find x if $\log_x 2{\cdot}69 = 2$.
3 Solve for x: (i) $3^{2x-1} = 5^x$; (ii) $7^{4x+2} = 9^{3x-1}$.
4 Solve the equation $2^{x^2} = \frac{1}{4}8^x$.
5 Find x if $9^{x^2} = 3^{5x-2}$.
6 Solve the equation $5^{2x} - 5^{1+x} + 6 = 0$.
7 Solve the equation $4^{2x} = 2^{6x-1}$.
8 Find x if $\log_x 8 - \log_{x^2} 16 = 1$.
9 Find x if $\log_x 3 + \log_3 x = 2{\cdot}5$.
10 Solve the simultaneous equations $2^{x+y} = 6$, $3^{x-y} = 4$.

Exercises 1

1 Solve the simultaneous equations $x + 2y = 7$, $x^2 + 2y^2 = 17$. [WJC]

2 Express $(2x^2 + 8x + 7)/(x^2 + 4x + 5)$ in the form

$$a - \frac{b}{(x + c)^2 + d}$$

and state the values of a, b, c and d.

3 Solve the equation $2^x.3^{1-x} = 6$.

4 Show that $(a + \sqrt{b})^2 = a^2 + b + 2a\sqrt{b}$. Hence evaluate the square root of $9 + 4\sqrt{5}$ in the form $c + \sqrt{d}$.

5 Given the simultaneous equations $x^2 - 6xy + 11y^2 = 3a^2$, $x^2 - 2xy - 3y^2 = 5a^2$, derive an equation in x and y only, and hence solve the equations for x and y in terms of a. [JMB, part]

6 Express in partial fractions $(8x + 15)/(x^2 + 4)(x - 3)$.

7 If $X = \dfrac{a}{(1 + t)^2}$ and $Y = \dfrac{a(1 - t)}{1 + t}$, show that $(Y + a)^2 = 4aX$.

8 Solve completely the equation $(x^2 + x)^2 = 5x^2 + 5x - 6$. [LU, part]

9 Express in partial fractions $\dfrac{18y - 10y^2}{(3 - y)(1 - y)^2}$.

10 Find the range of values of x for which $\dfrac{x(x - 2)}{x + 6} > 2$.

11 For what values of x is $\dfrac{x(x - 1)}{2x + 3} > 0$?

12 By putting $y = x + \dfrac{1}{x}$, solve the equation $2x^4 - 9x^3 + 14x^2 - 9x + 2 = 0$.

13 Solve the equation $\sqrt{(2x-1)} - \sqrt{(x-1)} = 1$.

14 If $p^2 = qr$, show that $\log_q p + \log_r p = 2 \log_q p \log_r p$.

15 Solve the equation $\log_3 x + \log_x 3 = \frac{10}{3}$.

16 Solve the inequality $\dfrac{x-2}{x-3} > \dfrac{x-3}{x+4}$.

17 Verify that $(l^2 + m^2 + n^2)(x^2 + y^2 + z^2) - (lx + my + nz)^2 = (ly - mx)^2 + (mz - ny)^2 + (nx - lz)^2$. Deduce that $(lx + my + nz)^2 \leqslant (l^2 + m^2 + n^2) \times (x^2 + y^2 + z^2)$.

18 If a, b and c are positive and unequal show that

$$(a + b + c)^2 < 3(a^2 + b^2 + c^2)$$

19 If a, b, c are positive and unequal show that

$$(a + b + c)\left(\frac{1}{a} + \frac{1}{b} + \frac{1}{c}\right) > 9$$

20 Solve the equation $27^{x-3} = 3 \times 9^{x-2}$.

21 By making the substitution $y = x + \dfrac{1}{x}$, solve the equation $x^4 + 8x^3 + 17x^2 + 8x + 1 = 0$. [WJC, part]

22 Solve for x: $\log_{10}\left(\dfrac{x^2 + 24}{x}\right) = 1$.

23 Solve for x: $2^x \times 3^{x+1} = 5^{2x+1}$.

24 If $\dfrac{u^2}{v} + \dfrac{v^2}{u} = 12$ and $\dfrac{1}{u} + \dfrac{1}{v} = \dfrac{1}{3}$ find the values of uv and hence solve the equations.

25 Solve the equation $\sqrt{(2x+3)} - \sqrt{(x-2)} = 2$.

26 If $x = \dfrac{a(1+t^2)}{1-t^2}$, $y = \dfrac{2bt}{1-t^2}$, show that $\dfrac{x^2}{a^2} - \dfrac{y^2}{b^2} = 1$.

27 If a and b are two real numbers such that $a + b = 1$, prove that $4ab \leqslant 1$. Hence or otherwise show that $a^2 + b^2 \geqslant \frac{1}{2}$. [JMB, part]

28 Find y if $\left|\dfrac{y-3}{y+1}\right| < 2$.

29 Use the result of question 17 to show that $(a^3 + b^3 + c^3)^2 \leqslant (a^2 + b^2 + c^2) \times (a^4 + b^4 + c^4)$.

30 Solve the equation $\log_{10}(x^2 + 9) - 2 \log_{10} x = 1$.

31 Solve the equations $x + 2y = 3$, $3x^2 + 4y^2 + 12x = 7$.

32 Solve the equations $3^{2x+y} = 12$, $2^{x-y} = 4$.

33 For what values of x is $\dfrac{(x-2)(x-3)}{x-4} \geqslant 0$?

34 Solve the equation $2^{2+2x} + 3 \times 2^x - 1 = 0$.

35 If $a^2 + b^2 = 23ab$ show that $\log a + \log b = 2 \log\left(\dfrac{a+b}{5}\right)$.

36 Solve the inequality $\dfrac{y+1}{3y-7} > 1$.

37 If $x = \sqrt[3]{p} + \dfrac{1}{\sqrt[3]{p}}$, $y = \sqrt{p} + \dfrac{1}{\sqrt{p}}$, show that $y^2 - 2 = x(x^2 - 3)$.

38 By putting $\alpha = \log a$, $\beta = \log b$, $\gamma = \log c$ in the identity $\alpha(\beta - \gamma) + \beta(\gamma - \alpha) + \gamma(\alpha - \beta) = 0$, show that

$$\left(\frac{b}{c}\right)^{\log a} \cdot \left(\frac{c}{a}\right)^{\log b} \cdot \left(\frac{a}{b}\right)^{\log c} = 1$$

where the logarithms are taken to any base.

39 Show that $(x - 2)$ is a factor of $6x^4 - 7x^3 - 27x^2 + 28x + 12$ and hence solve the equation $6x^4 - 7x^3 - 27x^2 + 28x + 12 = 0$. [AEB]

40 When $f(x) \equiv x^3 + ax^2 + bx + c$ is divided by $x - 3$, the remainder is 30. If $x^2 - 4$ is a factor of $f(x)$, find the values of a, b and c. [AEB]

41 For what values of x is $|x - 2| > 5|x - 3|$? [LU]

42 Find the set of values of x for which $\dfrac{1}{x - 2} < \dfrac{1}{x + 2}$. [LU]

43 Given that $y = \log_a x^3$ and $z = \log_x a$, show that $yz = 3$. Hence find the numerical values of y and z when $\log_a(3\log_a x) - \log_a(\log_x a) = \log_a(27)$. [AEB]

44 Given that $\log_2(x - 5y + 4) = 0$ and $\log_2(x + 1) - 1 = 2\log_2 y$, find the values of x and y. [AEB]

45 When the expression $x^3 + ax^2 + bx + c$ is divided by $x^2 - 4$, the remainder is $18 - x$; and when it is divided by $x + 3$, the remainder is 21. Find the remainder when the expression is divided by $x + 1$. [AEB]

46 Solve the simultaneous equations

$$\frac{x}{3} + \frac{y}{4} = 1$$

$$\frac{3}{x} - \frac{2}{y} = \frac{7}{12}$$

[C]

47 If $2\log_y x + 2\log_x y = 5$ show that $\log_y x$ is either $\frac{1}{2}$ or 2. Hence find all pairs of values of x and y which satisfy simultaneously the equation above and the equation $xy = 27$. [JMB]

48 The variables x and y are connected by the relation $y = ax^n$, where a and n are constants; $y = 3$ when $x = 4$ and $y = 2$ when $x = 9$. Find the exact values of n and a. [C]

49 Find all the solutions of the system of equations

$$5y + 1 = 2xy$$
$$3x + y - 7 = xy$$

[O]

50 Find the set of values of x for which $\dfrac{(2x - 1)(x + 4)}{x - 5} > 0$. [LU]

2

Finite sequences and series

2.1 Sequences and series

Sequences

A sequence, or progression, is a set of numbers in some definite order, the successive terms (or numbers) of the sequence being formed according to some rule.

For the sequence of positive integers 1, 2, 3, 4, ..., the rth term is the integer r; for the sequence 1, 4, 9, 16, ..., the rth term is the number r^2.

It is usual to denote the rth term of a general sequence by u_r, and the sequence by $u_1, u_2, u_3 \ldots u_r \ldots$ The rule defining a sequence is often given in the form of some formula for u_r in terms of r although this is not necessarily so. (See Example 2.) Thus, for our first sequence $u_r = r$; for our second sequence $u_r = r^2$.

Example 1 Find u_r in terms of r for the sequences

(i) 3, 5, 7, 9, ...
(ii) $1, \frac{1}{4}, \frac{1}{9}, \frac{1}{16}, \frac{1}{25}, \ldots$
(iii) 1, 4, 3, 16, 5, 36, 7, 64, ...
(iv) $-1, 1, -1, 1, -1, 1, \ldots$
(v) $1, -2, 3, -4, 5, -6, \ldots$

(i) By inspection, we see that the terms can be written

$$2 \times 1 + 1, \quad 2 \times 2 + 1, \quad 2 \times 3 + 1, \quad 2 \times 4 + 1, \ldots$$

$$u_r = 2r + 1$$

(ii) By inspection, we see that the terms can be written

$$\frac{1}{1^2}, \frac{1}{2^2}, \frac{1}{3^2}, \frac{1}{4^2}, \ldots$$

$$u_r = \frac{1}{r^2}$$

(iii) If r is odd, $u_r = r$, if r is even $u_r = r^2$. Since $2r$ is always even and $2r + 1$ always odd, $u_{2r+1} = 2r + 1$, $u_{2r} = 4r^2$ adequately describe the sequence.

(iv) The odd terms are -1, the even terms $+1$. Therefore, $u_{2r+1} = -1$,

$u_{2r} = +1$. However, the sequence may be described quite adequately by *one* formula in this case: $u_r = (-1)^r$.

(v) By using the result (iv), we have

$$u_r = (-1)^r(-r) = (-1)^{r+1}r$$

Example 2 Find the first five terms of the sequence defined as follows: the first two terms are 1 and 3 respectively; each later term is formed by multiplying its predecessor by 3 and subtracting the next previous term, i.e.

$$\begin{aligned} u_r &= 3u_{r-1} - u_{r-2} \\ u_1 &= 1 \\ u_2 &= 3 \\ u_3 &= 3u_2 - u_1 = 8 \\ u_4 &= 3u_3 - u_2 = 24 - 3 = 21 \\ u_5 &= 3u_4 - u_3 = 3 \times 21 - 8 = 63 - 8 = 55 \end{aligned}$$

The first five terms are thus 1, 3, 8, 21, 55. (Note: this rule adequately defines a sequence although it would not be easy to find a formula for u_r in terms of r.)

Series

A series is obtained by forming the sum of the terms of a sequence. A finite series is obtained if a finite number of terms of the sequence are summed. The sum of the first n terms of the sequence $u_1, u_2, \ldots$ is generally denoted by S_n:

$$S_n = u_1 + u_2 + u_3 + \ldots + u_n \tag{2.1}$$

S_n is the sum of the first n terms of the series $u_1 + u_2 + u_3 + \ldots$, or, as it is sometimes put, S_n is the sum to n terms of the series $u_1 + u_2 + u_3 + \ldots$. The rth term of the series is u_r, the corresponding term of the sequence from which the series is derived by summation.

$$S_n = u_1 + u_2 + u_3 + \ldots + u_n$$

is often denoted by

$$\sum_{r=1}^{n} u_r \equiv u_1 + u_2 + \ldots + u_n \tag{2.2}$$

Σ is the Greek capital letter sigma and the symbol above means evaluate u_r for all values of r from 1 to n and sum the results. The specific form of u_r may be inserted. Thus, with u_r (the general term of the series) $= r$

$$\sum_{r=1}^{n} u_r = \sum_{r=1}^{n} r = 1 + 2 + 3 + \ldots + n$$

In the same way

$$\sum_{r=m}^{n} u_r = u_m + u_{m+1} + \ldots + u_n \quad (n > m) \tag{2.3}$$

Example 3 The sum of the first n terms of a series is given by the formula

$S_n = n^2 + 3n$ for all values of n. Find an expression for the rth term of the series.

If u_r denotes the rth term

$$u_r \equiv S_r - S_{r-1} \tag{2.4}$$
$$\equiv [u_1 + u_2 + \ldots + u_{r-1} + u_r] - [u_1 + u_2 + \ldots + u_{r-1}]$$
$$= r^2 + 3r - [(r-1)^2 + 3(r-1)]$$
$$= r^2 + 3r - [r^2 + r - 2]$$

Therefore

$$u_r = 2r + 2$$

Example 4 Evaluate (i) $\sum_{r=1}^{4} r^2$, (ii) $\sum_{r=3}^{7} 2^r$.

(i) $\sum_{r=1}^{4} r^2 = 1^2 + 2^2 + 3^2 + 4^2 = 1 + 4 + 9 + 16 = 30$

(ii) $\sum_{r=3}^{7} 2^r = 2^3 + 2^4 + 2^5 + 2^6 + 2^7$

$= 8 + 16 + 32 + 64 + 128 = 248$

Exercises 2a

1 Evaluate the first five terms of the sequences whose rth terms (u_r) are (i) $3r - 1$, (ii) $(-\frac{1}{3})^{r-1}$, (iii) $2^r + r^2$.

2 Find a formula for u_r for the sequences
(i) 1, 8, 27, 64, 125 ...
(ii) 1, -4, 9, -16, 25 ...
(iii) $\frac{1}{2}$, 2, 8, 32, 128 ...

3 A sequence is defined by the rule $u_1 = 1$, $u_2 = 2$ and $u_r = u_{r-1} + u_{r-2}$ for $r \geqslant 3$. Find the first seven terms of this sequence. [This is the Fibonacci sequence; as r increases, the ratio u_{r+1}/u_r approaches the value τ, the *golden ratio* of Pythagoras.]

4 If $u_1 = -1$, $u_2 = -5$ and $u_r = a + br$, find a, b and u_5.

5 Evaluate u_r for the sequences
(i) 0, 7, 26, 63, 124 ...
(ii) 2, 4, 6, 8, 10 ...
(iii) 2, 11, 32, 71, 134 ...
(iv) 3, 9, 27, 81, 243 ...
(v) 2, -4, 8, -16, 32 ...
(vi) 5, 5, 35, 65, 275 ...
(vii) 6, -36, 216, -1296 ....

6 Find the first six terms of the sequence defined by $u_1 = 0$, $u_2 = 2$ and $u_r = u_{r-1} - u_{r-2}$ for $r > 2$. Hence evaluate $\sum_{r=1}^{6} u_r$.

7 $u_1 = 0$, $u_2 = 3$, $u_3 = 12$, and $u_r = a + br + cr^2$. Find a, b, c and
(i) $\sum_{r=1}^{5} u_r$ (ii) $\sum_{r=4}^{7} u_r$.

8 Evaluate (i) S_8 for the series $1 + 3 + 6 + 9 + 12 + \ldots$, (ii) S_5 for the series $3 + 9 + 27 + 81 + \ldots$.

9 The sum of the first n terms of a series is given by $S_n = n^3 - 2n$ for all values of n. Find u_r.

10 If $u_r = \log_{10} r$, show that $\sum_{r=1}^{10} u_r = \sum_{r=1}^{10} \log_{10} r = \log_{10} 3\,628\,800$.

2.2 The arithmetic sequence and series

If the consecutive terms of a sequence differ by a constant number, their terms are said to form an arithmetic sequence or an arithmetic progression. Thus, for example, the numbers 1, 3, 5, 7, ... are in arithmetic progression, the difference between consecutive terms being 2.

An arithmetic sequence is completely defined by its first term (conventionally denoted for the general arithmetic sequence by a) and the common difference (the difference between consecutive terms) denoted by d. The general arithmetic sequence is then

$$a,\ a + d,\ a + 2d,\ a + 3d,\ \ldots,\ a + (r - 1)d,\ \ldots \tag{2.5}$$

and the rth term of the sequence is

$$u_r = a + (r - 1)d \tag{2.6}$$

If the numbers $u_1, u_2, u_3, u_4, \ldots, u_{r-1}, u_r$ are in arithmetic progression, $u_2, u_3, \ldots, u_{r-1}$ are said to form $(r - 2)$ arithmetic means between u_1 and u_r. This is simply an extension of the usual notion of an arithmetic mean (or average). The *three* quantities $a - d$, a, $a + d$ are in arithmetic progression and a is the arithmetic mean of the other two.

The sum of the terms of an arithmetic sequence form an arithmetic series. For the general sequence (2.5), the sum of the first n terms is

$$S_n = a + (a + d) + (a + 2d) + \ \ldots\ + a + (n - 1)d \tag{2.7}$$

For this particular series, we can obtain a closed formula in terms of n for S_n

$$S_n = a + (a + d) + (a + 2d) + \ \ldots\ + [a + (n - 2)d] + [a + (n - 1)d]$$

and

$$S_n = [a + (n - 1)d] + [a + (n - 2)d] + \ldots + (a + d) + a$$

On addition, since corresponding pairs add to $2a + (n - 1)d$,

$$\begin{aligned} 2S_n &= [2a + (n - 1)d] + [2a + (n - 1)d] + \ldots\ n \text{ times} \\ &= n[2a + (n - 1)d] \end{aligned}$$

therefore

$$S_n = \frac{n}{2}[2a + (n - 1)d] \tag{2.8}$$

This result can be written in another useful form. Since a is the first term, $l = a + (n - 1)d$ is the last term of the arithmetic series above and

$$a + l = 2a + (n - 1)d$$

we have

$$S_n = \frac{n}{2}(a + l) \tag{2.9}$$

Example 1 Find three numbers in arithmetic progression whose sum is 21 and whose product is 315.

Let the numbers be $a - d$, a, $a + d$. Then $a - d + a + a + d = 21$, therefore $3a = 21$, so that $a = 7$.

$$a(a - d)(a + d) = a(a^2 - d^2) = 315$$
$$a^2 - d^2 = 45 \quad (\text{since } a = 7)$$
$$d^2 = 4 \quad \text{so that} \quad d = \pm 2$$

and the required numbers are 5, 7, 9.

Example 2 Find six arithmetic means between -3 and 18.

We require eight terms in arithmetic progression, -3 being the first and 18 the eighth. If their common difference is d, then

$$18 = -3 + 7d \quad \text{so that} \quad d = 3$$

The arithmetic means are 0, 3, 6, 9, 12, 15.

Example 3 Evaluate the nth term and the sum of the first n terms of the arithmetic series: $3 + 7 + 11 + 15 + \ldots$. Evaluate u_{11} and S_{20}.

The first term is 3 and the common difference 4. By (2.6), we have

$$u_n = 3 + (n - 1)4 = 4n - 1$$

For S_n we have, by (2.8),

$$S_n1 = \frac{n}{2}[2(3) + (n - 1)4] = \frac{n}{2}(4n + 2) = 2n^2 + n$$

therefore

$$u_{11} = 4(11) - 1 = 43$$
$$S_{20} = \tfrac{20}{2}[2(3) + 19(4)] = 820.$$

Example 4 The first two terms of an arithmetic series are -2 and 3. How many terms are needed for the sum to equal 306?

The first two terms a, $a + d$ are -2 and 3, so that $a = -2$ and $d = 5$. The sum of the first n terms is thus

$$S_n = \frac{n}{2}[-4 + 5(n - 1)]$$

$$= \frac{n}{2}[5n - 9]$$

therefore, if $S_n = 306$,

$$5n^2 - 9n = 612$$
$$5n^2 - 9n - 612 = 0$$
$$(5n + 51)(n - 12) = 0$$
$$n = 12 \quad \text{or} \quad -\tfrac{51}{5}$$

Thus, 12 terms are required.

Example 5 Obtain a formula for $\sum_{r=1}^{n} r$ in terms of n. (The sum of the first n positive integers.)

$$\sum_{r=1}^{n} r = 1 + 2 + 3 + \ldots + n$$

is the sum of the first n terms of an arithmetic series, whose first term is 1 and common difference 1. Therefore

$$\sum_{r=1}^{n} r = \frac{n}{2}[2(1) + (n-1)1]$$

$$\sum_{r=1}^{n} r = \frac{n(n+1)}{2} \qquad (2.10)$$

This result can be used to obtain the sum of any arithmetic series. For the series of Example 3

$$\begin{aligned} S_n &= 3 + 7 + 11 + 15 \ldots \text{ to } n \text{ terms} \\ &= \sum_{r=1}^{n} (4r - 1) \qquad (u_r = 4r - 1) \\ &= 4\sum_{r=1}^{n} r - \sum_{r=1}^{n} 1 \\ &= \frac{4n(n+1)}{2} - n \qquad \left(\sum_{r=1}^{n} 1 = 1 + 1 + 1 + \ldots n \text{ times} = n\right) \\ &= 2n^2 + 2n - n = 2n^2 + n \end{aligned}$$

as before.

Exercises 2b

1 Find three numbers in arithmetic progression whose sum is 3 and whose product is -15.
2 The sum of three numbers in arithmetic progression is 18 and the sum of their squares is 206. Find the numbers.
3 Find 12 arithmetic means between -5 and 60.
4 Find the sum of the first 16 terms of the series $3\frac{1}{2} + 4\frac{3}{4} + 6 + 7\frac{1}{4} + \ldots$.
5 The first term of an arithmetic series is 7, the last is 70 and the sum is 385. Find the number of terms in the series and the common difference.
6 Find the sum of the first n terms of the series $-1 + (-3) + (-5) + (-7) + \ldots$.
7 Evaluate (i) $\sum_{r=1}^{8} (3r + 2)$, (ii) $\sum_{r=1}^{n} (5r - 7)$, (iii) $\sum_{r=1}^{n} (2 - 3r)$.
8 The third term of an arithmetic progression is 18, the seventh term is 30. Find the sum of the first 33 terms.
9 Sum the first $2n$ terms of the series
(i) $5 + 11 + 17 + 23 \ldots$

(ii) $a + 3b + 2a + 6b + 3a + 9b + \ldots$
(iii) $3a - 2b + 4a - 4b + 5a - 6b + \ldots$

10 Evaluate (i) the sum of the positive integers less than 100; (ii) the sum of the positive integers less than 100 which are multiples of 3; (iii) the sum of the positive integers less than 100 which are not multiples of 7.

2.3 The finite geometric sequence and series

If the consecutive terms of a sequence are all in the same ratio, the terms are said to form a geometric sequence or a geometric progression. Thus, for example, the numbers 1, 2, 4, 8, 16 ... are in geometric progression, the ratio of any pair of consecutive terms being 2.

A geometric sequence is completely defined by its first term (conventionally denoted for the general geometric progression by a) and the common ratio (the ratio of consecutive terms) denoted by r. The general geometric sequence is then

$$a, ar, ar^2, ar^3, \ldots, ar^{n-1}, \ldots \tag{2.11}$$

The nth term of the sequence is

$$u_n = ar^{n-1} \tag{2.12}$$

If the numbers $u_1, u_2, u_3, \ldots, u_{n-1}, u_n$ are in geometric progression, $u_2, u_3, \ldots, u_{n-1}$ are said to form $(n-2)$ geometric means between u_1 and u_n. This is simply an extension of the usual meaning for the geometric mean, G, of two numbers c, d, namely, $G = \sqrt{cd}$. The *three* quantities a/r, a, ar are in geometric progression and a is the geometric mean of the other two.

The sum of the terms of a geometric sequence form a geometric series. For the general sequence (2.11), the sum of the first n terms is

$$S_n = a + ar + ar^2 + \ldots + ar^{n-1} \tag{2.13}$$

As with the arithmetic series, we can obtain a closed formula for S_n in terms of n:

$$S_n = a + ar + ar^2 + \ldots + ar^{n-2} + ar^{n-1}$$

therefore

$$rS_n = ar + ar^2 + \ldots + ar^{n-1} + ar^n$$

On subtraction, all terms cancel except the first and last; therefore

$$S_n(1 - r) = a - ar^n$$

$$S_n = \frac{a(1 - r^n)}{1 - r} = \frac{a(r^n - 1)}{r - 1} \tag{2.14}$$

Example 1 Find three numbers in geometrical progression whose sum is 28 and whose product is 512.

Let the numbers be a/r, a, ar, then

$$\frac{a}{r}.a.ar = a^3 = 512 \qquad a = 8$$

Therefore
$$\frac{8}{r} + 8 + 8r = 28$$
$$8r^2 + 8r + 8 = 28r$$
$$2r^2 - 5r + 2 = 0$$
$$(2r - 1)(r - 2) = 0$$
$$r = \tfrac{1}{2} \quad \text{or} \quad r = 2$$

The required numbers are, therefore, $\frac{8}{2}$, 8, 8×2, i.e. 4, 8, 16.

Example 2 Find four geometric means between 2 and 486.

We require six numbers in geometric progression such that the first is 2 and the sixth 486. Let r be the common ratio. Then by (2.12)
$$2r^5 = 486$$
$$r^5 = 243$$
$$r = \sqrt[5]{243} = 3$$

Therefore, the required geometric means are 2×3, 2×3^2, 2×3^3, 2×3^4, i.e. 6, 18, 54, 162.

Example 3 Find the sum of the first n terms of the series $1 - \frac{1}{2} + \frac{1}{4} - \frac{1}{8} + \frac{1}{16} - \frac{1}{32} + \ldots$.

The series is a geometric series first term 1 and common ratio $-\frac{1}{2}$. Therefore, by (2.14),
$$S_n = \frac{1[1 - (-\frac{1}{2})^n]}{1 - (-\frac{1}{2})}$$
$$S_n = \tfrac{2}{3}[1 - (-\tfrac{1}{2})^n] = \tfrac{2}{3} - \tfrac{2}{3}(-\tfrac{1}{2})^n$$

Example 4 The first and last terms of a geometric series are 2 and 2048 respectively. The sum of the series is 2730. Find the number of terms and the common ratio.

Let the number of terms be n and the common ratio r. Then by (2.12)
$$2r^{n-1} = 2048$$
and by (2.14)
$$\frac{2(r^n - 1)}{r - 1} = 2730$$

Therefore, $r^{n-1} = 1024$ and $(r^n - 1)/(r - 1) = 1365$, and these are the simultaneous equations we must solve for r and n. Substituting from the first into the second, we obtain
$$\frac{1024r - 1}{r - 1} = 1365$$
$$1024r - 1 = 1365r - 1365$$
$$1364 = 341r$$
$$r = 4$$
$$4^{n-1} = 1024$$
$$n - 1 = 5 \qquad \text{i.e. } n = 6$$

The number of terms is 6 and the common ratio 4.

Exercises 2c

1 Find three numbers in geometric progression whose sum is 13 and whose product is -64.

2 The product of three numbers in geometric progression is 1, their sum is $-\frac{7}{3}$. Find the numbers.

3 Find three geometric means between 5 and 80.

4 The third term of a geometric sequence is -1, the seventh term is -81. Find the ninth term.

5 The second term of a geometric sequence is 24, the fifth term is 81. Find the seventh term.

6 Find the sum of the first eight terms of the series $\frac{1}{2} + \frac{1}{3} + \frac{2}{9} + \ldots$.

7 Evaluate $1 + \sqrt{3} + 3 + 3\sqrt{3} + \ldots + 81\sqrt{3}$.

8 The first term of a geometric series is 3, the last term 768. If the sum of the terms is 1533 find the common ratio and the number of terms.

9 The pth, qth and rth terms of an arithmetic sequence are in geometric progression. Show that the common ratio is $\dfrac{q-r}{p-q}$ or $\dfrac{p-q}{q-r}$.

10 The sum of the first n terms of a geometric series is 127 and the sum of their reciprocals is $\frac{127}{64}$. The first term is 1. Find n and the common ratio.

2.4 The infinite geometric series

Consider the geometric series (with common ratio $\frac{1}{2}$) $1 + \frac{1}{2} + \frac{1}{4} + \frac{1}{8} + \ldots$

$$S_n = \frac{1 - (\frac{1}{2})^n}{1 - \frac{1}{2}} = 2[1 - (\tfrac{1}{2})^n] = 2 - (\tfrac{1}{2})^{n-1}$$

Thus the sum of the first four terms is $1\frac{7}{8}$, the sum of the first eight terms is $1\frac{127}{128}$. From these results we see that, as we add more and more terms, the sum of the series gets nearer and nearer to 2. Indeed, the difference between the S_n and 2 is just $(\frac{1}{2})^{n-1}$, and as n increases so this number decreases and approaches zero. We see that S_n tends to 2 as n tends to infinity, since we can make S_n as near to 2 as we please by choosing n large enough. The limit of S_n as n tends to infinity is 2, which we write $\lim_{n\to\infty} S_n = 2$, or $S_n \to 2$ as $n \to \infty$.

We say that the series above is convergent to sum 2, or 'the sum to infinity' of the series is 2.

For the general geometric series $a + ar + ar^2 + \ldots$

$$S_n = \frac{a(1 - r^n)}{1 - r} = \frac{a}{1 - r} - \frac{a}{1 - r}r^n$$

Now if $-1 < r < 1$, r^n decreases as n increases. We say that the limiting value of r^n is zero. Thus, as n increases, S_n approaches the limiting value (denoted by S) $\dfrac{a}{1-r}$. We say that the series converges to the sum $\dfrac{a}{1-r}$, the sum to infinity of the series.

Thus, if $-1 < r < 1$, the sum to infinity of the geometric series $a + ar + ar^2 + \ldots$ is

$$\lim_{n\to\infty} S_n = S = \frac{a}{1-r} \tag{2.15}$$

(The condition $-1 < r < 1$ is often written in the form $|r| < 1$. $|r|$, the modulus of r, is the positive number having the same magnitude as r. Thus $|\frac{1}{2}| = \frac{1}{2}$, $|-\frac{1}{4}| = \frac{1}{4}$, $|-3| = 3$, etc. See Section 1.4.)

The result (2.15) is only valid if $|r| < 1$. For the series $1 + 2 + 4 + 8 + \ldots$ for which $r = 2$, S_n increases indefinitely as n increases, and so S_n has no finite limit.

Example 1 To what sum does the following series converge: $1 - \frac{1}{3} + \frac{1}{9} - \frac{1}{27} + \ldots$?

This is a geometric series with common ratio $-\frac{1}{3}$. $|-\frac{1}{3}| = \frac{1}{3}$ is less than 1. Therefore, by (2.15) the series converges to sum

$$S = \frac{1}{1-(-\frac{1}{3})} = \frac{1}{\frac{4}{3}} = \frac{3}{4}$$

Example 2 Express 0·777 recurring as a fraction.

$$0{\cdot}777 \text{ recurring} = \tfrac{7}{10} + \tfrac{7}{100} + \tfrac{7}{1000} + \tfrac{7}{10\,000} + \ldots$$

that is, it is the limit of the sum of the geometric series whose first term is $\frac{7}{10}$ and whose common ratio is $\frac{1}{10}(<1)$. Therefore, by (2.15)

$$0{\cdot}777 \text{ recurring} = \frac{\frac{7}{10}}{1-\frac{1}{10}} = \frac{7}{9}$$

Example 3 For what values of x does the series

$$x + \frac{x}{1+x} + \frac{x}{(1+x)^2} + \frac{x}{(1+x)^3} + \ldots$$

converge? And to what sum does it then converge?

The series is a geometric series with common ratio $\dfrac{1}{1+x}$. The series will thus converge if

$$\left|\frac{1}{1+x}\right| < 1$$

which we write as

$$|x+1| > 1$$

Thus

$$x + 1 > 1 \quad \text{i.e.} \quad x > 0$$

or

$$x + 1 < -1 \quad \text{i.e.} \quad x < -2$$

Thus, the series converges if $x > 0$ or $x < -2$. The limit of the sum of the series is then

$$\frac{x}{1-\dfrac{1}{1+x}} = \frac{x}{\dfrac{x}{1+x}}$$
$$= 1 + x$$

We also observe that the series will converge if $x = 0$, for then each and every term is zero. The sum will of course also be zero.

Exercises 2d

1 Find the sum to which the following series converge:
(i) $1 - x + x^2 - x^3 + \ldots \quad (-1 < x < 1)$
(ii) $\frac{1}{5} + \frac{1}{25} + \frac{1}{125} + \frac{1}{625} + \ldots$
(iii) $\frac{1}{6} + \frac{1}{6}(\frac{5}{6})^2 + \frac{1}{6}(\frac{5}{6})^4 + \ldots$

2 Show that the series

$$1 + \frac{2x}{x^2+4} + \left(\frac{2x}{x^2+4}\right)^2 + \left(\frac{2x}{x^2+4}\right)^3 + \ldots$$

is always convergent and find the limit of its sum.

3 Express 0·232323 recurring as a fraction.

4 Show that if θ is an acute angle between 0 and $\pi/2$, the limit of the sum of the series $\cos\theta + \cos\theta\sin^2\theta + \cos\theta\sin^4\theta + \ldots$ is $\sec\theta$.

5 Determine the range, or ranges of values of x for which $|3x - 5| > 7$. If x has a value which satisfies this condition show that the infinite series

$$1 + \left(\frac{7}{3x-5}\right) + \left(\frac{7}{3x-5}\right)^2 + \left(\frac{7}{3x-5}\right)^3 + \ldots$$

has sum $\dfrac{3x-5}{3x-12}$.

Exercises 2

1 The sum of the squares of three positive numbers in arithmetic progression is 155. The sum of the numbers is 21. Find the numbers.

2 The sum of the first n terms of a geometric series is 364. The sum of their reciprocals is $\frac{364}{243}$. If the first term is 1, find n and the common ratio.

3 The sum of the geometric progression a, ar, ar^2, ar^{n-1} is S. The product of these n terms is P. Find R, the sum of the reciprocals of these n terms, and show that $\left(\dfrac{S}{R}\right)^n = P^2$. [AEB]

4 If a and r are both positive, prove that the series $\log a + \log ar + \log ar^2 + \ldots + \log ar^{n-1}$ is an arithmetic series and find the sum of the terms.

5 The first term of an arithmetic series is $3p + 5$, where p is a positive integer. The last term is $17p + 17$ and the common difference is 2. Find, in terms of p (i) the number of terms, (ii) the sum of the series. Show that the sum of the series is divisible by 14, only when p is odd.

[AEB]

6 Find the sum of each of the following:
(i) All the odd numbers with three digits of which the first digit is not zero.
(ii) All the odd numbers less than 1000.
(iii) All the numbers less than 100 which end in 5 or 7.

7 If m is a positive integer prove that

$$(1 + x + x^2 + \dots + x^{2m})(1 - x + x^2 - x^3 + \dots + x^{2m}) = (1 + x^2 + x^4 + \dots + x^{4m})$$

8 Write down the nth term of the arithmetic series with first term a and common difference d.
In each of a set of n separate arithmetic series, the first term is 1. The common difference of the first series is 1, of the second 2, of the third 2^2, and so on. Find, in its simplest form, a formula for the sum of the nth terms of the n series. [JMB, part]

9 The third, sixth and seventh terms of a geometric progression (whose common ratio is neither 0 nor 1) are in arithmetic progression. Prove that the sum of the first three is equal to the fourth term.

10 The first term of an arithmetic series is 3, the common difference is 4 and the sum of all the terms is 820. Find the number of terms and the last term. [JMB]

11 The sum to infinity of a geometric series is S. The sum to infinity of the squares of the terms is $2S$. The sum to infinity of the cubes is $\frac{64}{13}S$. Find S and the first three terms of the original series. [LU]

12 Find how many terms of the series $1 + \dfrac{1}{5} + \dfrac{1}{5^2} + \dfrac{1}{5^3} + \dots$ must be taken so that the sum will differ from the sum to infinity by less than 10^{-6}.

13 A ball when dropped from any given height loses 20 per cent of its previous height at each rebound. If it is dropped from a height of 40 m, find how often it will rise to a height of over 8 m. How far does the ball travel before coming to rest?

14 Find the sum of all the positive integers less than 1000 that are not multiples of 3.

15 The three real, distinct and non-zero numbers a, b, c are such that a, b, c are in arithmetic progression, and a, c, b are in geometric progression. Find the numerical value of the common ratio of the geometric progression. Hence find an expression, in terms of a, for the sum to infinity of the geometric series whose first terms are a, c, b. [JMB]

16 (i) In an arithmetical progression the sum of the squares of five consecutive terms equals 20 times the square of the middle term and the product of the five terms equals 80. Find the middle term.
(ii) A nail, 4 cm long, is driven into wood by blows of a hammer. The first blow drives it in $2\frac{1}{2}$ cm and each successive blow drives it in two fifths of the previous distance (except the last for which the distance is less). Find how many blows must be used.
If the blows are reduced, the ratio of successive distances being maintained, find the least initial distance through which the first blow must drive the nail in order that the nail may ultimately be driven home. [LU]

17 The nth term of a series is $(an + 5r^n)$ where a and r are constant with $r \neq 0$ or 1. Find the sum of the first n terms. The nth term of the series $18 + 36 + 64 + \ldots$ is of the form stated above. Find a and r and the sum of the first 10 terms.

18 The first term of a geometric series is 18 and the sum to infinity is 20. Find the common ratio and the sum of the first six terms. Find also in its simplest form the ratio of the nth term to the sum of all the subsequent terms of the infinite series. [JMB]

19 If S_n denotes the sum of the first n terms of a geometric progression whose first term is a and whose common ratio is r, show that

(i) $S_n(S_{3n} - S_{2n}) = (S_{2n} - S_n)^2$

(ii) $r^{m-n} = \dfrac{S_{m+p} - S_m}{S_{n+p} - S_n}$

20 For what values of x do both the series

$$1 - x + x^2 - x^3 + x^4 + \ldots$$

$$1 + \frac{x}{1+x} + \left(\frac{x}{1+x}\right)^2 + \ldots$$

converge?

If for any value of x in this range, the limits of the sum of the two series are S_1 and S_2, show that $S_1S_2 = 1$.

21 Write down the sum of the natural numbers from m to $n\,(n > m)$ inclusive. The natural numbers are arranged in groups thus: $1 + (2 + 3) + (4 + 5 + 6) + (7 + 8 + 9 + 10) + \ldots$, so that the rth group contains r numbers. Find (i) the first number in the rth group, (ii) the sum of the numbers in the rth group. Show that the sum of the numbers in the $(2r - 1)$th group is $r^4 - (r - 1)^4$. [WJC]

22 Show that the series $c + \dfrac{c^2}{1+c} + \dfrac{c^3}{(1+c)^2} + \dfrac{c^4}{(1+c)^3} + \ldots$ converges for all values of c greater than $-\frac{1}{2}$.

If c_1 and c_2 are two possible values of c for which the series converges and S_1 and S_2 are the corresponding sums, show that if $c_1 > c_2$, $S_1 > S_2$.

23 Find for what values of x the series

(i) $\dfrac{1}{1+x} - \dfrac{1-x}{(1+x)^2} + \dfrac{(1-x)^2}{(1+x)^3} - \dfrac{(1-x)^3}{(1+x)^4} + \ldots$

(ii) $\dfrac{1}{1+3x} + \dfrac{1+x}{(1+3x)^2} + \dfrac{(1+x)^2}{(1+3x)^3} + \dfrac{(1+x)^3}{(1+3x)^4} + \ldots$

converge, and prove that the sum to infinity of the first series is $\frac{1}{2}$. [LU]

24 If $T_n = a^{n-1}$, $a \neq 1$, and $S_n = T_1 + T_2 + \ldots + T_n$, find, in terms of a and n, in their simplest form,

(i) $T_1 + T_2 + T_3 + \ldots + T_n$
(ii) $T_1T_2T_3 \ldots T_n$
(iii) $S_1 + S_2 + S_3 + \ldots + S_n$ [AEB]

25 If p, q, r, s, are successive terms of an arithmetic sequence show that

$\frac{1}{qrs}, \frac{1}{rsp}, \frac{1}{spq}, \frac{1}{pqr}$ are also successive terms in an arithmetic sequence.

26 The pth, qth and rth terms of a sequence are P, Q and R respectively. Show that

(i) if the sequence is arithmetic,

$$P(q-r)+Q(r-p)+R(p-q)=0$$

(ii) if the sequence is geometric,

$$(q-r)\log P+(r-p)\log Q+(p-q)\log R=0$$

27 The sum of the first p, q, r terms of an arithmetic series are P, Q, R respectively. Show that

$$Pqr(q-r)+Qpr(r-p)+Rpq(p-q)=0.$$

28 An arithmetic series and a geometric series have r as the common difference and the common ratio respectively. The first term of the arithmetic series is 1 and the first term of the geometric series is 2. If the fourth term of the arithmetic series is equal to the sum of the third and fourth terms of the geometric series, find three possible values of r. When $|r|<1$ find, in the form $p+q\sqrt{2}$, (i) the sum to infinity of the geometric series, (ii) the sum of the first ten terms of the arithmetic series. [AEB]

29 The sum of three real distinct quantities in geometric progression is p, and the sum of their squares is q. Show that the middle one is equal to $\frac{p^2-q}{2p}$, and that p^2/q must be in one or other of the ranges: $\frac{1}{3}<\frac{p^2}{q}<1$, $1<\frac{p^2}{q}<3$. [WJC]

30 The first term of a geometric series is 2 and the second term is x. State the set of values of x for which the series is convergent. Show that when convergent the series converges to a sum greater than 1. If $x=\frac{1}{2}$, find the smallest positive integer n such that the sum of the first n terms differs from the sum to infinity by less than 2^{-10}. [LU]

31 The sum of the first n terms of a series is $2n^2+n$. Find the nth term of the series and show that the series is an arithmetic progression. State the values of the first term and the common difference. Find the least value of n for which the sum of the first n terms of the series is greater than 1000. [C]

32 A geometric series with first term 3 converges to the sum of 2. Find the fifth term in the series. [LU]

33 (i) The sum to infinity of a geometric progression is 3. When the terms of this geometric progression are squared a new geometric progression is obtained whose sum to infinity is 1·8. Find the first term and the common ratio of each series.

(ii) Express the recurring decimal $0{\cdot}3\dot{2}\dot{1}$ in the form p/q, where p and q are integers with no common factor. [AEB]

34 Show that when the $2n$ terms of the series

$$1^2-2^2+3^2-4^2+\ldots+(2r-1)^2-(2r)^2+\ldots+(2n-1)^2-(2n)^2$$

are bracketed in pairs, an arithmetic progression is formed. Hence find the sum of the given series. [C]

35 The sum of the first twenty terms of an arithmetic progression is 45, and the sum of the first forty terms is 290. Find the first term and the common difference. Find the number of terms in the progression which are less than 100. [JMB]

3
The binomial theorem

3.1 The binomial theorem for a positive integral index

In this section, we shall obtain a formula for the expansion, in terms of powers of x, of $(1 + x)^n$, where n is a positive integer.

First, we observe that by ordinary multiplication

$$\begin{aligned}(1 + x) &= 1 + x \\ (1 + x)^2 &= (1 + x)(1 + x) = 1 + 2x + x^2 \\ (1 + x)^3 &= (1 + 2x + x^2)(1 + x) = 1 + 3x + 3x^2 + x^3 \\ (1 + x)^4 &= (1 + x)^3(1 + x) = 1 + 4x + 6x^2 + 4x^3 + x^4 \\ (1 + x)^5 &= (1 + x)^4(1 + x) = 1 + 5x + 10x^2 + 10x^3 + 5x^4 + x^5\end{aligned}$$

and so on.

From these few results, we notice that in each case the first and last coefficients are unity. Further, we notice that each of the other coefficients in $(1 + x)^{n+1}$ is the sum of the corresponding coefficient and the preceding one in the expansion of $(1 + x)^n$. Thus we can lay out the coefficients for successive powers in the form of a triangle (Pascal's triangle) using these two rules. The last two rows which are obtained in this way are, as is readily verified, the coefficients in the expansion of $(1 + x)^6$ and $(1 + x)^7$ respectively.

1	1							
1	2	1						
1	3	3	1					
1	4	6	4	1				
1	5	10	10	5	1			
1	6	15	20	15	6	1		
1	7	21	35	35	21	7	1	etc.

In order to be able to write down the expansion oı $(1 + x)^n$ in powers of x, we need a formula for the coefficient of x^r in the expansion of $(1 + x)^n$. We shall denote this by the symbol $\binom{n}{r}$ so that

$$(1 + x)^n = 1 + \binom{n}{1}x + \binom{n}{2}x^2 + \ldots + \binom{n}{r}x^r + \ldots + \binom{n}{n-1}x^{n-1} + x^n \quad (3.1)$$

Our problem is then to find a formula for $\binom{n}{r}$ in terms of n and r.

It can be seen that the coefficients in the row corresponding to $(1 + x)^n$ follow the pattern

$$1,\ n,\ \frac{n(n-1)}{1.2},\ \frac{n(n-1)(n-2)}{1.2.3},\ \ldots$$

Thus

$$\binom{6}{3} = \frac{6(6-1)(6-2)}{1.2.3} = 5.4 = 20$$

$$\binom{4}{2} = \frac{4(4-1)}{1.2} = \frac{4.3}{1.2} = 6$$

and so on. We shall show that

$$\binom{n}{r} = \frac{n(n-1)(n-2)\ldots(n-r+1)}{r(r-1)(r-2)\ldots 3.2.1} \tag{3.2}$$

is the appropriate result for the general case. There are r factors in both numerator and denominator. In the numerator the factors begin at n and decrease by one each time to $n - r + 1$; in the denominator they begin at r and decrease by one each time to 1. Equations (3.1) and (3.2) constitute the binomial theorem when n is a positive integer.

Before proving this result, we shall verify its correctness for some of the cases already considered.

Example 1 Use (3.2) to evaluate the coefficients of (i) x, x^2, x^3, x^4, x^5 in $(1 + x)^5$ and (ii) $x, x^2, x^3, x^4, x^5, x^6$ in $(1 + x)^6$.

(i) $\binom{5}{1} = \frac{5}{1} = 5 \qquad \binom{5}{2} = \frac{5.4}{2.1} = 10$

$\binom{5}{3} = \frac{5.4.3}{3.2.1} = 10 \qquad \binom{5}{4} = \frac{5.4.3.2}{4.3.2.1} = 5$

which are the results we had before.

Note $\binom{5}{5} = \frac{5.4.3.2.1}{5.4.3.2.1} = 1$, so that the result is true for the last coefficient as well.

(ii) $\binom{6}{1} = \frac{6}{1} = 6 \qquad \binom{6}{2} = \frac{6.5}{2.1} = 15 \qquad \binom{6}{3} = \frac{6.5.4}{3.2.1} = 20$

$\binom{6}{4} = \frac{6.5.4.3}{4.3.2.1} = 15 \qquad \binom{6}{5} = \frac{6.5.4.3.2}{5.4.3.2.1} = 6$

which are the results we had before. Also

$$\binom{6}{6} = \frac{6.5.4.3.2.1}{6.5.4.3.2.1} = 1$$

Indeed, we see that

$$\binom{n}{n} = \frac{n(n-1)\ldots 3.2.1}{n(n-1)\ldots 3.2.1} = 1 \quad \text{for all } n$$

Thus (3.1) and (3.2) would give

$$(1 + x)^5 = 1 + 5x + 10x^2 + 10x^3 + 5x^4 + x^5$$

and

$$(1 + x)^6 = 1 + 6x + 15x^2 + 20x^3 + 15x^4 + 6x^5 + x^6$$

which we know to be correct. It is readily verified that (3.1) and (3.2) also give the correct expansion for $(1 + x)^2$, $(1 + x)^3$, $(1 + x)^4$.

We also observe that our formula for $\binom{n}{r}$ satisfies the result that a coefficient in $(1 + x)^{n+1}$ is the sum of the corresponding coefficient and the preceding one in the expansion of $(1 + x)^n$. In symbols, this is

$$\binom{n+1}{r} = \binom{n}{r} + \binom{n}{r-1} \tag{3.3}$$

and with $\binom{n}{r}$ as given by (3.2)

$$\begin{aligned}\binom{n}{r} + \binom{n}{r-1} &= \frac{n(n-1)(n-2)\ldots(n-r+1)}{r(r-1)\ldots 3.2.1} + \frac{n(n-1)\ldots(n-r+2)}{(r-1)(r-2)\ldots 2.1}\\ &= \frac{n(n-1)(n-2)\ldots(n-r+1) + r(n)(n-1)\ldots(n-r+2)}{r(r-1)\ldots 3.2.1}\\ &= \frac{n(n-1)(n-2)\ldots(n-r+2)(n-r+1+r)}{r(r-1)\ldots 3.2.1}\\ &= \frac{(n+1)(n)(n-1)\ldots(n-r+2)}{r(r-1)(r-2)\ldots 3.2.1}\\ &= \frac{(n+1)(n+1-1)(n+1-2)\ldots(n+1-r+1)}{r(r-1)\ldots 3.2.1}\\ &= \binom{n+1}{r}\end{aligned}$$

which proves the result.

The expansion of $(a + x)^n$

Assuming for the moment the validity of (3.1) and (3.2), we notice that $(a + x)^n$ can be expanded in terms of powers of x and a with the same coefficients.

$$\begin{aligned}(a + x)^n &= \left[a\left(1 + \frac{x}{a}\right)\right]^n = a^n\left(1 + \frac{x}{a}\right)^n\\ &= a^n\left[1 + \binom{n}{1}\frac{x}{a} + \binom{n}{2}\frac{x^2}{a^2} + \ldots \binom{n}{r}\frac{x^r}{a^r} + \ldots + \frac{x^n}{a^n}\right]\end{aligned}$$

therefore

$$(a+x)^n = a^n + \binom{n}{1}xa^{n-1} + \binom{n}{2}x^2a^{n-2} + \ldots + \binom{n}{r}x^ra^{n-r} + \ldots + x^n \tag{3.4}$$

Example 2 Expand $(2+x)^5$ in powers of x.

Using the coefficients as given by Example 1,

$$\begin{aligned}(2+x)^5 &= 2^5 + 5.2^4x + 10.2^3x^2 + 10.2^2x^3 + 5.2x^4 + x^5\\ &= 32 + 80x + 80x^2 + 40x^3 + 10x^4 + x^5\end{aligned}$$

The evaluation of $\binom{n}{r}$

$$\begin{aligned}\binom{n}{r} &= \frac{n(n-1)\ldots(n-r+1)}{r(r-1)\ldots 3.2.1}\\ &= \frac{n(n-1)\ldots(n-r+1)}{r(r-1)\ldots 3.2.1}\,\frac{(n-r)(n-r-1)\ldots 3.2.1}{(n-r)(n-r-1)\ldots 3.2.1}\end{aligned}$$

therefore

$$\binom{n}{r} = \frac{n!}{r!(n-r)!} \tag{3.5}$$

where $n!$ (factorial n) is used to denote the product of the integers from n down to 1. Thus $2! = 2.1 = 2$, $3! = 3.2.1 = 6$, $4! = 4.3.2.1 = 24$, etc. From (3.5) we see that

$$\binom{n}{n-r} = \frac{n!}{(n-r)![n-(n-r)]!} = \frac{n!}{r!(n-r)!}$$

therefore

$$\binom{n}{n-r} = \binom{n}{r} \tag{3.6}$$

a result which merely expresses the symmetry of the coefficients which was apparent for the numerical cases considered earlier.

We use this symmetry to define

$$\binom{n}{0} = \binom{n}{n} = 1 \tag{3.7}$$

Then

$$\begin{aligned}(1+x)^n &= \binom{n}{0} + \binom{n}{1}x + \binom{n}{2}x^2 + \ldots + \binom{n}{r}x^r + \ldots + \binom{n}{n}x^n\\ &= \sum_{r=0}^{n}\binom{n}{r}x^r \end{aligned} \tag{3.8}$$

using the notation of the previous chapter.

Example 3 Evaluate (i) $\binom{20}{2}$, (ii) $\binom{20}{17}$, (iii) $\binom{8}{5}$.

(i) By (3.2)

$$\binom{20}{2} = \frac{20.19}{2.1} = 190$$

(ii) We first use (3.6)

$$\binom{20}{17} = \binom{20}{3} = \frac{20.19.18}{3.2.1} = 60.19 = 1140$$

(Had we used (3.2) for $\binom{20}{17}$, we should have 17 factors in the numerator and denominator.)

(iii)
$$\binom{8}{5} = \binom{8}{3} = \frac{8.7.6}{3.2.1} = 56$$

Example 4 Expand $(1-2x)^4$ in powers of x.

We write $(1-2x)$ as $[1+(-2x)]$

$$\begin{aligned}(1-2x)^4 &= [1+(-2x)]^4\\ &= 1+\binom{4}{1}(-2x)+\binom{4}{2}(-2x)^2+\binom{4}{3}(-2x)^3+(-2x)^4\\ &= 1+\binom{4}{1}(-2x)+\binom{4}{2}(-2x)^2+\binom{4}{3}(-2x)^3+(-2x)^4\\ &= 1-8x+24x^2-32x^3+16x^4\end{aligned}$$

Example 5 Expand $(x+3y)^6$.

$$\binom{6}{1} = \binom{6}{5} = 6 \qquad \binom{6}{2} = \binom{6}{4} = \frac{6.5}{2.1} = 15 \qquad \binom{6}{3} = \frac{6.5.4}{3.2.1} = 20$$

By (3.4)

$$\begin{aligned}(x+3y)^6 &= x^6+6x^5(3y)+15x^4(3y)^2+20x^3(3y)^3\\ &\qquad +15x^2(3y)^4+6x(3y)^5+(3y)^6\\ &= x^6+18x^5y+135x^4y^2+540x^3y^3+1215x^2y^4\\ &\qquad +1458xy^5+729y^6\end{aligned}$$

Exercises 3a

Assume the validity of (3.1) and (3.2).

1 Obtain the expansions in powers of x of (i) $(1+x)^2$, (ii) $(1+x)^3$, (iii) $(1+x)^4$.

2 Use the rules described at the beginning of this section to obtain the coefficients in the expansions of $(1+x)^8$ and $(1+x)^9$, i.e. obtain the next two rows of Pascal's triangle.

3 Use (3.1) and (3.2) to verify the results obtained in question 2.

4 Evaluate (i) $\binom{18}{16}$, (ii) $\binom{11}{9}$, (iii) $\binom{13}{8}$, (iv) $\binom{9}{7}$.

5 Expand (i) $(1+3x)^4$, (ii) $(1-x)^5$, (iii) $(1-2x)^7$.

6 Expand (i) $(3+x)^4$, (ii) $(2-x)^6$.

7 Expand (i) $(2x + 3y)^3$, (ii) $(2x - 5y)^4$.
8 Obtain the fourth term in the expansion, in ascending powers of x, of $(2 + 3x)^{11}$.
9 Calculate a if the coefficient of x^3 in $(a + 2x)^5$ is 320.
10 Calculate the coefficient of x^3y^4 in $(2x - 3y)^7$.

3.2 Proof of the binomial theorem when *n* is a positive integer

The results of the previous section suggest that the results (3.1) and (3.2) are true for all positive integer values of n. We shall now prove this by a method of proof known as mathematical induction.

We assume that for a particular value of n, say $n = N$, the results (3.1) and (3.2) are correct. (This is not so unreasonable since we have seen that this is so when $n = 2$, $n = 3$, $n = 4$, $n = 5$, and $n = 6$ among others.)

Thus we assume

$$(1 + x)^N = 1 + \binom{N}{1}x + \binom{N}{2}x^2 + \dots + \binom{N}{r}x^r + \dots + \binom{N}{N-1}x^{N-1} + x^N$$

Then

$$(1 + x)^{N+1} = (1 + x)(1 + x)^N$$

$$= (1 + x)\left(1 + \binom{N}{1}x + \binom{N}{2}x^2 + \dots + \binom{N}{r}x^r + \dots + x^N\right)$$

Therefore

$$(1 + x)^{N+1} = 1 + \left(\binom{N}{1} + 1\right)x + \left(\binom{N}{2} + \binom{N}{1}\right)x^2$$
$$+ \dots + \left(\binom{N}{r} + \binom{N}{r-1}\right)x^r + \dots + x^{N+1}$$

by ordinary multiplication.

Now

$$\binom{N}{1} + 1 = N + 1 = \binom{N+1}{1}$$

and by (3.3) we have

$$\binom{N}{r} + \binom{N}{r-1} = \binom{N+1}{r}$$

Thus our initial assumption inevitably leads to

$$(1 + x)^{N+1} = 1 + \binom{N+1}{1}x + \binom{N+1}{2}x^2 + \dots + \binom{N+1}{r}x^r$$
$$+ \dots + \binom{N+1}{N}x^N + x^{N+1}$$

Thus, we have proved that if the results (3.1) and (3.2) are true for a given value $n = N$, then they are true for the next value of n, i.e. $n = N + 1$. But we have seen that the result is true for $n = 2$, therefore it is true for $n = 3$, therefore it is true for $n = 4$, and so on for all other positive integer values of n. Thus we have shown that (3.1) and (3.2) are true for all positive integer values of n. We have also proved (3.4) and (3.8) for such values of n.

Example 1 Expand $(1 + x + x^2)^3$ in powers of x.

$$\begin{aligned}(1 + x + x^2)^3 &= [1 + (x + x^2)]^3 \\ &= 1 + 3(x + x^2) + 3(x + x^2)^2 + (x + x^2)^3 \\ &= 1 + 3x + 3x^2 + 3(x^2 + 2x^3 + x^4) \\ &\quad + (x^3 + 3x^4 + 3x^5 + x^6) \\ &= 1 + 3x + 6x^2 + 7x^3 + 6x^4 + 3x^5 + x^6\end{aligned}$$

Example 2 Find the coefficient of x^8 in $\left(x^2 + \dfrac{2y}{x}\right)^{10}$

The $(r + 1)$th term in the expansion of $\left(x^2 + \dfrac{2y}{x}\right)^{10}$ is

$$\binom{10}{r}(x^2)^{10-r}\left(\frac{2y}{x}\right)^r = \binom{10}{r}2^r y^r x^{20-2r-r} = \binom{10}{r}2^r y^r x^{20-3r}$$

Thus, for the term in x^8,

$$20 - 3r = 8 \qquad \text{i.e.} \quad r = 4$$

Therefore, the required coefficient is $\binom{10}{4}2^4 y^4$

$$= \frac{10.9.8.7}{4.3.2.1}16y^4 = 210.16y^4 = 3360y^4$$

Example 3 Find the values of a if the coefficient of x^2 in the expansion of $(1 + ax)^4(2 - x)^3$ is 6.

$$\begin{aligned}(1 + ax)^4 &= 1 + 4(ax) + 6(ax)^2 + 4(ax)^3 + (ax)^4 \\ (2 - x)^3 &= [2 + (-x)]^3 = 2^3 + 3.2^2(-x) + 3.2(-x)^2 + (-x)^3 \\ &= 8 - 12x + 6x^2 - x^3\end{aligned}$$

Therefore, the coefficient of x^2 in the expansion of $(1 + ax)^4(2 - x)^3$ is the coefficient of x^2 in

$$(1 + 4ax + 6a^2x^2 + 4a^3x^3 + a^4x^4) \times (8 - 12x + 6x^2 - x^3)$$

This coefficient is $6 - 48a + 48a^2$. If this is equal to 6,

$$\begin{aligned}48a^2 - 48a + 6 &= 6 \\ 8a^2 - 8a &= 0 \\ 8a(a - 1) &= 0 \\ a = 0 \quad &\text{or} \quad a = 1\end{aligned}$$

Example 4 Expand $(x + 5y)^5$. Hence evaluate $(1{\cdot}05)^5$ correct to three decimal places

$$\binom{5}{1} = \binom{5}{4} = 5 \qquad \binom{5}{2} = \binom{5}{3} = \frac{5.4}{2.1} = 10$$

Therefore

$$\begin{aligned}(x + 5y)^5 &= x^5 + 5x^4(5y) + 10x^3(5y)^2 + 10x^2(5y)^3 + 5x(5y)^4 + (5y)^5 \\ &= x^5 + 25x^4y + 250x^3y^2 + 1250x^2y^3 + 3125xy^4 + 3125y^5\end{aligned}$$

With $x = 1$, $y = 0{\cdot}01$, we obtain

$$\begin{aligned}(1{\cdot}05)^5 &= 1 + 25(0{\cdot}01) + 250(0{\cdot}0001) + 1250(0{\cdot}000\,001) \\ &\quad + 3125(0{\cdot}000\,000\,01) + 3125(0{\cdot}000\,000\,0001) \\ &\approx 1 + 0{\cdot}25 + 0{\cdot}025 + 0{\cdot}001\,25\end{aligned}$$

where we have omitted the last two terms since they do not affect the first five decimal figures. Therefore

$$(1{\cdot}05)^5 \approx 1{\cdot}276\,25$$

or to three decimal places

$$(1{\cdot}05)^5 = 1{\cdot}276$$

Exercises 3b

1 Expand in powers of x, $(1 + 2x + 2x^2)^3$.
2 Expand in powers of x, $(1 - x + 2x^2)^4$.
3 Find the coefficient of x^3 in the expansion of $(1 + x + 2x^2)^6$.
4 Expand $(1 + 2x + x^2)^3$. (You might be able to obtain this result in an easier way.)
5 Find the coefficient of x^6 in the expansion $\left(\dfrac{1}{x^2} - x\right)^{18}$.
6 Find the term independent of y in the expansion of $\left(\dfrac{x^4}{y^3} + \dfrac{y^2}{2x}\right)^{10}$.
7 Find the term independent of x in the expansion of $\left(2x - \dfrac{3}{x^2}\right)^6$.
8 Find the coefficient of x^{22} in the expansion of $(1 - 3x)(1 + x^3)^{10}$.
9 Find a if the coefficient of x in the expansion of

$$(1 + ax)^8(1 + 3x)^4 - (1 + x)^3(1 + 2x)^4$$

is zero. What is the coefficient of x^2?
10 Expand $(x + 2y)^7$. Hence evaluate $(1{\cdot}02)^7$ to four significant figures.

3.3 The binomial theorem when *n* is not a positive integer

It can be shown (although the proof is beyond the scope of this book) that if $-1 < x < 1$ and n has any value

$$(1+x)^n = 1 + nx + \frac{n(n-1)}{1.2}x^2 + \frac{n(n-1)(n-2)}{1.2.3}x^3 + \ldots \tag{3.9}$$

The expansion (3.9) is known as the binomial theorem. If n is, in fact, a positive integer, the coefficients after the coefficient of x^n are all zero since they each contain the factor $(n-n)$. Thus, for this case, we see that the expansion terminates with x^n. We obtain, of course, the same result as was given by (3.1) and (3.2) and for this particular case the requirement that $-1 < x < 1$ is not necessary.

If n has a value other than a positive integer, the expansion will not terminate and the requirement $-1 < x < 1$ is absolutely essential. This latter requirement is often written, following the notation of the first chapter, $|x| < 1$. Although the expansion does not terminate in this case, it can still be used with great effect to approximate (as accurately as we like) the value of $(1+x)^n$ when x has a value such that $|x| < 1$.

Example 1 Obtain the first five terms in the expansion $(1+x)^{1/2}$. Hence evaluate $\sqrt{1{\cdot}03}$ to five significant figures.

For this case, $n = \frac{1}{2}$ and so by (3.9)

$$\begin{aligned}(1+x)^{1/2} &= 1 + (\tfrac{1}{2})x + \frac{\frac{1}{2}(\frac{1}{2}-1)}{1.2}x^2 + \frac{\frac{1}{2}(\frac{1}{2}-1)(\frac{1}{2}-2)}{1.2.3}x^3 \\ &\quad + \frac{\frac{1}{2}(\frac{1}{2}-1)(\frac{1}{2}-2)(\frac{1}{2}-3)}{1.2.3.4}x^4 + \ldots \\ &= 1 + \tfrac{1}{2}x - \tfrac{1}{8}x^2 + \tfrac{1}{16}x^3 - \tfrac{5}{128}x^4 + \ldots\end{aligned}$$

With $x = 0{\cdot}03$, which is certainly between -1 and $+1$,

$$(1{\cdot}03)^{1/2} = \sqrt{1{\cdot}03} = 1 + \tfrac{1}{2}(0{\cdot}03) - \tfrac{1}{8}(0{\cdot}0009) + \tfrac{1}{16}(0{\cdot}000\,027) + \ldots$$

and since we only require the result to four decimal places, we need in fact only consider the first three terms. Therefore

$$\begin{aligned}(1{\cdot}03)^{1/2} &= 1 + 0{\cdot}015 - 0{\cdot}000\,1125 + 0{\cdot}000\,0017 + \ldots \\ &\approx 1{\cdot}014\,8892 \\ \sqrt{1{\cdot}03} &= 1{\cdot}0149 \quad \text{correct to 5 significant figures}\end{aligned}$$

Example 2 Expand (i) $(2+x)^{-1}$, (ii) $\dfrac{1}{1-x}$.

$$\text{(i)}\ (2+x)^{-1} = \frac{1}{2+x} = \frac{1}{2\left(1+\frac{x}{2}\right)} = \frac{1}{2}\left(1+\frac{x}{2}\right)^{-1}$$

$$= \frac{1}{2}\left[1 + (-1)\frac{x}{2} + \frac{(-1)(-2)}{1.2}\left(\frac{x}{2}\right)^2 + \frac{(-1)(-2)(-3)}{1.2.3}\left(\frac{x}{2}\right)^3 + \ldots\right] \quad \text{provided } \left|\frac{x}{2}\right| < 1$$

Therefore

$$(2+x)^{-1} = \frac{1}{2} - \frac{x}{4} + \frac{x^2}{8} - \frac{x^3}{16} + \frac{x^4}{32} + \dots \qquad \text{provided } |x| < 2$$

(ii) $$\frac{1}{1-x} = [1+(-x)]^{-1}$$

$$= 1 + (-1)(-x) + \frac{(-1)(-2)}{1.2}(-x)^2 + \frac{(-1)(-2)(-3)}{1.2.3}(-x)^3 + \dots$$

Therefore

$$\frac{1}{1-x} = 1 + x + x^2 + x^3 + x^4 + \dots \qquad \text{provided } |x| < 1$$

(cf. (2.15) with $a = 1$)

If we ignore the requirement $|x| < 1$ and put $x = 2$ in the left-hand side, we obtain -1. On the right-hand side, we obtain $1 + 2 + 4 + 8 + \dots$, which is quite meaningless and is certainly not -1. The point to realise is that the expansion is only valid of $-1 < x < 1$.

Example 3 Show that if x is so small that x^3 and higher powers of x can be neglected, then $\dfrac{(1+2x)^{3/2} - 4(1+x)^{1/2}}{1+x^2} = -3 + x + 5x^2$.

$$(1+2x)^{3/2} = 1 + (\tfrac{3}{2})2x + \frac{(\frac{3}{2})(\frac{1}{2})(2x)^2}{1.2} \qquad \text{as far as terms in } x^2$$

$$(1+x)^{1/2} = 1 + \tfrac{1}{2}x + \frac{(\frac{1}{2})(-\frac{1}{2})}{1.2}x^2 \qquad \text{as far as terms in } x^2$$

$$(1+x^2)^{-1} = 1 - x^2 \qquad \text{as far as terms in } x^2$$

Therefore

$$\frac{(1+2x)^{3/2} - 4(1+x)^{1/2}}{1+x^2} = (1-x^2)[1 + 3x + \tfrac{3}{2}x^2 - 4(1 + \tfrac{1}{2}x - \tfrac{1}{8}x^2)]$$

Simplifying and retaining only terms which involve x^2 and lower powers of x, we have

$$\frac{(1+2x)^{3/2} - 4(1+x)^{1/2}}{1+x^2} = (1-x^2)(-3 + x + 2x^2) = -3 + x + 5x^2$$

Example 4 Expand $\dfrac{3-x}{(1-2x)(1+x^2)}$ in ascending powers of x as far as the term in x^3.

We first resolve $\dfrac{3-x}{(1-2x)(1+x^2)}$ into partial fractions:

$$\begin{aligned}\frac{3-x}{(1-2x)(1+x^2)} &= \frac{2}{1-2x}+\frac{1+x}{1+x^2}\\ &= 2(1-2x)^{-1}+(1+x)(1+x^2)^{-1}\\ &= 2(1+2x+4x^2+8x^3+\ldots)\\ &\quad +(1+x)(1-x^2+x^4-x^6+x^8+\ldots)\\ &= (2+4x+8x^2+16x^3+\ldots)\\ &\quad +1+x-x^2-x^3+\ldots\\ &= 3+5x+7x^2+15x^3+\ldots\end{aligned}$$

This expression is valid only if both $-2x$ and x^2 lie between -1 and $+1$:

$$-1<-2x<1 \quad \text{if} \quad -\tfrac{1}{2}<x<\tfrac{1}{2}$$
$$-1<x^2<1 \quad \text{if} \quad -1<x<1$$

Therefore, for the expansion to be valid, x must lie between $-\frac{1}{2}$ and $\frac{1}{2}$, i.e. $|x|<\frac{1}{2}$.

Exercises 3c

1 Obtain the first four terms of the expansion of $(1+3x)^{1/3}$ in ascending powers of x. Hence evaluate $\sqrt[3]{1{\cdot}03}$ to five significant figures.

2 Evaluate $(0{\cdot}95)^{1{\cdot}3}$. Use logarithms to check your result.

3 Show that (i) $\dfrac{1}{(1-x)^2}=1+2x+3x^2+4x^3+\ldots$ and (ii) $\dfrac{1}{(1-x)^3}=1+3x+\dfrac{3.4x^2}{2}+\dfrac{4.5x^3}{2}+\ldots.$

4 Obtain the expansions in ascending powers of x of

(i) $\dfrac{1}{1+x}$ (ii) $\dfrac{1}{(1+x)^2}$ (iii) $\dfrac{1}{(1+x)^3}$.

5 Use the results above to expand in ascending powers of x

(i) $\dfrac{1}{1+3x}$ (ii) $\dfrac{1}{(6-x^3)^2}$ (iii) $\dfrac{1}{(3+x)^3}$.

6 Show that $\sqrt{(9+x^2)}\approx 3+\frac{1}{6}x^2-\frac{1}{216}x^4$. For what values of x is the expansion valid?

7 Show that

$$\frac{1}{a+bx}\approx\frac{1}{a}-\frac{b}{a^2}x+\frac{b^2}{a^3}x^2-\frac{b^3}{a^4}x^3+\ldots.$$

For what values of x is the expansion valid?

8 Show that, if x is so small that x^3 and higher powers of x may be neglected,

$$\frac{(1+x)^{3/2}-(1+\frac{1}{2}x)^3}{\sqrt{(1-x)}}\approx-\tfrac{3}{8}x^2.$$

9 Obtain the expansion of $\dfrac{3-4x}{1-3x+2x^2}$ in ascending powers of x as far as the term in x^4.

10 Show that if x is so small that x^3 and higher powers of x can be neglected $\sqrt{\left(\frac{1+x}{1-x}\right)} = 1 + x + \frac{1}{2}x^2$. By putting $x = \frac{1}{7}$, show that $\sqrt{3} \approx \frac{196}{113}$.

3.4 Mathematical induction

In Section 3.1, we proved the binomial theorem for n a positive integer using the method of mathematical induction. This method is very useful where particular cases suggest that some result is true quite generally. The following examples are further illustrations of the method.

Example 1 Show that if n is a positive integer,

$$1 + 2 + 3 + 4 + \ldots + n = \frac{n(n+1)}{2} \qquad \text{(cf. (2.10))}$$

We observe that the result is true for the particular cases $n = 1, n = 2$, namely

$$1 = \frac{1(1+1)}{2} \qquad 1 + 2 = 3 = \frac{2(2+1)}{2}$$

We assume the result is true for a particular value, say $n = N$. Thus

$$1 + 2 + 3 + \ldots + N = \frac{N(N+1)}{2}$$

Then for the next value of n, $n = N + 1$

$$\begin{aligned} 1 + 2 + 3 + \ldots + N + (N+1) &= \frac{N(N+1)}{2} + (N+1) \\ &= (N+1)\left(\frac{N}{2} + 1\right) \\ &= \tfrac{1}{2}(N+1)(N+2) \\ &= \tfrac{1}{2}(N+1)[(N+1)+1] \end{aligned}$$

Thus, if the result is true for any particular value of n, it is also true for the next value of n. But we have seen that it is true for $n = 1$, therefore it is true for $n = 2$, therefore it is true for $n = 3$ and so on for all positive integer values of n.

Example 2 Show that for all positive integer values of n, $5^{2n} + 3n - 1$ is an integer multiple of 9.

We first observe that for $n = 1$

$$5^2 + 3.1 - 1 = 27$$

which is a multiple of 9. We assume the result to be true for $n = N$, that is

$$5^{2N} + 3N - 1 = 9M$$

where M is some integer. Then for $n = N + 1$, we obtain

$$\begin{aligned} 5^{2(N+1)} + 3(N+1) - 1 &= 25.5^{2N} + 3N + 2 \\ &= 25(9M - 3N + 1) + 3N + 2 \\ &\qquad \text{(since } 5^{2N} = 9M - 3N + 1\text{)} \end{aligned}$$

$$= 9.25M - 75N + 25 + 3N + 2$$
$$= 9.25M - 72N + 27$$
$$= 9(25M - 8N + 3)$$

This proves that $5^{2(N+1)} + 3(N+1) - 1$ is a multiple of 9 if $5^{2N} + 3N - 1$ is a multiple of 9.

The required result is true when $n = 1$, therefore it is true when $n = 2$ and so on for all positive integer values of n.

Exercises 3d

Use the method of mathematical induction to prove that if n is a positive integer:

1 $2^n > n$.
2 $1^2 + 2^2 + 3^2 + \dots + n^2 = \frac{1}{6}n(n+1)(2n+1)$.
3 $1 + 3 + 5 + \dots + (2n-1) = n^2$.
4 $n(n+1)(n+2)$ is an integer multiple of 6.
5 $7^{2n+1} + 1$ is an integer multiple of 8.

Exercises 3

1 Expand $(1 - x + 2x^2)^5$ in ascending powers of x as far as the term in x^4.

2 Use the binomial theorem to evaluate $\sqrt{99}$ to five significant figures.

3 Expand $(1-x)^{-1} - 2(1-2x)^{-1/2} + (1-3x)^{-1/3}$ in ascending powers of x, up to and including the term in x^3.

4 If $N = x^3 + t$ and t/x^3 is so small that its fourth and higher powers may be neglected, show that

$$\sqrt[3]{N} = x + \frac{t}{3x^2} - \frac{t^2}{9x^5} + \frac{5t^3}{81x^8}$$

Hence evaluate $\sqrt[3]{64{\cdot}032}$ correct to six decimal places. [AEB]

5 Show that $\dfrac{1}{\sqrt{(1-x)}} - \sqrt{(1+x)} = \dfrac{x^2}{2} + \dfrac{x^3}{4}$ if x^4 and higher powers of x may be neglected.

6 If x is small show that $(1 - x^2)^{-1/3} \approx 1 + \frac{1}{3}x^2 + \frac{2}{9}x^4$.

7 Express $\dfrac{7 - 23x + 48x^2}{(2+x)(1-3x)^2}$ in partial fractions and find the coefficient of x^n when the expression is expanded in ascending powers of x when $-\frac{1}{3} < x < \frac{1}{3}$. [LU, part]

8 Expand $\sqrt{4-x}$ as a series in ascending powers of x up to and including the term in x^2. If terms in x^n, $n \geqslant 3$, can be neglected, find the quadratic approximation to $\sqrt{\dfrac{4-x}{1-2x}}$. State the range of values of x for which this approximation is valid. [AEB]

9 If x^4 and higher powers of x can be neglected show that

$$\sqrt{\left(\frac{1-x}{1+x+x^2}\right)} = 1 - x + \tfrac{1}{2}x^3$$

10 In the expansion in powers of x of the function $(1+x)(a-bx)^{12}$ the coefficient of x^8 is zero. Find in its simplest form the value of a/b.

11 Show that $\binom{n+2}{r} = \binom{n}{r-2} + 2\binom{n}{r-1} + \binom{n}{r}$ where $n > r > 2$.

12 Use the method of induction to show that

$$\sum_{r=1}^{n} r^3 = \frac{n^2(n+1)^2}{4}$$

13 If x is small show that

$$\frac{\sqrt{(1+6x)} - \dfrac{1}{\sqrt{(1-6x)}}}{\sqrt{(1+3x)} - \dfrac{1}{\sqrt{(1-3x)}}} \approx 4 + 6x$$

14 Use the method of induction to show that

$$\frac{1}{1.2} + \frac{1}{2.3} + \frac{1}{3.4} + \ldots + \frac{1}{n(n+1)} = \frac{n}{n+1}$$

15 Expand $(1+4x^2)^{1/2}$ in ascending powers of x up to the term in x^6. Hence evaluate $\sqrt{1{\cdot}04}$.

16 Evaluate the term independent of x in the binomial expansion of

$$\left(x^2 - \frac{1}{2x}\right)^9$$

17 Show that if x is so small that x^4 and higher powers can be neglected then $\dfrac{1+2x+3x^2}{(1-x)(1+x^2)}$ can be expressed in the form $A + Bx + Cx^2 + Dx^3$ and find A, B, C, D.

18 Write down the first few terms in the expansion in ascending powers of x of $(1+4x)^{1/2}$, and simplify the coefficients. Hence by putting $x = -\frac{1}{100}$, calculate $\sqrt{6}$ correct to four decimal places. [JMB, part]

19 If x is small show that $(1+x)^{3/2} \approx 1 + \dfrac{3x}{2} + \dfrac{3x^2}{8} - \dfrac{1}{16}x^3$. Evaluate $\sqrt{11}$.

20 Find p and q if the coefficients of x and x^3 in the expansion of $(1+px+qx^2+4x^3)(1+x)^6$ are both zero.

21 If $|x| < 1$, prove that the sum to infinity of the series

$$1 + 5x + 9x^2 + \ldots + (4n+1)x^n \quad \text{is} \quad \frac{1+3x}{(1-x)^2}$$

[JMB, part]

22 Show that $3^{4n+2} + 2.4^{3n+1}$ is exactly divisible by 17 if n is a positive integer.

23 Use the binomial theorem to show that the expansion of

$$\frac{1}{(1-x)^{1/2}} \quad \text{is} \quad 1 + \frac{1}{2}x + \frac{1.3}{2.4}x^2 + \frac{1.3.5}{2.4.6}x^3 + \ldots \quad (|x| < 1)$$

Also show that

$$1+\frac{4}{10}+\frac{4.12}{10.20}+\frac{4.12.20}{10.20.30}+\ldots$$
$$=2\left(1+\frac{1}{10}+\frac{1.3}{10.20}+\frac{1.3.5}{10.20.30}+\ldots\right)$$

[JMB, part]

24 If $\dfrac{1+x}{(1-x)^2}$ is expanded in ascending powers of x, where $-1<x<1$, show that the coefficient of x^{n-1} is $2n-1$ and that the sum of the terms after the nth term is $\dfrac{(2n+1)x^n}{1-x}+\dfrac{2x^{n+1}}{(1-x)^2}$. [LU, part]

25 Show that n^4+4n^2+11 is a multiple of 16 for all odd positive integers n.

26 Obtain the first four terms in the expansion in ascending powers of x of

$$\frac{(1+4x)^{1/4}}{(1-x)^3}$$ [LU]

27 When terms in x^n, $n \geqslant 4$, are omitted

$$\frac{3ax}{4}+\sqrt{(4+ax)}-\frac{2}{\sqrt{(1-ax)}}=-x^2+bx^3$$

Find the values of a and b. [AEB]

28 Express in partial fractions $\dfrac{3x-5}{x^2-2x-3}$. State the values of $|x|$ for which this expression can be expanded as a series of ascending powers of x and obtain the first three terms of this expansion. [LU]

29 Show that the first three terms in the expansion in ascending powers of x of $(1+8x)^{1/4}$ are the same as the first three terms in the expansion of $\dfrac{1+5x}{1+3x}$. Use the corresponding approximation

$$(1+8x)^{1/4}\approx\frac{1+5x}{1+3x}$$

to obtain an approximation to $(1{\cdot}16)^{1/4}$ as a rational fraction in its lowest terms. [JMB]

30 If x is so small that terms in x^n, $n \geqslant 3$, can be neglected and

$$\frac{3+ax}{3+bx}=(1-x)^{1/3}$$

find the values of a and b. Hence, without the use of tables, find an approximation in the form p/q, where p and q are integers, for $\sqrt[3]{0{\cdot}96}$. [AEB]

4

Complex numbers

4.1 Introduction

In Section 1.1, we discussed the necessity for the introduction of other types of real number apart from the natural integers which are used for counting. The solution of an equation such as $5x = 4$ requires the introduction of rational numbers; the solution of an equation such as $3x + 4 = 0$ requires the introduction of negative numbers. Then we can say that without exception every linear equation has one and only one solution.

Some quadratic equations, such as $x^2 = 2$, require the introduction of the irrational numbers into our number system before we can give a meaning to their solution. There are other equations, such as $x^2 + 16 = 0$, which do not have a solution within the system of real numbers. To solve the last equation, we require a number whose square is -16 and such a number does not exist within the system of real numbers. Thus, we have a breakdown of the general rule that quadratic equations have two solutions. (If the two solutions are equal, we say that the equation has a repeated root, i.e. *two* equal roots.)

This situation is not peculiar to the last-mentioned equation. Indeed, the equations $x^2 + 1 = 0$, $x^2 + 2x + 5 = 0$, whose solutions we might formally write as

$$x = \pm\sqrt{-1} \qquad x = \frac{-2 \pm \sqrt{4 - 4(5)(1)}}{2} \quad \text{by (1.2)}$$

that is

$$x = \pm\sqrt{-1} \qquad x = \frac{-2 \pm \sqrt{-16}}{2}$$

have no solutions within the realm of real numbers. In each case, we require the existence of a number whose square is negative before we can give a meaning to the solutions.

This situation calls for yet another extension of our number system. We can achieve this extension by the introduction of a new number, generally denoted by i, whose square is -1. The number i has the property

$$i^2 = -1 \tag{4.1}$$

This new number i having the property (4.1) need give no cause for concern.

It is new in the sense that up to now it has been outside our experience, but so at one stage were fractions, and negative numbers.

The solution of the equation $x^2 + 1 = 0$ is then

$$x = \pm\sqrt{-1} \qquad \text{i.e.} \quad x = \pm i$$

The solution of the equation $x^2 + 16 = 0$ is then

$$x = \pm\sqrt{-16} = \pm\sqrt{16(-1)} = \pm\sqrt{16}\sqrt{-1} = \pm 4i$$

The solution of the equation $x^2 + 2x + 5 = 0$ is then

$$x = \frac{-2 \pm \sqrt{-16}}{2} = \frac{-2 \pm 4i}{2} = -1 \pm 2i$$

A number of the form $a + ib$, where a and b are real numbers, is called a complex number. Our real numbers can be regarded as complex numbers for which b is zero. Thus, we need only consider complex numbers since our real numbers will be contained within the system of complex numbers. This will require us to consider carefully the rules for the addition and multiplication of complex numbers, so that these rules, when applied to real numbers of the form $(a + i0)$, give us the correct results within the real number system. This we shall do in the next section. For the present, we observe that higher powers of i can be reduced to ± 1 or $\pm i$.

Thus, since $i^2 = -1$

$$i^3 = i^2.i = -i \qquad i^4 = i^3.i = -i.i = -(-1) = 1$$
$$i^5 = i^4.i = i$$

and so on.

Exercises 4a

1 Express the solution of the following equations in the form $a + bi$:
(i) $3x - 7 = 0$ (ii) $x^2 - 9 = 0$
(iii) $x^2 + 30 = 0$ (iv) $x^2 + 3x + 10 = 0$
(v) $x^2 + 49 = 0$ (vi) $x^2 + 2x + 8 = 0$
(vii) $x^2 + 4x + 40 = 0$ (viii) $x^2 - x + 1 = 0$

2 Show that (i) $i^7 = -i$, (ii) $i^{13} = i$, (iii) $1 + i - 3i^2 + i^7 = 4$.

3 Show that $i^9 + 2i^{11} + i^{13} = 0$.

4 Show that the cubic equation $x^3 - 1 = 0$ has *three* solutions:

$$x = 1 \qquad x = -\frac{1}{2} + \frac{i\sqrt{3}}{2} \qquad x = -\frac{1}{2} - \frac{i\sqrt{3}}{2}$$

[Hint: $x^3 - 1 \equiv (x - 1)(x^2 + x + 1)$.]

5 Show that the quartic equation $x^4 - 1 = 0$ has *four* solutions:

$$x = 1 \qquad x = -1 \qquad x = i \qquad x = -i$$

[Hint: $x^4 - 1 \equiv (x^2 - 1)(x^2 + 1)$.]

4.2 The rules for the manipulation of complex numbers

In this section, we shall consider the rules governing the addition, subtraction, multiplication and division of complex numbers. These rules are defined so that they become the rules of algebra for real numbers when applied to complex numbers of the form $a + i0$. First, we define what is meant by the equality of two complex numbers $a + ib$ and $c + id$, where a, b, c, d are real numbers.

We say

$$a + ib = c + id \quad \text{if and only if} \quad a = c \quad \text{and} \quad b = d \tag{4.2}$$

This definition is very reasonable and is consistent with treating i as an ordinary algebraic quantity. For if $a + ib = c + id$, we would expect $a - c = i(d - b)$ which yields, on squaring both sides,

$$(a - c)^2 = i^2(d - b)^2$$
$$(a - c)^2 = -(d - b)^2$$

Now $a - c$ and $d - b$ are both real numbers and so their squares are positive or zero. If both squares are positive, the above states that a positive number is equal to a negative number, which can be rejected as absurd. If one is positive and the other zero, then either a positive number is equal to zero or zero is equal to a negative number. Both these situations are absurd and can be rejected. Indeed, the only possible and sensible situation is that both the squares are zero, i.e. $(a - c)^2 = 0$ and $(d - b)^2 = 0$, which results in $a = c$ and $b = d$. Thus we see that the equality of two complex numbers implies two relations of equality among real numbers.

Most of the mathematical literature speaks of a as being the real part and b as being the imaginary part of the complex number $a + ib$. We write this as

$$a = \text{Re}(a + ib) \qquad b = \text{Im}(a + ib) \tag{4.3}$$

The terminology, which perhaps indicates the hesitation with which complex numbers were first used, can be quite convenient provided we do not interpret it too literally. There is nothing imaginary about i, although it is perhaps a little unfamiliar to most readers at this stage. Put into words, (4.2) states that two complex numbers are equal if their real parts *and* their imaginary parts are equal.

The addition of two complex numbers is defined by

$$a + ib + c + id = a + c + i(b + d) \tag{4.4}$$

This rule, if applied to the numbers $a + i0$ and $c + i0$ (i.e. to the real numbers a and c), yields $a + c + i0$ (i.e. the real number $a + c$) and so is a natural extension of the rule for adding real numbers. Similarly for subtraction:

$$a + ib - (c + id) = a - c + i(b - d) \tag{4.5}$$

These two rules may be stated thus: to add (or subtract) two complex numbers add (or subtract) their real and their imaginary parts, respectively.

Example 1 Express in the form $a + ib$ (i) $6 + 3i + 7 - i$, (ii) $7 + 2i + 3 - 4i - (5 + i)$.

(i) $6 + 3i + 7 - i = 13 + 2i$
(ii) $7 + 2i + 3 - 4i - (5 + i) = 10 - 2i - (5 + i) = 5 - 3i$

The multiplication of two complex numbers is defined by the rule

$$(a + ib)(c + id) = ac - bd + i(bc + ad) \tag{4.6}$$

This rule, if applied to the numbers $a + i0$ and $c + i0$ (i.e. the real numbers a and c), yields $ac - 0 + i0 = ac + i0$ (i.e. the real number ac) and so is a natural extension of the rule for multiplying real numbers. The rule is consistent with treating i like any other algebraic quantity and this in practice is how we normally multiply complex numbers. On this basis

$$\begin{aligned}(a + ib)(c + id) &= ac + i^2bd + ibc + iad \\ &= ac - bd + i(bc + ad) \qquad (\text{since } i^2 = -1)\end{aligned}$$

which is the result (4.6).

Example 2 Express in the form $a + ib$ (i) $(2 + 3i)^2$, (ii) $(2 + i)(2 - i) + (3 + 2i)(3 - 2i)$.

(i) $(2 + 3i)^2 = (2 + 3i)(2 + 3i) = 2.2 - 3.3 + i(2.3 + 2.3) = -5 + 12i$

(ii) $(2 + i)(2 - i) + (3 + 2i)(3 - 2i)$
$= 4 - 1(-1) + i[(2.1 + 2(-1)] + 9 - 2(-2) + i[3.2 + 3(-2)]$
$= 4 + 1 + i0 + 9 + 4 + i0$
$= 18 + i0$

Before defining division for complex numbers, it is useful to introduce the notion of a complex conjugate. Two complex numbers which differ only in the sign of their imaginary part are called complex conjugates. The complex conjugate of $a + ib$ is thus $a - ib$; $a - ib$ is the complex conjugate of $a + ib$. Two numbers which are complex conjugates have the property that their sum and product are both real numbers. For

$$a + ib + a - ib = 2a + i0 \tag{4.7}$$

$$\begin{aligned}(a + ib)(a - ib) &= a^2 - b(-b) + i[ab + a(-b)] \\ &= a^2 + b^2 + i0\end{aligned} \tag{4.8}$$

The $i0$ may, of course, be omitted from (4.7) and (4.8). It has been written in to emphasise that the real numbers are just a subset (with imaginary part zero) of the complex numbers.

The division of two complex numbers is defined by

$$\frac{a + ib}{c + id} = \frac{ac + bd}{c^2 + d^2} + i\frac{bc - ad}{c^2 + d^2} \tag{4.9}$$

This rule, if applied to the numbers $a + i0$ and $c + i0$, gives

$$\frac{ac + 0}{c^2 + 0} + i\frac{0 + 0}{c^2 + 0} = \frac{a}{c} + i0$$

so that (4.9), complicated as it may appear, is just a natural extension of the rule for division for real numbers.

As with (4.6), the rule is consistent with treating i as an ordinary algebraic quantity. The following working, which in practice is the way in which we normally carry out division, illustrates this.

$$\frac{a+ib}{c+id}=\frac{(a+ib)(c-id)}{(c+id)(c-id)}$$

on multiplying top and bottom by the conjugate of the denominator. Therefore, using (4.8),

$$\begin{aligned}\frac{a+ib}{c+id}&=\frac{(a+ib)(c-id)}{c^2+d^2}\\&=\frac{ac+bd+i(bc-ad)}{c^2+d^2}\\&=\frac{ac+bd}{c^2+d^2}+i\frac{(bc-ad)}{c^2+d^2}\end{aligned}$$

which is the result (4.9).

Example 3 Evaluate in the form $a+ib$: (i) $\dfrac{1+i}{2-i}$, (*ii*) $\dfrac{3-4i}{5+2i}$.

(i) $\dfrac{1+i}{2-i}=\dfrac{(1+i)(2+i)}{(2-i)(2+i)}=\dfrac{2-1+i(2+1)}{4+1}=\dfrac{1}{5}+i\dfrac{3}{5}$

(ii) $\dfrac{3-4i}{5+2i}=\dfrac{(3-4i)(5-2i)}{(5+2i)(5-2i)}=\dfrac{15-8+i(-20-6)}{25+4}=\dfrac{7}{29}-i\dfrac{26}{29}$

Example 4 Evaluate X and Y if $X+iY=\dfrac{3-i}{1+i}$.

We could find X and Y by (4.9). However, by way of illustration, we proceed as follows:

$$X+iY=\frac{3-i}{1+i}$$

$$3-i=(1+i)(X+iY)=X-Y+i(X+Y)$$

Therefore, by (4.2)

$$\begin{aligned}3&=X-Y\\-1&=X+Y\end{aligned}$$

We solve these two equations for X and Y. On addition, we obtain $2=2X$, i.e. $X=1$, and on subtraction $Y=-2$. Therefore

$$\frac{3-i}{1+i}=1-2i$$

It is readily verified that (4.9) gives the same result. Rules (4.2), (4.4), (4.5), (4.6) and (4.9) define the algebraic operations for complex numbers. Readers should verify that under these rules complex numbers obey the rules I to V described in Section 1.1.

Exercises 4b

1 Express in the form $a + ib$: (i) $i^3(1 + i) + i^5(3 - i) + i^7(2 + i)$, (ii) $2i(1 - 3i) + 3(4 - i)$.
2 Evaluate in the form $a + ib$: (i) $(3 + i)(4 - 2i)$, (ii) $(6 + 2i)(1 - 3i)$.
3 Find the solutions of the equation $x^2 + 6x + 18 = 0$ in the form $a + ib$. Verify their correctness by substitution into the equation.
4 Evaluate (i) $(1 + i)^2$, (ii) $(1 + i)^3$, (iii) $(1 + i)^4$.
5 Use (3.1) and (3.2) with $x = i$ to calculate $(1 + i)^2$, $(1 + i)^3$, $(1 + i)^4$. Verify that the results are in agreement with your answers to question 4. (This suggests that the binomial theorem remains valid when complex numbers are involved.)
6 Express in the form $a + ib$: (i) $\dfrac{2 - 5i}{7 + 3i}$, (ii) $\dfrac{1 - 5i}{3 + 2i}$.
7 Show that (i) $(\cos\theta + i\sin\theta)^2 = \cos 2\theta + i\sin 2\theta$
(ii) $(\cos\theta + i\sin\theta)^{-1} = \cos\theta - i\sin\theta$.
8 Find real numbers x and y such that $(x + iy)^2 = 40 + 42i$. Hence evaluate $\sqrt{(40 + 42i)}$. Calculate also $\sqrt{(35 - 12i)}$.
9 Evaluate in the form $a + ib$:

(i) $(x - i)^5$ (ii) $\dfrac{3 - 4i}{(2 + i)^2}$ (iii) $\left(\dfrac{1 - i}{1 + i}\right)^2$

(iv) $\dfrac{1 + 2i}{i^3(1 - 3i)}$ (v) $\dfrac{(2 + i)^3}{(3 - i)^2}$ (vi) $\dfrac{1 - 2i}{(4 - 3i)^2}$

10 If $\dfrac{a + ib}{c + id} = X + iY$ show, following the method of Example 4, that X and Y satisfy the equations $cX - dY = a$, $dX + cY = b$. Solve these equations and show that the results are consistent with (4.9).

4.3 The geometrical representation of complex numbers

It was shown in Section 1.1 that the real numbers could be represented by the points of a line. For convenience, we recall this representation. *Figure 4.1* shows a line, on which we choose a point 0 as origin. Positive numbers are represented by points to the right of 0, or displacements to the right of 0, and negative numbers by points or displacements to the left of 0.

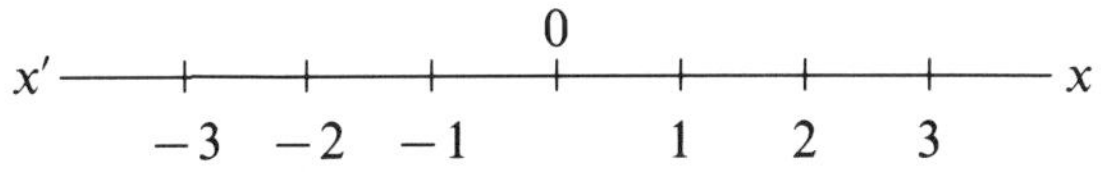

Figure 4.1

The real numbers, integers, rationals, irrationals and transcendental numbers completely fill the line, from which it is clear that we shall not be able to represent the complex numbers by further points of the line. A complex number of the form $x + iy$ is specified by the two real numbers x and y. The natural way in which to represent a complex number is thus by a point in the plane whose cartesian co-ordinates are the numbers x and y. The complex number $x + iy$ is thus represented by the point P (x, y) or by the vector or displacement OP (*Figure 4.2*). Our real numbers which are, of course, just particular complex numbers (with imaginary part zero) are confined to the line $x'Ox$, but complex numbers can be anywhere in the complex plane. This representation, which is not necessarily the only possible representation of complex numbers, was originally due to J. R. Argand and for this reason the complex plane of *Figure 4.2* is often called the Argand diagram.

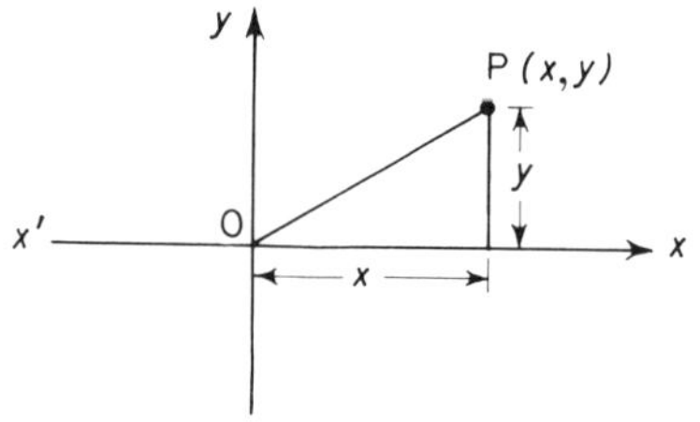

Figure 4.2

The length of OP, where ***OP*** is the vector representing the complex number $x + iy$, is known as the modulus of the complex number. By Pythagoras' theorem, we see that $OP = \sqrt{(x^2 + y^2)}$. The modulus of a complex number is represented by the symbol $|x + iy|$. Thus

$$|x + iy| = \sqrt{(x^2 + y^2)} \tag{4.10}$$

It is conventional to represent a complex number by the single bold letter z, i.e. $z = x + iy$ and so also $|z| = |x + iy| = \sqrt{(x^2 + y^2)}$.

Example 1 Represent the complex numbers (i) $2 + i$, (ii) i, (iii) $-i$, (iv) $-3 - 2i$ by points in the complex plane (Argand diagram).

In *Figure 4.3*, the numbers (i), (ii), (iii) and (iv) are represented by the points A, B, C, D respectively or by the vectors ***OA***, ***OB***, ***OC***, ***OD***.

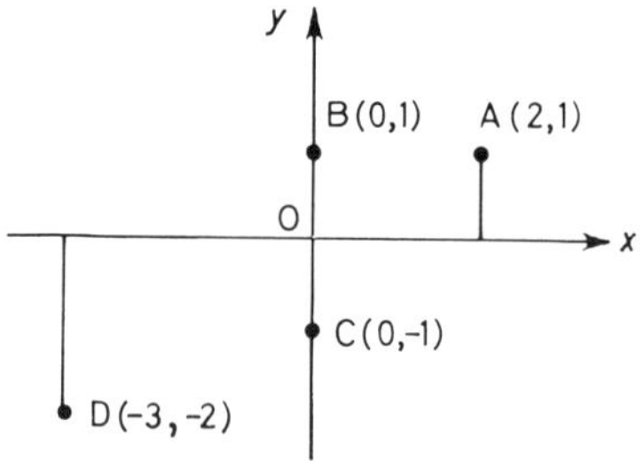

Figure 4.3

Example 2 If $z = 1 + i$, mark the points (i) $1 + z$, (ii) $\dfrac{1}{1+z}$ in the complex plane.

(i) $1 + z = 1 + 1 + i = 2 + i$

(ii) $\dfrac{1}{1+z} = \dfrac{1}{2+i} = \dfrac{2-i}{2^2+1} = \dfrac{2}{5} - i\dfrac{1}{5}$

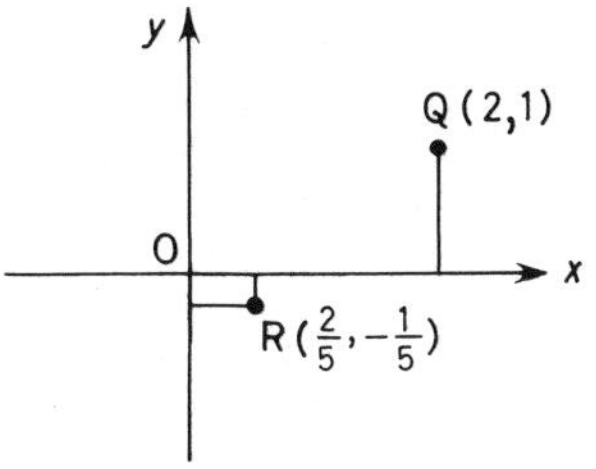

Figure 4.4

Thus the numbers (i) and (ii) are represented by the points Q and R in *Figure 4.4* or by the vectors ***OQ***, ***OR***.

Exercises 4c

1 Mark the points in the complex plane corresponding to the complex numbers (i) $2 + 3i$, (ii) $4 - i$, (iii) $-3 - 6i$, (iv) $-1 + i$.
2 Find the modulus of each of the complex numbers in question 1.
3 $z = 2 - 3i$. Mark the points (i) z, (ii) iz, (iii) i^2z in the complex plane.
4 Show that $|z| = |iz|$, where z is any complex number.
5 z is the complex number $x + 7i$. If $|z| = 25$, find x.

4.4 The geometry of complex numbers

The representation of complex numbers described in Section 4.3 enables us to give a geometrical interpretation of the rules for the addition and multiplication of complex numbers.

Consider the two complex numbers $2 + 3i$ and $3 + i$. These may be represented by the vectors ***OP*** and ***OQ*** (*Figure 4.5*).

Figure 4.5

Their sum is the complex number $5 + 4i$, which may be represented by the vector ***OR***, and we see from *Figure 4.6* that ***OR*** is the vector sum of and ***OQ***. (That is, starting from O draw OP, from P draw a line parallel to OQ and of magnitude OQ. We then arrive at the point R; OR is the diagonal of the parallelogram with adjacent sides OP, OQ.)

If we consider the complex number $4 + 6i$, multiplication by 2, $\frac{1}{2}$, -1 and i produces the complex numbers $8 + 12i$, $2 + 3i$, $-4 - 6i$ and $-6 + 4i$, respectively. These numbers and the original number are represented by the vectors ***OA***, ***OB***, ***OC***, ***OD*** and ***OP***, respectively (*Figure 4.7*).

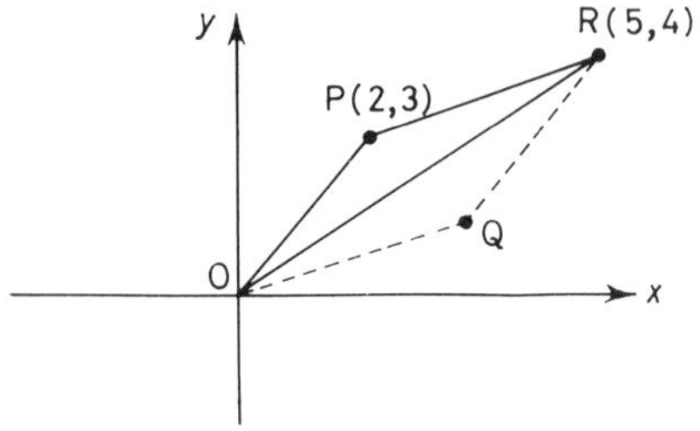

Figure 4.6

Thus we see that multiplication by a positive number (2 or $\frac{1}{2}$ in our case) dilates or shrinks a vector: $\boldsymbol{OA} = 2\boldsymbol{OP}$; $\boldsymbol{OB} = \frac{1}{2}\boldsymbol{OP}$. This is in agreement with our results for real numbers, which were all represented along the one direction $x'Ox$. Multiplication merely increases or decreases a number. Multiplication by -1 rotates a vector through 180° but does not change its length: $\boldsymbol{OC} = -\boldsymbol{OP}$. This again corresponds with results for real numbers. Multiplication of 4 by -1 produces the number -4, which lies to the left of O and is obtained by rotating the vector of length 4 units drawn in the direction of the positive x-axis through 180°. Multiplication by $(-1)^2$, i.e. by 1, will rotate a vector through $180° + 180° = 360°$ and so will leave it unchanged.

Of particular interest is the interpretation of multiplication by i. From *Figure 4.7*, we see that the vector $\boldsymbol{OD}$ is perpendicular to the vector $\boldsymbol{OP}$ but has the same length as this vector. Thus multiplication by i rotates a vector through 90° in an anticlockwise direction. Of course, multiplication by i^2 will rotate a vector through $90° + 90° = 180°$. But this is equivalent to multiplication by -1. This is consistent with (4.1), namely, $i^2 = -1$.

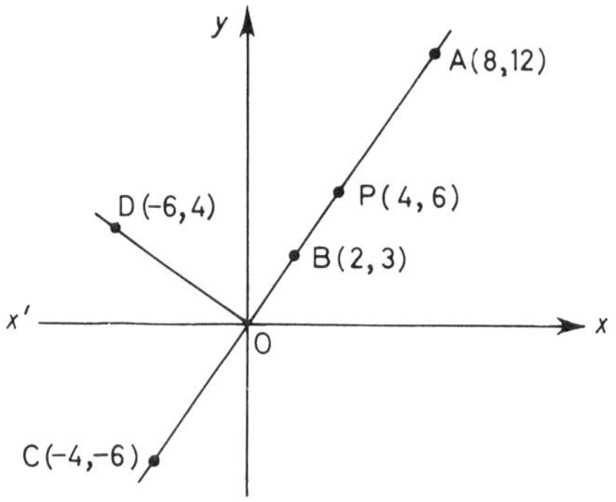

Figure 4.7

The results of this section have been obtained by reference to particular examples. However, the results are perfectly general. The addition of complex numbers is equivalent to vector addition. Multiplication by a positive number merely increases (or decreases) the length of a vector representing a number; multiplication by -1 and i rotate the vector representing a number through 180° and 90° respectively.

Example 1 If $z = x + iy$, the conjugate complex number $x - iy$ is conventionally represented by $\bar{z}$. What is the geometrical representation of $\bar{z}$?

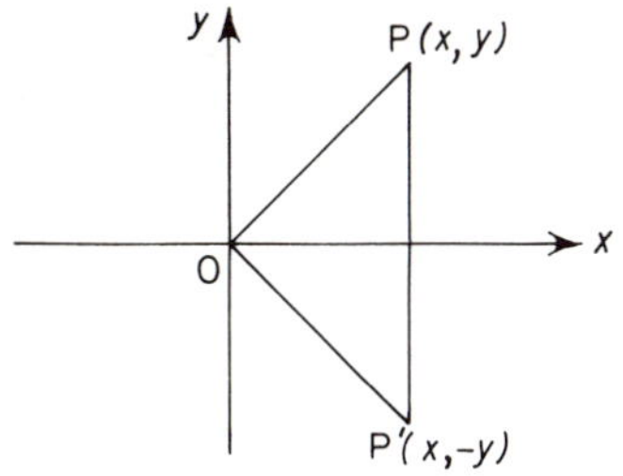

Figure 4.8

With reference to *Figure 4.8* we see that $\bar{z}$ is represented by the reflection of the point representing z in the real axis.

Example 2 Where must the complex number represented by z lie if $|z| = 1$? Find the cartesian equation of all points satisfying this condition.

If $|z| = 1$, the length of the vector representing z is unity. Thus, the point lies on a circle whose centre is the origin and radius is 1 (see *Figure 4.9*).

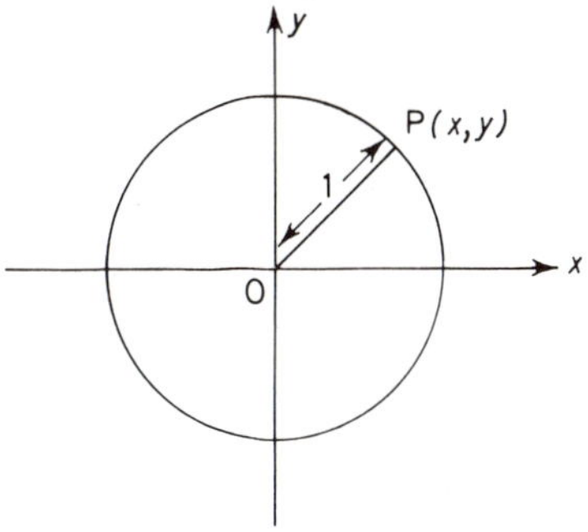

Figure 4.9

If $z = x + iy$, $|z|^2 = x^2 + y^2 = 1$ and this is the equation required.

Example 3 What is the geometrical significance of $|z - (2 + 3i)|$?

If z is the complex number $x + iy$, then

$$z - (2 + 3i) = (x - 2) + i(y - 3)$$

Therefore

$$|z - (2 + 3i)| = \sqrt{[(x - 2)^2 + (y - 3)^2]}$$

and so represents the distance between the points in the complex plane representing z and $2 + 3i$ (see *Figure 4.10*; cf. Section 17.5).

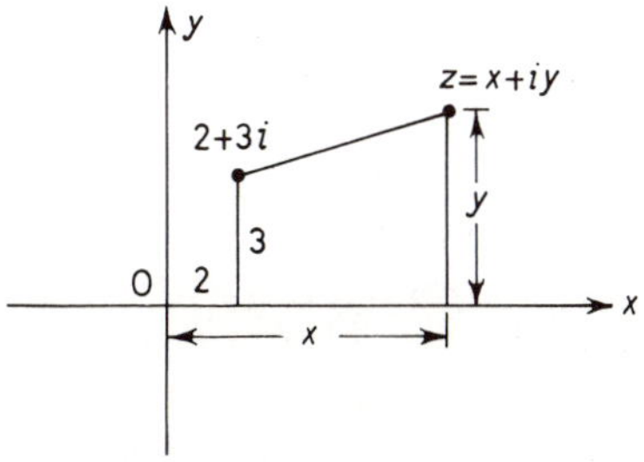

Figure 4.10

Example 4 Find the cartesian equation of the locus of the points in the complex plane such that $|z - i| = |z + i|$. What is the geometrical interpretation of this locus?

If $z = x + iy$,

$$z - i = x + i(y - 1) \qquad z + i = x + i(y + 1)$$

Therefore, if $|z - i|^2 = |z + i|^2$,

$$x^2 + (y - 1)^2 = x^2 + (y + 1)^2$$
$$x^2 + y^2 - 2y + 1 = x^2 + y^2 + 2y + 1$$
$$4y = 0$$

which is the x-axis. The locus is the locus of points equidistant from the points $i(0, 1)$ and $-i(0, -1)$, i.e. the perpendicular bisector of the line joining these points, which is the x-axis (see *Figure 4.11*).

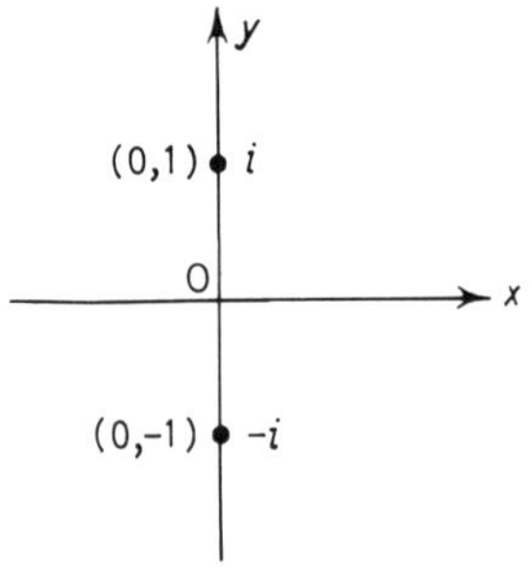

Figure 4.11

It is important to realise that, as far as complex numbers are concerned, the notion that 'one complex number is greater than a second complex number' is *meaningless*. This applies to the notion that a complex number may be greater than or less than zero. There is no classification of complex numbers into positive and negative, this latter notion applying only to the real numbers which either lie to the right or left of O. However, the complex number i, for example, is neither to the right nor to the left of O, indeed it is directly above O. This perhaps serves to remove some of the mystery that surrounds complex numbers. The idea that the square of a number is always positive is derived from the usual rule of signs as they apply to the real numbers. But the notion of sign has no place as far as complex numbers are concerned, so it is not surprising that the squares of such numbers should be negative.

In Example 1 of this section, we introduced the notion of the complex conjugate of a complex number, $z = x + iy$. Its conjugate is $\bar{z} = x - iy$, so that

$$\text{Re}(z) = \text{Re}(\bar{z})$$

and

$$\text{Im}(z) = -\text{Im}(\bar{z})$$

We say that z and $\bar{z}$ are a conjugate pair. Such pairs are of common occurrence. They will arise as the solutions of a quadratic equation which has real coefficients but not real roots. For example, as we have seen in Section 4.1, the equation $x^2 + 16 = 0$ has two roots, $z_1 = 4i$ and $z_2 = -4i$, from which we see that $z_1 = \bar{z}_2$ and $z_2 = \bar{z}_1$. The equation $x^2 + 2x + 5 = 0$ has also two roots, $z_1 = -1 + 2i$ and $z_2 = -1 - 2i$, and again we see that $z_1 = \bar{z}_2$ and $z_2 = \bar{z}_1$.

This result can be taken a stage further and applied to the roots of cubic, quartic or higher-degree equations. Thus, we have the important result that if $z = a + ib$ is a root of the equation

$$P(x) = c_n x^n + c_{n-1} x^{n-1} + \ldots + c_1 x + c_0 = 0$$

where the c's are real, then $\bar{z} = a - ib$ is also a root, i.e. *complex roots occur in conjugate pairs.*

We first note that

$$\begin{aligned}(x - z)(x - \bar{z}) &= (x - a - ib)(x - a + ib)\\ &= x^2 - 2ax + a^2 + b^2\end{aligned}$$

is wholly real.

Thus, in line with (1.12), if we divide $P(x)$ by the quadratic $(x - \bar{z})(x - \bar{z})$, we shall obtain a quotient $Q(x)$ and a remainder of first degree, $lx + m$. We shall show that this remainder is zero.

$$P(x) \equiv (x - z)(x - \bar{z})Q(x) + lx + m$$

therefore

$$P(z) = P(a + ib) = 0$$

so that

$$\begin{aligned}l(a + ib) + m &= 0\\ la + m + ibl &= 0\end{aligned}$$

On equating the real parts and the imaginary parts of this expression to zero, we obtain

$$la + m = 0 \quad \text{and} \quad bl = 0$$

Now, since $b \neq 0$ (we are dealing with a genuine complex root) $l = 0$ and so $m = 0$. Therefore

$$P(x) = (x - a - ib)(x - a + ib)Q(x)$$

Thus, $P(a - ib) = 0$, which proves the result.

Example 5 Solve the equation $x^4 + 2x^3 + 7x^2 + 8x + 12 = 0$ given that $2i$ is one root.

The equation has real coefficients. Thus, if $2i$ is a root so is $-2i$. Therefore, $(x - 2i)(x - (-2i))$, i.e. $x^2 + 4$, is a factor of $x^4 + 2x^3 + 7x^2 + 8x + 12$. By division we obtain

$$x^4 + 2x^3 + 7x^2 + 8x + 12 = (x^2 + 4)(x^2 + 2x + 3)$$

Thus, the other roots are the roots of $x^2 + 2x + 3 = 0$. These are

$$x = \frac{-2 \pm \sqrt{-8}}{2}$$

Therefore, the solutions are $2i$, $-2i$, $-1 + i\sqrt{2}$, $-1 - i\sqrt{2}$.

In this case, the other two roots are also complex and form a pair of complex conjugates.

Exercises 4d

1 If $z = 3 - 4i$, evaluate (i) $\bar{z}$, (ii) $z + 3$, (iii) $z - 3i$. Interpret the results geometrically in the complex plane.

2 If $z = 2 - i$, evaluate (i) $2z$, (ii) $-3z$, (iii) $4iz$. Interpret the results geometrically in the complex plane.

3 Evaluate (i) $3(4 + 5i)$, (ii) $2i(4 + 5i)$. Interpret the results on the Argand diagram.

4 Evaluate $(3 + 2i)(4 + 5i)$ and interpret the result on the Argand diagram. (Connect this with the results of question 3.)

5 If $z = x + iy$, show that $|z|^2 = z\bar{z}$.

6 If $z = 3 - 2i$, evaluate $\dfrac{1}{z}$ and interpret the result geometrically.

7 Show that $\left|\dfrac{1}{z}\right| = \dfrac{1}{|z|}$.

8 If $z = \bar{z}$, find the locus of the point represented by z.

9 If $|z - 2| = |z + 2|$, find the locus of the point represented by z.

10 If $|z - 3i| = 2|z - 3|$, find the locus of the point represented by z.

11 If $z = x + iy$, show that (i) $\mathrm{Re}(z + \bar{z}) = 2\mathrm{Re}(z)$, (ii) $\mathrm{Im}(z - \bar{z}) = 2\mathrm{Im}(z)$.

12 If $z = u + iv$, show that $z\bar{z} = |z|^2 = u^2 + v^2$.

13 If $z = -\bar{z}$, find the locus of the point represented by z.

14 If $z\bar{z} = 16$, find the locus of the point represented by z.

15 Given that one root of the equation $x^4 + x^3 + 3x^2 + x + 2 = 0$ is i, find the other roots.

4.5 The cube roots of unity

We have already seen (question 4 of Exercises 4a) that the cubic equation $x^3 = 1$, i.e. $x^3 - 1 = 0$, has three roots:

$$x^3 - 1 = (x - 1)(x^2 + x + 1)$$

Therefore, if $x^3 - 1 = 0$,

$$x - 1 = 0 \quad \text{or} \quad x^2 + x + 1 = 0$$

$$x = 1 \quad \text{or} \quad x = \frac{-1 \pm \sqrt{(1 - 4)}}{2} = -\frac{1}{2} \pm i\frac{\sqrt{3}}{2}$$

The three roots of the equation are thus 1, $-\dfrac{1}{2} + i\dfrac{\sqrt{3}}{2}$ and $-\dfrac{1}{2} - i\dfrac{\sqrt{3}}{2}$.

By direct multiplication, we see that the square of the second (third) root equals the third (second) root.

$$\left(-\frac{1}{2}+i\frac{\sqrt{3}}{2}\right)^2=\left(-\frac{1}{2}+i\frac{\sqrt{3}}{2}\right)\left(-\frac{1}{2}+i\frac{\sqrt{3}}{2}\right)$$
$$=\frac{1}{4}-\frac{3}{4}+i\left[\frac{\sqrt{3}}{2}\left(-\frac{1}{2}\right)+\frac{\sqrt{3}}{2}\left(-\frac{1}{2}\right)\right]$$
$$=-\frac{1}{2}-i\frac{\sqrt{3}}{2}$$

and

$$\left(-\frac{1}{2}-i\frac{\sqrt{3}}{2}\right)^2=\left(-\frac{1}{2}-i\frac{\sqrt{3}}{2}\right)\left(-\frac{1}{2}-i\frac{\sqrt{3}}{2}\right)$$
$$=\frac{1}{4}-\frac{3}{4}+i\left[\left(\frac{-\sqrt{3}}{2}\right)\left(-\frac{1}{2}\right)+\left(-\frac{\sqrt{3}}{2}\right)\left(-\frac{1}{2}\right)\right]$$
$$=-\frac{1}{2}+i\frac{\sqrt{3}}{2}$$

Thus we may denote the three cube roots of unity by 1, w, w^2, where

$$w=-\frac{1}{2}+i\frac{\sqrt{3}}{2}\quad\text{or}\quad-\frac{1}{2}-i\frac{\sqrt{3}}{2}$$

Then,

$$|w|=\sqrt{\left[\left(\frac{1}{2}\right)^2+\left(\frac{\sqrt{3}}{2}\right)^2\right]}=1$$

and if we represent the three roots by points in the complex plane, we see that they are equally spaced round a circle whose centre is the origin and whose radius is unity (*Figure 4.12*). We also observe that the sum of the three cube roots

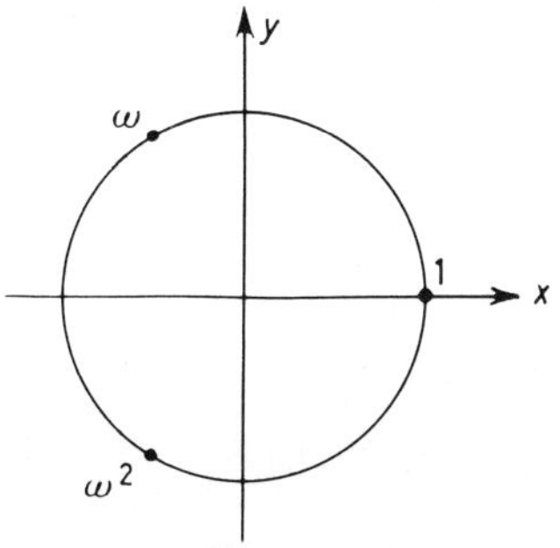

Figure 4.12

$$1 + w + w^2 = 1 + \left(-\frac{1}{2} + i\frac{\sqrt{3}}{2}\right) + \left(-\frac{1}{2} - i\frac{\sqrt{3}}{2}\right) = 0$$

Therefore

$$1 + w + w^2 = 0 \tag{4.11}$$

Example 1 If w is one of the complex cube roots of unity, show that $(1 + w^2)^4 = w$.

By (4.11)

$$1 + w^2 = -w$$

therefore

$$(1 + w^2)^4 = (-w)^4 = w^4 = w^3 . w$$
$$(1 + w^2)^4 = w \quad \text{since} \quad w^3 = 1$$

Example 2 If w is one of the complex cube roots of unity, show that $(a + wb + w^2c)(a + w^2b + wc) = a^2 + b^2 + c^2 - ab - bc - ca$.

$$\begin{aligned}(a + wb + w^2c)(a + w^2b + wc) &= a^2 + w^3b^2 + w^3c^2 + ab(w + w^2)\\ &\quad + bc(w^2 + w^4) + ca(w^2 + w)\\ &= a^2 + b^2 + c^2 + ab(-1)\\ &\quad + bc(w^2 + w) + ca(-1)\\ &= a^2 + b^2 + c^2 - ab - bc - ca\end{aligned}$$

Exercises 4e

If w is one of the complex cube roots of unity show that:

1 $(1 + w - w^2)^3 - (1 - w + w^2)^3 = 0$

2 $(1 - w + w^2)(1 + w - w^2) = 4$

3 $(1 + w)^6 = 1$

4 $a^3 + b^3 = (a + b)(a + wb)(a + w^2b)$

5 $6xy = (x + y)^2 + (wx + w^2y)^2 + (wy + w^2x)^2$

Exercises 4

1 Express in the form $a + ib$:

(i) $\dfrac{3 - i}{2 + i}$ (ii) $\dfrac{4 + 3i}{2 - i}$ (iii) $\left(\dfrac{1 + i}{1 - i}\right)^2$

2 If $(x + iy)^2 = a + ib$ show that $x^2 - y^2 = a$, $2xy = b$. Hence evaluate $\sqrt{(8 + 6i)}$.

3 Find the solutions of the equation $x^2 + 7x + 20 = 0$ in the form $a + ib$. Find the sum and product of the roots.

4 If $z^3 = 1$, find the possible values for $1 + z + z^2$.

5 If $z = \cos\theta + i\sin\theta$, find the modulus of $\dfrac{z + 1}{z - 1}$ if $0 < \theta < \pi/2$.

6 If $z = x + iy$, find the real and imaginary parts of (i) z^3 and (ii) $\dfrac{1}{z^2}$.

7 Show how to represent geometrically the sum of two complex numbers z_1 and z_2. What is the meaning of $|z_1 + z_2|$?

8 Use the result of question 7 to show that $|z_1 + z_2| \leqslant |z_1| + |z_2|$.

9 Find the locus of a point z which moves so that $\left|\dfrac{z+1}{z-1}\right| = 3$.

10 If w is one of the complex cube roots of unity evaluate
(i) $(1-w)(1-w^2)$
(ii) $(a+b)(wa+w^2b)(w^2a+bw)$.

11 Solve the equation $x^2 + 4x + 20 = 0$ giving the roots in the form $p \pm iq$, where p and q are real. [JMB, part]

12 If $\alpha + i\beta = \sqrt{[(A+iB)(C+iD)]}$ show that $\alpha^2 - \beta^2 = AC - BD$, $2\alpha\beta = AD + BC$ and $2\alpha^2 = AC - BD + \sqrt{[(A^2+B^2)(C^2+D^2)]}$.

13 Express in the form $X + iY$:

(i) $(2-i)^3$ (ii) $\dfrac{2-3i}{1+3i}$ (iii) $\dfrac{2-i}{3+i}$ (iv) $\dfrac{(4-i)^2}{3-i}$

14 Find two real numbers x and y so that

$$x(3+4i) - y(1+2i) + 5 = 0$$

15 If $(x+iy)^3 = a + ib$ show that $a^2 + b^2 = (x^2+y^2)^3$.

16 If w is one of the complex cube roots of unity show that
(i) $(1+w^2)^{12} = 1$
(ii) $(1-w)(1-w^2)(1-w^4)(1-w^5)(1-w^7)(1-w^8) = 27$.

17 Show that the square roots of unity are equally spaced round the unit circle. Show that this is the case for the cube and fourth roots of unity. Where do you suppose the fifth or the sixth roots of unity lie in the complex plane? Verify by actual multiplication that the complex sixth root in the first quadrant is indeed a sixth root of unity. What is this complex sixth root?

18 (i) Find the real and imaginary parts of $\dfrac{2+3i}{3+4i}$.
(ii) If the complex number $4+7i$ is represented by the point P on the Argand diagram, write down the complex numbers which are represented by (a) the reflection of P in the x-axis; (b) the reflection of P in the line $y = x$; (c) the reflection of P in the line $y = -x$. [JMB, part]

19 (a) Show that $1 + i$ is a fifth root of $-4 - 4i$.
(b) Show that if a, b, c, d and $(a+ib)/(c+id)$ are real, then $ad = bc$. Hence show that if $z = x + iy$ and $\dfrac{z^2+2z}{z^2+4}$ is real, the point represented by z lies on the real axis or on a certain circle. [JMB]

20 Show that

$$(a+b+c)(a+wb+w^2c)(a+w^2b+wc) = a^3 + b^3 + c^3 - 3abc$$

where w is a complex cube root of unity.

21 Solve the equation $\left(\dfrac{z+1}{z-1}\right)^2 = i$.

22 P represents the complex number z; Q represents $z + iz$. Show that OPQ is a right-angled triangle where O is the origin.

23 The points A, B, C, D on the Argand diagram correspond to the complex numbers a, b, c, d respectively. Prove that
(i) if $a - b + c - d = 0$, then ABCD is a parallelogram
(ii) if also $a + ib - c - id = 0$, then ABCD is a square. [JMB, part]

24 P represents the complex number z; Q represents the complex number $z + \frac{1}{z}$. Show that if P moves on the circle $|z| = 2$, Q moves on the ellipse $\frac{x^2}{25} + \frac{y^2}{9} = \frac{1}{4}$.

25 ABCD is a square in the complex plane. If A represents $3 + 2i$ and D represents $4 + 3i$, what complex numbers are represented by B and C?

26 Write down, or obtain, the non-real cube roots of unity, w_1 and w_2, in the form $a + ib$, where a and b are real. A regular hexagon is drawn in an Argand diagram such that two adjacent vertices represent w_1 and w_2, respectively, and the centre of the circumscribing circle of the hexagon is the point (1, 0). Determine, in the form $a + ib$, the complex numbers represented by the other four vertices of the hexagon and find the product of these four complex numbers. [JMB]

27 In an Argand diagram, the point P represents the complex number z, where $z = x + iy$. Given that $z + 2 = \lambda i(z + 8)$, where λ is a real parameter, find the cartesian equation of the locus of P as λ varies. If also $z = \mu(4 + 3i)$, where μ is real, prove that there is only one possible position for P. [JMB]

28 (i) Given that $z_1 = 3 + 4i$ and $z_2 = -1 + 2i$, represent z_1, z_2, $(z_1 + z_2)$, and $(z_2 - z_1)$ in the Argand diagram. Express $(z_1 + z_2)/(z_2 - z_1)$ in the form $a + ib$, where a, b are real. Find the magnitude of the angle between the vectors representing $(z_1 + z_2)$ and $(z_2 - z_1)$.
(ii) One root of the equation $z^3 - 6z^2 + 13z + k = 0$, where k is real, is $2 + i$. Find the other roots and the value of k. [LU]

29 Find real values of a and b such that $(a + ib)^2 = i$. Hence, or otherwise, solve the equation $z^2 + 2z + 1 - i = 0$, giving your solutions in the form $z = p + iq$. [O]

30 (i) Given that $3 + 4i = (x + iy)^2$, find x and y. Find also the square roots of i in the form $a + ib$, where a and b are real.
(ii) Show that $1 + i$ is a root of the equation $z^3 - 4z^2 + 6z - 4 = 0$, and find the other roots of this equation. [LU]

5
The quadratic function and the quadratic equation

5.1 The general quadratic equation

We have already used the formula (1.2) in order to solve a quadratic equation. The result (1.2) can be obtained as follows.

For the general quadratic equation $ax^2 + bx + c = 0$, where a, b, c are any real numbers with a at least non-zero, we have after division by a and a slight rearrangement of the terms

$$x^2 + \frac{b}{a}x = \frac{-c}{a}$$

The addition of the quantity $b^2/4a^2$ to both sides makes the left hand side a perfect square, namely $[x + (b/2a)]^2$. Thus

$$\left(x + \frac{b}{2a}\right)^2 = \frac{b^2}{4a^2} - \frac{c}{a} = \frac{b^2 - 4ac}{4a^2}$$

Therefore

$$x + \frac{b}{2a} = \pm\sqrt{\left(\frac{b^2 - 4ac}{4a^2}\right)} = \frac{\pm\sqrt{(b^2 - 4ac)}}{2a}$$

$$x = \frac{-b \pm \sqrt{(b^2 - 4ac)}}{2a} \tag{5.1}$$

Expression (5.1) enables us not only to solve quadratic equations but also to investigate the dependence of the roots on the relative values of a, b and c. In particular, the types of root which arise depend on the quantity $b^2 - 4ac$ whose square root occurs in (5.1). This quantity is called the discriminant of the equation and is often denoted by the symbol D:

$$D = b^2 - 4ac \tag{5.2}$$

If $b^2 - 4ac > 0$, its square root will be a real number and we shall obtain two real distinct roots of the equation.

If $b^2 - 4ac = 0$, so is its square root, and both roots of the equation will be real and equal to $-b/2a$.

If $b^2 - 4ac < 0$, its square root is that of a negative number. Such a square root cannot be a real number. Indeed, we have seen that it is a complex number. In this case, we say that the equation has no real roots or the equation has complex roots.

Example 1 Find the values of a for which the equation $(3a+1)x^2+(a+2)x+1=0$ has equal roots.

The discriminant

$$\begin{aligned} D &= (a+2)^2 - 4(3a+1).1 \\ &= a^2+4a+4-12a-4 \\ &= a^2-8a \end{aligned}$$

For equal roots, $D = 0$; therefore

$$\begin{aligned} a^2 - 8a &= 0 \\ a(a-8) &= 0 \\ a = 0 \quad \text{or} \quad a &= 8 \end{aligned}$$

Example 2 Show that the roots of the equation $(x-a)(x-b)=k^2$ are always real if a, b and k are real.

We first write the equation in the form

$$x^2-(a+b)x+ab-k^2=0$$

The equation has real roots provided its discriminant is not a negative number.

The discriminant $D=(a+b)^2-4(ab-k^2)$. We require to show that $D \geqslant 0$. Now

$$\begin{aligned} D &= (a+b)^2-4(ab-k^2) \\ &= a^2+2ab+b^2-4ab+4k^2 \\ &= a^2-2ab+b^2+4k^2 \\ &= (a-b)^2+4k^2 \end{aligned}$$

We have been able to express D as the sum of the squares of two real numbers. This proves $D \geqslant 0$, which is the condition for real roots. (Note: a standard technique for proving a number to be non-negative is to express it as the sum of the squares of two or more real numbers.)

Example 3 Find the values of λ for which the roots of the equation $x^2-(3\lambda+1)x+\lambda^2-1=5\lambda$ are real.

The equation can be written in the form

$$x^2-(3\lambda+1)x+\lambda^2-5\lambda-1=0$$

The discriminant

$$\begin{aligned} D &= (3\lambda+1)^2-4(1)(\lambda^2-5\lambda-1) \\ &= 9\lambda^2+6\lambda+1-4\lambda^2+20\lambda+4 \\ &= 5\lambda^2+26\lambda+5 \end{aligned}$$

For real roots, $D \geqslant 0$, that is

$$5\lambda^2 + 26\lambda + 5 \geqslant 0$$

therefore

$$(5\lambda + 1)(\lambda + 5) \geqslant 0$$

This inequality is satisfied if both factors have the same sign. If this sign is positive, $\lambda \geqslant -\frac{1}{5}$; if this sign is negative, $\lambda \leqslant -5$. Thus the equation has real roots if $\lambda \leqslant -5$ or $\lambda \geqslant -\frac{1}{5}$. (See also Section 1.4.)

Exercises 5a

1 Find a if the equation $(5a + 1)x^2 - 8ax + 3a = 0$ has equal roots.
2 If the equation $(7p + 1)x^2 + (5p - 1)x + p = 1$ has equal roots, find p.
3 For what values of k does the equation $x^2 - (4 + k)x + 9 = 0$ have real roots?
4 Find the greatest value of λ for which the equation $(\lambda - 1)x^2 - 2x + (\lambda - 1) = 0$ has real roots.
5 Show that the equation $x^2 - 2px + p^2 - q^2 = 0$ has real roots provided p and q are both real.
6 Show that the equation $x^2 - 2ax + 3a^2 + b^2 = 0$ cannot have real roots if a and b are real.
7 Show that the roots of the equation $x^2 + 2x = (2a + 2b + 1)(2a + 2b - 1)$ are integers if a and b are integers.
8 The equation $x^2 + 2px + p^2 + q^2 = r^2$ has real roots. Show that $r^2 \geqslant q^2$.
9 Find the values of a if the equation $(a + 3)x^2 - (11a + 1)x + a = 2(a - 5)$ has equal roots.
10 Show that the roots of the equation $x^2 - 2x = (b - c)^2 - 1$ are rational if b and c are rational numbers.

5.2 The quadratic function

In this section, we shall examine the values of the quadratic function $y = ax^2 + bx + c$ as x takes on different real values.

First, we notice that for large positive or large negative values of x, the dominant part of the right-hand side is ax^2. Since x^2 is always positive, we see that y will have the same sign as a for large positive or large negative values of x.

$y = ax^2 + bx + c$ may be written as

$$y \equiv a\left(x^2 + \frac{bx}{a} + \frac{c}{a}\right)$$

$$\equiv a\left(x^2 + \frac{bx}{a} + \frac{b^2}{4a^2} + \frac{c}{a} - \frac{b^2}{4a^2}\right)$$

$$\equiv a\left[\left(x + \frac{b}{2a}\right)^2 - \frac{b^2 - 4ac}{4a^2}\right]$$

that is

$$y \equiv a\left[\left(x+\frac{b}{2a}\right)^2 - \frac{D}{4a^2}\right] \tag{5.3}$$

$$\equiv a\left[\left(x+\frac{b}{2a}\right)^2 - \left(\frac{\sqrt{(b^2-4ac)}}{2a}\right)^2\right]$$

$$\equiv a\left[x+\frac{b}{2a}-\frac{\sqrt{(b^2-4ac)}}{2a}\right]\left[x+\frac{b}{2a}+\frac{\sqrt{(b^2-4ac)}}{2a}\right]$$

Therefore

$$y = ax^2+bx+c \equiv a(x-\alpha)(x-\beta) \tag{5.4}$$

where

$$\alpha = \frac{-b+\sqrt{(b^2-4ac)}}{2a} \quad \text{and} \quad \beta = \frac{-b-\sqrt{(b^2-4ac)}}{2a}$$

are the roots of the equation $ax^2+bx+c=0$.

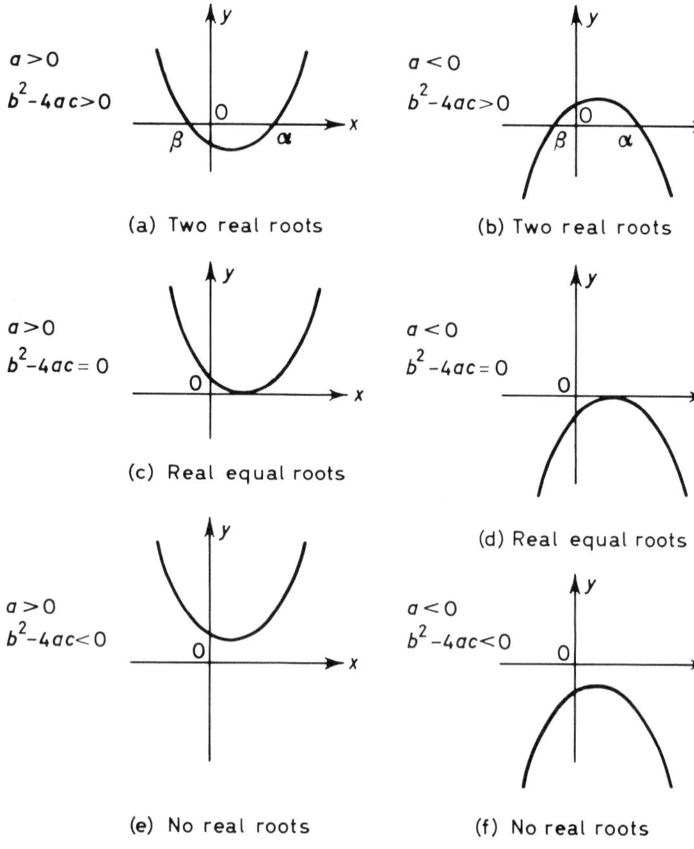

Figure 5.1

Thus we see that $y \equiv a(x-\alpha)(x-\beta)$ is zero for the two values $x=\alpha$ and $x=\beta$, as we would expect.

We may suppose without any loss of generality that one root, say α, is not less than β, i.e. $\alpha \geqslant \beta$, and we also note that for any value of x the sign of y is the same as the sign of $a(x-\alpha)(x-\beta)$.

Then $(x-\alpha)(x-\beta)$ is positive when $x>\alpha$ since both factors are then positive, and $(x-\alpha)(x-\beta)$ is also positive for $x<\beta$ since both factors are then negative. For $\beta<x<\alpha$, however, $(x-\alpha)(x-\beta)$ is negative since one factor is then negative and the other positive. To sum up, we have the important result:

THEOREM *The sign of $y=ax^2+bx+c$ is the same as the sign of a except for those values of x which lie between the roots of the equation $ax^2+bx+c=0$.*

This result combined with the procedure for determining the nature of the roots enables us to make a sketch of the graph of the function $y=ax^2+bx+c$. There are six cases to consider (see *Figure 5.1*).

Example 1 Show that $3x^2+6x+20$ is always positive.

$$\begin{aligned}3x^2+6x+20 &= 3(x^2+2x+\tfrac{20}{3})\\ &= 3(x^2+2x+1+\tfrac{17}{3})\\ &= 3[(x+1)^2+\tfrac{17}{3}]\end{aligned}$$

which, being three times the sum of two positive quantities, is always positive.

Example 2 Show that if $a>0$, $y=ax^2+bx+c$ has a minimum value when $x=-b/2a$, and determine that minimum value.

We have seen that

$$\begin{aligned}y &= a\left[\left(x+\frac{b}{2a}\right)^2+\frac{4ac-b^2}{4a^2}\right]\\ &= a\left(\frac{4ac-b^2}{4a^2}\right)+a\left(x+\frac{b}{2a}\right)^2\end{aligned}$$

Now $a\left(x+\dfrac{b}{2a}\right)^2$ is never negative if a is positive, and has a minimum value of zero when $x=-b/2a$. Thus y has a minimum value of

$$a\left(\frac{4ac-b^2}{4a^2}\right)=\frac{4ac-b^2}{4a} \quad \text{when} \quad x=-\frac{b}{2a}$$

Example 3 If x is real, show that $y=\dfrac{x^2+x+1}{x+1}$ is either $\geqslant 1$ or $\leqslant -3$.

We have

$$y=\frac{x^2+x+1}{x+1}$$

This we rewrite as a quadratic in x so that

$$x^2+x+1=xy+y$$
$$x^2+x(1-y)+1-y=0$$

Since x is always real, the discriminant (see (5.2)) of this equation is greater than or equal to zero:

$$(1-y)^2 - 4(1-y) \geqslant 0$$
$$(1-y)(-3-y) \geqslant 0$$
$$(y-1)(y+3) \geqslant 0$$

Now, the roots of the equation $(y-1)(y+3) = 0$ are $y = 1$ and $y = -3$, and the coefficient of y^2 for the function $(y-1)(y+3)$ is 1. Therefore (see theorem page 81),

$$(y-1)(y+3) \geqslant 0 \quad \text{(the same sign as 1)}$$

for $y \geqslant 1$ or $y \leqslant -3$. Therefore, $y \geqslant 1$ or $\leqslant -3$ whatever the value of x.

Exercises 5b

1 Show that $x^2 + 4x + 13 > 0$ for all values of x.
2 Show that $16x^2 - 24x + 10 > 0$ for all values of x.
3 Prove that $6x - 4 - 9x^2$ can never be greater than -3.
4 Show that if $a < 0$, $y = ax^2 + bx + c$ has a maximum value when $x = -b/2a$. What is this maximum value of y?
5 Find the maximum value of $5 + 6x - x^2$.
6 Find the minimum value of $12x^2 + 24x + 13$.
7 Show that if x is real, $\dfrac{x^2 - 12}{2x - 7}$ can have no real values between 3 and 4.
8 If x is real show that $\dfrac{(x-2)^2 + 16}{2(x+2)}$ can take on any real value which does not lie between $-4(\sqrt{2} + 1)$ and $4(\sqrt{2} - 1)$.
9 If p and q are both real and $q > 4$, show that $\dfrac{x^2 + px + p}{x^2 + qx + q}$ cannot be between p/q and $\dfrac{p-4}{q-4}$ when x is real.
10 Find the possible values of k if $\dfrac{x^2 + 3x - 4}{5x - k}$ may be capable of taking on all values when x is real.

5.3 The relation between the roots of a quadratic equation and the coefficients

We have seen that the roots of the quadratic equation $ax^2 + bx + c = 0$ are the two numbers

$$\frac{-b + \sqrt{(b^2 - 4ac)}}{2a} \quad \text{and} \quad \frac{-b - \sqrt{(b^2 - 4ac)}}{2a}$$

If we denote these roots by α and β in some order, we see that

$$\alpha + \beta = -\frac{b}{2a} + \frac{\sqrt{(b^2 - 4ac)}}{2a} - \frac{b}{2a} - \frac{\sqrt{(b^2 - 4ac)}}{2a} = -\frac{b}{a}$$

and

$$\alpha\beta = \left[-\frac{b}{2a} + \frac{\sqrt{(b^2 - 4ac)}}{2a}\right]\left[-\frac{b}{2a} - \frac{\sqrt{(b^2 - 4ac)}}{2a}\right]$$

$$= \left(-\frac{b}{2a}\right)^2 - \left(\frac{b^2 - 4ac}{4a^2}\right) = \frac{4ac}{4a^2} = \frac{c}{a}$$

Thus we have the following important results.

If α and β are the roots of the quadratic equation $ax^2 + bx + c = 0$, the sum of the roots equals $-b/a$:

$$\alpha + \beta = -\frac{b}{a} \tag{5.5}$$

and the product of the roots equals c/a:

$$\alpha\beta = \frac{c}{a} \tag{5.6}$$

This same result would have been obtained by using (5.4). For then we have

$$ax^2 + bx + c \equiv a(x - \alpha)(x - \beta)$$
$$ax^2 + bx + c \equiv ax^2 - a(\alpha + \beta)x + a\alpha\beta.$$

These two expressions are identical and, on equating the coefficient of x and the term independent of x, we obtain

$$b = -a(\alpha + \beta) \quad \text{and} \quad c = a\alpha\beta$$

$$\alpha + \beta = -\frac{b}{a} \qquad \alpha\beta = \frac{c}{a} \quad \text{as before}$$

(We notice that in the identity the coefficients of x^2 are both equal to a.)

Example 1 If α and β are the roots of the equation $ax^2 + bx + c = 0$, obtain in terms of a, b and c the values of (i) $\alpha^2 + \beta^2$, (ii) $\dfrac{\alpha}{\beta} + \dfrac{\beta}{\alpha}$, (iii) $\alpha^3 + \beta^3$.

We express (i), (ii) and (iii) in terms of $\alpha + \beta$ and $\alpha\beta$.

For (i)

$$\alpha^2 + \beta^2 = \alpha^2 + 2\alpha\beta + \beta^2 - 2\alpha\beta$$
$$= (\alpha + \beta)^2 - 2\alpha\beta$$
$$= \left(-\frac{b}{a}\right)^2 - \frac{2c}{a}$$

Therefore

$$\alpha^2 + \beta^2 = \frac{b^2}{a^2} - \frac{2c}{a} = \frac{b^2 - 2ac}{a^2}$$

For (ii)

$$\frac{\alpha}{\beta}+\frac{\beta}{\alpha}=\frac{\alpha^2+\beta^2}{\alpha\beta}=\frac{b^2-2ac}{a^2}\bigg/\frac{c}{a} \quad \text{by the previous result}$$

$$\frac{\alpha}{\beta}+\frac{\beta}{\alpha}=\frac{b^2-2ac}{ac}$$

For (iii)

$$\alpha^3+\beta^3=(\alpha+\beta)(\alpha^2-\alpha\beta+\beta^2)$$

$$=-\frac{b}{a}\left(\frac{b^2-2ac}{a^2}-\frac{c}{a}\right) \quad \text{by (i)}$$

$$\alpha^3+\beta^3=-\frac{b}{a}\frac{(b^2-3ac)}{a^2}=\frac{3abc-b^3}{a^3}$$

Example 2 If one root of the equation $px^2+qx+r=0$ is three times the other root, show that $3q^2=16pr$.

Suppose the roots are α and 3α. Then from (5.5) and (5.6)

$$\alpha+3\alpha=4\alpha=-\frac{q}{p} \quad \text{and} \quad \alpha.3\alpha=3\alpha^2=\frac{r}{p}$$

We eliminate α from these two equations. From the first equation, $\alpha=-(q/4p)$. Therefore

$$3\left(\frac{-q}{4p}\right)^2=\frac{r}{p}$$

$$\frac{3q^2}{16p^2}=\frac{r}{p}$$

$$3q^2=16pr$$

Example 3 The roots of the equation $x^2+px+q=0$ are γ and δ. Form the quadratic equation whose roots are $\gamma+\delta$ and $1/\gamma+1/\delta$.

Let the required equation be $x^2+Px+Q=0$. By (5.5)

$$\frac{P}{1}=-\left(\gamma+\delta+\frac{1}{\gamma}+\frac{1}{\delta}\right)=-\left(\gamma+\delta+\frac{\gamma+\delta}{\gamma\delta}\right)$$

$$=-\left(-p-\frac{p}{q}\right)=p+\frac{p}{q}=\frac{p}{q}(1+q)$$

By (5.6)

$$\frac{Q}{1}=(\gamma+\delta)\left(\frac{1}{\gamma}+\frac{1}{\delta}\right)=(\gamma+\delta)\left(\frac{\gamma+\delta}{\gamma\delta}\right)$$

$$=\frac{(\gamma+\delta)^2}{\gamma\delta}=\frac{(-p)^2}{q}=\frac{p^2}{q}$$

Therefore, the required equation is

$$x^2+(p/q)(1+q)x+p^2/q=0$$

that is

$$qx^2 + p(1 + q)x + p^2 = 0$$

Exercises 5c

1 If α and β are the roots of the equation $3x^2 - 7x - 1 = 0$ find the values of (i) $(\alpha - \beta)^2$, (ii) $\alpha^2 + \beta^2$, (iii) $\alpha^4 + \beta^4$.

2 If α and β are the roots of the equation $5x^2 - 3x - 1 = 0$, form the equations with integral coefficients which have the roots (i) $1/\alpha^2$ and $1/\beta^2$, (ii) α^2/β and β^2/α.

3 Find the condition that the roots of the equation $px^2 + qx + r = 0$ should be (i) equal in magnitude and opposite in sign, (ii) reciprocals.

4 One root of the equation $px^2 + qx + r = 0$ is twice the other root. Show that $2q^2 - 9rp = 0$.

5 γ and δ are the roots of the equation $px^2 + qx + r = 0$. Find in terms of p, q and r (i) $\gamma - \delta$, (ii) $\gamma^2 - \delta^2$, (iii) $\gamma^3 - \delta^3$ $[= (\gamma - \delta)(\gamma^2 + \gamma\delta + \delta^2)]$, (iv) $\gamma^4 - \delta^4$.

6 One root of the equation $x^2 - px + q = 0$ is the square of the other. Show that $p^3 - q(3p + 1) - q^2 = 0$ provided $q \neq 1$.

7 If α and β are the roots of the equation $ax^2 + bx + c = 0$, form the equation whose roots are α/β^2 and β/α^2.

8 If one root of the equation $px^2 + qx + r = 0$ is four times the other show that $4q^2 - 25pr = 0$.

9 Find the relationship which must exist between a, b and c if the roots of equation $ax^2 + bx + c = 0$ are in the ratio p/q.

10 Form the quadratic equation for which the sum of the roots is 5 and the sum of the squares of the roots is 53.

Exercises 5

1 Find in their simplest rational forms the quadratic equations whose roots are (i) $3 \pm \sqrt{5}$, (ii) $-2 \pm 3\sqrt{2}$, (iii) $a \pm 2b$.

2 Prove that if a, b and c are real the roots of the equation $(a^2 + b^2)x^2 + 2(a^2 + b^2 + c^2)x + (b^2 + c^2) = 0$ are also real.

3 Show that if x is real, $\dfrac{(x - 1)(x - 5)}{(x - 2)(x - 4)}$ cannot lie between 1 and 4. Can it attain these two values and if so for what values of x?

4 If the roots of the equation $px^2 - 6qx - (9p - 10q) = 0$ are $2\alpha - 3$ and $2\beta - 3$, find the equation whose roots are α and β. [LU, part]

5 For what values of k has the equation $(x + 1)(x + 2) = k(3x + 7)$ equal roots?

6 Show that the roots of the equation $(a - b + 1)x^2 + 2x + (b - a + 1) = 0$ are both real if a and b are real.

7 If α and β are the roots of the equation $2p^2x^2 + 2pqx + q^2 - 3p^2 = 0$, show that $\alpha^2 + \beta^2$ is independent of p and q.

8 If the roots of the quadratic equation $x^2 - 3px + p^2 = 0$ are α and β, where $\alpha > \beta$, find the values of $\alpha^2 + \beta^2$ and $\alpha - \beta$ when p is positive. Find, in terms of p, a quadratic equation whose roots are α^3/β and $-\beta^3/\alpha$. [AEB]

9 Prove that, for real values of x, the function $\dfrac{x+2}{x^2+3x+6}$ cannot be greater than $\frac{1}{3}$, nor less than $-\frac{1}{5}$. Find for what values of x, if any, it attains these values. [SUJB]

10 Show that the roots of the equation $2bx^2+2(a+b)x+3a=2b$ are real when a and b are real. If one root is double the other show that $a=2b$ or $4a=11b$.

11 For what values of k does the equation $10x^2+4x+1=2kx(2-x)$ have real roots?

12 Show that the value of the expression $\dfrac{(x+2)^2}{x+1}$ cannot lie between 0 and 4 if x is real.

13 Find the ranges of values of k for which the equation $x^2+(k-3)x+k=0$ has (i) real distinct roots, (ii) roots of the same sign. [JMB, part]

14 Find the condition that the equation $\dfrac{a}{x-a}+\dfrac{b}{x-b}=5$ shall have roots equal in magnitude but opposite in sign.

15 If a and b are real, prove that the roots of the equation $(3a-b)x^2+(b-a)x-2a=0$ are real.

16 α and β are the roots of the equation $x^2+px+q=0$. Form the equation whose roots are $\alpha+\beta$ and $\alpha-\beta$.

17 Show that for all real values of α and β the value of the function

$\dfrac{x^2-\alpha\beta}{2x-\alpha-\beta}$ cannot lie between α and β. [JMB, part]

18 If $y=\dfrac{x^2+2x+\lambda}{2x-3}$ and x is real, find the greatest value of λ for which y can take all real values. [AEB]

19 Show that if the roots of the equation $3x^2+6x-1+m(x-1)^2=0$ are real, then m is not greater than $\frac{3}{2}$. Find m if one root is the negative of the other.

20 Find the condition that the quadratic equations $l_1x^2+m_1x+n_1=0$ and $l_2x^2+m_2x+n_2=0$ have a common root.

21 Find the values of k for each of which the quadratic equations $x^2+kx-6k=0$ and $x^2-2x-k=0$ have a common root. [JMB, part]

22 If the quadratic equations $x^2+ax+b=0$ and $x^2+bx+a=0$ $(a\neq b)$ have a common root, show that the solutions of $2x^2+(a+b)x=(a+b)^2$ are $x=1$ and $x=-\frac{1}{2}$. [LU, part]

23 If α and α' are the roots of the equation $(x-\beta)(x-\beta')=\gamma$, show that β and β' are the roots of the equation $(x-\alpha)(x-\alpha')+\gamma=0$.

24 If the roots of the equation $x^2+bx+c=0$ are α and β and the roots of the equation $x^2+\lambda bx+\lambda^2c=0$ are γ and δ, prove that: (i) $(\alpha\gamma+\beta\delta)(\alpha\delta+\beta\gamma)=2\lambda^2c(b^2-2c)$; (ii) the equation whose roots are $\alpha\gamma+\beta\delta$ and $\alpha\delta+\beta\gamma$ is $x^2-\lambda b^2x+2\lambda^2c(b^2-2c)=0$. [JMB, part]

25 Find the limits between which k must lie in order that $\dfrac{kx^2-6x+4}{4x^2-6x+k}$ may be capable of all values when x is real.

26 (a) If α and β are the roots of the equation $ax^2+bx+c=0$, form, without solving this equation, an equation whose roots are α^2/β and β^2/α.

(b) Find the set of possible values of p if $\dfrac{x^2+p}{x-1}$ can take all real values when x is real. Find the set of possible values of $\dfrac{x^2+p}{x-1}$ when $p = 3$. [AEB]

27 The roots of the equation $9x^2 + 6x + 1 = 4kx$, where k is a real constant, are denoted by α and β. (a) Show that the equation whose roots are $1/\alpha$ and $1/\beta$ is $x^2 + 6x + 9 = 4kx$. (b) Find the set of values of k for which α and β are real. (c) Find also the set of values of k for which α and β are real and positive. [LU]

28 The roots of the equation $x^2 + px + q = 0$ are α and β. (i) Given that the roots differ by $2\sqrt{3}$ and that the sum of the reciprocals of the roots is 4, find the possible values of p and q. (ii) Find an equation whose roots are $\alpha + \dfrac{2}{\beta}$ and $\beta + \dfrac{2}{\alpha}$, expressing the coefficients in terms of p and q. [C]

29 Given that $f(x) \equiv x^2 + (k+2)x + 2k$, show that the roots of the equation $f(x) = 0$ are real for all real values of k. (a) Find the roots of $f(x-k) = 0$. (b) Find the value of k for which the equation $f(x-k) - 2x = 0$ has roots $x = 0$ and $x = 7$. For this value of k, find the minimum value of $f(x-k) - 2x$. (c) Given that $2c = 2 - k$, show that the roots of the equation $f(x-k) + c^2 = 0$ are equal, and find the value of these equal roots when $k = 1$. [LU]

30 Given that the roots of the equation $x^2 - x - 1 = 0$ are α and β, find, in its simplest form, the quadratic equation with numerical coefficients whose roots are $\dfrac{1+\alpha}{2-\alpha}$ and $\dfrac{1+\beta}{2-\beta}$. [JMB]

6

Properties of the trigonometric functions

6.1 The measurement of angle

When a line OP rotates from a position OX to some other position OP, the angle POX is said to be positive if the sense of rotation is anticlockwise, and negative if the sense of rotation is clockwise. Thus, in *Figure 6.1a*, $\angle POX = 115^\circ$ and in *Figure 6.1b*, $\angle POX = -49^\circ$.

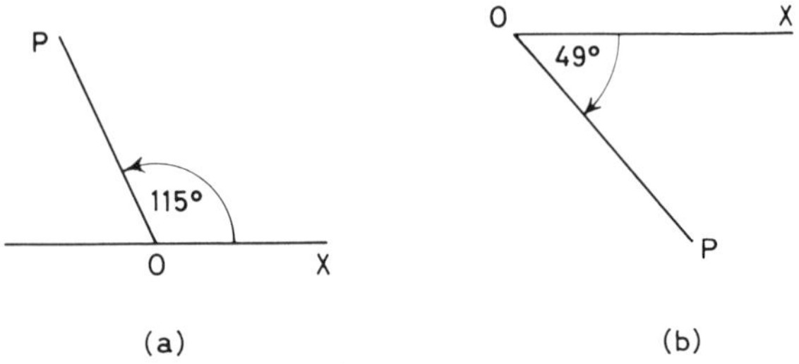

Figure 6.1

Angles are generally measured in degrees (360° = one revolution) or radians, this latter unit being defined as follows. Let AB be an arc of a circle, centre O, equal in length to the radius r of the circle (*Figure 6.2a*). Then $\angle AOB$ is one radian. If CD (*Figure 6.2b*) is an arc of length s, then $\angle COD$ is defined to be

$$\theta = \frac{s}{r} \text{ radians} \tag{6.1}$$

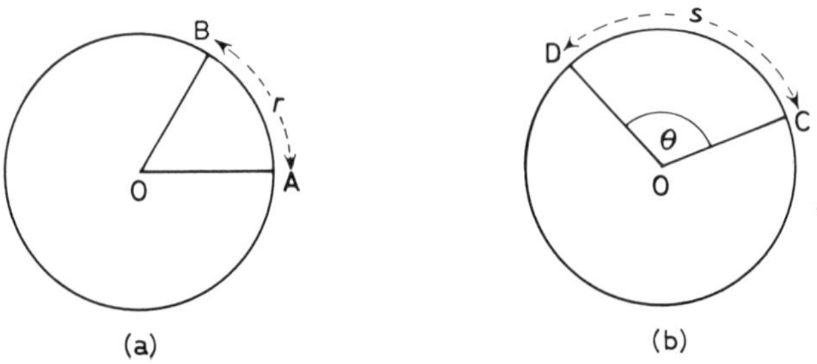

Figure 6.2

It follows from (6.1) that one complete revolution is equivalent to $2\pi r/r = 2\pi$ radians. Thus, we have the relationship between the two units:

$$2\pi \text{ radians} = 360°$$

$$\pi \text{ radians} = 180° \tag{6.2}$$

$$1 \text{ radian} = \left(\frac{180}{\pi}\right)^{\circ} = 57{\cdot}2958° = 57°\ 17'\ 45'' \tag{6.3}$$

$$1° = \frac{\pi}{180} \text{ radians} \tag{6.4}$$

Example 1 Express the following angles in radians: (i) 37°, (ii) −143° 10′.

(i) By (6.4)

$$37° = \frac{37 \times \pi}{180} \text{ radian} = 0.6458 \text{ radian}$$

(ii)

$$-143°\ 10' = -143\tfrac{1}{6}° = -143{\cdot}166°$$

Therefore

$$-143°\ 10' = -\frac{143{\cdot}166 \times \pi}{180} \text{ radians} = -2{\cdot}499 \text{ radians}$$

Example 2 Express the following angles in degrees and minutes correct to the nearest minute: (i) 2·1 radians, (ii) $\pi/12$ radian.

(i) By (6.3)

$$2{\cdot}1 \text{ radians} = \frac{2{\cdot}1 \times 180}{\pi} \text{ degrees} = 120{\cdot}316°$$

$$= 120°\ 19' \quad \text{(correct to the nearest minute)}$$

(ii)

$$\frac{\pi}{12} \text{ radian} = \left(\frac{\pi}{12} \times \frac{180}{\pi}\right)^{\circ} = 15°$$

Although an understanding of the relationship between the units concerned is desirable, in practice the conversions of the examples above are best carried out with the aid of tables or a calculator.

Exercises 6a

1 Express the following angles in degrees and minutes correct to the nearest minute: (i) 3·2 radians, (ii) −1·58 radians, (iii) $\pi/5$ radian.

2 Express in radians (i) 235°, (ii) 14° 4′, (iii) −128° 10′.

3 Which is the larger of the following pairs? (i) 129° or 2·16 radians, (ii) 19° or $\frac{1}{3}$ radian.

4 Verify the correctness of the following useful equivalents:

(i) $\frac{\pi}{2}$ radians = 90° (ii) $\frac{\pi}{4}$ radian = 45° (iii) $\frac{\pi}{3}$ radians = 60°

(iv) $\frac{\pi}{6}$ radian = 30° (v) $\frac{2\pi}{3}$ radians = 120°

5 For which of the following angles will the positions of OP coincide? $\angle$POX = (i) $-120°$, (ii) $135°$, (iii) $600°$, (iv) $240°$, (v) $-225°$, (vi) $30°$.

6.2 The trigonometric ratios for an acute angle

For an acute angle θ, the trigonometric ratios are defined as follows (see *Figure 6.3*):

$$\sin\theta = \frac{y}{r} \qquad \cos\theta = \frac{x}{r} \qquad \tan\theta = \frac{y}{x} \tag{6.5}$$

$$\operatorname{cosec}\theta = \frac{r}{y} \qquad \sec\theta = \frac{r}{x} \qquad \cot\theta = \frac{x}{y} \tag{6.6}$$

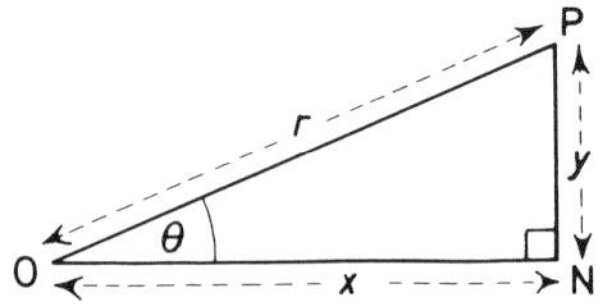

Figure 6.3

Thus, we have immediately the following relationships between the six ratios:

$$\operatorname{cosec}\theta = \frac{1}{\sin\theta} \qquad \sec\theta = \frac{1}{\cos\theta} \qquad \cot\theta = \frac{1}{\tan\theta} \tag{6.7}$$

Also

$$\tan\theta = \frac{y}{x} = \frac{y}{r}\bigg/\frac{x}{r} = \frac{\sin\theta}{\cos\theta}$$

and

$$\cot\theta = \frac{x}{y} = \frac{x}{r}\bigg/\frac{y}{r} = \frac{\cos\theta}{\sin\theta} \tag{6.8}$$

Furthermore, $x^2 + y^2 = r^2$, therefore

$$\frac{x^2}{r^2} + \frac{y^2}{r^2} = 1$$

that is

$$\cos^2\theta + \sin^2\theta = 1 \tag{6.9}$$

This may be written in either of the forms

$$\sin^2\theta = 1 - \cos^2\theta \quad \text{or} \quad \cos^2\theta = 1 - \sin^2\theta \tag{6.10}$$

In addition, we have

$$1 + \tan^2\theta = 1 + \frac{y^2}{x^2} = \frac{x^2 + y^2}{x^2} = \frac{r^2}{x^2}$$

therefore

$$1 + \tan^2\theta = \sec^2\theta \tag{6.11}$$

Also

$$1 + \cot^2\theta = 1 + \frac{x^2}{y^2} = \frac{y^2 + x^2}{y^2} = \frac{r^2}{y^2}$$

therefore

$$1 + \cot^2\theta = \operatorname{cosec}^2\theta \tag{6.12}$$

The relationships (6.5) to (6.12), which are true generally for any acute angle, enable us to calculate all the trigonometric ratios if one is known, and are of value in rewriting trigonometric expressions in alternative and simpler forms.

Example 1 If $\sin\theta = 1/\sqrt{3}$ and $0° \leqslant \theta \leqslant 90°$, find the values of the other trigonometric ratios of the angle θ.

From (6.10)

$$\cos^2\theta = 1 - \sin^2\theta = 1 - \tfrac{1}{3} = \tfrac{2}{3}$$

$$\cos\theta = \sqrt{\tfrac{2}{3}}$$

By (6.8)

$$\tan\theta = \frac{\sin\theta}{\cos\theta} = \frac{1}{\sqrt{2}}$$

and by (6.7)

$$\sec\theta = \sqrt{\tfrac{3}{2}} \qquad \operatorname{cosec}\theta = \sqrt{3} \qquad \cot\theta = \sqrt{2}$$

Example 2 Show that $\sin^2\theta + (1 + \cos\theta)^2 = 2(1 + \cos\theta)$.

The left-hand side (LHS) $= \sin^2\theta + (1 + \cos\theta)^2$

$= \sin^2\theta + 1 + 2\cos\theta + \cos^2\theta$

$= 2 + 2\cos\theta$

(since $\cos^2\theta + \sin^2\theta = 1$)

$= 2(1 + \cos\theta)$

$=$ right-hand side (RHS), as required.

Example 3 Show that $\left(\dfrac{1 + \sin x}{1 + \cos x}\right)\left(\dfrac{1 + \sec x}{1 + \operatorname{cosec} x}\right) = \tan x$

$$\text{LHS} = \left(\frac{1 + \sin x}{1 + \cos x}\right)\left[\frac{1 + \dfrac{1}{\cos x}}{1 + \dfrac{1}{\sin x}}\right] = \frac{1 + \sin x}{1 + \cos x}\left[\frac{\dfrac{\cos x + 1}{\cos x}}{\dfrac{\sin x + 1}{\sin x}}\right]$$

$$= \frac{1 + \sin x}{1 + \cos x} \cdot \frac{1 + \cos x}{\cos x} \cdot \frac{\sin x}{1 + \sin x} = \frac{\sin x}{\cos x} = \tan x$$

Trigonometric identities may often be proved by reducing one expression (usually the more complicated) to the second. In other cases we may proceed by showing that

(i) LHS − RHS = 0, or

(ii) $\dfrac{\text{LHS}}{\text{RHS}} = 1.$

Example 4 Show that $\dfrac{\sin\phi}{1-\cos\phi} = \dfrac{1+\cos\phi}{\sin\phi}$.

$$\frac{\text{LHS}}{\text{RHS}} = \frac{\sin\phi}{1-\cos\phi}\cdot\frac{\sin\phi}{1+\cos\phi} = \frac{\sin^2\phi}{1-\cos^2\phi} = \frac{\sin^2\phi}{\sin^2\phi} = 1$$

therefore

$$\text{LHS} = \text{RHS}$$

$$\frac{\sin\phi}{1-\cos\phi} = \frac{1+\cos\phi}{\sin\phi}$$

Exercises 6b

1 If $\cos\theta = \frac{12}{13}$ and $0^\circ \leqslant \theta \leqslant 90^\circ$, evaluate $\sin\theta$, $\tan\theta$, $\cot\theta$, $\sec\theta$, $\operatorname{cosec}\theta$.
2 If $\tan\theta = \frac{3}{2}$ and $0^\circ \leqslant \theta \leqslant 90^\circ$, evaluate $\sin\theta$ and $\cos\theta$.
3 If $x = a\cos\theta$, simplify (i) $a^2 - x^2$, (ii) $\left(1 - \dfrac{x^2}{a^2}\right)^{5/2}$
4 If $x = a\tan\theta$, simplify (i) $\dfrac{1}{a^2+x^2}$, (ii) $\sqrt{\left(1+\dfrac{x^2}{a^2}\right)}$.
5 If $c = \cos\theta$, express in terms of c (i) $3\sin^2\theta - 2\cos\theta$, (ii) $\tan^2\theta + 2\cos\theta$, (iii) $\operatorname{cosec}\theta + \sin\theta$.
6 If $6\cos^2\theta + 2\sin^2\theta = 5$, show that $\tan^2\theta = \frac{1}{3}$.
7 If $a\cos^2\theta + b\sin^2\theta = c$, show that $\tan^2\theta = \dfrac{c-a}{b-c}$.
8 If $\cot^2\theta + 3\operatorname{cosec}^2\theta = 7$, show that $\tan\theta = \pm 1$.

Show that

9 $\tan\theta + \cot\theta = \sec\theta\operatorname{cosec}\theta$
10 $4 - 3\cos^2\theta = 3\sin^2\theta + 1$
11 $\dfrac{1+\cos\theta}{1-\cos\theta}\cdot\dfrac{\sec\theta - 1}{\sec\theta + 1} = 1$
12 $(1+\sin\theta)^2 + \cos^2\theta = 2(1+\sin\theta)$
13 $(\sin\theta + \cos\theta)^2 + (\sin\theta - \cos\theta)^2 = 2$
14 $(\operatorname{cosec} x - \cot x)(\operatorname{cosec} x + \cot x) = 1$
15 $\cos^4 A - \sin^4 A = \cos^2 A - \sin^2 A$
16 $\dfrac{1-\sin\theta}{1+\sin\theta} = (\sec\theta - \tan\theta)^2$
17 $\dfrac{1+\tan A - \sec A}{\sec A + \tan A - 1} = \dfrac{1+\sec A - \tan A}{\sec A + \tan A + 1}$
18 $\dfrac{\cos\theta}{1+\sin\theta} + \dfrac{1+\sin\theta}{\cos\theta} = 2\sec\theta$
19 If $x' = x\cos\theta + y\sin\theta$ and $y' = x\sin\theta - y\cos\theta$, show that $x'^2 + y'^2 = x^2 + y^2$.

20 If $x = r\sin\theta\cos\phi$, $y = r\sin\theta\sin\phi$, $z = r\cos\theta$, show that $x^2 + y^2 + z^2 = r^2$.

6.3 The trigonometric ratios for any angle

The definitions of the trigonometric ratios given in the previous section can only apply to acute angles, since they involve the ratio of the sides of a right-angled triangle containing the given angle. In this section, we shall define the trigonometric ratios in such a way as to be applicable to angles of any size. These new definitions, if applied to acute angles, will, of course, yield the same results as before.

We shall measure angles from a fixed line X′OX on the plane. Y′OY is a line in the plane perpendicular to X′OX (*Figure 6.4*). This pair of lines divides the plane into four quadrants, XOY, YOX′, X′OY′, Y′OX (the first, second, third and fourth quadrants respectively). Let OP be any line in the plane through O and let ∠POX = θ. This is defined in accordance with the sign convention of Section 6.1.

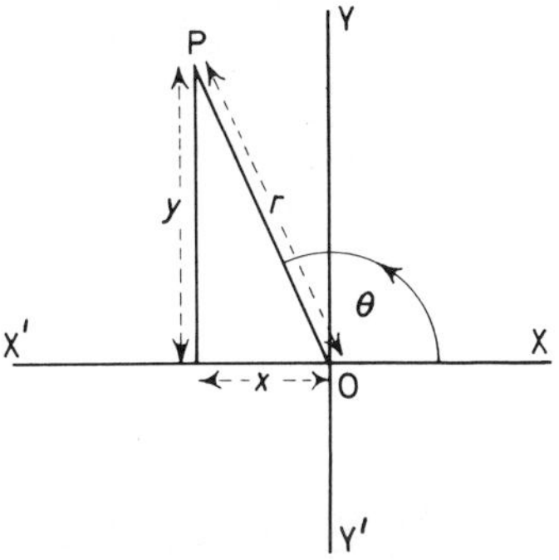

Figure 6.4

Let x and y be the cartesian co-ordinates of P, referred to the axes X′OX, Y′OY, also defined with the usual sign conventions. Let OP = r be measured as positive for all positions of P. Then the trigonometric ratios for ∠POX are defined as

$$\sin\theta = \frac{y}{r} \qquad \cos\theta = \frac{x}{r} \qquad \tan\theta = \frac{y}{x} \tag{6.13}$$

and

$$\operatorname{cosec}\theta = \frac{1}{\sin\theta} \qquad \sec\theta = \frac{1}{\cos\theta} \qquad \cot\theta = \frac{1}{\tan\theta} \tag{6.14}$$

If these definitions are applied to an acute angle, i.e. one which lies in the first quadrant, the results are identical with those given in Section 6.2 (see *Figure 6.5*).

For other angles, however, we must take account of the signs of x and y. If P is in the first quadrant, then x, y and r are all positive, so that the sine,

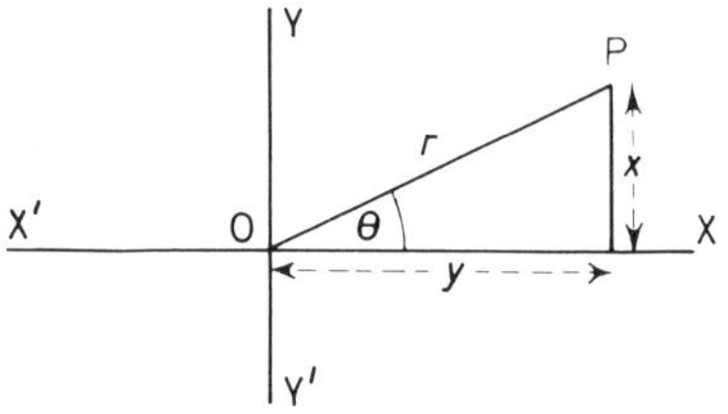

Figure 6.5

cosine and tangent are all positive. If P is in the second quadrant, y is positive and x is negative, so that the sine is positive but the cosine and tangent are negative. If P is in the third quadrant, x and y are negative, so that the sine and cosine are negative but the tangent is positive. If P is in the fourth quadrant, x is positive and y is negative, so that the sine and tangent are negative but the cosine is positive. These results may be memorised with the aid of the following diagram which shows which of three ratios (sine, cosine and tangent) are *positive* in each quadrant.

sin	all
tan	cos

With the definitions (6.13) and (6.14) it is clear that the identities (6.9) to (6.12) will remain valid for angles in any of the four quadrants since the relation $x^2 + y^2 = r^2$ is true whatever the signs of x and y.

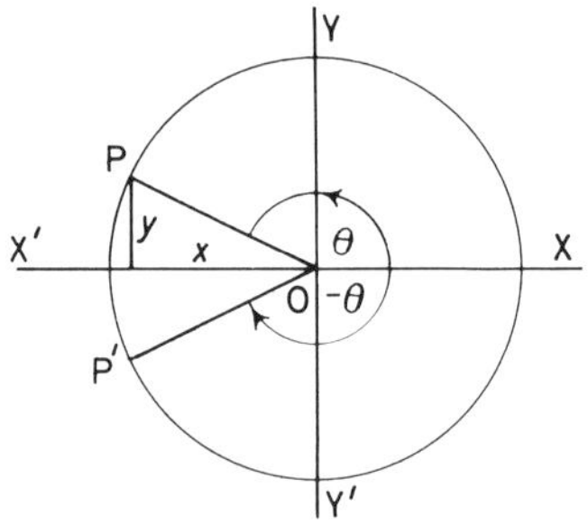

Figure 6.6

The definitions (6.13) and (6.14) enable us to define the trigonometric ratios for any angles. Tables of the trigonometric functions only exist for angles in the range 0° to 90°. These tables are, however, quite sufficient, for the trigonometric ratios of any angle may be expressed in terms of the trigonometric ratios of an acute angle. The following general relationships proved below are of value in such transformations.

First, we observe that if $\angle$POX $= \theta$ and $\angle$P′OX′ $= -\theta$, then P and P′ are the mirror images of each other in the line X′OX (*Figure 6.6*). Thus, OP = OP′ = r, say.

Also, the abscissae of P and P′ are the same, but their ordinates though equal in magnitude are opposite in sign, thus by (6.13)

$$\left.\begin{aligned} \sin(-\theta) &= -\frac{y}{r} = -\sin\theta \\ \cos(-\theta) &= \frac{x}{r} = \cos\theta \\ \tan(-\theta) &= -\frac{y}{x} = -\tan\theta \end{aligned}\right\} \quad (6.15)$$

If P and P′ are the ends of a diameter of a circle radius r, and $\angle$POX $= \theta$, then $\angle$P′OX $= 180° + \theta$. But we can see (*Figure 6.7*) that in this case the abscissae and ordinates of P and P′ are equal in magnitude but opposite in sign.

Thus by (6.13)

$$\left.\begin{aligned} \sin(180° + \theta) &= \frac{-y}{r} = -\sin\theta \\ \cos(180° + \theta) &= \frac{-x}{r} = -\cos\theta \\ \tan(180° + \theta) &= \frac{-y}{-x} = \frac{y}{x} = \tan\theta \end{aligned}\right\} \quad (6.16)$$

If we replace θ by $-\theta$ in (6.16), we obtain, after using (6.15),

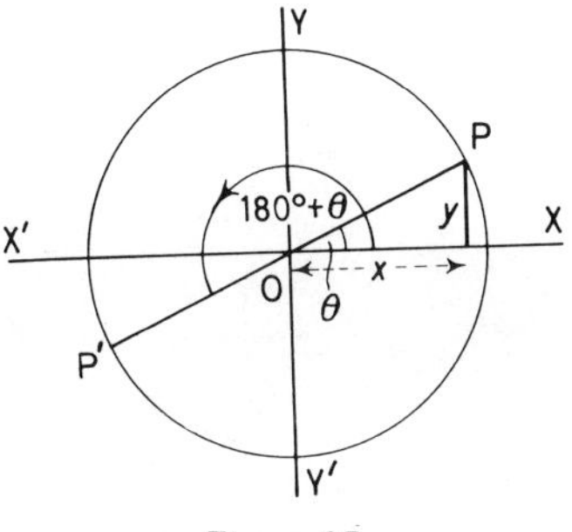

Figure 6.7

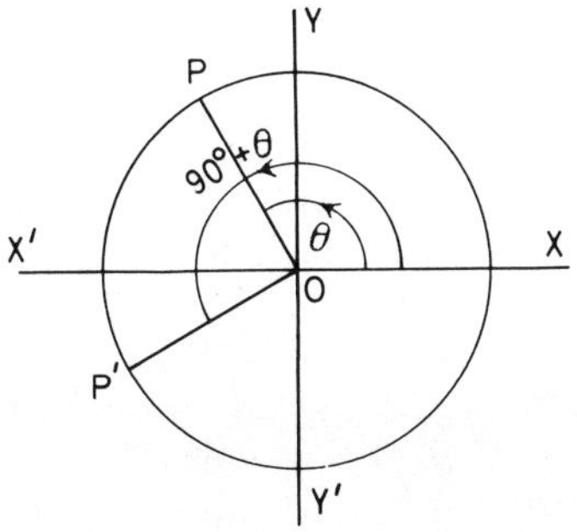

Figure 6.8

$$\left.\begin{aligned} \sin(180^\circ - \theta) &= -\sin(-\theta) = \sin\theta \\ \cos(180^\circ - \theta) &= -\cos(-\theta) = -\cos\theta \\ \tan(180^\circ - \theta) &= \tan(-\theta) = -\tan\theta \end{aligned}\right\} \quad (6.17)$$

If $\angle$POX $= \theta$ and $\angle$P′OX $= 90^\circ + \theta$, then P and P′ lie on the ends of perpendicular radii of a circle centre O and radius r (*Figure 6.8*).

Thus, if P is the point (x,y), P′ is the point with co-ordinates $(-y,x)$. Thus

$$\left.\begin{aligned} \sin(90^\circ + \theta) &= \frac{x}{r} = \cos\theta \\ \cos(90^\circ + \theta) &= \frac{-y}{r} = -\sin\theta \\ \tan(90^\circ + \theta) &= \frac{-y}{x} = -\cot\theta \end{aligned}\right\} \quad (6.18)$$

If we replace θ by $-\theta$ in (6.18), we obtain after using (6.15)

$$\left.\begin{aligned} \sin(90^\circ - \theta) &= \cos(-\theta) = \cos\theta \\ \cos(90^\circ - \theta) &= -\sin(-\theta) = \sin\theta \\ \tan(90^\circ - \theta) &= -\cot(-\theta) = \cot\theta \end{aligned}\right\} \quad (6.19)$$

If $\angle$POX $= \theta$ and $\angle$P′OX $= 360^\circ + \theta$, then P and P′ coincide and if P is the point (x,y) P′ is also the point (x,y) (*Figure 6.9*). Thus

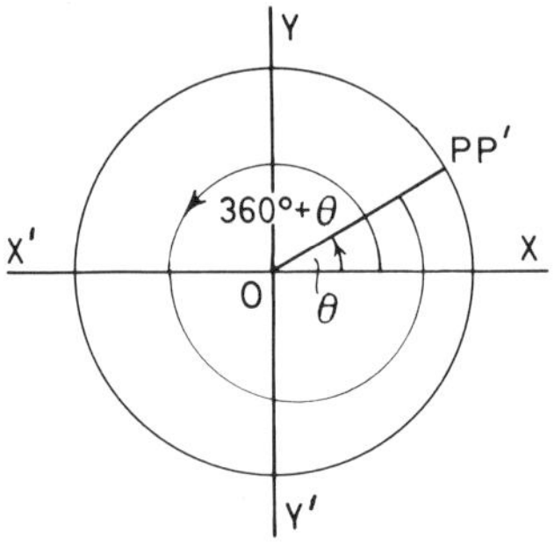

Figure 6.9

$$\left.\begin{aligned} \sin(360^\circ + \theta) &= \sin\theta \\ \cos(360^\circ + \theta) &= \cos\theta \\ \tan(360^\circ + \theta) &= \tan\theta \end{aligned}\right\} \quad (6.20)$$

If we replace θ by $-\theta$ in (6.20), we obtain

$$\left.\begin{aligned} \sin(360^\circ - \theta) &= \sin(-\theta) = -\sin\theta \\ \cos(360^\circ - \theta) &= \cos(-\theta) = \cos\theta \\ \tan(360^\circ - \theta) &= \tan(-\theta) = -\tan\theta \end{aligned}\right\} \quad (6.21)$$

Example 1 Evaluate (i) sin 150°, (ii) cos 210°, (iii) tan 300°, (iv) cos 420°.

$$\left.\begin{aligned}
&\text{(i)}\ \sin 150^\circ = \sin(180^\circ - 150^\circ)\\
&\qquad = \sin 30^\circ = 0{\cdot}5\\
&\text{(ii)}\ \cos 210^\circ = \cos(180^\circ + 30^\circ)\\
&\qquad = -\cos 30^\circ = -0.866\\
&\text{(iii)}\ \tan 300^\circ = -\tan(360^\circ - 300^\circ)\\
&\qquad = -\tan 60^\circ = -1{\cdot}732\\
&\text{(iv)}\ \cos 420^\circ = \cos(360^\circ + 60^\circ)\\
&\qquad = \cos 60^\circ = 0{\cdot}5
\end{aligned}\right\}\ \text{From tables}$$

Example 2 Express in terms of the trigonometric ratios of positive acute angles (i) $\cos -170^\circ$, (ii) $\tan 210^\circ$, (iii) $\cos -300^\circ$, (iv) $\sin -500^\circ$.

(i) $\cos -170^\circ = \cos 170^\circ = -\cos(180^\circ - 170^\circ) = -\cos 10^\circ$
(ii) $\tan 210^\circ = \tan(180^\circ + 30^\circ) = \tan 30^\circ$
(iii) $\cos -300^\circ = \cos 300^\circ = \cos(360^\circ - 300^\circ) = \cos 60^\circ$
(iv) $\sin -500^\circ = -\sin 500^\circ = -\sin(360^\circ + 140^\circ)$
$= -\sin 140^\circ = -\sin(180^\circ - 140^\circ) = -\sin 40^\circ$

Exercises 6c

1 Evaluate (i) $\sin 160^\circ$, (ii) $\cos -400^\circ$, (iii) $\tan 520^\circ$, (iv) $\sin -200^\circ$.
2 Express in terms of the trigonometric ratios of positive acute angles (i) $\cos 190^\circ$, (ii) $\tan -410^\circ$, (iii) $\cos 300^\circ$, (iv) $\sin -740^\circ$.
3 Show that $\sin(270^\circ - \theta) = -\cos\theta$, $\cos(270^\circ - \theta) = -\sin\theta$, $\tan(270^\circ - \theta) = \cot\theta$.
4 Show that $\sin(270^\circ - \theta) + \sin(270^\circ + \theta) = -2\cos\theta$.
5 Show that $\cos 210^\circ \cos 150^\circ - \sin 210^\circ \sin 150^\circ = 1$.
6 Evaluate (i) $\sin(3\pi/2)$, (ii) $\cos(-9\pi/4)$, (iii) $\tan(11\pi/3)$, (iv) $\sin(-8\pi/3)$.
7 Evaluate (i) $\sin 180^\circ$, (ii) $\sin 270^\circ$, (iii) $\sin 360^\circ$, (iv) $\cos 180^\circ$, (v) $\cos 270^\circ$, (vi) $\cos 360^\circ$.
8 If $\sin\theta = 1/\sqrt{3}$ and θ is obtuse find $\cos\theta$ and $\tan\theta$.
9 $\tan\theta = \frac{3}{4}$ and θ is in the third quadrant; calculate $\sin\theta$ and $\cos\theta$.
10 If A, B and C are angles of a triangle, show that (i) $\sin(90^\circ + A) = -\cos(B + C)$, (ii) $\sin\frac{1}{2}(A + B) = \cos\frac{1}{2}C$.

6.4 The graphs of the trigonometric functions

For angles in the range 0° to 90°, the trigonometric ratios have been tabulated. Thus, the graphs of these ratios may be plotted very accurately for acute angles. The results of the previous section then enable us to plot the graphs for other values of the angle.

Less accurately, we may obtain the graph of the function $y = \sin x$ as follows. Consider a circle of unit radius with centre O. Let X′OX and Y′OY be two perpendicular axes. Then, if $\angle$POX $= x$ and P is on the circumference of the circle, $\sin x$ is equal to the projection of OP on the axis Y′OY. In *Figure 6.10*, several positions of OP corresponding to different angles are shown together with the projection of OP on Y′OY. These enable us to obtain the graph shown.

From *Figure 6.10*, the maximum value of $\sin x$ is seen to be 1 and the minimum value -1. The graph crosses the x-axis at $x = 0^\circ, \pm 180^\circ, \pm 360^\circ$, etc.

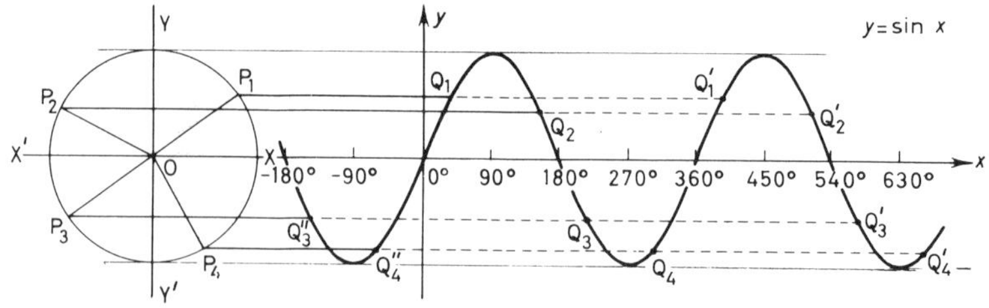

Figure 6.10

The graph for the function $y = \cos x$ can be obtained from *Figure 6.10* by means of the relation

$$\cos x = \sin(90^\circ + x) \quad \text{from (6.18)}$$

Thus the ordinate of the sine curve at $x + 90^\circ$ is the ordinate of the cosine curve at x. The cosine curve is thus the sine curve moved 90° along to the *left*. This is shown in *Figure 6.11*, from which we see that $\cos x$ has a maximum and a minimum value of $+1$ and -1, respectively. The curve crosses the x-axis at $x = \pm 90^\circ$, $\pm 270^\circ$, $\pm 450^\circ$, etc.

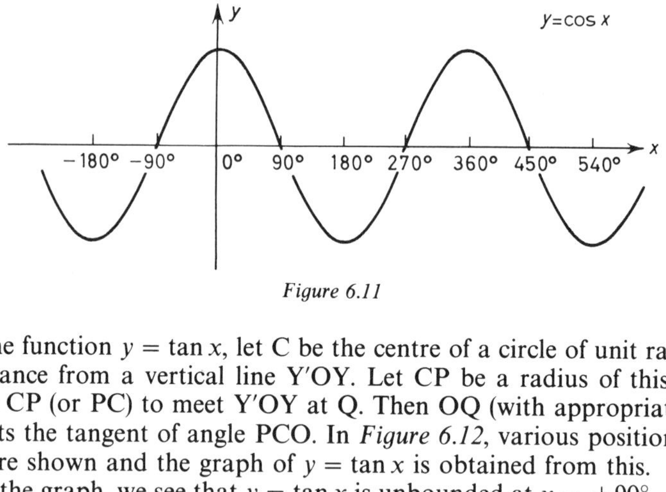

Figure 6.11

For the function $y = \tan x$, let C be the centre of a circle of unit radius at unit distance from a vertical line Y′OY. Let CP be a radius of this circle. Produce CP (or PC) to meet Y′OY at Q. Then OQ (with appropriate sign) represents the tangent of angle PCO. In *Figure 6.12*, various positions of P and Q are shown and the graph of $y = \tan x$ is obtained from this.

From the graph, we see that $y = \tan x$ is unbounded at $x = \pm 90^\circ$, $\pm 270^\circ$, etc. and crosses the x-axis at $x = 0^\circ$, $\pm 180^\circ$, $\pm 360^\circ$, etc.

The graphs above, or more accurately the tables of the trigonometric functions, enable us to evaluate the trigonometric ratio of any angle. The ratios for the particular angles 30°, 60° and 45° can, however, be obtained exactly from considerations of an equilateral triangle and an isosceles right-angled triangle. Thus in *Figure 6.13*, OPN is an isosceles right-angled triangle in which ON = PN = 1 unit. Then, $\angle$OPN = 45° and, by Pythagoras' theorem, OP = $\sqrt{2}$ units. Thus

$$\sin 45^\circ = \frac{ON}{OP} = \frac{1}{\sqrt{2}} \qquad \cos 45^\circ = \frac{PN}{OP} = \frac{1}{\sqrt{2}} \qquad \tan 45^\circ = \frac{ON}{PN} = 1 \quad (6.22)$$

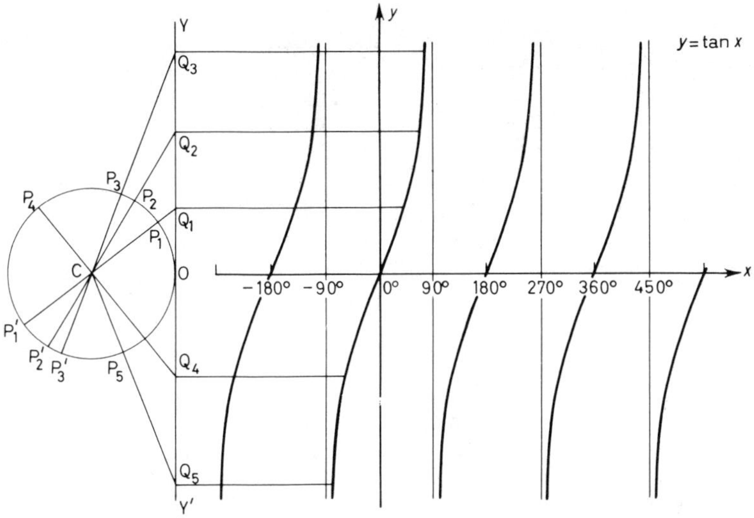

Figure 6.12

In *Figure 6.14*, OPQ is an equilateral triangle of side 1 unit, ON being an altitude. Thus, $PN = NQ = \frac{1}{2}$ unit and $\angle OPN = 60°$. $\angle PON = 30°$. Now $ON^2 = OP^2 = 1 - \frac{1}{4} = \frac{3}{4}$. Therefore

$$ON = \frac{\sqrt{3}}{2}$$

and so

$$\left.\begin{aligned} \sin 30° &= \frac{PN}{OP} = \cos 60° = \frac{1}{2} \\ \cos 30° &= \frac{ON}{OP} = \sin 60° = \frac{\sqrt{3}}{2} \\ \tan 30° &= \frac{PN}{ON} = \cot 60° = \frac{1}{\sqrt{3}} \\ \tan 60° &= \cot 30° = \sqrt{3} \end{aligned}\right\} \qquad (6.23)$$

The results (6.22) and (6.23) are often useful and are worth the trouble of memorising.

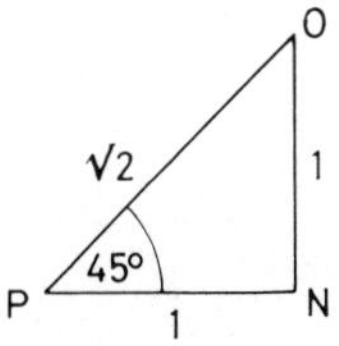

Figure 6.13

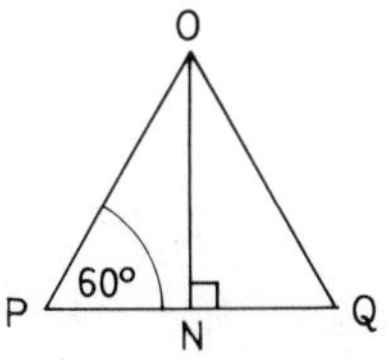

Figure 6.14

Exercises 6d

1 Plot the graphs of (i) $y = \sec\theta$, (ii) $y = \operatorname{cosec}\theta$, (iii) $y = \cot\theta$ for all values of θ in range $-720°$ to $+720°$.
2 Plot the graph of (i) $y = \sin 2x$, (ii) $y = \cos(x + 45°)$, (iii) $y = \tan(2x + 30°)$.
3 Plot the graph of (i) $y = 3\cos x + 4\sin x$ and on the same scale (ii) $y = 5\sin(x + 36° 52')$. What do you notice?
4 Plot the graph of (i) $y = \cos x - \sin x$ and on the same scale (ii) $y = (1/\sqrt{2})\cos(x + 45°)$. What do you notice?
5 Without using tables write down the values of (i) sin 120°, (ii) tan −135°, (iii) tan 315°, (iv) cos −240°, (v) $\sin(3\pi/4)$, (vi) $\tan(-2\pi/3)$, (vii) $\tan(-5\pi/4)$, (viii) $\cos(5\pi/4)$.

6.5 The addition formulae

In this section, we shall obtain formulae for the trigonometric ratios of the sum and difference of two angles in terms of the trigonometric ratios of those angles.

We shall prove these results to be valid for any pair of angles A and B. We first obtain an expression for $\cos(A - B)$ in terms of $\cos A$, $\cos B$, $\sin A$ and $\sin B$. Suppose P and Q are two points on the circumference of a circle centre O and radius r and such that ∠POX = A, ∠QOX = B. Then, from the definitions of the sine and cosine (6.13), P is the point $(r\cos A, r\sin A)$ and Q is the point $(r\cos B, r\sin B)$.

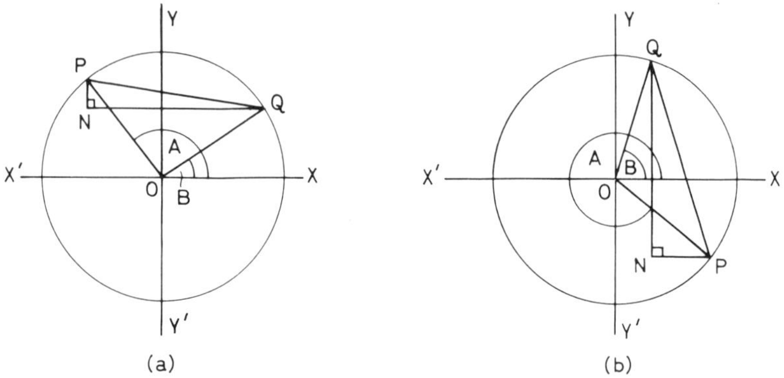

Figure 6.15

Figures 6.15a and *b* show two possible configurations. In *Figure 6.15a*, ∠POQ = $A - B$; in *Figure 6.15b*, ∠POQ = $360° - (A - B)$. In both cases (and all other cases, as readers should easily verify), cos POQ = $\cos(A - B)$ by virtue of (6.20) and (6.21).

Thus, on applying the cosine rule (8.3) to the triangle OPQ,

$$\begin{aligned} PQ^2 &= OP^2 + OQ^2 - 2OP.OQ\cos POQ \\ &= r^2 + r^2 - 2r^2\cos(A - B) = 2r^2 - 2r^2\cos(A - B) \end{aligned}$$

The triangle PQN obtained by drawing parallels to the axes through P and Q has ∠PNQ = 90°, PN = $r\sin A - r\sin B$, QN = $r\cos A - r\cos B$. Therefore, by Pythagoras,

$$\begin{aligned}PQ^2 &= PN^2 + QN^2\\ &= r^2 \sin^2 A - 2r^2 \sin A \sin B + r^2 \sin^2 B + r^2 \cos^2 A\\ &\quad - 2r^2 \cos A \cos B + r^2 \cos^2 B\\ &= r^2(\sin^2 A + \cos^2 A) + r^2(\sin^2 B + \cos^2 B)\\ &\quad - 2r^2(\cos A \cos B + \sin A \sin B)\\ &= 2r^2 - 2r^2(\cos A \cos B + \sin A \sin B)\end{aligned}$$

On equating these two expressions for PQ^2, we obtain

$$\cos(A - B) = \cos A \cos B + \sin A \sin B \qquad (6.24)$$

If in (6.24) we replace B by $-B$ (i.e. $-B$ by $+B$), we obtain

$$\cos(A + B) = \cos(A)\cos(-B) + \sin A \sin(-B)$$

which by (6.15) gives

$$\cos(A + B) = \cos A \cos B - \sin A \sin \mathrm{B} \qquad (6.25)$$

In (6.24) replace A by $(90° - A)$, then

$$\begin{aligned}\cos(90° - A - B) &= \cos[90° - (A + B)]\\ &= \cos(90° - A)\cos B + \sin(90° - A)\sin B\end{aligned}$$

Therefore, by (6.19) we have

$$\sin(A + B) = \sin A \cos B + \cos A \sin B \qquad (6.26)$$

In (6.26) replace B by $-B$, then

$$\sin(A - B) = \sin A \cos(-B) + \cos A \sin(-B)$$

and so by (6.15)

$$\sin(A - B) = \sin A \cos B - \cos A \sin B \qquad (6.27)$$

From these results, we obtain by division the corresponding formulae for $\tan(A + B)$ and $\tan(A - B)$

$$\tan(A + B) = \frac{\sin(A + B)}{\cos(A + B)} = \frac{\sin A \cos B + \cos A \sin B}{\cos A \cos B - \sin A \sin B}$$

On dividing numerator and denominator by $\cos A \cos B$, we obtain

$$\tan(A + B) = \frac{\tan A + \tan B}{1 - \tan A \tan B} \qquad (6.28)$$

In the same way

$$\tan(A - B) = \frac{\sin A \cos B - \cos A \sin B}{\cos A \cos B + \sin A \sin B} = \frac{\tan A - \tan B}{1 + \tan A \tan B} \qquad (6.29)$$

Example 1 Given $\sin 45° = \cos 45° = 1/\sqrt{2}$, $\sin 30° = 1/2$, $\cos 30° = \sqrt{3}/2$, calculate $\sin 15°$.

By (6.27)

$$\begin{aligned}\sin 15° &= \sin(45° - 30°) = \sin 45° \cos 30° - \cos 45° \sin 30°\\ &= \frac{1}{\sqrt{2}}\cdot\frac{\sqrt{3}}{2} - \frac{1}{\sqrt{2}}\cdot\frac{1}{2} = \frac{\sqrt{3} - 1}{2\sqrt{2}}\end{aligned}$$

Example 2 Show that $\tan(45^\circ + A) = \dfrac{1+\tan A}{1-\tan A}$.

By (6.28)

$$\tan(45^\circ + A) = \frac{\tan 45^\circ + \tan A}{1 - \tan 45^\circ \tan A} = \frac{1+\tan A}{1-\tan A}$$

since $\tan 45^\circ = 1$.

Example 3 Show that $\sin(x+y)\sin(x-y) = \sin^2 x - \sin^2 y$.

$$\begin{aligned} \text{LHS} &= (\sin x\cos y + \cos x\sin y)(\sin x\cos y - \cos x\sin y) \\ &= \sin^2 x\cos^2 y - \cos^2 x\sin^2 y \\ &= \sin^2 x(1-\sin^2 y) - (1-\sin^2 x)\sin^2 y \\ &= \sin^2 x - \sin^2 y = \text{RHS} \end{aligned}$$

Exercises 6e

1 If $\sin A = \frac{4}{5}$ and $\cos B = \frac{5}{13}$ evaluate without using tables (i) $\sin(A+B)$, (ii) $\cos(A-B)$, (iii) $\tan(A+B)$, if A and B are acute. Is $(A+B)$ an acute angle?
2 If $\tan(x+y) = \frac{4}{3}$ and $\tan x = \frac{1}{2}$, evaluate $\tan y$.
3 Evaluate without tables (i) $\cos 75^\circ$, (ii) $\sin 75^\circ$.
4 Using (6.25) show that (i) $\cos(90^\circ + A) = -\sin A$, (ii) $\sin(90^\circ + A) = \cos A$.
5 Show that (i) $\sin(180^\circ - A) = \sin A$, (ii) $\cos(180^\circ - A) = -\cos A$.
6 Simplify (i) $\sin 40^\circ\cos 30^\circ - \cos 40^\circ\sin 30^\circ$
(ii) $\cos 50^\circ\cos 60^\circ - \sin 50^\circ\sin 60^\circ$.
7 Simplify (i) $\cos 40^\circ\cos 30^\circ + \sin 40^\circ\sin 30^\circ$
(ii) $\sin 150^\circ\cos 160^\circ + \cos 150^\circ\sin 160^\circ$.
8 Simplify (i) $\dfrac{\tan 30^\circ + \tan 40^\circ}{1 - \tan 30^\circ\tan 40^\circ}$, (ii) $\dfrac{\tan 60^\circ - \tan 30^\circ}{1 + \tan 60^\circ\tan 30^\circ}$.
9 Show that $\sin x + \sin(x + \frac{2}{3}\pi) + \sin(x + \frac{4}{3}\pi) = 0$.

Show that

10 $\tan A + \tan B = \dfrac{\sin(A+B)}{\cos A\cos B}$

11 $\cot(x+y) = \dfrac{\cot x\cot y - 1}{\cot x + \cot y}$

12 $(\cos\theta + \cos\phi)^2 + (\sin\theta + \sin\phi)^2 = 2 + 2\cos(\theta - \phi)$

13 $\tan(x + \frac{1}{4}\pi) = \dfrac{\cos x + \sin x}{\cos x - \sin x}$

14 $\cos(A+B)\cos(A-B) = \cos^2 A - \sin^2 B$
15 $\tan(\frac{1}{4}\pi + x)\tan(\frac{1}{4}\pi - x) = 1$
16 If $\tan\lambda = \mu$, show that $\dfrac{\sin\theta + \mu\cos\theta}{\cos\theta - \mu\sin\theta} = \tan(\theta + \lambda)$.
17 Given $\sin(x-\alpha) = \cos(x-\alpha)$, show that

$$\tan\alpha = \frac{\tan x - 1}{\tan x + 1} = \tan(x - \tfrac{1}{4}\pi)$$

18 Express in terms of the sines and cosines of A, B and C (i) $\sin(A+B+C)$, (ii) $\cos(A+B+C)$.

19 Show that

$$\tan(A + B + C) = \frac{\tan A + \tan B + \tan C - \tan A \tan B \tan C}{1 - \tan A \tan B - \tan B \tan C - \tan C \tan A}$$

20 If A, B and C are the angles of a triangle show that
(i) $\cos A + \cos(B - C) = 2 \sin B \sin C$
(ii) $\cos \tfrac{1}{2}C + \sin \tfrac{1}{2}(A - B) = 2 \sin \tfrac{1}{2}A \cos \tfrac{1}{2}B$.

6.6 Multiple and submultiple angle formulae

If we put $B = A$ in (6.26) and (6.25), we obtain

$$\sin 2A = \sin A \cos A + \cos A \sin A$$

$$\sin 2A = 2 \sin A \cos A \tag{6.30}$$

and

$$\cos 2A = \cos^2 A - \sin^2 A \tag{6.31}$$

Since $\cos^2 A = 1 - \sin^2 A$ and $\sin^2 A = 1 - \cos^2 A$, this result can be put in either of the forms

$$\cos 2A = 2\cos^2 A - 1 \tag{6.32}$$

or

$$\cos 2A = 1 - 2\sin^2 A \tag{6.33}$$

Also

$$\tan 2A = \frac{\tan A + \tan A}{1 - \tan A \tan A}$$

$$\tan 2A = \frac{2 \tan A}{1 - \tan^2 A} \tag{6.34}$$

We can use these important results to obtain expressions for $\sin 3A$, $\cos 3A$, $\tan 3A$, etc.

$$\begin{aligned} \sin 3A = \sin(2A + A) &= \sin 2A \cos A + \cos 2A \sin A \\ &= 2 \sin A \cos^2 A + (\cos^2 A - \sin^2 A) \sin A \\ &= 2 \sin A \cos^2 A + \sin A \cos^2 A - \sin^3 A \\ &= 3 \sin A \cos^2 A - \sin^3 A \\ &= 3 \sin A(1 - \sin^2 A) - \sin^3 A \end{aligned}$$

Therefore

$$\sin 3A = 3 \sin A - 4 \sin^3 A \tag{6.35}$$

$$\begin{aligned} \cos 3A = \cos(2A + A) &= \cos 2A \cos A - \sin 2A \sin A \\ &= (2\cos^2 A - 1)\cos A - 2\sin^2 A \cos A \\ &= 2\cos^3 A - \cos A - 2\cos A(1 - \cos^2 A) \end{aligned}$$

Therefore

$$\cos 3A = 4\cos^3 A - 3\cos A \tag{6.36}$$

$$\tan 3A = \tan(2A + A) = \frac{\tan 2A + \tan A}{1 - \tan 2A \tan A}$$

$$= \frac{\dfrac{2\tan A}{1-\tan^2 A} + \tan A}{1 - \tan A\dfrac{2\tan A}{1-\tan^2 A}}$$

$$= \frac{2\tan A + \tan A - \tan^3 A}{1 - \tan^2 A - 2\tan^2 A}$$

Therefore

$$\tan 3A = \frac{3\tan A - \tan^3 A}{1 - 3\tan^2 A} \tag{6.37}$$

If we replace A by $\frac{1}{2}x$ in (6.30), (6.31), (6.32), (6.33) and (6.34), we have

$$\sin x = 2\sin\tfrac{1}{2}x\cos\tfrac{1}{2}x \tag{6.38}$$

$$\cos x = \cos^2\tfrac{1}{2}x - \sin^2\tfrac{1}{2}x \tag{6.39}$$

$$= 2\cos^2\tfrac{1}{2}x - 1 \tag{6.40}$$

$$= 1 - 2\sin^2\tfrac{1}{2}x \tag{6.41}$$

$$\tan x = \frac{2\tan\frac{1}{2}x}{1 - \tan^2\frac{1}{2}x} \tag{6.42}$$

These results enable us to express $\sin x$, $\cos x$ and $\tan x$ in terms of $\tan\frac{1}{2}x$. For with $\tan\frac{1}{2}x = t$, we have immediately from (6.42)

$$\tan x = \frac{2t}{1 - t^2} \tag{6.43}$$

$$\sin x = 2\sin\tfrac{1}{2}x\cos\tfrac{1}{2}x$$

$$= \frac{2\sin\frac{1}{2}x\cos\frac{1}{2}x}{\cos^2\frac{1}{2}x + \sin^2\frac{1}{2}x} \quad (\text{since } \cos^2\tfrac{1}{2}x + \sin^2\tfrac{1}{2}x = 1)$$

On dividing the numerator and denominator by $\cos^2\frac{1}{2}x$, we have

$$\sin x = \frac{2\dfrac{\sin\frac{1}{2}x}{\cos\frac{1}{2}x}}{1 + \dfrac{\sin^2\frac{1}{2}x}{\cos^2\frac{1}{2}x}}$$

therefore

$$\sin x = \frac{2t}{1 + t^2} \tag{6.44}$$

Also

$$\cos x = \frac{\cos^2 \frac{1}{2}x - \sin^2 \frac{1}{2}x}{\cos^2 \frac{1}{2}x + \sin^2 \frac{1}{2}x} = \frac{1 - \dfrac{\sin^2 \frac{1}{2}x}{\cos^2 \frac{1}{2}x}}{1 + \dfrac{\sin^2 \frac{1}{2}x}{\cos^2 \frac{1}{2}x}}$$

therefore

$$\cos x = \frac{1 - t^2}{1 + t^2} \tag{6.45}$$

Example 1 If $\tan\theta = \frac{24}{7}$ and θ is acute, calculate $\tan\frac{1}{2}\theta$.
We have from (6.43) with $t = \tan\frac{1}{2}\theta$

$$\frac{24}{7} = \frac{2t}{1 - t^2}$$

$$24 - 24t^2 = 14t$$
$$24t^2 + 14t - 24 = 0$$
$$12t^2 + 7t - 12 = 0$$
$$(4t - 3)(3t + 4) = 0$$
$$t = \tfrac{3}{4} \quad \text{or} \quad t = -\tfrac{4}{3}$$

But since θ is acute, so is $\frac{1}{2}\theta$, so that $\tan\frac{1}{2}\theta$ is positive. Therefore, we have $\tan\frac{1}{2}\theta = \frac{3}{4}$.

Example 2 Show that $\dfrac{\sin 2\theta}{1 + \cos 2\theta} = \tan\theta$.

$$\text{LHS} = \frac{2\sin\theta\cos\theta}{1 + 2\cos^2\theta - 1} = \frac{2\sin\theta\cos\theta}{2\cos^2\theta} = \frac{\sin\theta}{\cos\theta} = \tan\theta.$$

Example 3 Prove that $\cos^4 A - \sin^4 A = \cos 2A$.

$$\begin{aligned}\text{LHS} = \cos^4 A - \sin^4 A &= (\cos^2 A - \sin^2 A)(\cos^2 A + \sin^2 A)\\ &= \cos^2 A - \sin^2 A\\ &\quad (\text{since } \cos^2 A + \sin^2 A = 1)\end{aligned}$$

Therefore

$$\cos^4 A - \sin^4 A = \cos 2A = \text{RHS}$$

Exercises 6f

1 If $\tan\theta = \frac{4}{3}$, calculate the possible values of $\tan\frac{1}{2}\theta$.
2 Show that $\tan 22\frac{1}{2}° = \sqrt{2} - 1$, without using tables.
3 If $\cos A = \frac{4}{5}$, find without tables $\sin 2A$, $\cos\frac{1}{2}A$ and $\tan\frac{1}{2}A$.
4 Given $\cos 30° = \sqrt{3}/2$, show that $\sin 15° = \frac{1}{2}\sqrt{(2 - \sqrt{3})}$.
5 From the values of cos 30° and sin 30° deduce those of cos 60° and sin 60°.

Show that

6 $\dfrac{\sin 2A}{1 - \cos 2A} = \cot A$

7 $\tan 2\theta - \tan\theta = \tan\theta \sec 2\theta$

8 $\dfrac{1 + \cos x + \cos 2x}{\sin x + \sin 2x} = \cot x$

9 $\dfrac{\sin 16\theta}{\sin\theta} = 16\cos\theta\cos 2\theta\cos 4\theta\cos 8\theta$

10 $(\cos\theta - \sin\theta)^2 = 1 - \sin 2\theta$

11 $\dfrac{1 + \tan^2 A}{(1 + \tan A)^2} = \dfrac{1}{1 + \sin 2A}$

12 $\sin^2 2\phi + 2\cos^2\phi\cos 2\phi = 2\cos^2\phi$

13 $\tan^2(\frac{1}{4}\pi - \frac{1}{2}\theta) = \dfrac{1 - \sin\theta}{1 + \sin\theta}$

14 If $\tan^2 x = 1 + 2\tan^2 y$, show that $\cos 2x + \sin^2 y = 0$.

15 Express in terms of $t = \tan\frac{1}{2}\theta$ (i) $1 + \sin\theta$, (ii) $1 + \sin\theta + \cos\theta$, (iii) $\sec\theta - \tan\theta$.

6.7 The factor formulae

The sums and differences of sines and cosines may be expressed as products of sines and cosines and vice versa.

From (6.26) and (6.27), we have

$$\sin(A + B) = \sin A\cos B + \cos A\sin B$$
$$\sin(A - B) = \sin A\cos B - \cos A\sin B$$

so that on addition

$$2\sin A\cos B = \sin(A + B) + \sin(A - B) \tag{6.46}$$

and on subtraction

$$2\cos A\sin B = \sin(A + B) - \sin(A - B) \tag{6.47}$$

Similarly from (6.24) and (6.25)

$$\cos(A - B) = \cos A\cos B + \sin A\sin B$$
$$\cos(A + B) = \cos A\cos B - \sin A\sin B$$

so that

$$2\cos A\cos B = \cos(A + B) + \cos(A - B) \tag{6.48}$$
$$2\sin A\sin B = \cos(A - B) - \cos(A + B) \tag{6.49}$$

These formulae enable us to express a product of sines and cosines as a sum or a difference.

If we put $A + B = C$ and $A - B = D$, so that $A = \frac{1}{2}(C + D)$ and $B = \frac{1}{2}(C - D)$, we obtain

$$\left.\begin{aligned}
\sin C + \sin D &= 2\sin\tfrac{1}{2}(C + D)\cos\tfrac{1}{2}(C - D)\\
\sin C - \sin D &= 2\cos\tfrac{1}{2}(C + D)\sin\tfrac{1}{2}(C - D)\\
\cos C + \cos D &= 2\cos\tfrac{1}{2}(C + D)\cos\tfrac{1}{2}(C - D)\\
\cos C - \cos D &= -2\sin\tfrac{1}{2}(C + D)\sin\tfrac{1}{2}(C - D)
\end{aligned}\right\} \tag{6.50}$$

These formulae enable us to express a sum or a difference of two sines or cosines as a product. Note the *minus* sign in the last result.

Example 1 Show that $\sin 2A \cos 4A + \sin 3A \cos 9A = \frac{1}{2}(\sin 12A - \sin 2A)$.

By (6.46)

$$\begin{aligned}\sin 2A \cos 4A &= \tfrac{1}{2}\sin(2A + 4A) + \tfrac{1}{2}\sin(2A - 4A)\\ &= \tfrac{1}{2}\sin 6A + \tfrac{1}{2}\sin(-2A)\\ &= \tfrac{1}{2}\sin 6A - \tfrac{1}{2}\sin 2A\end{aligned}$$

Similarly

$$\begin{aligned}\sin 3A \cos 9A &= \tfrac{1}{2}\sin(3A + 9A) + \tfrac{1}{2}\sin(3A - 9A)\\ &= \tfrac{1}{2}\sin 12A - \tfrac{1}{2}\sin 6A\end{aligned}$$

On addition, we have

$$\sin 2A \cos 4A + \sin 3A \cos 9A = \tfrac{1}{2}(\sin 12A - \sin 2A)$$

as required.

Example 2 Show that $\sin 7x + \sin x - 2\sin 2x \cos 3x = 4\cos^2 3x \sin x$.

$$\begin{aligned}\text{LHS} &= \sin 7x + \sin x - 2\sin 2x \cos 3x\\ &= 2\sin\tfrac{1}{2}(7x + x)\cos\tfrac{1}{2}(7x - x) - 2\sin 2x \cos 3x\\ &= 2\sin 4x \cos 3x - 2\sin 2x \cos 3x\\ &= 2\cos 3x(\sin 4x - \sin 2x)\\ &= 2\cos 3x . 2\cos\tfrac{1}{2}(4x + 2x)\sin\tfrac{1}{2}(4x - 2x)\\ &= 4\cos^2 3x \sin x\end{aligned}$$

Example 3 If $\sin\theta + \sin\phi = a$ and $\cos\theta + \cos\phi = b$, show that

$$\cos^2\tfrac{1}{2}(\theta - \phi) = \tfrac{1}{4}(a^2 + b^2)$$

We have

$$2\sin\tfrac{1}{2}(\theta + \phi)\cos\tfrac{1}{2}(\theta - \phi) = a \qquad 2\cos\tfrac{1}{2}(\theta + \phi)\cos\tfrac{1}{2}(\theta - \phi) = b$$

On squaring and adding, we obtain

$$4\cos^2\tfrac{1}{2}(\theta - \phi)[\sin^2\tfrac{1}{2}(\theta + \phi) + \cos^2\tfrac{1}{2}(\theta + \phi)] = a^2 + b^2$$

therefore

$$\cos^2\tfrac{1}{2}(\theta - \phi) = \frac{a^2 + b^2}{4}$$

Exercises 6g

Show that

1 $\sin 50^\circ + \sin 40^\circ = \sqrt{2}\cos 5^\circ$
2 $\cos 70^\circ + \cos 20^\circ = \sqrt{2}\cos 25^\circ$
3 $\sin 70^\circ + \sin 50^\circ = \sqrt{3}\cos 10^\circ$
4 $\cos 75^\circ - \cos 15^\circ = -1/\sqrt{2}$
5 $\cos 5A + \cos 3A = 2\cos A \cos 4A$
6 $\sin 6x - \sin 2x = 2\cos 4x \sin 2x$
7 $\cos A - \cos 13A = 2\sin 7A \sin 6A$

8 $2\sin 3\theta\cos\theta = \sin 4\theta + \sin 2\theta$

9 $2\sin 7\theta\sin\theta = \cos 6\theta - \cos 8\theta$

10 $2\cos 5\theta\cos\theta = \cos 6\theta + \cos 4\theta$

11 $\dfrac{\cos 7\theta + \cos 3\theta}{\sin 7\theta + \sin 3\theta} = \cot 5\theta$

12 $\dfrac{\sin\theta + \sin 5\theta}{\cos\theta + \cos 5\theta} = \tan 3\theta$

13 $\dfrac{\cos 70^\circ - \cos 50^\circ}{\sin 70^\circ - \sin 50^\circ} = -\sqrt{3}$

14 $\dfrac{\sin\theta + \sin\phi}{\cos\theta + \cos\phi} = \tan\frac{1}{2}(\theta + \phi)$

15 $\dfrac{\sin\theta - \sin\phi}{\cos\theta + \cos\phi} = \tan\frac{1}{2}(\theta - \phi)$

16 $\dfrac{\cos 70^\circ + \cos 20^\circ}{\sin 70^\circ + \sin 20^\circ} = 1$

17 $\dfrac{\sin\beta + \sin 3\beta + \sin 5\beta}{\cos\beta + \cos 3\beta + \cos 5\beta} = \tan 3\beta$

18 $\dfrac{\cos 2\alpha + \cos 5\alpha + \cos 8\alpha}{\sin 2\alpha + \sin 5\alpha + \sin 8\alpha} = \cot 5\alpha$

19 $\tan\frac{1}{2}(A - B) + \tan\frac{1}{2}(A + B) = \dfrac{2\sin A}{\cos A + \cos B}$

20 $\cos^2(\alpha + \beta) + \cos^2(\alpha - \beta) = 1 + \cos 2\alpha\cos 2\beta$

21 $4\cos A\cos(A + \frac{2}{3}\pi)\cos(A + \frac{4}{3}\pi) = \cos 3A$

22 $\sin A\cos 3A - \sin 3A\cos 5A = \frac{1}{2}(\sin 4A - \sin 8A)$

23 $\sin 4A\sin 5A + \sin 2A\sin 11A = \sin 7A\sin 6A$

24 $\cos x + 2\cos 2x + \cos 3x = 4\cos^2\frac{1}{2}x\cos 2x$

25 $\sin\alpha + \sin(\alpha + 3x) + \sin(\alpha + 5x) + \sin(\alpha + 8x)$
$= 4\sin(\alpha + 4x)\cos\frac{3}{2}x\cos\frac{5}{2}x$

6.8 The function $a\cos\theta + b\sin\theta$

Expressions of the form $a\cos\theta + b\sin\theta$ arise in many practical problems. We shall show that this expression may be written in a form involving either the sine or the cosine of some other angle. We saw this in two particular cases (Exercises 6d, questions 3 and 4).

If we set $a\cos\theta + b\sin\theta \equiv R\sin(\theta + \alpha)$, then R and α can be found by the following reasoning. We require $a\cos\theta + b\sin\theta$ to be identical with

$$R\sin\theta\cos\alpha + R\cos\theta\sin\alpha\,[\equiv R\sin(\theta + \alpha) \quad \text{by (6.26)}]$$

This will be so if $R\cos\alpha = b$ and $R\sin\alpha = a$. If we square and add these equations, we obtain

$$R^2(\cos^2\alpha + \sin^2\alpha) = a^2 + b^2$$
$$R = \pm\sqrt{(a^2 + b^2)}$$

On division, we obtain

$$\frac{R\sin\alpha}{R\cos\alpha} = \frac{a}{b}$$

that is

$$\tan\alpha = \frac{a}{b}$$

Therefore

$$a\cos\theta + b\sin\theta \equiv \pm\sqrt{(a^2+b^2)}\sin(\theta+\alpha) \tag{6.51}$$

where

$$\tan\alpha = \frac{a}{b}$$

If we chose R to be positive, then α is determined by the signs of $\sin\alpha$ and $\cos\alpha$ (which are those of a and b respectively) *and* $\tan\alpha = a/b$.

$$\sin\alpha = \frac{a}{\sqrt{(a^2+b^2)}} \qquad \cos\alpha = \frac{b}{\sqrt{(a^2+b^2)}}$$

Alternatively, if we set

$$\begin{aligned} a\cos\theta + b\sin\theta &\equiv R\cos(\theta-\beta) \\ &\equiv R\cos\theta\cos\beta + R\sin\theta\sin\beta \end{aligned}$$

then $R\cos\beta = a$ and $R\sin\beta = b$. Therefore, on squaring and adding, we have (for positive R)

$$R = \sqrt{(a^2+b^2)}$$

and on dividing, we have

$$\tan\beta = \frac{b}{a}$$

Therefore

$$a\cos\theta + b\sin\theta \equiv \sqrt{(a^2+b^2)}\cos(\theta-\beta) \tag{6.52}$$

where

$$\tan\beta = \frac{b}{a}$$

Example 1 Express in the form $R\sin(\theta+\alpha)$, $2\cos\theta + 3\sin\theta$.

By (6.51)

$$2\cos\theta + 3\sin\theta = \sqrt{(2^2+3^2)}\sin(\theta+\alpha)$$

where $\tan\alpha = \frac{2}{3}$, $\sin\alpha = 2/\sqrt{13}$ and $\cos\alpha = 3/\sqrt{13}$, so that α is acute (see *Figure 6.16*).

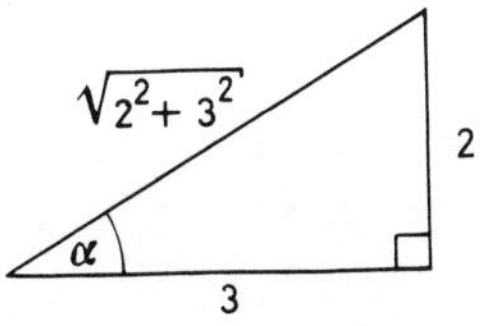

Figure 6.16

From tables, $\alpha = 33°41'$, therefore

$$2\cos\theta + 3\sin\theta = \sqrt{13}\sin(\theta + 33°41')$$

Example 2 Express in the form $R\cos(\theta - \alpha)$, $\cos\theta - 2\sin\theta$.
By (6.52)

$$\cos\theta - 2\sin\theta = \sqrt{(1^2 + 2^2)}\cos(\theta - \alpha)$$

where $\tan\alpha = -2/1 = -2$, $\sin\alpha = -2/\sqrt{5}$, $\cos\alpha = 1/\sqrt{5}$, so that α lies in the fourth quadrant (*Figure 6.17*). From tables, $\alpha = -63°26'$, therefore

$$\cos\theta - 2\sin\theta = \sqrt{5}\cos[\theta - (-63°26')]$$
$$= \sqrt{5}\cos(\theta + 63°26')$$

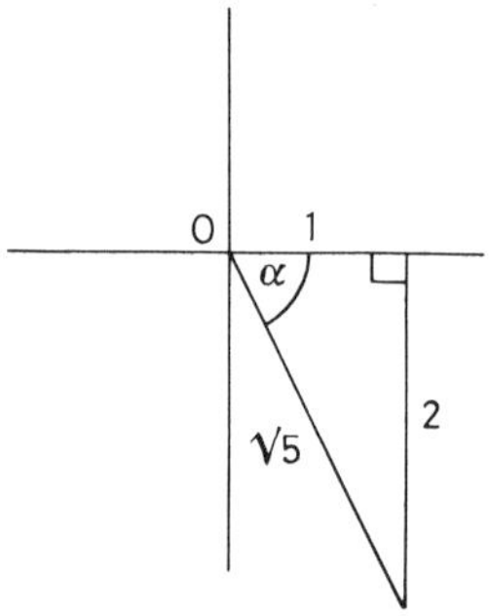

Figure 6.17

Example 3 Express in the form $R\sin(\theta - \alpha)$, $\cos\theta - \sin\theta$.
We notice that by (6.27)

$$R\sin(\theta - \alpha) \equiv R\sin\theta\cos\alpha - R\cos\theta\sin\alpha$$

If this is to be identical with $\cos\theta - \sin\theta$, we need to choose R and α so that

$$R\cos\alpha = -1 \quad \text{and} \quad -R\sin\alpha = 1$$

On squaring and adding, we obtain

$$R^2(\cos^2\alpha + \sin^2\alpha) = (-1)^2 + (1)^2 = 2$$

therefore

$$R = \sqrt{2} \quad \text{(Note: we take the positive root)}$$

On dividing, we have $\tan\alpha = 1$, but since $\sin\alpha = \cos\alpha = -1/\sqrt{2}$, α is in the *third* quadrant. From tables, $\alpha = 225°$ (see *Figure 6.18*). Therefore

$$\cos\theta - \sin\theta = \sqrt{2}\sin(\theta - 225°)$$

Example 4 Sketch the graph of the function $y = a\cos\theta + b\sin\theta$ and state the maximum and minimum values of y.
We have seen (6.51) that

$$y = a\cos\theta + b\sin\theta = \sqrt{(a^2 + b^2)}\sin(\theta + \alpha)$$

where $\tan\alpha = a/b$.

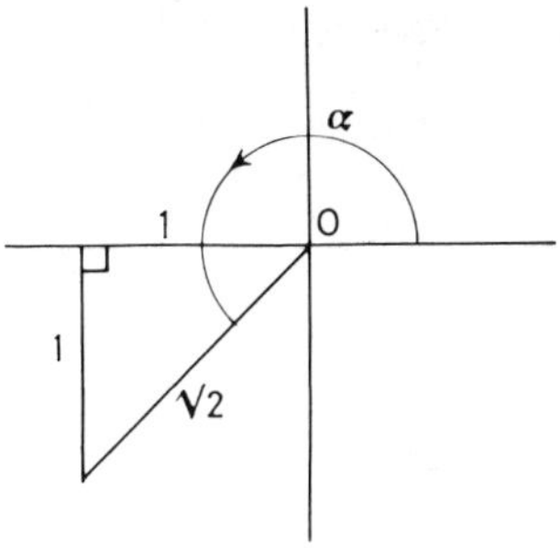

Figure 6.18

The graph of $y = \sin(\theta + \alpha)$ is obtained from the graph of $y = \sin\theta$ by displacing the latter graph a distance α to the left. Multiplication by $\sqrt{(a^2 + b^2)}$ produces the graph of y shown in *Figure 6.19*.

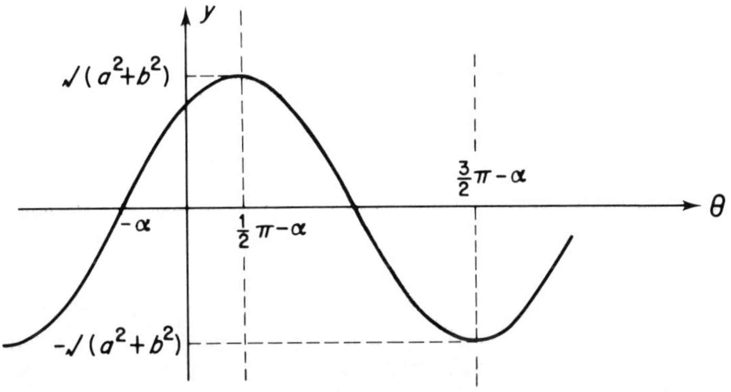

Figure 6.19

The maximum and minimum values of $\sin(\theta + \alpha)$ are 1 and -1 respectively. Hence, the maximum and minimum values of y are $\sqrt{(a^2 + b^2)}$ and $-\sqrt{(a^2 + b^2)}$ respectively. When $\theta = -\alpha$, y is zero and one value of θ which produces the maximum value of y is $\frac{1}{2}\pi - \alpha$. One value of θ which produces the minimum value of y is $\frac{3}{2}\pi - \alpha$.

Exercises 6h

1 Express $\cos\theta + \sin\theta$ in the form (i) $A\sin(\theta + \alpha)$, (ii) $B\cos(\theta - \beta)$.
2 Express $3\cos\theta - 4\sin\theta$ in the form $R\cos(\theta + \alpha)$.
3 Express $2\sin\theta - 3\cos\theta$ in the form $R\sin(\theta - \alpha)$.
4 Express $3\cos 2\theta + 4\sin 2\theta$ in the form $R\sin(2\theta + \alpha)$. Hence state the maximum value of $3\cos 2\theta + 4\sin 2\theta$.
5 By expressing $\sin\theta + 3\cos\theta$ in the form $R\sin(\theta + \text{a})$ calculate the maximum value of this expression. Find an acute angle θ for which this maximum is attained.

6.9 The inverse trigonometric functions

If $\sin y = x$, we say that y is the number of radians in the angle whose sine is x. This we write as

$$y = \arcsin x$$

y is called the inverse sine of x. The statements

$$\sin y = x \quad \text{and} \quad y = \arcsin x \tag{6.53}$$

are equivalent. *Figure 6.20* shows the graph of the function $y = \arcsin x$. It is easily derived from the graph $x = \sin y$. From the graph we see that to any value of x there correspond many possible values for y. This ambiguity can be avoided (see Section 9.1) by confining our attention to the value which

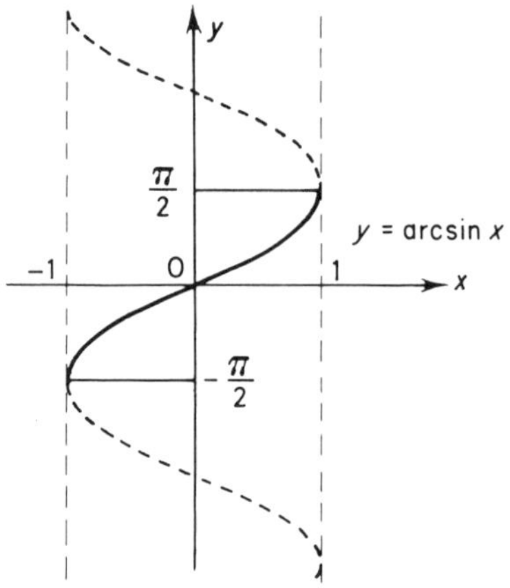

Figure 6.20

lies in the range $-\pi/2$ to $\pi/2$. This value is called the principal value of the inverse sine.

In the same way, if $x = \cos y$ then $y = \arccos x$ is the inverse cosine of x: y is the number of radians in the angle whose cosine is x. The principal value of the inverse cosine is the value of y in the range 0 to π. The graph of arccose x is shown in *Figure 6.21*.

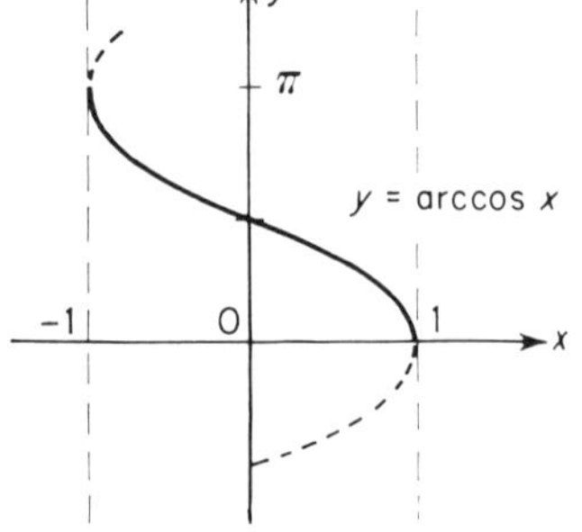

Figure 6.21

In the same way, $y = \arctan x$ is called the inverse tangent of x and means that $x = \tan y$. The principal value of y is the value in the range $-\pi/2$ to $\pi/2$. The graph of $\arctan x$ is shown in *Figure 6.22*.

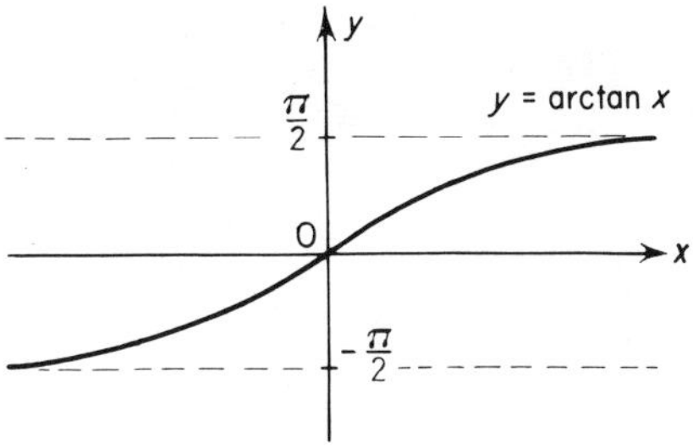

Figure 6.22

The notation $\sin^{-1} x$, $\cos^{-1} x$ and $\tan^{-1} x$ is also used for the inverse functions, but we discourage its use because of possible confusion with the reciprocals of the trigonometric functions, namely, $(\sin x)^{-1}$, $(\cos x)^{-1}$ and $(\tan x)^{-1}$. To be certain of avoiding confusion the reciprocal functions should always be written

$$\frac{1}{\sin x} \qquad \frac{1}{\cos x} \qquad \frac{1}{\tan x}$$

It follows as a direct consequence of the definitions above that

$$\sin(\arcsin x) = \cos(\arccos x) = \tan(\arctan x) = x \tag{6.54}$$

and

$$\arcsin(\sin y) = \arccos(\cos y) = \arctan(\tan y) = y \tag{6.55}$$

provided y lies in the appropriate principal value range.

Although they are not used so much as the inverse functions already defined, it is possible to define the inverse functions arcsec x, arccosec x, arccot x. The principal value arcsec x is taken to lie in the range $-\pi/2$ to $\pi/2$, while the principal values arcsec x and arccot x are taken to lie in the range 0 to π.

Example 1 Show that $\arctan x = \arcsin\left(\dfrac{x}{\sqrt{(1+x^2)}}\right)$

With $\arctan x = \alpha$, $x = \tan\alpha$. Now by (6.55) $\alpha = \arcsin(\sin\alpha)$, so we express $\sin\alpha$ in terms of x. Since

$$\operatorname{cosec}^2\alpha = 1 + \cot^2\alpha = 1 + \frac{1}{x^2} = \frac{1+x^2}{x^2}$$

therefore

$$\sin^2\alpha = \frac{x^2}{1+x^2}$$

so that

$$\sin\alpha = \frac{x}{\sqrt{(1+x^2)}}$$

The sign of the square root is positive in accordance with our conventions for the principal value. If x is positive, α will lie between 0 and $\pi/2$ so that

$\sin\alpha$ is positive. If x is negative, α will lie between $-\pi/2$ and 0 and so $\sin\alpha$ will also be negative. Therefore

$$\alpha = \arctan x = \arcsin\left(\frac{x}{\sqrt{(1+x^2)}}\right)$$

Example 2 Show that $\arcsin(-x) = -\arcsin x$.

We consider the cases x positive, x negative and x zero separately. If x is positive and $\alpha = \arcsin x$, then α lies in the range 0 to $\pi/2$ and $\sin\alpha = x$. Therefore

$$-x = -\sin\alpha = \sin(-\alpha)$$
$$\arcsin(-x) = -\alpha = -\arcsin x$$

If x is negative, put $x = -y$ so that y is positive. Then $\arcsin(-y) = -\arcsin y$ by the above. Rearranging this we have

$$\arcsin y = -\arcsin(-y)$$

that is

$$\arcsin(-x) = -\arcsin x$$

If x is zero, $\arcsin 0 = 0 = \arcsin(-0)$. Thus, we have proved the result for all values of x.

Example 3 Show that $\arctan x + \arctan y = \arctan\left(\dfrac{x+y}{1-xy}\right)$.

Let $\arctan x = \alpha$, $\arctan y = \beta$, so that $\tan\alpha = x$, $\tan\beta = y$. Then

$$\arctan x + \arctan y = \alpha + \beta = \arctan[\tan(\alpha+\beta)] \quad \text{by (6.55)}$$
$$= \arctan\left(\frac{\tan\alpha + \tan\beta}{1 - \tan\alpha\tan\beta}\right) \quad \text{by (6.28)}$$

Therefore

$$\arctan x + \arctan y = \arctan\left(\frac{x+y}{1-xy}\right) \tag{6.56}$$

Exercises 6i

1 Write down the value of (i) $\arccos(1/\sqrt{2})$, (ii) $\arctan 1$, (iii) $\arctan\sqrt{3}$.
2 Evaluate (i) $\arctan(-1)$, (ii) $\arcsin(-\frac{1}{2})$, (iii) $\arccos(\sqrt{3}/2)$.
3 Show that $2\arctan x = \arctan\dfrac{2x}{1-x^2}$.
4 Show that $\arcsin\frac{1}{2} = \arccos(\sqrt{3}/2)$.
5 Show that $\arccos(-x) = \pi - \arccos(x)$. (Consider the three cases x positive, x negative, x zero and proceed as in Example 2.)
6 Show that if $\arccos x$, $\arccos y$ and $\arccos x + \arccos y$ are all in the range 0 to $\frac{1}{2}\pi$ then

$$\arccos x + \arccos y = \arccos\{xy - \sqrt{[(1-x^2)(1-y^2)]}\}$$

7 Find x if $\arctan x + \arctan(1-x) = \arctan\frac{4}{3}$.
8 Show that $2\arcsin x = \arcsin[2x\sqrt{(1-x^2)}]$ if $\arcsin x < \frac{1}{4}\pi$.
9 Show that $2\arcsin\frac{5}{13} = \arctan\frac{120}{119}$.
10 Show that $\arctan\frac{1}{4} + \arctan\frac{2}{9} = \arctan\frac{1}{2}$.

6.10 Small angles

From the definition of the radian, we see that if s is the length of an arc of a circle radius r which subtends an angle θ at the centre then

$$s = r\theta \quad \text{(measured in radians)} \tag{6.57}$$

(Compare with equation (6.1).)

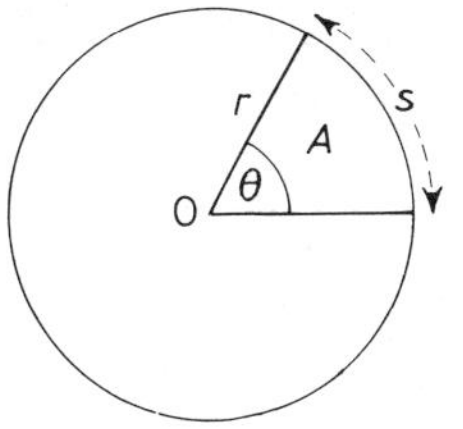

Figure 6.23

Also, if A is the area of the sector defined by the arc s and the bounding radii (see *Figure 6.23*) then

$$\frac{A}{\text{area of circle}} = \frac{\theta}{2\pi}$$

$$A = \pi r^2 \frac{\theta}{2\pi} = \tfrac{1}{2} r^2 \theta \tag{6.58}$$

We shall use these results to obtain useful approximations to $\sin\theta$, $\cos\theta$ and $\tan\theta$ when θ is small.

In *Figure 6.24* PR is a chord of a circle radius r subtending an acute angle θ at the centre O of this circle. PT is the tangent at P.

Figure 6.24

It is clear that area $\triangle$POR < area sector POR < area $\triangle$POT. Thus since OP = OR = r and PT = $r \tan\theta$

$$\tfrac{1}{2} r^2 \sin\theta < \tfrac{1}{2} r^2 \theta < \tfrac{1}{2} r^2 \tan\theta$$

$$\sin\theta < \theta < \tan\theta \tag{6.59}$$

Thus $1 < \dfrac{\theta}{\sin\theta} < \dfrac{1}{\cos\theta}$ or $1 > \dfrac{\sin\theta}{\theta} > \cos\theta$, which is equivalent.

Now let $\theta \to 0$, then $\cos\theta \to 1$, so that we have

$$\frac{\sin\theta}{\theta} \to 1 \quad \text{as} \quad \theta \to 0$$

Thus $\dfrac{\sin\theta}{\theta} \approx 1$ for small values of θ, that is

$$\sin\theta \approx \theta \quad \text{if } \theta \text{ is small} \tag{6.60}$$

In practice, this approximation is valid to about four decimal places for values of θ less than 6°, although to use the approximation we emphasise that the units for θ must be *radians*.

Since $\cos\theta = 1 - 2\sin^2\frac{1}{2}\theta$ and since $\sin\frac{1}{2}\theta \approx \frac{1}{2}\theta$ for small values of θ, we have

$$\cos\theta \approx 1 - \tfrac{1}{2}\theta^2 \quad \text{if } \theta \text{ is small} \tag{6.61}$$

This approximation is more accurate than the coarser approximation

$$\cos\theta \approx 1 \quad \text{if } \theta \text{ is small} \tag{6.62}$$

From (6.60) and (6.62) we obtain by division

$$\tan\theta \approx \theta \quad \text{if } \theta \text{ is small} \tag{6.63}$$

Example 1 Find an approximate value of θ if $\sin\theta = 0{\cdot}48$.

Since $\sin\theta$ is nearly 0·5, θ must be approximately $\frac{1}{6}\pi$ radians. We can improve on this first approximation by letting $\theta = \frac{1}{6}\pi - \alpha$, where α is small. Then

$$0{\cdot}48 = \sin(\tfrac{1}{6}\pi - \alpha) = \sin\tfrac{1}{6}\pi\cos\alpha - \cos\tfrac{1}{6}\pi\sin\alpha$$

Thus since

$$\cos\tfrac{1}{6}\pi = \sqrt{3}/2 \qquad \sin\tfrac{1}{6}\pi = \tfrac{1}{2} \qquad \cos\alpha \approx 1 \qquad \sin\alpha \approx \alpha$$

we have

$$0{\cdot}48 \approx \frac{1}{2} - \frac{\sqrt{3}}{2}\alpha$$

therefore

$$\alpha \approx \left(\frac{2}{\sqrt{3}}\right) \times 0{\cdot}02 = 0{\cdot}0231 \text{ radian}$$

$$\alpha \approx 1°19' \quad \text{so that} \quad \theta \approx 28°41'$$

Example 2 The diameter of a circular flat disc is 1 m. At what distance will it subtend an angle of 15 minutes?

Let AB be a diameter of the circular flat disc; C its centre (*Figure 6.25*). Then

$$AB = 1\text{ m} = 2OC\tan AOC$$

But

$$\tan AOC \approx \angle AOC = \frac{1}{8}\frac{\pi}{180} \text{ radians}$$

therefore

$$OC \approx \frac{720}{\pi}\text{ m} = 229{\cdot}2\text{ m}$$

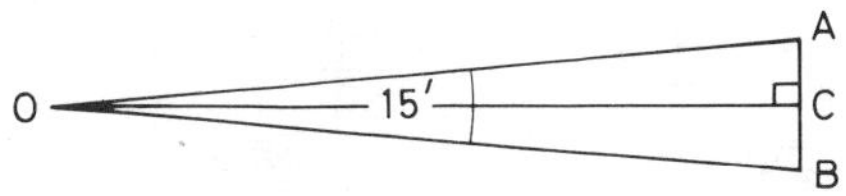

Figure 6.25

Exercises 6j

1 Use tables to evaluate $\sin\theta$ for $\theta = 1°, 2°, 3°, 4°, 5°, 10°$. Convert θ to radians and compare the values.
2 Calculate $\tan\theta$ for the values of θ in question 1. Compare the values of $\sin\theta$, θ and $\tan\theta$ (see (6.59)).
3 Find an approximate value for θ if $\sin\theta = 0{\cdot}51$.
4 A hill 15 km away has an angle of elevation of 30′. Find its approximate height in metres.
5 Find the angle subtended by a building 50 m high at a distance of 8·8 km.

Exercises 6

In the following questions (and also in subsequent chapter-end exercises), both types of notation occur for the inverse trigonometric functions, namely, $\arcsin x$ etc. and $\sin^{-1} x$ etc. This has been done to familiarise readers with the form of notation preferred by some examining boards.

1 Show that $\tan^2\theta - \sin^2\theta = \sin^2\theta\tan^2\theta$.
2 Evaluate (i) $\cos 386°$, (ii) $\sin -429°$, (iii) $\tan -819°$, (iv) $\sin 881°$.
3 If $90° < x < 180°$ and $\sin x = 0{\cdot}8$ evaluate (i) $\cos x$, (ii) $\tan x$, (iii) $\sin 2x$, (iv) $\cos 2x$.
4 If $\sin\theta = s$ and θ is acute express all the other trigonometric ratios of θ in terms of s.
5 Show that $\sin^4\theta + \cos^4\theta = 1 - 2\sin^2\theta\cos^2\theta$.
6 Show that $(\sec^2\theta + \tan\theta)(\operatorname{cosec}^2\theta - \cot\theta) = 1 + \tan^2\theta + \cot^2\theta$.
7 Prove without tables that $\arcsin\frac{3}{5} - \arccos\frac{63}{65} = 2\arctan\frac{1}{5}$.
8 Show that $\sin A + 2\sin 5A + \sin 9A = 4\cos^2 2A\sin 5A$.
9 If A, B, C are the angles of a triangle, show that
(i) $\sin A + \sin B + \sin C = 4\cos\frac{1}{2}A\cos\frac{1}{2}B\cos\frac{1}{2}C$
(ii) $\sin 2A + \sin 2B + \sin 2C = 4\sin A\sin B\sin C$.
10 If $\tan\theta = 1/p$ and $\tan\phi = 1/q$ and $pq = 2$, show that $\tan(\theta + \phi) = p + q$.
11 Show that

(a) $\sin 2A + \cos 2A = \dfrac{(1 + \tan A)^2 - 2\tan^2 A}{1 + \tan^2 A}$

(b) $\dfrac{\cos(2P - 2Q) + \cos 3Q}{\sin(2P - 3Q) + \sin 3Q} = \cot P.$ [LU, part]

12 Show that $\dfrac{1 + \sin x + \cos x}{1 + \sin x - \cos x} = \cot\frac{1}{2}x.$
13 Show that $\sin\theta + \sin(\theta + x) + \sin(\theta + 2x) + \sin(\theta + 3x)$
$= 4\cos(x/2)\cos x\sin(\theta + 3x/2).$
14 Given that $\sin 2\alpha + \sin 2\beta = p$, $\cos 2\alpha + \cos 2\beta = q$, prove that $p/q = \tan(\alpha + \beta)$. Prove also that $\dfrac{4p}{p^2 + q^2 + 2q} = \dfrac{\sin(\alpha + \beta)}{\cos\alpha\cos\beta}$ and deduce an expression for $\tan\alpha\tan\beta$ in terms of p and q. [JMB, part]

15 Express the function $4\cos x - 6\sin x$ in the form $R\cos(x+\alpha)$, where R is positive and $0 < \alpha < 360°$. Hence find the co-ordinates of the point on the graph of $y = 4\cos x - 6\sin x$ for $0° \leqslant x \leqslant 360°$, where y has its minimum value. Sketch the graph. State the values of x where the curve crosses the x-axis. [AEB]

16 Given that $\sin(150° - \theta) = p\sin\theta$, show that $\cot\theta = 2p - \sqrt{3}$. By choosing a suitable value of θ find $\cot 75°$ and using $\theta = 135°$, find $\sin 15°$, giving both your answers in surd form. [JMB]

17 If $\sin\theta = \dfrac{1-x}{1+x}$ show that $\tan(\frac{1}{4}\pi - \frac{1}{2}\theta) = \sqrt{x}$.

18 Show that $\dfrac{1+\sin x - \cos x}{1+\sin x + \cos x} + \dfrac{1+\sin x + \cos x}{1+\sin x - \cos x} = \dfrac{2}{\sin x}$.

19 If θ is not a multiple of $\pi/2$, and if x, y, z are given as sums of the following infinite series

$$x = 1 + \cos^2\theta + \cos^4\theta + \ldots$$
$$y = 1 + \sin^2\theta + \sin^4\theta + \ldots$$
$$z = 1 + \cos^2\theta\sin^2\theta + \cos^4\theta\sin^4\theta + \ldots$$

prove that (i) $x + y = xy$, (ii) $x + y + z = xyz$. [JMB, part]

20 Given that $\tan 3\theta = 2$ evaluate without using tables

$$\frac{\sin\theta + \sin 3\theta + \sin 5\theta}{\cos\theta + \cos 3\theta + \cos 5\theta}$$

[JMB, part]

21 If $\sin(\alpha+\beta) = \lambda\sin(\alpha-\beta)$ show that $\tan\alpha = \dfrac{\lambda+1}{\lambda-1}\tan\beta$.

22 If $\sin\alpha + \sin\beta = p$ and $\cos\alpha + \cos\beta = q$ show that

(i) $\sin(\alpha+\beta) = \dfrac{2pq}{p^2+q^2}$ (ii) $\cos(\alpha+\beta) = \pm\left(\dfrac{q^2-p^2}{p^2+q^2}\right)$

23 If α, β, γ are all greater than $\pi/2$ and less than 2π and $\sin\alpha = \frac{1}{2}$, $\tan\beta = \sqrt{3}$, $\cos\gamma = 1/\sqrt{2}$, find the value of $\tan(\alpha+\beta+\gamma)$ in surd form. [WJC]

24 (i) Express $\tan 3A$ and $\operatorname{cosec} 4A + \cot 4A$ in terms of $\tan A$.
(ii) If $\cos\theta + \cos 3\theta = k\cos\phi$ and $\sin\theta + \sin 3\theta = k\sin\phi$, show that $\cos\theta = \pm\frac{1}{2}k$ and find the values of $\tan\phi$ and $\cos 2\phi$ in terms of k. [LU]

25 Show that

(i) $\dfrac{\sin x\sin y}{\cos x + \cos y} = \dfrac{2\tan\frac{1}{2}x\tan\frac{1}{2}y}{1 - \tan^2\frac{1}{2}x\tan^2\frac{1}{2}y}$

(ii) $\dfrac{\cos x\cos y}{\cos x + \cos y} = \dfrac{(1-\tan^2\frac{1}{2}x)(1-\tan^2\frac{1}{2}y)}{2(1-\tan^2\frac{1}{2}x\tan^2\frac{1}{2}y)}$

26 If θ is an acute angle such that $\cos\theta = 1 - x$, where x is so small that x^2 is negligible compared with unity, prove that $\cos 2\theta = 1 - 4x$ and $\cos 3\theta = 1 - 9x$ approximately. [LU, part]

27 Show that $\tan(B-C) + \tan(C-A) + \tan(A-B) = \tan(B-C)\tan(C-A)\tan(A-B)$.

28 Evaluate $\arcsin(1/\sqrt{5}) + \arcsin(1/\sqrt{10})$.

29 (i) Prove that $a \sin n\theta - 2(a-1)\sin(n-1)\theta\cos\theta + (a-2)\sin(n-2)\theta = 2\sin\theta\cos(n-1)\theta$.

(ii) Prove that, if $0 < x < 1/\sqrt{2}$, $2\sin^{-1}x = \sin^{-1}2x\sqrt{(1-x^2)}$. (Note $\sin^{-1}x$ means 'the principal value of the inverse sine of x', and '$\sqrt{(1-x^2)}$' means the positive square root.) State the corresponding formula if $1/\sqrt{2} < x < 1$.

Express $2\cos^{-1}x$ as an inverse cosine, considering all values of x between 0 and 1. [SUJB]

30 If $\tan 2\phi - \sin 2\phi = x$ and $\tan 2\phi + \sin 2\phi = y$, show that (i) $x/y = \tan^2\phi$, (ii) $(x^2 - y^2)^2 = 16xy$. [LU]

31 Prove that $4\arctan\frac{1}{5} - \arctan\frac{1}{239} = \pi/4$.

32 Find x if $\arctan 2x + \arctan 3x = \pi/4$.

33 Express the functions $6\cos^2\theta + 8\sin\theta\cos\theta$ in terms of $\cos 2\theta$ and $\sin 2\theta$. Deduce an expression for the function in the form $A + 5\cos(2\theta - \alpha)$, where A and α are constants. Hence write down the greatest and least values of the function and find, correct to the nearest minute, one value of θ corresponding to each. [JMB]

34 If $\sin\theta + \sin\phi = p$ and $\cos\theta + \cos\phi = q$ prove that

(i) $$\tan\theta + \tan\phi = \frac{8pq}{(p^2+q^2) - 4p^2}$$

(ii) $$\cos 2\theta + \cos 2\phi = \frac{(q^2-p^2)(p^2+q^2-2)}{p^2+q^2}$$

35 If $\sin x = \alpha\sin\theta$, where $\alpha \approx 1$, show that $x \approx \theta + (\alpha - 1)\tan\theta$.

36 Show that $\sin 3A = 3\sin A - 4\sin^3 A$. Deduce that $\sin^3 A + \sin^3(120° + A) + \sin^3(240° + A) = -\frac{3}{4}\sin 3A$. [JMB, part]

37 If $\sin\theta + \sin\omega = a$, $\cos\theta + \cos\omega = b$, and $\cos\theta\cos\omega = c$, show that $(a^2+b^2)(a^2+b^2-4c) = 4a^2$.

38 Express the function $f(\theta) = \sin\theta(2\sin\theta + \cos\theta)$ in terms of $\sin 2\theta$ and $\cos 2\theta$. Hence show that $f(\theta)$ can be written in the form $1 - R\cos(2\theta + \alpha)$, where R is positive and α is acute. State the values of R and $\tan\alpha$. Deduce the maximum value of $f(\theta)$ and the smallest positive value of θ for which this maximum value occurs. [JMB]

39 If $\tan(x+y) = a$ and $\tan(x-y) = b$, express $\dfrac{\sin 2x + \sin 2y}{\sin 2x - \sin 2y}$ in terms of a and b. Show that $\tan 2y = \dfrac{a-b}{1+ab}$ and by using this result, or otherwise, obtain an expression for $\tan(x+3y)$ in terms of a and b. [AEB]

40 Prove that $\sin(\alpha+\beta)\sin(\alpha-\beta) = \sin^2\alpha - \sin^2\beta$. By using this result or otherwise, prove that

$$\sin(\alpha+\beta+\gamma)\sin(\beta+\gamma-\alpha)\sin(\gamma+\alpha-\beta)\sin(\alpha+\beta-\gamma) = (a+b+c)(b+c-a)(c+a-b)(a+b-c) - 4a^2b^2c^2$$

where $a = \sin\alpha$, $b = \sin\beta$, $c = \sin\gamma$. [JMB]

7
Trigonometric equations

7.1 The general expression for angles with a given trigonometric ratio

In this chapter we shall consider equations in which the trigonometric ratios of the unknown quantity occur. We shall show that the solution of such equations can be reduced to the solution of one or more equations of the type $\sin x = \alpha$, $\cos x = \alpha$, or $\tan x = \alpha$, where α is known and x is to be found. We first consider these particular equations.

As an example, consider the equation $\tan x = 1$. One solution is $x = 45°$, but this is not the only solution. From (6.16)

$$\tan(\theta + 180°) = \tan\theta$$

so that

$$\tan(45° + 180°) = \tan 45° = 1$$

Thus, $x = 225°$ is also a solution. Again, $\tan(225° + 180°) = \tan 225° = \tan 45° = 1$, so that $x = 405°$ is also a solution. It is clear that we can proceed indefinitely in this way and obtain as solutions $x = 45°$, $x = 45° + 180°$, $x = 45° + 2 + 180°$, $x = 45° + 3 \times 180°$, etc. From (6.15 and 6.17)

$$\tan(\theta - 180°) = -\tan(180° - \theta) = \tan\theta$$

so that

$$\tan(45° - 180°) = 1 \qquad \tan(45° - 2 \times 180°) = 1$$

Therefore, $x = 45° - 180°$, $x = 45° - 2 \times 180°$, $x = 45° - 3 \times 180°$, etc., are also solutions. All the solutions above may be expressed in the one form

$$x = 45° + n180°$$

where n is an integer, either positive, negative or zero. By giving n different values, we obtain the different solutions of the equation.

For the equation $\tan x = \alpha$ suppose that θ is any angle that $\tan\theta = \alpha$. (In practice, θ is found with the aid of tables of tangents.) Then any angle $x = \theta + n180°$, where n is an integer, will satisfy $\tan x = \tan\theta = \alpha$.

Thus all solutions of the equation $\tan x = \alpha$ are of the form

$$x = \theta + n180° \qquad (7.1)$$

where n is an integer, positive, negative or zero.

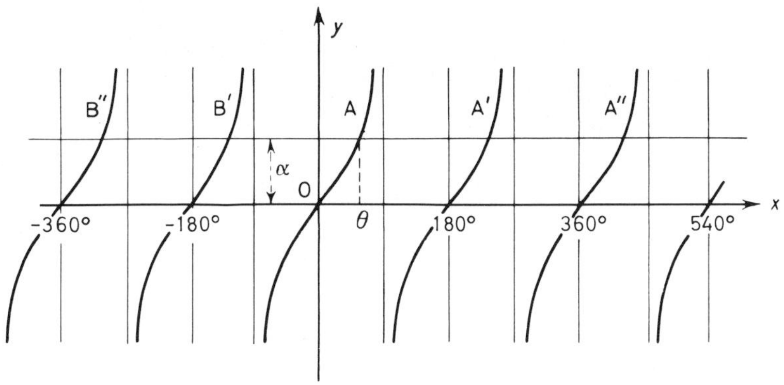

Figure 7.1

Reference to the graph $y = \tan x$ in *Figure 7.1* may help to clarify this. The horizontal line $y = \alpha$ is drawn on the same scale and intersects the tangent curve at points A, A′, B′, A″, B″, etc.

If the abscissa of A is θ, that of A′ is $\theta + 180°$, of A″, $\theta + 2 \times 180°$, etc. and that of B′, $\theta - 180°$, of B″, $\theta - 2 \times 180°$, etc. If the angles are expressed in radians, (7.1) assumes the form

$$x = \theta + n\pi \tag{7.2}$$

where n is an integer, positive, negative or zero.

For the equation $\cos x = \alpha$, we first observe that if α is numerically greater than one, no solution will exist. If α is numerically less than one, we proceed as follows. *Figure 7.2* shows the graph of $y = \cos x$ and $y = \alpha$ drawn on the same scale. The abscissae of the points of intersection will give the solutions of the equation $\cos x = \alpha$.

If θ is an angle (the smallest) for which $\cos \theta = \alpha$, the abscissa of A is θ. The abscissae of A′, A″, A‴, etc. are $360° - \theta$, $360° + \theta$, $2 \times 360° - \theta$, $2 \times 360° + \theta$, etc. The abscissae of B′, B″, B‴, etc. are $-\theta$, $-360° + \theta$, $-360° - \theta$, etc. These are all particular cases of the formula $n360° \pm \theta$, where n is an integer. Thus, the general solution of the equation $\cos x = \alpha$ $(= \cos \theta)$ is

$$x = n360° \pm \theta \tag{7.3}$$

or $x = 2n\pi \pm \theta$ (if angles are measured in radians), where n is an integer, positive, negative or zero.

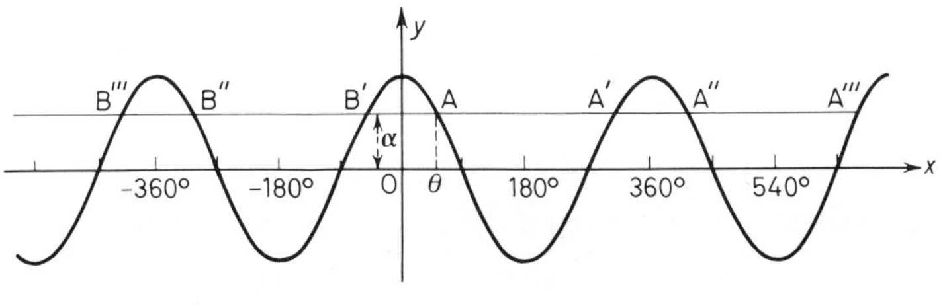

Figure 7.2

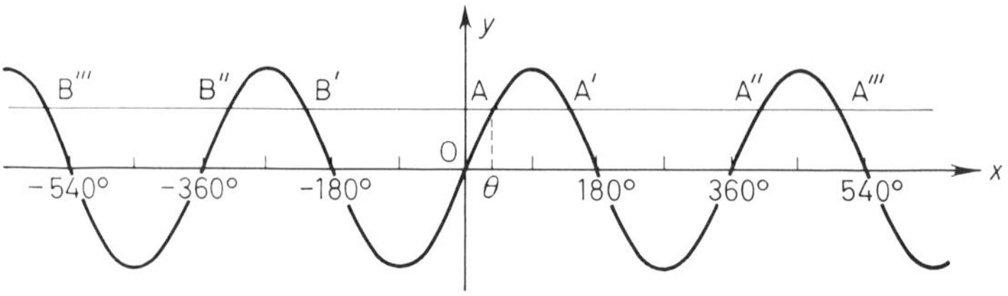

Figure 7.3

We treat the equation $\sin x = \alpha$ in the same way. *Figure 7.3* shows the graph $y = \sin x$ and $y = \alpha(-1 < \alpha < 1)$. From the graph we see that if $\sin\theta = \alpha$ (θ is the abscissa of A), the other solutions are given by the abscissae of A′, B′, A″, B″, etc., i.e. $180° - \theta$, $-180° - \theta$, $360° + \theta$, $-360° + \theta$, $540° - \theta$, $-540° - \theta$, etc. These are particular cases of the formula $n180° + (-1)^n\theta$, where n is an integer.

Thus, if $\sin\theta = \alpha$, the general solution of the equation $\sin x = \alpha$ is

$$x = n180° + (-1)^n\theta \tag{7.4}$$

or $x = n\pi + (-1)^n\theta$ (if angles are measured in radians), where n is an integer, positive, negative or zero.

Example 1 Find the solutions of the equation $\sin x = 0{\cdot}515$ which lie in the range 0° to 360°. State the general solution.

From tables, $\sin 31° = 0.515$, hence $x = 31°$ is a solution. Thus by (7.4), the general solution is $x = n180° + (-1)^n 31°$. With $n = 0$ or 1, we obtain the solutions in the range 0° to 360°, that is $x = 31°$ or $x = 180° - 31° = 149°$.

Example 2 Solve the equation $\tan x = -\sqrt{3}$, giving the general solution and the solutions which lie in the range 0° to 360°.

From tables, or by (6.23), $\tan 60° = \sqrt{3}$. Thus by (6.17)

$$\tan(180 - \theta) = -\tan\theta \quad \text{so that} \quad \tan 120° = -\sqrt{3}$$

Therefore, one solution is $\theta = 120°$. By (7.1) the general solution is

$$x = 120° + n180°$$

With $n = 0$ or 1, we obtain solutions in the range 0° to 360°: $x = 120°$ or $x = 300°$.

Example 3 Find in radians the general solution of the equation

$$\cos 2\theta = \cos(\theta - \pi/4)$$

From (7.3) the general solution is such that

$$2\theta = 2n\pi \pm (\theta - \pi/4)$$

where n is an integer. Thus, with the positive sign, we have

$$2\theta = 2n\pi + \theta - \pi/4$$
$$\theta = 2n\pi - \pi/4$$

With the negative sign, we have

$$2\theta = 2n\pi - \theta + \pi/4$$

$$\theta = \frac{2n\pi}{3} + \frac{\pi}{12}$$

Therefore

$$\theta = 2n\pi - \frac{\pi}{4} \quad \text{or} \quad \frac{2n\pi}{3} + \frac{\pi}{12}$$

where n is an integer.

Example 4 Solve the equation $\sin 2\theta = \cos 3\theta$, giving the general solution and solutions in the range 0° to 360°.

Since $\cos 3\theta = \sin(90° - 3\theta)$ the equation is

$$\sin 2\theta = \sin(90° - 3\theta)$$

so that by (7.4)

$$2\theta = n180° + (-1)^n(90° - 3\theta) \quad \text{(where } n \text{ is an integer)}$$

Therefore

$$\theta[2 + 3(-1)^n] = n180° + (-1)^n 90°$$

$$\theta = \frac{n180° + (-1)^n 90°}{2 + 3(-1)^n} \quad \text{(where } n \text{ is an integer)}$$

Solutions in the range 0° to 360° are obtained by giving n particular values:

With $n = 0$ $\quad \theta = \frac{90}{5} = 18°$

With $n = 2$ $\quad \theta = \frac{450}{5} = 90°$

With $n = 4$ $\quad \theta = \frac{810}{5} = 162°$

With $n = 6$ $\quad \theta = \frac{1170}{5} = 234°$

With $n = 8$ $\quad \theta = \frac{1530}{5} = 306°$

Other positive values of n (including all the odd values) lead to solutions outside the range 0° to 360°.

With $n = -1$

$$\theta = \frac{-180° - 90°}{-1} = 270°$$

but other negative values of n lead to solutions outside the range 0° to 360°. The required solutions are thus 18°, 90°, 162°, 234°, 270°, 306°.

Exercises 7a

Find the general solutions of the following equations. (Find *all* solutions in the range 0° to 360°.*)

1 $\sin x = 0{\cdot}831$

2 $\cos x = 0{\cdot}7125$

3 $\tan x = 2{\cdot}1155$

4 $\cos x = -0{\cdot}5577$

5 $\sin x = -0{\cdot}4775$

6 $\tan x = -0{\cdot}300$

7 $\sin 3x = 0.500$

8 $\tan 6x = -1.23$

9 $\cos \frac{1}{5}x = \sqrt{3}/2$

10 $\cos(2x - 30°) = 0{\cdot}564$

* It is suggested that, at first reading, only solutions in the range 0° to 360° be considered.

11 $\sin(x + 18°3') = 0{\cdot}813$

12 $\sin 4x = \sin 2x$

13 $\tan 6\theta = \tan\theta$

14 $\cos 3x = -\cos x$

15 $\cos 2x = \sin 3x$

16 $\tan 2x = \cot 3x$

17 $\sin 2\theta \cos 3\theta = 0$

18 $\tan 2\theta \cot 4\theta = 0$

19 (i) $\cos\theta = -\frac{1}{2}$, (ii) $\tan\theta = \sqrt{3}$

20 From the result of question 19 find the general value of θ and the values in the range 0° to 360° which satisfy simultaneously $\tan\theta = \sqrt{3}$, $\cos\theta = -\frac{1}{2}$.

7.2 Trigonometric equations involving different ratios of the same angle

Trigonometric equations which involve more than one ratio of the same angle are generally solved by obtaining an equation which involves just one trigonometric ratio. This generally calls for the use of some of the identities (6.5) to (6.12). The aforementioned equation is then solved for the particular ratio involved. This will result in one or more equations of the type considered in the previous sections. The following examples will illustrate some of the techniques commonly used.

Example 1 Find the general solution and all solutions in the range 0° to 360° of the equation $2\cos x - 3\sin x = 0$.

We have $2\cos x - 3\sin x = 0$, therefore

$$\frac{\sin x}{\cos x} = \frac{2}{3}$$

provided $\cos x \neq 0$, which is so for our solutions but we must check this point. Therefore

$$\tan x = \tfrac{2}{3} = \tan 33°41' \quad \text{(from tables)}$$
$$x = 33°41' + n180°$$

Therefore, the solutions in the range 0° to 360° are, with $n = 0$ and 1, 33°41′ and 213°41′.

Example 2 Find x in the range 0° to 360° if $2\cos^2 x = 2 - \sin x$.

Since $\cos^2 x = 1 - \sin^2 x$, we have

$$2 - \sin x = 2 - 2\sin^2 x \quad \text{(which involves only } \sin x\text{)}$$

therefore

$$2\sin^2 x - \sin x = 0$$
$$\sin x(2\sin x - 1) = 0^*$$

which means that

$$\sin x = 0 \quad \text{or} \quad 2\sin x - 1 = 0 \qquad \text{i.e.} \quad \sin x = \tfrac{1}{2}$$
$$x = 0,\ 180°,\ 360°, \ldots;\ n180°$$

or

$$x = 30°,\ 150°, \ldots;\ n180° + (-1)^n 30°$$

The required solutions are thus 0°, 30°, 150°, 180°, 360°.

* Note that we do not divide by $\sin x$ since we should then lose the solutions which result from $\sin x = 0$.

Example 3 Find θ in the range 0° to 360° if $2\sin 2\theta + 3\operatorname{cosec} 2\theta = 7$.

We have $2\sin 2\theta + \dfrac{3}{\sin 2\theta} = 7$, which involves only $\sin 2\theta$. Therefore

$$2\sin^2 2\theta - 7\sin 2\theta + 3 = 0$$
$$(2\sin 2\theta - 1)(\sin 2\theta - 3) = 0$$
$$2\sin 2\theta - 1 = 0 \quad \text{or} \quad \sin 2\theta - 3 = 0$$

that is

$$\sin 2\theta = \tfrac{1}{2} \quad \text{or} \quad \sin 2\theta = 3$$

The second alternative gives no real values for θ. If $\sin 2\theta = \frac{1}{2}$, $2\theta = 30°$, $180° - 30°$, $360° + 30°$, $540° - 30°$, $720° + 30°$. Since we require values of θ in the range 0° to 360°, we must find all values for 2θ in the range 0° to 720°. From the above, we see that the required values for θ are 15°, 75°, 195°, 255°.

Example 4 Solve for θ, $3\sec^2\theta = 2\tan\theta + 4$.
We have $3\sec^2\theta - 2\tan\theta - 4 = 0$, therefore

$$3(1 + \tan^2\theta) - 2\tan\theta - 4 = 0 \quad \text{(which involves only } \tan\theta)$$
$$3\tan^2\theta - 2\tan\theta - 1 = 0$$
$$(\tan\theta - 1)(3\tan\theta + 1) = 0$$
$$\tan\theta = 1 \quad \text{or} \quad \tan\theta = -\tfrac{1}{3}$$

because $\tan\theta = 1 = \tan 45°$, then

$$\theta = 45° + n180°$$

From tables, $\tan\theta = -\frac{1}{3} = \tan(-18°\,26')$, therefore

$$\theta = n180° - 18°\,26'$$

Example 5 Solve for x in the range 0° to 360°

$$3\cos^2 x - 3\sin x\cos x + 2\sin^2 x = 1$$

After division by $\cos^2 x$, we have

$$3 - 3\tan x + 2\tan^2 x = \sec^2 x$$

(We must check that our solutions do *not* give $\cos x = 0$.)

$$3 - 3\tan x + 2\tan^2 x = 1 + \tan^2 x \quad \text{(which involves only } \tan x)$$
$$\tan^2 x - 3\tan x + 2 = 0$$
$$(\tan x - 1)(\tan x - 2) = 0$$
$$\tan x = 1 \quad \text{or} \quad \tan x = 2$$

that is

$$x = 45° + n180° \quad \text{or} \quad x = 63°\,26' + n180°$$

The required solutions are thus 45°, 225°, 63° 26′, 243° 26′.

Exercises 7b

Find the general solution, together with *all* the solutions in the range 0° to 360°, of the equations*

1 $2\tan^2 x - 3\tan x + 1 = 0$
2 $7\tan^2 x - 3\sec^2 x = 9$
3 $8\sin^2\theta - 6\cos\theta = 3$
4 $9\cos\theta = 4\sin\theta$
5 $4\cos x = 3\tan x + 3\sec x$
6 $3\sin^2\theta = \cos^2\theta$
7 $4\cos^2\theta + 5\sin^2\theta = 5$
8 $\tan x \operatorname{cosec} x = 5$
9 $\cot^2 x + 2\operatorname{cosec}^2 x = 6$
10 $9\sin^2 x + 10\sin x\cos x - 2\cos^2 x = 1$
11 $7\sec^2\theta = 6\tan\theta + 8$
12 $2\tan 2\theta + 3\sec 2\theta = 4\cos 2\theta$
13 $\sin 2\theta(1 + 2\cos 2\theta) = 0$
14 $6\sin\theta = \tan\theta$
15 $3\sin 3\theta - \operatorname{cosec} 3\theta + 2 = 0$

7.3 Trigonometric equations involving multiple angles

If the equation for x involves trigonometric ratios of $2x$, $3x$, etc., we still seek an equation involving a single trigonometric ratio of a single angle be this x, $2x$, $3x$, etc. This will generally require us to use the identities of Sections 6.6 and 6.7.

Example 1 Find x if $\tan 2x + 3\tan x = 0$.

From (6.34) we have

$$\frac{2\tan x}{1-\tan^2 x} + 3\tan x = 0 \quad \text{(which involves only } \tan x\text{)}$$

Therefore

$$2\tan x + 3\tan x - 3\tan^3 x = 0$$
$$3\tan^3 x - 5\tan x = 0$$
$$\tan x(3\tan^2 x - 5) = 0$$
$$\tan x = 0 \quad \text{or} \quad 3\tan^2 x - 5 = 0$$

which give

$$\tan x = 0 \quad \text{or} \quad \tan x = \sqrt{\tfrac{5}{3}} \quad \text{or} \quad \tan x = -\sqrt{\tfrac{5}{3}}$$

Thus $x = 0°, 180°, -180°$, etc. In general, $x = n180°$ if $\tan x = 0$.
When $\tan x = \sqrt{\frac{5}{3}} = 1{\cdot}291 = \tan 52°\,14'$, then

$$x = n180° + 52°\,14'$$

When $\tan x = -\sqrt{\frac{5}{3}} = -1{\cdot}291 = \tan(-52°\,14')$, then

$$x = n180° - 52°\,14'$$

The general solution is thus $n180°$ or $n180° \pm 52°\,14'$.

* It is suggested that, at first reading, only solutions in the range 0° to 360° be considered.

Example 2 Find all values of θ in the range from 0° to 360° for which $\cos 2\theta - \cos\theta - 2 = 0$.

We have $2\cos^2\theta - 1 - \cos\theta - 2 = 0$, which involves only $\cos\theta$. Therefore

$$2\cos^2\theta - \cos\theta - 3 = 0$$
$$(2\cos\theta - 3)(\cos\theta + 1) = 0$$
$$2\cos\theta - 3 = 0 \quad \text{or} \quad \cos\theta + 1 = 0$$
$$\cos\theta = \tfrac{3}{2} \quad \text{or} \quad \cos\theta = -1$$

$\cos\theta = -1$ gives $\theta = 180°$; $\cos\theta = \frac{3}{2}$ gives no real solution for θ.

Example 3 Find θ if $\sin\theta + \sin 5\theta = \sin 3\theta$.

We have by (6.50)

$$\sin 5\theta + \sin\theta = 2\sin\tfrac{1}{2}(\theta + 5\theta)\cos\tfrac{1}{2}(5\theta - \theta)$$
$$= 2\sin 3\theta\cos 2\theta$$

Therefore, the equation becomes

$$2\sin 3\theta\cos 2\theta = \sin 3\theta$$
$$\sin 3\theta(2\cos 2\theta - 1) = 0$$
$$\sin 3\theta = 0 \quad \text{or} \quad \cos 2\theta = \tfrac{1}{2}$$

If $\sin 3\theta = 0°$, $3\theta = 0$, 180°, 360°, 540°, etc. (in general $n180°$). If $\cos 2\theta = \frac{1}{2}$, $2\theta = \pm 60°$, $360° \pm 60°$, $720° \pm 60°$ (in general $n360° \pm 60°$). Thus, the general solution of the equation is $\theta = n60°$ or $\theta = n180° \pm 30°$, where n is an integer.

Example 4 Find all values of x in the range 0° to 360° for which $\sin 3x \sin x = 2\cos 2x + 1$.

By (6.49)

$$\sin 3x \sin x = \tfrac{1}{2}[\cos(3x - x) - \cos(3x + x)]$$

Therefore, the equation becomes

$$\cos 2x - \cos 4x = 4\cos 2x + 2$$
$$\cos 4x + 3\cos 2x + 2 = 0$$

But $\cos 4x = 2\cos^2 2x - 1$, therefore

$$2\cos^2 2x + 3\cos 2x + 1 = 0 \quad \text{(which involves only } \cos 2x\text{)}$$
$$(2\cos 2x + 1)(\cos 2x + 1) = 0$$
$$2\cos 2x = -1 \quad \text{or} \quad \cos 2x = -1$$

If $\cos 2x = -\frac{1}{2} = \cos 120°$, $2x = n360° \pm 120°$, therefore

$$x = n180° \pm 60°$$

If $\cos 2x = -1 = \cos 180°$, $2x = n360° \pm 180°$, therefore

$$x = n180° \pm 90°$$

Thus, the solutions in the range 0° to 360° are 90°, 270°, 60°, 120°, 240°, 300°.

Exercise 7c

Find the general solution, together with *all* solutions in the range 0° to 360°, of the equations*

1 $\cos 2x + \sin^2 x = 1$
2 $2\cos^2\theta - 3\sin 2\theta - 2 = 0$
3 $2\cos 2\theta + 2\sin\theta\cos\theta = 1$
4 $\tan 2\theta - 1 = 6\cot 2\theta$
5 $\cos\theta + \cos 5\theta = \cos 2\theta$
6 $\sin x + \sin 3x = \sin 2x + \sin 4x$
7 $\cos\frac{3}{2}x\cos\frac{1}{2}x = 1 + \cos x$
8 $2\cos 3x\cos x = \cos 2x + \sin 2x + 1$
9 $\sin 3\theta = \cos 2\theta - 1 + \sin\theta$
10 $\tan 2\theta\tan 4\theta = 1$

7.4 The equation $a\cos\theta \pm b\sin\theta = c$

Equations of the type $a\cos\theta \pm b\sin\theta = c$, where a, b, c are constants, may be solved by first expressing $a\cos\theta \pm b\sin\theta$ in the form $R\cos(\theta \mp \alpha)$ (see (6.52)) or $R\sin(\theta \pm \beta)$ (see (6.51)). If we always write the equation with a positive, we need only use the first form. Alternatively, by expressing $\cos\theta$ and $\sin\theta$ in terms of $t = \tan\frac{1}{2}\theta$ (see (6.44) and (6.45)), we can obtain an equation for $\tan\frac{1}{2}\theta$ from which θ may be found.

Example 1 Solve the equation $3\cos\theta - 4\sin\theta = -2{\cdot}5$.
We have

$$\begin{aligned} 3\cos\theta - 4\sin\theta &= \sqrt{(3^2+4^2)}(\tfrac{3}{5}\cos\theta - \tfrac{4}{5}\sin\theta) \\ &= 5(\tfrac{3}{5}\cos\theta - \tfrac{4}{5}\sin\theta) \\ &= 5(\cos\theta\cos\alpha - \sin\theta\sin\alpha) \\ &\qquad\qquad (\text{where } \tan\alpha = \tfrac{4}{3}, \text{ cf. } (6.52)) \\ &= 5\cos(\theta+\alpha) \end{aligned}$$

Now, $\alpha = \arctan\frac{4}{3} = 53°8'$, therefore

$$5\cos(\theta + 53°8') = -2{\cdot}5$$
$$\cos(\theta + 53°8') = -\tfrac{1}{2} \quad (= \cos 120°)$$
$$\theta + 53°8' = n360° \pm 120°$$
$$\theta = n360° + 66°52' \quad \text{or} \quad \theta = n360° - 173°8'$$

If we tackle this problem by the second method, we have with $t = \tan\frac{1}{2}\theta$

$$\frac{3(1-t^2)}{1+t^2} - \frac{4.2t}{1+t^2} = -2{\cdot}5$$
$$3 - 3t^2 - 8t = -2{\cdot}5 - 2{\cdot}5t^2$$
$$6 - 6t^2 - 16t = -5 - 5t^2$$
$$t^2 + 16t - 11 = 0$$

* It is suggested that, at first reading, only solutions in the range 0° to 360° be considered.

Therefore

$$t = \frac{-16 \pm \sqrt{(256+44)}}{2} = \frac{-16 \pm \sqrt{300}}{2} = \frac{-16 \pm 17{\cdot}321}{2}$$

$$t = \tan\tfrac{1}{2}\theta = 0{\cdot}6605 \quad \text{or} \quad -16{\cdot}6605$$

that is

$$\tfrac{1}{2}\theta = n180^\circ + 33^\circ 26' \quad \text{or} \quad \tfrac{1}{2}\theta = n180^\circ - 86^\circ 34'$$

Thus $\theta = n360^\circ + 66^\circ 52'$ or $\theta = n360^\circ - 173^\circ 8'$, as before.

An alternative method is to proceed as follows. We have

$$4\sin\theta = 3\cos\theta + 2{\cdot}5$$

On squaring both sides

$$16\sin^2\theta = 9\cos^2\theta + 15\cos\theta + 6{\cdot}25$$
$$64\sin^2\theta = 36\cos^2\theta + 60\cos\theta + 25$$
$$64(1-\cos^2\theta) = 36\cos^2\theta + 60\cos\theta + 25$$
$$100\cos^2\theta + 60\cos\theta - 39 = 0$$

Therefore

$$\cos\theta = \frac{-60 \pm \sqrt{(3600 + 15\,600)}}{200} = \frac{-60 \pm \sqrt{19\,200}}{200} = \frac{-6 \pm 13{\cdot}856}{20}$$

that is

$$\cos\theta = 0{\cdot}3928 \quad \text{or} \quad \cos\theta = -0{\cdot}9928$$
$$\theta = n360^\circ \pm 66^\circ 52' \quad \text{or} \quad \theta = n360^\circ \pm 173^\circ 8'$$

In fact, as is readily verified, only the solutions

$$\theta = n360^\circ + 66^\circ 52' \quad \text{and} \quad \theta = n360^\circ - 173^\circ 8'$$

satisfy the equation $3\cos\theta - 4\sin\theta = -2{\cdot}5$. The two other results are solutions of the equation $3\cos\theta + 4\sin\theta = -2{\cdot}5$, equivalent to $-4\sin\theta = 3\cos\theta + 2{\cdot}5$ which *on squaring* gives $16\sin^2\theta = 9\cos^2\theta + 15\cos\theta + 6{\cdot}25$ as before.

Thus, if we adopt this method for solving this type of equation, we must check our solutions (cf. Section 1.2).

Exercises 7d

Find the general solution, together with *all* solutions in the range 0° to 360°, of the equations

1 $2\cos\theta + \sin\theta = 1$
2 $3\cos x + 4\sin x = 5$
3 $24\cos x - 7\sin x = 12{\cdot}5$
4 $2\cos\tfrac{1}{2}\theta + 3\sin\tfrac{1}{2}\theta = 2$

5 $5\cos 2x - \sin 2x = 2$
6 $\operatorname{cosec}\theta = 3 + 4\cot\theta$
7 $\sin 3x - \cos 3x = 1$
8 $5\cos 2\theta - \sqrt{2}\sin 2\theta = 3$
9 $\sin 3\theta - 4\cos 3\theta = 4$
10 $\cos\theta + \sin\theta + 1 = 0$ [Write $\cos\theta$ and $\sin\theta$ in terms of $t = \tan\frac{1}{2}\theta$ and notice how part of the solution is lost if this method of solution is used.]

Exercises 7

1 Find all solutions in the range 0° to 360° of the equations
(i) $\sin 2\theta = 0{\cdot}8799$
(ii) $\cos 4\theta = 0{\cdot}5659$
(iii) $\tan\theta = -1{\cdot}7699$
(iv) $\sin 3x = \sin 2x$
(v) $\cos 3x = \sin x$.

2 Find all angles in the range 0° to 360° which satisfy (i) $\tan\theta + \sec\theta = 2$, (ii) $\cos\theta - \cos 2\theta = \frac{1}{2}$.

3 Find x if $3\sin 2x = 7 - 8\cos^2 x$.

4 Find θ in the range 0° to 360° if $3(\cos 2\theta - 1) = 4\sin\theta\cos^2\theta$.

5 Find the values of x between 0° and 180° for which (i) $\sin 3x = \sin x$, (ii) $2\cos^2 x - \sin^2 x = 1$, (iii) $\sin 2x + \cos x = 0$.

6 Find the general value of θ if $3\tan^2\theta - 7\sec\theta + 5 = 0$.

7 If $3\cos\theta - 4\sin\theta - 2 = 0$, find all values for θ which lie in the range 0° to 360°.

8 Find the general solution and all solutions in the range 0° to 360° of the equations (i) $\sin\theta + \cos\theta = \sin 2\theta + \cos 2\theta$, (ii) $3(\sec\theta - \tan\theta) = 1$.

9 Solve completely the equation $\cos 3\theta = 2\cos 2\theta$. [First use (6.36) to express $\cos 3\theta$ in terms of $\cos\theta$.] [WJC, part]

10 Solve the following equations, giving all solutions within the range $0° < x < 360°$:
(i) $2\sin^2 x = 2 + \cos 2x$
(ii) $3\sin^2 x = 1 + \sin 2x$
(iii) $\sin(x + 30°) + \sin(x + 60°) = \cos(x + 45°) + \cos(x + 75°)$. [LU]

11 Find the general value of x if $3\tan^2 x - 5\sec x + 1 = 0$.

12 Solve the following equations for $0° \leqslant x \leqslant 180°$: (i) $6\sin^2 x - \cos x = 5$, (ii) $\tan 2x + 4\cot x = 0$. [AEB]

13 (i) Express $\tan 2x$ and $\tan 3x$ in terms of $\tan x$, and solve the equation $\tan x + \tan 2x + \tan 3x = 0$.
(ii) Solve the equation $\cos x + \cos 2x = \sin 3x$. (In each equation give all solutions between 0° and 180° inclusive.) [SUJB]

14 State and prove a formula for $\sin A - \sin B$ in terms of the angles $\frac{1}{2}(A + B)$ and $\frac{1}{2}(A - B)$. [You may, if you wish, quote formulae for $\sin(x \pm y)$.]

Prove that $\sin 7x\cos 2x - \sin 5x\cos 4x = \sin 2x\cos 3x$ and obtain a similar simplification of the expression $\sin 7x\sin 2x - \sin 5x\sin 4x$. Give all the solutions between 0° and 180° of the equation

$$\sin 7x\cos 2x = \sin 2x\cos 3x$$ [SUJB]

15 Find the values of θ in the range 0° to 360° which satisfy the equation $\tan\theta - 3\cot\theta = 2\tan 3\theta$.
16 Solve the equation $\cos x + \cos 5x = \cos 3x$.
17 Solve the equations (i) $3\sin x - 4\cos x = 2$, (ii) $24\sin x + 7\cos x = 5$.
18 Find the general solution of the equation $\tan 2x = 2\sin x$. [JMB, part]
19 Prove that $4\cos\theta\cos 3\theta + 1 = \dfrac{\sin 5\theta}{\sin\theta}$. Hence find all the values of θ in the range 0° to 180° inclusive for which $\cos\theta\cos 3\theta = -\frac{1}{2}$. [JMB, part]
20 Solve the simultaneous equations $\cos A + \cos B = \frac{2}{3}k$ and $\cos A\cos B = -\frac{1}{3}k^2$ to find $\cos A$ and $\cos B$ in terms of k. Find the range of values of k for which A and B exist. [AEB]
21 Find all values of θ in the range $0° < \theta < 360°$ for which $\tan^2\theta = 5 + \sec\theta$.
22 Solve the following equations, giving the general solutions:
(i) $\cos 2x + 3\cos x = -2$
(ii) $\cos 3x = \sin x$
(iii) $\sin(2x + 30°)\cos(2x - 20°) = \frac{1}{2}$. [SUJB]
23 Find the range $-180° < x < 180°$ the solutions of the equation $\cos 5x = \cos x$. [JMB, part]
24 Given that $5\cos\theta + 12\sin\theta = R\cos(\theta - \alpha)$, where R and α are independent of θ and R is positive, obtain the values of R and α. Hence find the values of θ between $-180°$ and 180° which satisfy the equation $5\cos\theta + 12\sin\theta = 3{\cdot}25$, giving the answers to the nearest minute. [JMB, part]
25 Find the complete solution of $16\tan x + 6\cot x + 17\sec x = 0$.
26 Solve the equation $5\cos\theta - 12\sin\theta + 10 = 0$.
27 Solve for x, $\arctan x + \arctan(x - 1) = \arctan 3$ [use (6.56)].
28 Find all angles between 0° and 360° for which

$$3\sin 3\theta + 2\cos 2\theta - \sin\theta = 2$$

29 Find the values of θ in the range 0° to 360° for which

$$\sin 5\theta + 2\cos 2\theta + \sin\theta = 0$$

30 Find x if $\tan^{-1}(2x + 1) - \tan^{-1}(2x - 1) = \tan^{-1}\frac{1}{8}$.
31 Find to the nearest minute the values of θ between 0° and 360° that satisfy the equation $4\cos 2\theta + \sin\theta = 4\sin^2\theta + 3$. [JMB, part]
32 Solve the equation $5\sin(x + 60°) - 3\cos(x + 30°) = 4$, giving all solutions between 0° and 360°. [LU, part]
33 If $\sin(\alpha + \theta)\sin(\beta + \phi) = \sin(\alpha + \phi)\sin(\beta + \theta)$, prove that either α and β or θ and ϕ differ by a multiple of π. [LU, part]
34 By putting $t = \tan\theta$, find the general solution of the equation

$$(1 - \tan\theta)(1 + \sin 2\theta) = 1 + \tan\theta$$

35 The acute angle θ satisfies the equation $\sin(2\theta + \alpha) = \sqrt{3}\cos(\theta - \alpha)$. If α is zero, show that $\theta = \pi/3$. If α is so small that its square may be neglected and $\theta = \pi/3 + \lambda$, prove that λ is approximately 4α. [JMB, part]
36 (a) Find two values of θ between 0° and 180° satisfying the equation $6\sin^2\theta = 5 + \cos\theta$.
(b) Find a value of x between 0 and π satisfying the equation

$$\sin(x + \pi/3) = \cos(x - \pi/3)$$

[JMB]

37 Show that the equation $a\cos\theta + b\sin\theta = c$ will have in general two solutions in the range 0° to 360° if $a^2 + b^2 > c^2$. α and β are two roots of the equation $8\cos\theta - \sin\theta = 4$ and both lie in the range 0° to 360°. Form the quadratic equations whose roots are $\sin\alpha$ and $\sin\beta$.

38 If α and β are two unequal values of θ which satisfy the equation $a\cos\theta + b\sin\theta = c$, show that

(i) $\sin\frac{1}{2}(\alpha+\beta)\sec\frac{1}{2}(\beta-\alpha) = b/c$

(ii) $\tan\frac{1}{2}\alpha\tan\frac{1}{2}\beta = \dfrac{c-a}{c+a}$.

39 Show that the equation

$$p\cos^2 x + 2q\cos x\sin x + r\sin^2 x = s \quad (r \neq s)$$

has a real solution only provided $q^2 \geqslant (s-p)(s-r)$ and that in this case there are in general two solutions which are in the range $0^\circ \leqslant x \leqslant 180^\circ$.
If θ and ϕ are these solutions, show that $\tan(\theta+\phi) = 2q/(p-r)$.

40 If $t = \tan\frac{1}{2}\theta$, write down expressions for $\sin\theta$ and $\cos\theta$ in terms of t. Use these to show that the equation

$$a\cos^2\theta + b\sin^2\theta + 2g\cos\theta + 2f\sin\theta + c = 0$$

can be written as a quartic equation in t. Write down an expression for $\tan\frac{1}{2}(\theta_1+\theta_2+\theta_3+\theta_4)$ in terms of $\tan\frac{1}{2}\theta_1$, $\tan\frac{1}{2}\theta_2$, $\tan\frac{1}{2}\theta_3$ and $\tan\frac{1}{2}\theta_4$. Deduce that the sum of the values of θ which satisfy the equation above is an even multiple of π.

41 Express $\cos x + 2\sin x$ in the form $R\cos(x-\alpha)$, where R is positive. Hence or otherwise solve the equation $\cos x + 2\sin x = 1{\cdot}52$ for $0 \leqslant x \leqslant 360^\circ$. [AEB]

42 Find the range of values of a for which the equation

$$\cos(x+90^\circ) + \cos x = a$$

has real solutions. For the case when $a = 0$, find all the solutions in the interval $0^\circ \leqslant x \leqslant 360^\circ$. Sketch the graph of $y = \cos(x+90^\circ) + \cos x$ for $0^\circ \leqslant x \leqslant 360^\circ$. [JMB]

43 Express the function $\cos\theta + 2\sin\theta$ in the form $R\sin(\theta+\alpha)$, where R is positive and $0^\circ < \alpha < 360^\circ$. Hence, or otherwise, find the values of θ between 0° and 360° which satisfy the equation $3\cos\theta + 6\sin\theta = 1$. [AEB]

44 (i) Given that $3\cos x + 2\sec x + 5 = 0$, find all the possible values of $\cos x$ and $\tan^2 x$.

(ii) By putting $\tan(\theta/2) = t$, or otherwise, find the general solution of the equation $\sin\theta + 7\cos\theta = 5$, giving your answers to the nearest tenth of a degree. [LU]

45 Prove that $\cos 3\theta \equiv 4\cos^3\theta - 3\cos\theta$. Solve the equation $\cos 3\theta + 2\cos\theta = 0$, giving all solutions between 0 and 2π. [O]

46 Prove the formulae

$$\sin\theta = \frac{2t}{1+t^2} \qquad \cos\theta = \frac{1-t^2}{1+t^2}$$

where $t = \tan\frac{1}{2}\theta$. Hence, or otherwise, find in degrees and minutes all the angles between 0° and 360° for which $16\sin x - 3\cos x + 11 = 0$. [C]

47 Write down expressions for $\tan\theta$ and $\sec\theta$ in terms of t, where $t = \tan\frac{1}{2}\theta$, and show that

$$\sec\theta + \tan\theta = \tan(45° + \tfrac{1}{2}\theta)$$

Find a solution in the interval $0° < \theta < 90°$ of the equation $\sec\theta + \tan\theta = \cot 2\theta$. [JMB]

48 Find the solutions in the range $0° \leqslant \theta \leqslant 180°$ of each of the following equations: (i) $\sin 2\theta = \cos\theta$, (ii) $2\cos 2\theta = 1 + 4\cos\theta$. [JMB]

49 Given that $\cos^4 x + \sin^4 x = \frac{5}{8}$, show, without the use of tables, that $\cos 4x = -\frac{1}{2}$. Hence, or otherwise, find all the values of x between 0 and π for which $\cos^4 x + \sin^4 x = \frac{5}{8}$. [C]

50 Show that $\sin 3A = 3\sin A - 4\sin^3 A$ (formulae for $\sin 2A$ and $\cos 2A$ may be quoted). Solve the equation $8\sin^3\theta - 6\sin\theta = 1$ for $0 < \theta < 360°$. [AEB]

8
The solution of triangles

8.1 The sine formula

The usual notation for a triangle is used. A, B, C denote the angles; a, b, c, the sides opposite these angles. R is the radius of the circle through the points A, B, C, the circumscribed circle, and $2s$ ($= a + b + c$) the perimeter of the triangle.

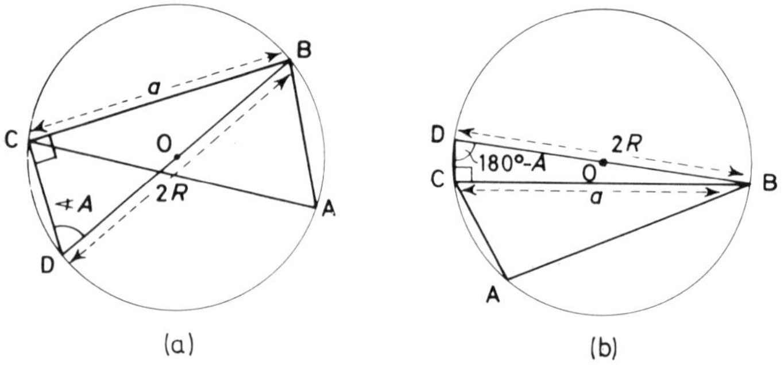

Figure 8.1

Let R be the radius and O the centre of the circumcircle of $\triangle$ABC. Draw the diameter BOD and join CD.

In $\triangle$BDC

$$\triangle \text{DCB} = 90^\circ \quad \text{(angle in a semicircle)}$$
$$\angle \text{BDC} = A \quad \text{or} \quad 180^\circ - A \quad \text{(see } \textit{Figure 8.1a} \text{ and } b\text{)}$$

since ABDC is a cyclic quadrilateral. Thus

$$\frac{\text{BC}}{\text{BD}} = \sin A \quad \text{or} \quad \sin(180^\circ - A)$$
$$\text{BC} = \text{BD} \sin A$$
$$a = 2R \sin A \quad \text{(BD is a diameter)}$$

Therefore

$$\frac{a}{\sin A} = 2R$$

Similarly

$$\frac{b}{\sin B} = 2R \qquad \frac{c}{\sin C} = 2R$$

Hence

$$\frac{a}{\sin A} = \frac{b}{\sin B} = \frac{c}{\sin C} = 2R \tag{8.1}$$

a result generally known as the sine formula.

8.2 The cosine formula

In $\triangle$ABC draw a perpendicular AL from A to meet BC, or BC produced at L. Consider these two cases separately:

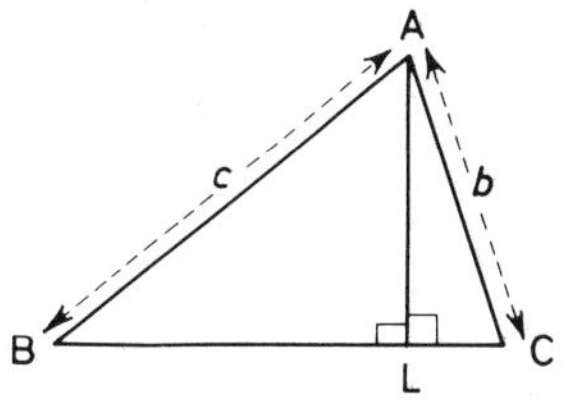

(a)

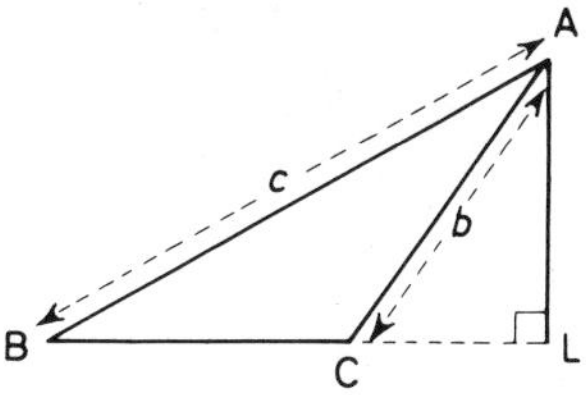

(b)

Figure 8.2a

BC = BL + LC

thence

BC = $c \cos B + b \cos C$

that is

$a = c \cos B + b \cos C$

Figure 8.2b

BC = BL − LC

thence

BC = $c \cos B - b \cos(180° - C)$

that is

$a = c \cos B + b \cos C$

This result is common to both cases.

By drawing the perpendiculars from B and C to the opposite sides, two similar results can be obtained. Collecting these together we have

$$\left.\begin{aligned} a &= b\cos C + c\cos B \\ b &= c\cos A + a\cos C \\ c &= a\cos B + b\cos A \end{aligned}\right\} \tag{8.2}$$

By multiplying these equations by $-a$, b, c, respectively and adding them we obtain

$$b^2 + c^2 - a^2 = 2bc \cos A$$

which gives

$$\left.\begin{aligned} a^2 &= b^2 + c^2 - 2bc\cos A \\ b^2 &= c^2 + a^2 - 2ca\cos B \\ c^2 &= a^2 + b^2 - 2ab\cos C \end{aligned}\right\} \quad (8.3)$$

Similarly

and

These results are known as the cosine formula.

Note that any one of the formulae (8.2) or (8.3) generates the others by a cyclic permutation of

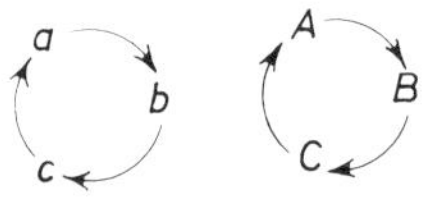

Example 1 The point P divides the side AB of a triangle ABC internally in the ratio $m{:}n$. If $\angle \mathrm{ACP} = \alpha$, $\angle \mathrm{BCP} = \beta$ and $\angle \mathrm{BPC} = \theta$, prove that $m\cot\alpha - n\cot\beta = (m+n)\cot\theta$.

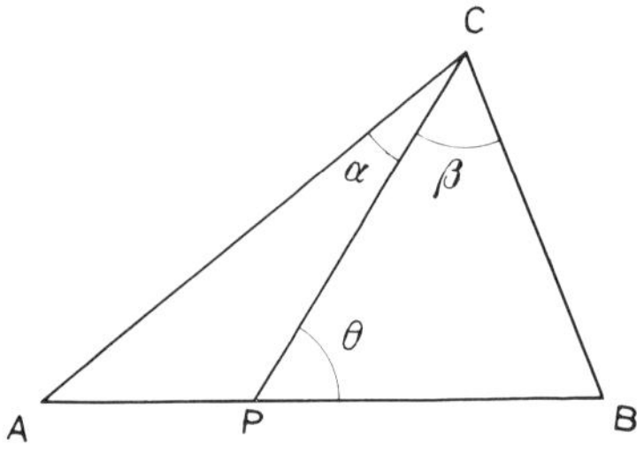

Figure 8.3

Referring to *Figure 8.3* and using the sine formula for the triangles ACP, BCP, we have

$$\frac{\mathrm{AP}}{\sin\alpha} = \frac{\mathrm{CP}}{\sin A} \quad \text{hence} \quad \frac{\mathrm{AP}}{\mathrm{CP}} = \frac{\sin\alpha}{\sin A} \quad \text{(i)}$$

$$\frac{\mathrm{BP}}{\sin\beta} = \frac{\mathrm{CP}}{\sin B} \quad \text{hence} \quad \frac{\mathrm{CP}}{\mathrm{BP}} = \frac{\sin B}{\sin\beta} \quad \text{(ii)}$$

From (i) and (ii)

$$\frac{m}{n} = \frac{\mathrm{AP}}{\mathrm{PB}} = \frac{\mathrm{AP}}{\mathrm{CP}}\cdot\frac{\mathrm{CP}}{\mathrm{PB}} = \frac{\sin\alpha}{\sin A}\cdot\frac{\sin B}{\sin\beta} \quad \text{(iii)}$$

Also

$$\theta = A + \alpha \quad \text{and} \quad \theta = 180^\circ - B - \beta \quad \text{(iv)}$$

Eliminating A and B from (iii) by means of (iv), we have

$$\frac{m}{n} = \frac{\sin\alpha}{\sin(\theta-\alpha)}\cdot\frac{\sin(180^\circ-\beta-\theta)}{\sin\beta}$$
$$= \frac{\sin\alpha}{\sin(\theta-\alpha)}\cdot\frac{\sin(\beta+\theta)}{\sin\beta}$$
$$= \frac{\sin\alpha(\sin\beta\cos\theta+\cos\beta\sin\theta)}{\sin\beta(\sin\theta\cos\alpha-\cos\theta\sin\alpha)}$$

Dividing above and below by $\sin\alpha\sin\beta\sin\theta$, we obtain

$$\frac{m}{n} = \frac{\cot\theta+\cot\beta}{\cot\alpha-\cot\theta}$$

therefore

$$m\cot\alpha - m\cot\theta = n\cot\theta + n\cot\beta$$
$$m\cot\alpha - n\cot\beta = (m+n)\cot\theta \qquad (8.4)$$

Example 2 Given a triangle whose sides are in the ratio 4:5:6, prove, without use of tables, that one angle is twice another angle.

Since the sides are in the ratio 4:5:6, their lengths are $4k$, $5k$, $6k$ (a, b, c say) where k is a constant.

Formulae (8.3) can be rewritten

$$\cos A = \frac{b^2+c^2-a^2}{2bc}$$

with similar results for $\cos B$ and $\cos C$. Hence

$$\cos A = \frac{25k^2+36k^2-16k^2}{60k^2} = \frac{3}{4}$$
$$\cos B = \frac{36k^2+16k^2-25k^2}{48k^2} = \frac{9}{16}$$
$$\cos C = \frac{16k^2+25k^2-36k^2}{40k^2} = \frac{1}{8}$$

Consider the smallest angle A:

$$\cos 2A = 2\cos^2 A - 1$$
$$= 2(\tfrac{3}{4})^2 - 1$$
$$= \tfrac{1}{8}$$
$$= \cos C$$

Hence

$$2A = 2n\pi \pm C$$

but since A and C are angles of a triangle, $2A = C$ is the only solution.

Example 3 A vertical tower stands on a river bank. From a point on the other bank directly opposite and at a height h above the water level, the angle of elevation of the top of the tower is α and the angle of depression of the reflection of the top of the tower is β. (Assume the water is smooth and the reflection of any object in the water surface will appear to be as far below the surface as the object is above it.) Prove that the height of the top of the tower above the water is $h\sin(\alpha+\beta)\operatorname{cosec}(\beta-\alpha)$ and the width of the river is $2h\cos\alpha\cos\beta\operatorname{cosec}(\beta-\alpha)$.

Refer to *Figure 8.4*. AB is the tower, O the observer, OP, OQ the horizontal and vertical through O. Let x be the height of the tower, and y the width of the river.

$$A'B = AB = x \qquad OP = BQ = y$$

In $\triangle$APO

$$AP = PO\tan\alpha$$

$$x - h = y\tan\alpha \qquad \text{(i)}$$

In $\triangle$A′PO

$$A'P = PO\tan\beta$$

$$x + h = y\tan\beta \qquad \text{(ii)}$$

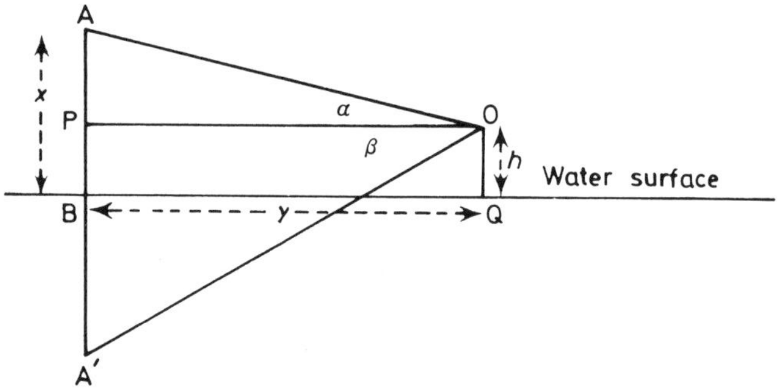

Figure 8.4

Subtracting (i) from (ii) to eliminate x, we obtain

$$\begin{aligned}2h &= y(\tan\beta - \tan\alpha)\\ &= y\left(\frac{\sin\beta}{\cos\beta} - \frac{\sin\alpha}{\cos\alpha}\right)\\ &= y\left(\frac{\sin\beta\cos\alpha - \sin\alpha\cos\beta}{\cos\alpha\cos\beta}\right)\\ &= y\left(\frac{\sin(\beta-\alpha)}{\cos\alpha\cos\beta}\right)\end{aligned}$$

whence

$$y = 2h\cos\alpha\cos\beta\operatorname{cosec}(\beta-\alpha)$$

To obtain the height of the tower x, multiply (i) by $\tan\beta$, (ii) by $\tan\alpha$ and subtract. Therefore

$$x\tan\beta - h\tan\beta - x\tan\alpha - h\tan\alpha = 0$$
$$x(\tan\beta - \tan\alpha) = h(\tan\alpha + \tan\beta)$$
$$x\left(\frac{\sin\beta}{\cos\beta} - \frac{\sin\alpha}{\cos\alpha}\right) = h\left(\frac{\sin\alpha}{\cos\alpha} + \frac{\sin\beta}{\cos\beta}\right)$$
$$x\left(\frac{\sin\beta\cos\alpha - \sin\alpha\cos\beta}{\cos\alpha\cos\beta}\right) = h\left(\frac{\sin\alpha\cos\beta + \cos\alpha\sin\beta}{\cos\alpha\cos\beta}\right)$$
$$x\sin(\beta - \alpha) = h\sin(\alpha + \beta)$$

whence

$$x = h\sin(\alpha + \beta)\operatorname{cosec}(\beta - \alpha).$$

Exercises 8a

1 The point P divides the base AB of a triangle ABC in the ratio $m:n$. If $\angle$BPC $= \theta$, prove that $n\cot A - m\cot B = (m+n)\cot\theta$.

2 Without using tables prove that there is a triangle whose angles are $\arccos\frac{23}{27}$, $\arccos\frac{7}{9}$, $\arccos(-\frac{1}{3})$.

3 The median AD of a triangle ABC makes angles β, γ respectively with AB, AC and $\angle$ADB $= \theta$. Show that $2\cot\theta = \cot\gamma - \cot\beta$. If AD $= 15$ m, $\beta = 35°$, $\gamma = 30°$ find B, C, a as accurately as the tables permit. [WJC]

4 If in any triangle ABC

$$\cos A\cos B + \sin A\sin B\sin C = 1$$

prove that $A = B = 45°$.

5 In a triangle ABC the angle C is 60°. Show that $c^2 = a^2 - ab + b^2$. If a, b are the roots of the equation $4x^2 - 10x + 3 = 0$ find the value of c and show that the length of the perpendicular from C to AB is $(3\sqrt{3})/16$. [JMB]

6 In the triangle ABC the perpendiculars AL, CN from A, C to the opposite sides intersect at H. R is the circumradius and O is the circumcentre of the triangle ABC.

(i) Prove that NA $= b\cos A$, and $\angle$NHA $= B$, and hence that AH $= 2R\cos A$.

(ii) Prove that $\angle$OAH $= C - B$ and hence by using the result in part (i) and applying the cosine rule to $\triangle$OAH, prove that OH^2 $= R^2(1 - 8\cos A\cos B\cos C)$.

7 Starting from the sine formula for a triangle ABC or otherwise, prove the formula $\dfrac{b-c}{a} = \dfrac{\sin\frac{1}{2}(B-C)}{\cos\frac{1}{2}A}$.

A, B and C are three towns. C is due south of A, and the bearing of B from A is 30° west of south. B is 50 miles from C, and 10 miles nearer to A than C. Calculate the bearing of B from C. [SUJB]

8 In the tetrahedron ABCD the three angles at A are each 60° and AB,

AC, AD are of lengths 2, 3, 4 cm respectively. Find the angles of the triangle BCD.

9 Prove that, in any triangle ABC, $\dfrac{a^2 - b^2}{c^2} = \dfrac{\sin(A - B)}{\sin(A + B)}$.

10 Where possible solve the following triangles:
(i) $c = 4{\cdot}13$, $b = 5{\cdot}62$, $\angle C = 61°23'$
(ii) $c = 5{\cdot}62$, $b = 5{\cdot}62$, $\angle C = 67°54'$
(iii) $a = 651$, $c = 792$, $\angle C = 73°22'$.

8.3 The area of a triangle

In triangle ABC, let AD be the perpendicular from A to BC. Let Δ denote the area of the triangle. Referring to *Figure 8.5*, we have, since $\text{AD} = c \sin B$,

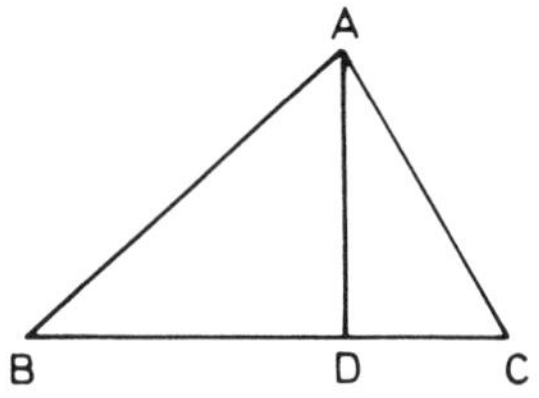

Figure 8.5

$$\Delta = \tfrac{1}{2}\text{BC}.\text{AD} = \tfrac{1}{2}ac \sin B \tag{8.5}$$

From (8.1)

$$\sin B = \frac{b}{2R}$$

and hence, on substituting in (8.5), we have

$$\Delta = \frac{abc}{4R} \tag{8.6}$$

The area of the triangle can be found in terms of the sides alone. From (8.5)

$$2ac \sin B = 4\Delta$$

and from the cosine formulae (8.3)

$$2ac \cos B = c^2 + a^2 - b^2$$

On squaring and adding, noting that $\cos^2 B + \sin^2 \text{B} = 1$, we have

$$4a^2c^2 = 16\Delta^2 + (c^2 + a^2 - b^2)^2$$

or

$$\begin{aligned} 16\Delta^2 &= 4a^2c^2 - (c^2 + a^2 - b^2)^2 \\ &= (2ac - c^2 - a^2 + b^2)(2ac + c^2 + a^2 - b^2) \\ &= [b^2 - (a - c)^2][(a + c)^2 - b^2] \\ &= (-a + b + c)(b + a - c)(a + b + c)(a - b + c) \end{aligned}$$

Now if we let $a + b + c = 2s$

$$16\Delta^2 = (2s - 2a)(2s - 2c)(2s)(2s - 2b)$$
$$\Delta^2 = s(s - a)(s - b)(s - c)$$
$$\Delta = \sqrt{[s(s - a)(s - b)(s - c)]} \qquad (8.7)$$

This is known as Hero's formula.

Example 1 Given that the sides of a triangle are of length $a = 3{\cdot}57$ m, $b = 2{\cdot}61$ m, $c = 4{\cdot}72$ m, find its area and the radius of its circumcircle.

$$2s = a + b + c = 3{\cdot}57 + 2{\cdot}61 + 4{\cdot}72$$
$$= 10{\cdot}90$$

therefore

$$s = 5{\cdot}45$$

Hence using formula (8.7)

number	log
5·45	0·7364
1·88	0·2742
2·84	0·4533
0·73	$\bar{1}$·8633
	1·3272
4·609	0·6636

$$\Delta = \sqrt{[5{\cdot}45(5{\cdot}45 - 3{\cdot}57)(5{\cdot}45 - 2{\cdot}61)(5.45 - 4.72)]}$$
$$= \sqrt{(5{\cdot}45 \times 1{\cdot}88 \times 2{\cdot}84 \times 0{\cdot}73)}$$
$$= 4{\cdot}609 \text{ m}^2$$

From (8.6)

$$\Delta = \frac{abc}{4R}$$
$$R = \frac{abc}{4\Delta}$$
$$= \frac{3{\cdot}57 \times 2{\cdot}61 \times 4{\cdot}72}{4 \times 4{\cdot}609}$$
$$R = 2{\cdot}385 \text{ m}$$

number	log
3·57	0·5527
2·61	0·4166
4·72	0·6739
	1·6432
18·436	1·2656
2·385	0·3776

The calculations above could easily be carried out using a pocket calculator.

Exercises 8b

1 ABC is a triangle with sides of lengths a, b, c opposite A, B, C, respectively. P is on the opposite side of BC to A and the triangle BCP is equilateral. Show that $\text{AP}^2 = a^2 + c^2 - ac\cos B + \sqrt{3}ac\sin B$. Deduce that $\text{AP}^2 = \frac{1}{2}(a^2 + b^2 + c^2) + 2\sqrt{3}\Delta$, where Δ is the area of ABC.

2 The median CC′ of a triangle ABC meets the side AB in the point C′. If θ is the angle AC′C, prove that

$$\frac{2c\cos\theta}{a^2 - b^2} = \frac{c\sin\theta}{2\Delta} = \frac{1}{\text{CC}'} \qquad \text{[JMB]}$$

3 If Δ is the area and R the radius of the circumcircle of the triangle ABC, prove that $\cos A + \cos(B - C) = 2\Delta/aR$.

4 If E is the middle point of the side CA of the triangle ABC and if Δ is the area of the triangle, prove that

$$\cot \mathrm{AEB} = \frac{\mathrm{BC}^2 - \mathrm{BA}^2}{4\Delta} \qquad \text{(LU)}$$

5 If p_1, p_2, p_3 are the lengths of the altitudes of a triangle and R the radius of its circumcircle, prove that (i) $8\Delta^3 = p_1 p_2 p_3 abc$ and (ii) $\Delta = \sqrt{(\tfrac{1}{2}Rp_1p_2p_3)}$.

8.4 Miscellaneous applications

We shall end the chapter with some typical examples.

Example 1 Show that, in any triangle ABC,

$$\tan \tfrac{1}{2}(B - C) = \frac{b - c}{b + c} \cot \tfrac{1}{2}A$$

Use this formula, rather than the cosine formula (8.3), to solve the triangle in which $b = 15{\cdot}32$, $c = 28{\cdot}6$ and $A = 39° 52'$.

From the sine formula (8.1), we have $b = 2R \sin B$, $c = 2R \sin C$. Hence, after cancelling $2R$,

$$\begin{aligned}\frac{b - c}{b + c} &= \frac{\sin B - \sin C}{\sin B + \sin C}\\ &= \frac{2 \cos \tfrac{1}{2}(B + C) \sin \tfrac{1}{2}(B - C)}{2 \sin \tfrac{1}{2}(B + C) \cos \tfrac{1}{2}(B - C)} \quad \text{by (6.50)}\\ &= \cot \tfrac{1}{2}(B + C) \tan \tfrac{1}{2}(B - C)\end{aligned}$$

Hence

$$\tan \tfrac{1}{2}(B - C) = \frac{b - c}{b + c} \tan \tfrac{1}{2}(B + C)$$

But since $\tfrac{1}{2}(B + C) = 90° - \tfrac{1}{2}A$, this can be rewritten

$$\tan \tfrac{1}{2}(B - C) = \frac{b - c}{b + c} \cot \tfrac{1}{2}A \qquad (8.8)$$

Since $b = 15{\cdot}32$, $c = 28{\cdot}6$, it is better to write the formula as

$$\tan \tfrac{1}{2}(C - B) = \frac{c - b}{c + b} \cot \tfrac{1}{2}A$$

Hence

$$\tan\tfrac{1}{2}(C-B) = \frac{28{\cdot}6 - 15{\cdot}32}{28.6 + 15{\cdot}32}\cot 19^\circ 56'$$
$$= \frac{13{\cdot}28}{43{\cdot}92}\cot 19^\circ 56'$$
$$= 0{\cdot}8337$$

number	log
13·28	1·1232
cot 19° 56′	0·4405
	1·5637
43·92	1·6427
0·8337	$\bar{1}$·9210
15·32	1·1853
sin 39° 52′	$\bar{1}$·8069
	0·9922
sin 30° 15′	$\bar{1}$·7022
19·50	1·2900

so that

$$\tfrac{1}{2}(C-B) = 39^\circ 49' \qquad \text{(i)}$$

Also

$$\tfrac{1}{2}(C+B) = 90^\circ - \tfrac{1}{2}A$$
$$= 90^\circ - 19^\circ 56'$$
$$= 70^\circ 4' \qquad \text{(ii)}$$

From (i) and (ii), $C = 109^\circ 53'$, $B = 30^\circ 15'$. Now

$$a = \frac{b\sin A}{\sin B} = \frac{15{\cdot}32\sin 39^\circ 52'}{\sin 30^\circ 15'}$$
$$= 19{\cdot}50$$

and the required solution is $B = 30^\circ 15'$, $C = 109^\circ 53'$, $a = 19{\cdot}50$.

Example 2 A, B are two points on the same level. The distance AB is c. The angles of elevation of the top T of a vertical tower from A and B are α and β respectively. If T′, the foot of the tower, is on the same level as A and B then $\angle$T′AB $= \gamma$ and $\angle$T′BA $= \delta$. Find an expression for the height of the tower in terms of c, γ, δ and either α or β. Also prove that $\sin\delta\tan\alpha = \sin\gamma\tan\beta$.

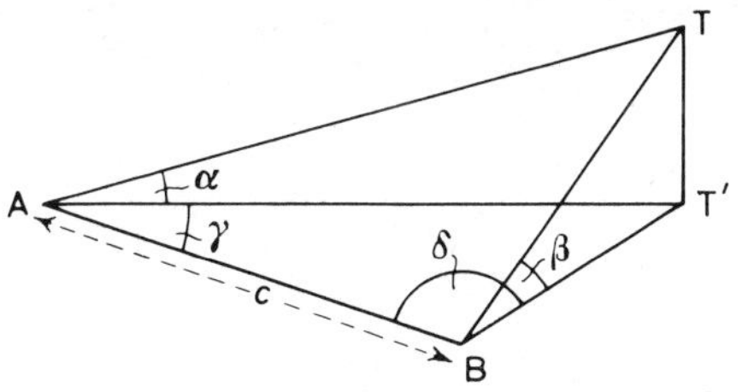

Figure 8.6

From the triangle ABT′ (*Figure 8.6*)

$$\frac{AT'}{\sin\delta} = \frac{BT'}{\sin\gamma} = \frac{AB}{\sin AT'B}$$

Since the angle AT′B is $180^\circ - \gamma - \delta$, then

$$\frac{AT'}{\sin\delta} = \frac{BT'}{\sin\gamma} = \frac{c}{\sin(\gamma+\delta)} \qquad \text{(i)}$$

From the right-angled triangle AT′T

$$TT' = AT' \tan\alpha$$

Hence from (i)

$$TT' = \frac{c\sin\delta\tan\alpha}{\sin(\gamma+\delta)} \qquad \text{(ii)}$$

Similarly, from the right-angled triangle BT′T

$$TT' = BT' \tan\beta$$

Hence from (i)

$$TT' = \frac{c\sin\gamma\tan\beta}{\sin(\gamma+\delta)} \qquad \text{(iii)}$$

Equating (ii) and (iii) and simplifying

$$\sin\delta\tan\alpha = \sin\gamma\tan\beta$$

Example 3 A point Q is in a direction $\theta°$ N of E from a point O. P is a point between O and Q such that OP = x. R is due north of Q and QR subtends angles α and β at O and P respectively. Prove that

$$QR = \frac{x\sin\alpha\sin\beta}{\sin(\beta-\alpha)\cos\theta}$$

Refer to *Figure* 8.7. In $\triangle$ORP

$$\frac{OP}{\sin ORP} = \frac{RP}{\sin\alpha}$$

but

$$\angle ORP = \beta - \alpha$$

hence

$$PR = \frac{x\sin\alpha}{\sin(\beta-\alpha)} \qquad \text{(i)}$$

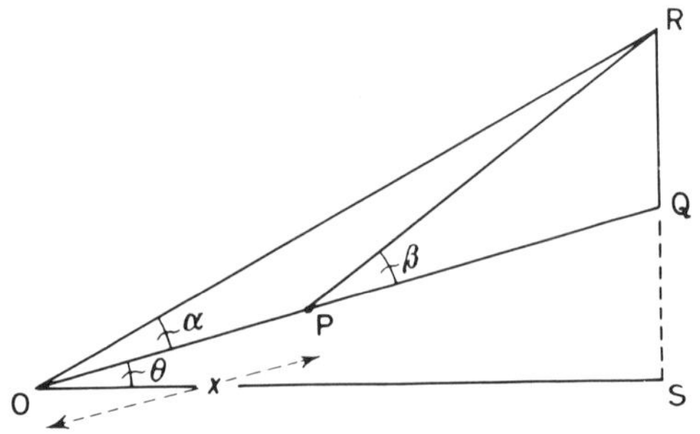

Figure 8.7

In $\triangle$PRQ

$$\frac{\text{QR}}{\sin\beta} = \frac{\text{PR}}{\sin \text{PQR}}$$

Now

$$\begin{aligned}\angle\text{PQR} &= 180^\circ - \angle\text{OQS} = 180^\circ - (90^\circ - \theta)\\ &= 90^\circ + \theta\end{aligned}$$

hence

$$\text{QR} = \frac{\text{PR}\sin\beta}{\sin(90^\circ + \theta)} \qquad \text{(ii)}$$

From (i) and (ii)

$$\text{QR} = \frac{x\sin\alpha\sin\beta}{\sin(\beta - \alpha)\cos\theta}$$

Example 4 ABCD is the square base of a pyramid. The top vertex of the pyramid is the point P, vertically above the point E, where AC meets BD. If AB = EP = $2a$, find: (i) the angle between the edge AP and the base of the pyramid; (ii) the angle between the planes ABP and ABCD; (iii) the angle between the planes ABP, BCP.

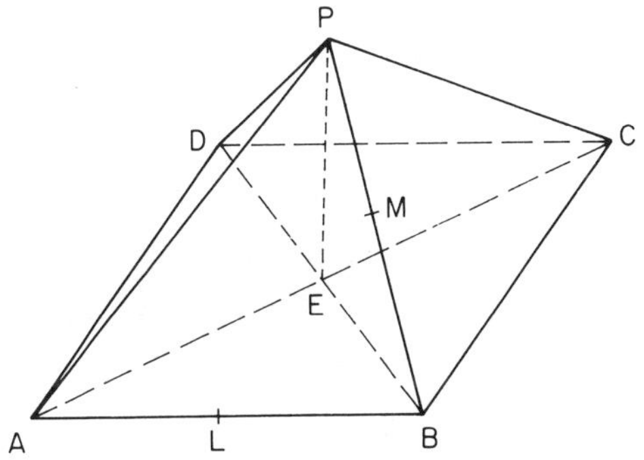

Figure 8.8

(i) The required angle is $\angle$PAE and from $\triangle$APE

$$\text{AE} = \sqrt{2}a \ (= \tfrac{1}{2}\text{AC}) \quad \text{and} \quad \text{PE} = 2a$$

therefore

$$\begin{aligned}\tan\text{PAE} &= 2a/\sqrt{2}a = \sqrt{2}\\ \angle\text{PAE} &= 54^\circ 44'\end{aligned}$$

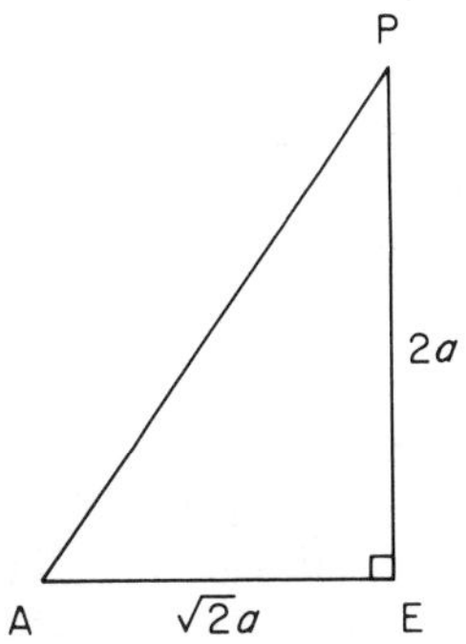

Figure 8.9

(ii) If L is the mid-point of AB, the required angle is ∠PLE, since PL in ABP and EL in ABCD are both perpendicular to AB, the line of intersection of the two planes. Therefore,

$$\tan \text{PLE} = 2$$
$$\angle \text{PLE} = 63^\circ 26'$$

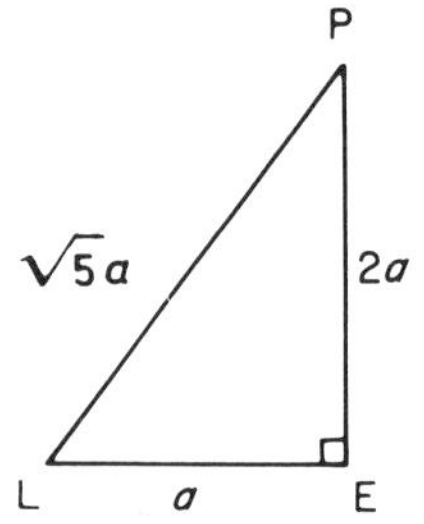

Figure 8.10

(iii) BP is the line of intersection of the two planes. If AM is the altitude from A to PB then CM and AM will both be perpendicular to PB and will lie in the planes BCP and ABP respectively. The angle required is ∠AMC.

$$\text{AM} = \text{MC} = \sqrt{\tfrac{10}{3}}a \quad (\text{from area of } \triangle\text{APB: AM}.\sqrt{6}a = \sqrt{5}a.2a)$$
$$\text{AC} = 2\sqrt{2}a$$
$$\sin \text{AME} = \sqrt{\tfrac{6}{10}}$$

Therefore

$$\angle \text{AME} = 50^\circ 46'$$
$$\angle \text{AMC} = 101^\circ 32'$$

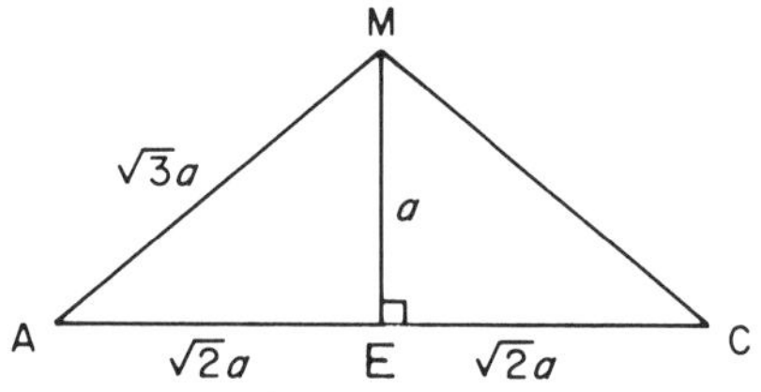

Figure 8.11

Exercises 8

1 The sides a, c of a triangle ABC are of length 3 cm and 7 cm respectively, and the angle C is 76°. Find the angles A and B and the side b.

2 Prove that for a triangle of area Δ

$$ab = 2\Delta \operatorname{cosec} C \qquad a^2 + b^2 = c^2 + 4\Delta \cot C$$

(You may assume the cosine formula for a triangle.) Find the remaining sides of a triangle in which one side is 5 m, the opposite angle is 45°, and the area is 15 m². [LU]

3 If in any triangle ABC $\sin\theta = \dfrac{2\sqrt{bc}}{b+c}\cos\tfrac{1}{2}A$, prove that $(b+c)\cos\theta = a$.

For the case $b = 123$, $c = 41{\cdot}2$, $A = 40^\circ\ 50'$, find the value of $\sin\theta$ and hence the value of a.

4 Two triangles ABC, PBC stand on the same side of the base BC of length 10 cm. If the angles ABC, PBC, ACB, PCB, are respectively 60°, 45°, 30°, 60°, calculate the distance AP.

5 In a triangle ABC, $b = 203$ mm, $c = 158$ mm, $B = 94^\circ\ 12'$. Calculate the values of a, A, C and find the area of the triangle.

6 If p is the altitude from A to BC of a triangle ABC, prove that

$$(b+c)^2 = a^2 + 2ap\cot\tfrac{1}{2}A$$
$$(b-c)^2 = a^2 - 2ap\tan\tfrac{1}{2}A$$

Hence, or otherwise, given $a = 80$, $p = 50$, $A = 37°$, calculate b, c. [WJC]

7 The length of the sides of a cyclic quadrilateral ABCD are given by AB = 3, BC = 4, CD = 5, DA = 6, Calculate the angles B and D and the length of the diagonal AC. [JMB, part]

8 A column is h m high. A man is standing at a horizontal distance a m from the base of the column, his eye level being at b m. He notices that a statue on top of the column subtends an angle θ at his eye. Find the height of the statue.

9 An observer O standing on top of a hill finds that the angles of depression to two points A and B on the same horizontal level are α and β respectively. If he is 300 m vertically above AB and the angle AOB is γ, find the distance AB in terms of α, β, γ.

10 In any triangle ABC assuming that $\dfrac{a}{\sin A} = \dfrac{b}{\sin B} = \dfrac{c}{\sin C}$, prove that $\dfrac{a+b-c}{a+b+c} = \tan\tfrac{1}{2}A\tan\tfrac{1}{2}B$. Calculate the value of c for the triangle in which $a + b = 185$ mm, $A = 72°\,14'$, $B = 45°\,42'$.

11 In a triangle ABC the angle A is $60°$, and the side a is the arithmetic mean of the sides b and c. Prove that the triangle is equilateral.

12 Prove that in any triangle ABC

(i) $a = b\cos C + c\cos B$

(ii) $\sin\tfrac{1}{2}(A-B) = \dfrac{a-b}{c}\cos\tfrac{1}{2}C$

(iii) $\sin^2 B + \sin^2 C = 1 + \cos(B-C)\cos A$. [JMB]

13 In a triangle ABC, the lengths of the sides BC, CA, AB, are in the ratio 8:5:9. A point P is taken on BC such that BP : PC = 1 : 3. Prove that angle ACP = 2 × angle APC.

14 In the triangle ABC, the sides AB, AC are equal and contain an angle 2θ. The circumscribed circle of the triangle has radius R. Show that the sum of the lengths of the perpendiculars from A, B, C to the opposite sides of the triangle is

$$2R(1 + 4\sin\theta - \sin^2\theta - 4\sin^3\theta)$$ [JMB, part]

15 The side BC of the triangle ABC is divided internally at a point A_1 such that $BA_1 = s - b$, where $s = \tfrac{1}{2}(a+b+c)$. Show that $A_1C = s - c$.

Points B_1 and C_1 are taken on CA and AB such that $CB_1 = s - c$ and $AC_1 = s - a$ respectively. Show that $\angle B_1A_1C_1 = \tfrac{1}{2}(B + C)$ and prove that if Δ_1 is the area of the triangle $A_1B_1C_1$

$$\Delta_1 = 2(s-b)(s-c)\sin\tfrac{1}{2}B\sin\tfrac{1}{2}C\cos\tfrac{1}{2}A$$

Deduce that if Δ is the area of the triangle ABC

$$s^2{\Delta_1}^3 = \Delta^4\sin A\sin B\sin C\sin\tfrac{1}{2}A\sin\tfrac{1}{2}B\sin\tfrac{1}{2}C$$ [JMB]

16 If the area of the triangle ABC is Δ and $2s = a + b + c$, prove that

$$\Delta = \frac{s^2}{\cot\frac{1}{2}A + \cot\frac{1}{2}B + \cot\frac{1}{2}C}$$

$$\left[\text{Hint: assume}\quad \tan\tfrac{1}{2}A = \sqrt{\left[\frac{(s-b)(s-c)}{s(s-a)}\right]}\right]$$

17 The medians AD, BE, CF of the triangle ABC meet at G. Prove that

$$\cot \text{AGF} + \cot \text{BGD} + \cot \text{CGE} = \cot A + \cot B + \cot C$$

18 A sloping plane bed of rock emerges at ground level in a horizontal line AB. At a point C on the same level as AB and such that BC = 1200 m and the angle ABC is 60° a vertical shaft CD of depth 300 m is sunk reaching the rock at D. Calculate the inclination of the plane of the rock to the horizontal. Another vertical shaft is sunk at M, the mid-point of BC, and reaches the rock at N. Given that AB is 1000 m calculate the inclination of AN to the horizontal. (Give answers to the nearest degree.) [JMB]

19 A plane is inclined at angle α to the horizontal and a line PQ on the plane makes an acute angle β with PR which is a line of greatest slope on the plane. Show that the inclination of PQ to the horizontal is given by $\sin\theta = \sin\alpha\cos\beta$. Show that the angle ϕ between the vertical plane through PQ and the vertical plane through PR is given by $\cos\phi\cos\theta = \cos\alpha\cos\beta$. [AEB]

20 D and E are points dividing the side BC of triangle ABC internally and externally in the ratio $p{:}q$. If $\angle \text{ADC} = \theta$ and $\angle \text{AEC} = \phi$ prove that $(p+q)\cot\theta = q\cot B - p\cot C$, and write down the corresponding result for $\cot\phi$. Hence prove that $\angle \text{DAE} = 90°$ only if $p{:}q = c{:}b$. [SUJB]

21 If A, B, C are the angles of a triangle and the products

$$\cos 2A \cos 2B \cos 2C \quad \text{and} \quad \sin 2A \sin 2B \sin 2C$$

have given values p, q respectively, prove that

$$p - q\cot 2A = \cos^2 2A$$

and deduce that $\tan 2A$, $\tan 2B$, $\tan 2C$ are the roots of the equation

$$(pt - q)(t^2 + 1) = t$$

Show that if $p = \frac{1}{2}$ and $q = 0$, then the angles of the triangle are in the ratios 1:3:4. [LU]

22 In the triangle ABC, AC = 5, BC = 7 and angle CAB is 60°. Prove that AB = 8. If D is a point on the circumcircle of the triangle on the side of BC away from A, and the angle CBD is 30°, show that $\sin \text{ABD} = \frac{13}{14}$ and find AD. [JMB]

23 Without using tables show that there is a triangle ABC whose angles are such that $\tan\frac{1}{2}A = \frac{3}{4}$, $\tan\frac{1}{2}B = \frac{9}{13}$, $\tan\frac{1}{2}C = \frac{1}{3}$.

24 Establish the sine rule for a triangle. A vertical tower AB stands on top of a hill which may be assumed to be a plane inclined at 8° to the horizontal. BCD is the line of greatest slope of the hill through B, the foot of the tower. The angles of elevation above the horizontal of A from C and D are 29° and 20° respectively and the length CD is 125 m. Find the height of the tower. [LU]

25 R is the radius of the circumcircle of a triangle ABC of area Δ. Show that $\Delta = \frac{1}{2}bc\sin A$ and that $R = abc/4\Delta$. D is the mid-point of BC and P and Q are the feet of the perpendiculars from D to AC and AB respectively. Find the area of the triangle DPQ in terms of a, Δ and R. Show that the area of the triangle APQ is

$$\frac{\Delta}{16b^2c^2}(3b^2 + c^2 - a^2)(3c^2 + b^2 - a^2)$$ [LU]

26 A point P is due south of a wireless mast and the angle of elevation of the top of the mast from P is α. A second point Q is due east of P and the angle of elevation of the top of the mast from Q is β. The horizontal distance between P and Q is c and the vertical height of P above Q is h. If x is the height of the top of the mast above P, prove that

$$x^2(\cot^2\beta - \cot^2\alpha) + 2hx\cot^2\beta + \mathrm{h}^2\cot^2\beta - c^2 = 0$$

Calculate the height to the nearest metre when $\alpha = 16°$, $\beta = 16°$, $h = 35$ m, $c = 340$ m. [JMB]

27 The base of a pyramid of vertex V is a square ABCD of side $2a$. Each of the slant edges is of length $a\sqrt{3}$. Find
(i) the angle between a slant face and the base
(ii) the perpendicular distance of D from the edge VA
(iii) the angle between two adjacent slant faces. [JMB]

28 If $\alpha + \beta + \gamma = 90°$ prove that

$$1 - \sin^2\alpha - \sin^2\beta - \sin^2\gamma - 2\sin\alpha\sin\beta\sin\gamma = 0$$

A convex quadrilateral ABCD is inscribed in a circle of which DA is a diameter. If $a = \mathrm{AB}$, $b = \mathrm{BC}$, $c = \mathrm{CA}$, $d = \mathrm{DA}$, prove that $d^3 - (a^2 + b^2 + c^2)d - 2abc = 0$. [JMB]

29 Points P, Q and R are taken on the sides of a triangle ABC (P on BC, Q on CA, R on AB) and lines are drawn through these points at right angles to the sides on which they lie. Prove that they will be concurrent if

$$\mathrm{BP}^2 - \mathrm{PC}^2 + \mathrm{CQ}^2 - \mathrm{QA}^2 + \mathrm{AR}^2 - \mathrm{RB}^2 = 0$$

Hence or otherwise, prove that the altitudes of any triangle are concurrent.

If the altitudes of the triangle ABC are AD, BE, and CF and if A′, B′, and C′ are the mid-points of BC, CA and AB respectively, prove that A′D.BC + B′E.CA + C′F.AB = 0, where the products are counted *plus* or *minus* according to the directions indicated by the order of the letters. [SUJB]

30 From a mountain peak P, 2000 m above sea level, observations are taken of two further peaks, A and B. The horizontal distance of A from P is 3 km, its angle of elevation from P is 10°, and its bearing from P is N 20°E. The horizontal distance of B from P is 1 km, its angle of depression from P is 15°, and its bearing from P is N 80° E. Find
(i) the horizontal distance of A from B
(ii) the heights of A and B above sea level
(iii) the angle of elevation of A from B.

31 A tent covers a rectangular piece of ground of length l and breadth b. Each of the long sides of the tent is a trapezium inclined at α to the ground

and each of its ends is an isosceles triangle inclined at β to the ground. Prove that
(i) the height of the tent is $\frac{1}{2}b \tan \alpha$
(ii) the length of the top edge is $l - b \tan \alpha \cot \beta$
Find the total area of the tent. [JMB]

32 The lengths of the sides AB, BC, CD, DA, of a quadrilateral ABCD are a, b, c, d respectively. The lengths of the diagonals AC, BD are x, y respectively. The sides AB, DC produced meet at an angle θ. Prove that $b^2 + d^2 = x^2 + y^2 - 2ac \cos \theta$.

Further if P, Q, R, S, are the mid-points of the sides in the above order and if $p = \text{PR}$, $q = \text{QS}$, prove that $p^2 - q^2 = bd \cos \phi - ac \cos \theta$, where ϕ is the angle of intersection of the sides BC, AD produced.

33 From a point on the side of a football field at a distance $2h$ from the corner flag the angle between the directions to the goal posts at the same end as the flag is α. Denoting the angle between the directions to the nearest post and the flag by θ, show that

$$\tan^2 \theta + \frac{d}{2h} \tan \theta + 1 - \frac{d}{2h} \cot \alpha = 0$$

where d is the distance between the posts. If, when the distance $2h$ is changed to h, the angle α changes to 2α, determine d in terms of h and $\tan \alpha$. [JMB]

34 A long square peg with cross-section of side a and with flat ends perpendicular to its axis is placed in a round cylindrical hole of diameter $d(> a\sqrt{2})$ and uniform depth h. The peg rests with its axis making an angle θ with the axis of the hole. The edge CD of the end face ABCD of the peg rests on the bottom of the hole, and the corners A and B are in contact with the wall of the hole. Also the long edges through C and D rest against the upper rim of the hole.

Show that the depth of the hole is given by the equation

$$h \tan \theta = \sqrt{(d^2 - a^2)} - a \cos \theta \qquad \text{[JMB]}$$

35 Two vertical cliff faces are at right angles and intersect in the line AOB, with B above A. A thin plane stratum of rock passes through O and intersects the cliff face in lines OL and OM respectively, each of which makes an acute angle θ with OB. Prove that
(i) the angle between the two lines of intersection is $\cos^{-1}(\cos^2 \theta)$
(ii) the angle of inclination of the stratum to the horizontal is $\tan^{-1}(\sqrt{2} \cot \theta)$. [JMB]

36 P is a point so that PA is 6 m and PA is perpendicular to the horizontal plane of a triangle ABC. When BC is 5 m and the angles of elevation of P from B and C are $\tan^{-1} 1$ and $\tan^{-1} \frac{6}{7}$ respectively, find the angle between the planes PBC and ABC. [AEB]

37 One face of a cube is ABCD, and the opposite face is A′B′C′D′, the lettering being such that AA′, BB′, CC′, DD′ are edges of the cube. Find (in the form $\sin^{-1} x$, $\cos^{-1} x$ or $\tan^{-1} x$) the angle between (i) the diagonal AC′ and the face ABCD, (ii) the diagonal AC′ and the plane BDD′B′. [O]

38 In the triangle ABC, the sides AB and AC are of lengths $2k$ and $3k$ respectively. The point P on the side AB at a distance x from A is joined to the point Q on the side AC at a distance y from A. If x and y vary in such a way that the area of the triangle APQ is always equal to half the area of the triangle ABC, show that $y = 3k^2/x$. State the range of possible values of x. Find, in terms of k, the minimum value of $x + y$. [JMB]

39 The points A, B, C lie on a circle centre O. Given that AB = 11 m, BC = 13 m, CA = 20 m, find the angles AOB, BOC, COA to the nearest tenth of a degree and the radius of the circle to the nearest tenth of a metre. [LU]

40 Points A, B and C are equally spaced on the circumference of a horizontal circle of radius 2 cm. Points P, Q and R are at heights 6 cm, 4 cm and 2 cm vertically above A, B and C respectively. Calculate the areas of the triangles ABC and PQR. Deduce that the angle between the plane PQR and the plane ABC is $\cos^{-1}\sqrt{\frac{3}{7}}$. The line PQ produced meets the plane ABC at the point S and the line PR produced meets the plane ABC at the point T. Calculate the length of ST. [AEB]

9
The fundamental ideas of the differential calculus

9.1 Functions

The calculus has at its foundations the notion that the value of one quantity depends on the value of another quantity. For example, the area A square units of a circle is related to the length r units of its radius by the formula $A = \pi r^2$. A temperature, F, measured in degrees Fahrenheit is related to the same temperature when measured in degrees Celsius, C, by the relationship $F = 32 + \frac{9}{5}C$. The volume, v, of a given mass of gas is related to the pressure, p, of the gas and under certain conditions the relationship may take the form $v = 60/p$. Readers will probably be able to think of other examples to add to the three given.

These examples lead us to the idea of a function. When the value of one quantity, say y, depends on the value of another quantity, say x, and to every possible value of x there corresponds *one* and *only one* value of y we say that y is a function of x. In the above examples, A is a function of r, F is a function of C, v is a function of p.

In the above examples, the quantities have been linked by a mathematical formula. This will often be the case but it is not essential. At a particular location, the wind speed may be measured using an anemometer, and on a given day the wind speed will vary with the time of day. It will almost certainly be impossible to find a mathematical formula connecting the two quantities, but since at any particular time there will be just one value for the wind speed, we can still say that wind speed is a function of time.

Unless we are dealing with a particular example, we shall generally use the symbols y and x to denote the two related quantities. The statement 'y is a function of x' is expressed mathematically by writing

$$y = \mathrm{f}(x) \tag{9.1}$$

The 'f' is used to indicate dependence on the bracketed quantity (x). Other letters can be used to distinguish between different functions, some commonly used symbols being $\mathrm{F}(x)$, $\phi(x)$, $\mathrm{g}(x)$.

As the value of x varies, the value of y varies in a way determined by the particular function. The value of y depends on the value of x. For this reason, y is called the dependent variable; x is called the independent variable. The value of y when x has the value 2, say, is denoted by $\mathrm{f}(2)$, and in general when x has the value a the value of y or $\mathrm{f}(x)$ is denoted by $\mathrm{f}(a)$. If $\mathrm{f}(x)$ is expressible as some formula involving x, these values can be calculated by substitution into this formula.

Example 1 If $f(x) = x^2 - 3x$, evaluate $f(2)$, $f(3)$, $f(-5)$, $f(a)$, $f(0)$.

$$f(2) = 2^2 - 3 \times 2 = -2 \qquad f(3) = 3^2 - 3 \times 3 = 0^*$$
$$f(-5) = (-5)^2 - 3 \times (-5) = 40 \qquad f(a) = a^2 - 3a$$
$$f(0) = 0 - 0 = 0^*$$

* Note that the values of $f(x)$ are not necessarily all different; the important point is that any one value of x gives *only one* value of $f(x)$.

The possible values for x constitute the *domain* of the function. The corresponding values for y constitute the *co-domain* of the function.

If the variables involved relate to physical quantities, the domain and co-domain will often be determined by physical considerations. For example $A = \pi r^2$, the domain is the set of positive real numbers: r can be any positive real number and A is also a positive real number. For $F = 32 + \frac{9}{5}C$, since the lowest possible temperature (the so-called absolute zero) is $-273\,°C$, the domain consists of the values of C which are not less than -273. We can write $F = g(C)$ say, where $g(C) = 32 + \frac{9}{5}C$. Since $g(-273) = -459{\cdot}4$, the co-domain consists of values of F which are not less than $-459{\cdot}4$.

If we think of the function $y = x^2$ without attaching any physical significance to the relationship, but merely regarding it as a rule by which we can obtain a value y corresponding to any value x, then the domain consists of all the real numbers. However, the co-domain is restricted to positive real numbers. For the function $y = \dfrac{1}{x-3}$, we see that x can take any value except the value 3. There is no way of finding the value of y corresponding to a value of $x = 3$. A function $f(x)$ is said to be defined for a certain value a, of x, if a definite value of $f(a)$ corresponds to this value of x.

Example 2 For what values of x are the following functions defined? What are their domains and co-domains? (i) $f(x) = 2x - 5$, (ii) $f(x) = \dfrac{1}{x-2}$.

(i) $y = f(x) = 2x - 5$ is defined for every value of x since to any value of x we obtain one value for y. The domain is all the real numbers. The co-domain is all the real numbers.

(ii) $y = \dfrac{1}{x-2}$ is defined for all values of x except $x = 2$ since, if we try to evaluate y when $x = 2$, we obtain $\dfrac{1}{0}$ which is meaningless since division by zero is not a valid operation. The domain is the real numbers except $x = 2$, the co-domain is the real numbers except $y = 0$ since no value of x makes y zero.

The function $v = \dfrac{60}{p}$ can be thought of as expressing the volume of a certain mass of gas as a function of the pressure of the gas. From this we can easily obtain the inverse relationship $p = \dfrac{60}{v}$ and we see that, in this case, this relationship expresses p as a function of v. Similarly from

$$F = 32 + \tfrac{9}{5}C$$

we obtain the inverse relationship

$$C = \tfrac{5}{9}(F - 32)$$

which expresses C as a function of F.

However, for the function $y = x^2$, the inverse relationship gives

$$x = \pm\sqrt{y}$$

and corresponding to a given value of y (this must be positive) there does not correspond a unique value for x.

Thus we have to think of the inverse relationship in terms of the *two* functions

$$x = +\sqrt{y} \qquad (y \geqslant 0)$$
$$x = -\sqrt{y} \qquad (y \geqslant 0)$$

Of course, for the first example we considered

$$A = \pi r^2$$

$$r = \sqrt{\frac{A}{\pi}}$$

since r must be positive on physical considerations.

Example 3 For the following functions find the inverse relationship. Is this relationship a function? (i) $y = 2x + 5$, (ii) $y = \dfrac{1}{x - 2}$, (iii) $T = \sqrt{\dfrac{l}{2}}$.

(i) If $y = 2x + 5$, then $x = \frac{1}{2}(y - 5)$ and this defines a function. Its domain consists of all real values for y and its co-domain consists of all the real numbers.

(ii) If $y = \dfrac{1}{x - 2}$, then $x = 2 + \dfrac{1}{y}$. This defines x as a function of y. The domain consists of all the real numbers except $y = 0$; the co-domain of all the real numbers except $x = 2$.

(iii) If $T = \sqrt{(l/2)}$, then $l = 2T^2$ and this defines l as a function of T. The domain consists of the positive real numbers for T, the co-domain of the positive real numbers for l.

If the functional relationship between y and x is expressed by a formula giving y in terms of x, we say that y is an *explicit* function of x. The functions just considered, $y = 2x + 5$, $y = 1/(x - 2)$, $T = \sqrt{(l/2)}$ are all explicit functions. It may be that the relationship between the quantities y and x is expressed by means of an equation; for example, $3y + 4x - 5 = 0$ or $x^3 + y^3 = 27$. In this situation, if for each possible value of x there is a *unique* value of y, we say that y is an *implicit* function of x. It is, in fact, possible to express y as an explicit function in both these cases. Thus, for the first example, $y = \frac{1}{3}(5 - 4x)$; and for the second example, $y = \sqrt[3]{(27 - x^3)}$. This will not always be possible. If $y^5 + xy + x^3 = 3$ is the equation which defines y as an implicit function of x, then to express y explicitly in terms of x we need to solve the quintic equation $y^5 + xy + x^3 - 3 = 0$, and this is not possible for a general value of x.

Example 4 Express y as an explicit function of x if (i) $xy = 3$, (ii) $2x - 3y + 2 = 0$, (iii) $x^2 + y^2 = 4$.

(i) $xy = 3$ therefore $y = 3/x$

(ii) $2x - 3y + 2 = 0$ therefore $y = \frac{1}{3}(2x + 2)$

(iii) $x^2 + y^2 = 4$ therefore $y^2 = 4 - x^2$

$$y = \pm\sqrt{(4 - x^2)}$$

and this relationship does not give a unique value for y. This is yet another case where we have to think in terms of two functions.

In the first,

$$y = u(x) = \sqrt{(4 - x^2)}$$

The domain consists of values of x in the range $-2 \leqslant x \leqslant 2$; the co-domain consists of values of y in the range $0 \leqslant y \leqslant 2$.

In the second,

$$y = v(x) = -\sqrt{(4 - x^2)}$$

The domain consists of values of x in the range $-2 \leqslant x \leqslant 2$; the co-domain consists of values of y in the range $-2 \leqslant y \leqslant 0$.

Exercises 9a

1 A rectangular enclosure is made using 100 m of fencing. The fencing is used on three sides only, the fourth side consisting of a stone wall. If the length of wall used for the enclosure is x m, find the area of the enclosure A m^2 as a function of x.

2 A cylindrical can, open at one end, is constructed so that its combined length and girth is 20 cm. If the height of the can is h cm, express the volume of the can as a function of h. If the radius of the can is r cm, express the surface area of the can as a function of r.

3 In triangle ABC, AB = AC = 10 cm. If $\angle$ABC = $x°$, express the height of the altitude from A to BC as a function of x.

4 A piece of wire of length 20 cm is bent into the shape of an isosceles triangle. If the length of one of the equal sides of the triangle is x cm, express the height of the triangle h cm as a function of x. What is the domain for this function? What is the co-domain?

5 If the piece of wire in question 4 is bent into the form of a sector of a circle of radius r cm and the bounding radii, find the angle θ of this sector, measured in radians, as a function of r.

6 $f(x) = x^3 - 3x$. Evaluate f(1), f(2), f(−1). What is the domain? What is the co-domain?

7 $\phi(x) = x^2 - 5x + 6$. Evaluate $\phi(0)$, $\phi(1)$. For what values of x is $\phi(x) = 0$?

8 $G(y) = \dfrac{1}{y^2 + 4}$. Evaluate G(0). What is (i) the domain, (ii) the co-domain?

9 $F(\theta) = \cos\theta - \sin\theta$. Evaluate F(0), F($\pi/2$). For what values of θ is $F(\theta) = 0$? If the domain is the values of θ for which $-2\pi \leqslant \theta \leqslant 2\pi$, find the co-domain.

10 For what values of x is the function $\phi(x) = 2x/(x - 1)(x - 2)$ defined?

11 For what values of x is the function $(x - 3)(x - 7)$ negative? For what values of x is the function $\sqrt{[(x - 3)(x - 7)]}$ defined?

12 Find x as a function of y if (i) $y = \dfrac{1}{2x-1}$, (ii) $y = 7x + 3$, (iii) $y = \sin(x + \frac{1}{4}\pi)$, (iv) $y = x^3$.

13 Define y as an explicit function of x (if possible) when (i) $xy + 4y = x^3$, (ii) $x + y + y^2 = x^2$, (iii) $x^5 + y^5 + xy = 3$.

14 The relationship between y and x is defined implicitly by (i) $xy + y = 2$, (ii) $xy^2 + y^4 + 1 = 0$. Evaluate y when $x = 1$ and when $x = 2$ (if possible) in each case.

15 The relationship between y and x is defined implicitly by $x^2 + y^2 = 25$. Find y when (i) $x = 3$, (ii) $x = 4$.

9.2 Graphical representation of a function

It is very helpful to represent the variation of a function by drawing its graph. The graph of the function $y = x^2$, with which readers are probably familiar, is shown in *Figure 9.1*.

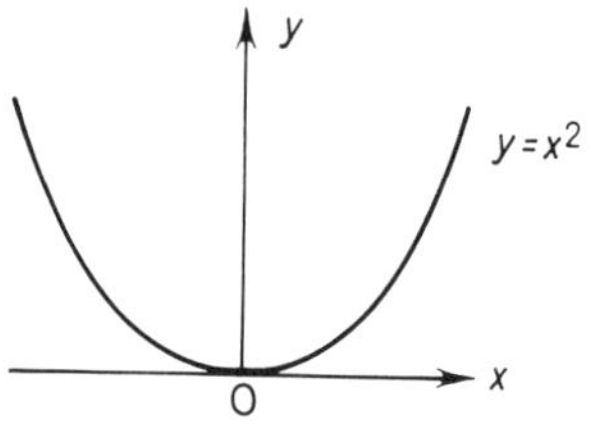

Figure 9.1

The graph of the function $y = x^2$ is a smooth continuous curve over the range of values of x for which the function is defined (in this case for all values of x). We say that the function $y = x^2$ is a continuous function of x.

Figure 9.2 shows the graph of the function $y = 1/x$. There is a break in the curve when $x = 0$. Indeed, the function is not defined when $x = 0$. The graph is discontinuous at this point, and we apply this same description, discontinuous, to the function at this point. The function is, of course, continuous for any range of values of x which does not include zero.

The word 'smooth' was applied to the curve in *Figure 9.1*. However,

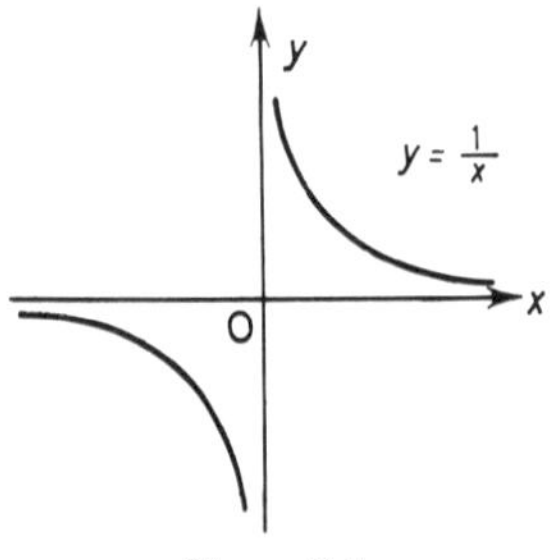

Figure 9.2

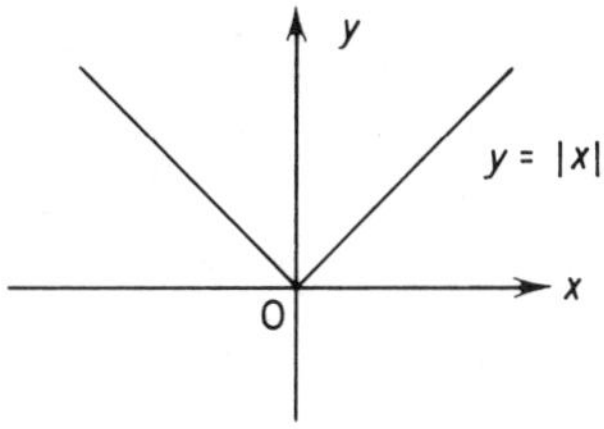

Figure 9.3

continuity does not necessarily imply a smooth curve. Consider the following function of x defined as 'y is the positive number having the same magnitude as x' (i.e. $y = |x|$, the modulus of x; see Section 1.4). This function may be expressed in the form $y = x$ if $x \geqslant 0$, $y = -x$ if $x \leqslant 0$. Its graph is shown in *Figure 9.3*. The graph is not smooth near the point on it where $x = 0$, but it is continuous.

For the function $y = x^2$ we have seen that there are two possible inverse functions. In the first, $x = \sqrt{y}$. In the second, $x = -\sqrt{y}$. The graphs of both these functions are shown in *Figure 9.4*. The first is shown with a solid line, the second with a broken line. It is interesting to compare the complete *Figure 9.4* with *Figure 9.1*.

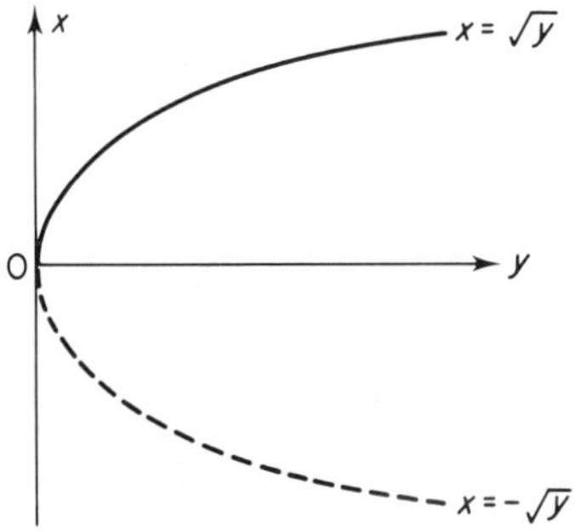

Figure 9.4

For the implicit relationship $x^2 + y^2 = 4$, we have seen that there are two possible explicit functions:

$$y = \mathrm{u}(x) = \sqrt{(4 - x^2)}$$
$$y = \mathrm{v}(x) = -\sqrt{(4 - x^2)}$$

The graphs of these are shown in *Figure 9.5*. One, u(x), gives the upper half of the circle centre O and radius 2; the other, v(x), the lower half of the same circle.

9.3 The rate of change of a function

If y is a function of x, as x changes y will in general change. We relate the change in y to the corresponding change in x by defining the average rate

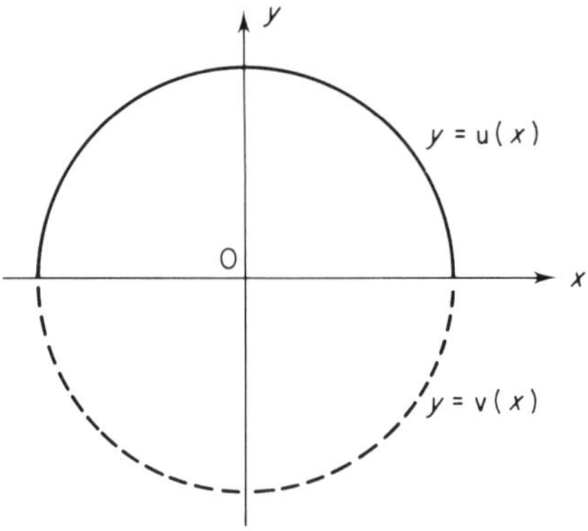

Figure 9.5

of change of the function to be the change in the function divided by the corresponding change in x. If x_1 and x_2 are two values of x, and the corresponding values of y are y_1 and y_2, then the average rate of change of the function as x changes from x_1 to x_2 is

$$\frac{y_2 - y_1}{x_2 - x_1} \tag{9.2}$$

Example 1 Find an expression for the average rate of change of the functions (i) $y = 2x + 5$, (ii) $y = x^2$ in the interval x_1 to x_2.

(i) By (9.2), the average rate of change for $y = 2x + 5$ is

$$\frac{(2x_2 + 5) - (2x_1 + 5)}{x_2 - x_1} = \frac{2(x_2 - x_1)}{x_2 - x_1} = 2$$

We notice that this is the same for each interval x_1 to x_2.

(ii) By (9.2) the average rate of change for $y = x^2$ is

$$\frac{x_2^2 - x_1^2}{x_2 - x_1} = \frac{(x_2 - x_1)(x_2 + x_1)}{x_2 - x_1} = x_1 + x_2$$

which is different for different intervals.

If we represent the function graphically, the average rate of change of the function in the interval x_1 to x_2 may be interpreted geometrically as being the gradient of the chord joining the points on the graph with abscissae x_1 and x_2. For the function $y = 2x + 5$, the graph is a straight line (*Figure 9.6a*) and the gradient of any chord is always 2. $\dfrac{\text{QN}}{\text{PN}} = \dfrac{\text{Q'N'}}{\text{P'N'}} = 2.$ But for the graph of $y = x^2$ (*Figure 9.6b*), the gradient of the chord PQ is different from the gradient of P′Q′, etc.

A practical application of this idea arises in connection with space–time graphs. Suppose a body moves so that the distance s travelled after time t is $s = \mathrm{f}(t)$. Then the average rate of change of s as t changes from t_1 to t_2, $(s_2 - s_1)/(t_2 - t_1)$, is just the average speed of the body in the interval t_1 to t_2 and is the gradient of the appropriate chord on the space–time graph.

Expression (9.2) expresses algebraically the gradient of the chord joining the

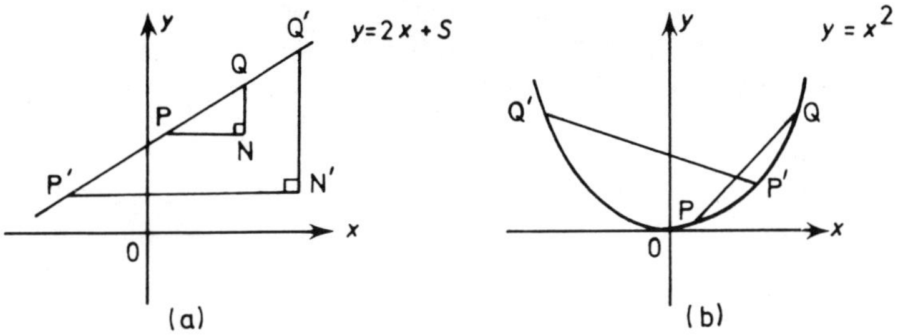

Figure 9.6

points with abscissae x_1 and x_2 on the graph of $y = f(x)$. Can the gradient of the tangent to the curve be given a similar algebraic interpretation? Geometrically, we feel no difficulty in drawing the tangent to a curve at a particular point, but in order to interpret this process algebraically we need to consider it in some detail.

Suppose we consider the definite problem of finding the gradient of the tangent at the point with abscissa 1 on the curve $y = x^2$. The gradient of the chord joining the points on this curve with abscissae 1 and x_2 is, from the result of Example 1(ii), equal to $1 + x_2$. Consider the chords PQ, PQ_1, PQ_2, PQ_3, where P has abscissa 1 and Q, Q_1, Q_2 and Q_3 have abscissae 1·5, 1·1, 1·01, 1·000 01. These chords (produced) are approaching nearer and nearer to being the tangent to the curve at P (*Figure 9.7* shows the curve in the region of P). Indeed, we can imagine that, as we try to draw the tangent at P, we rotate our ruler through the positions of these chords before finally drawing the tangent at P.

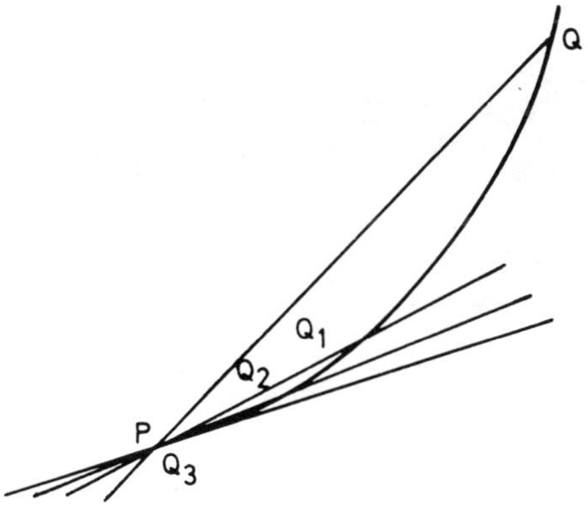

Figure 9.7

Geometrically, as Q approaches P, so PQ approaches the position of the tangent at P. Algebraically, the gradient of the chord is $1 + x_2$ (x_2 the abscissa of Q) and as x_2 gets nearer and nearer to the value 1, this expression will approach nearer and nearer to the gradient of the tangent at P. The gradients of the chords PQ, PQ_1, PQ_2, PQ_3 are 2·5, 2·1, 2·01 and 2·000 01. These are approaching the value 2, which we say is the gradient of the tangent at P.

Quite generally, the gradient of the chord joining the points R and S with abscissae x_1 and x_2 on the curve $y = x^2$, is $x_1 + x_2$. If we allow the abscissa of S to approach the abscissa of R, the chord RS approaches nearer and nearer to being the tangent at R. Algebraically, the expression $x_1 + x_2$ takes values closer and closer to the value $2x_1$ as x_2 takes values closer and closer to the value x_1, and so the gradient of the tangent to the curve $y = x^2$ at the point with abscissa x_1 is $2x_1$.

The same procedure enables us to find the gradient of the tangent at the point with abscissa x_1 on the graph of the function $y = f(x)$. The gradient of the chord joining the two points with abscissae x_1 and x_2 is $\dfrac{y_2 - y_1}{x_2 - x_1}$. This expression will depend on x_1 and x_2 and on the particular function $f(x)$ being considered. As x_2 takes values which approach x_1, so the chord approaches the tangent at the point with abscissa x_1. The gradient of this tangent is then the value approached by $(y_2 - y_1)/(x_2 - x_1)$ as x_2 takes values nearer and nearer to x_1.

An immediate application of the gradient of the tangent occurs in connection with space–time graphs. Just as the gradient of a chord on such a graph represents the average speed in a certain time interval, so the gradient of the tangent represents the speed at a particular instant.

Example 2 Find the gradient of the chord joining the points with abscissae 2 and x_2 on the curves (i) $y = 1/x$ and (ii) $y = 3/x^2$. What is the gradient of the tangents at the points with abscissa 2 on these two curves?

(i) By (9.2), the gradient of the chord of $y = 1/x$ is

$$\frac{\dfrac{1}{x_2} - \dfrac{1}{2}}{x_2 - 2} = \frac{2 - x_2}{2x_2(x_2 - 2)} = \frac{-1}{2x_2}$$

As x_2 approaches the value 2, $-1/2x_2$ approaches the value $\dfrac{-1}{2^2} = -\dfrac{1}{4}$, which is the gradient of the tangent as required.

(ii) By (9.2), the chord of $y = 3/x^2$ has gradient

$$\frac{\dfrac{3}{x_2^{\,2}} - \dfrac{3}{4}}{x_2 - 2} = \frac{3(4 - x_2^{\,2})}{4x_2^{\,2}(x_2 - 2)}$$

$$= \frac{3(2 - x_2)(2 + x_2)}{4x_2^{\,2}(x_2 - 2)}$$

$$= \frac{-3(2 + x_2)}{4x_2^{\,2}}$$

As x_2 approaches the value 2, $\dfrac{-3(2 + x_2)}{4x_2^{\,2}}$ approaches the value $\dfrac{-3 \times 4}{4 \times (2)^2} = -\dfrac{3}{4}$, which is the gradient of the tangent as required.

Example 3 The distance s m of a particle (which moves along a fixed line X′OX) from the point O, after time t sec, is given by $s = t + \dfrac{1}{t + 1}$. Find its speed after 1 sec.

The average speed of the particle in the time interval 1 to t sec is given by

$$\frac{t+\dfrac{1}{t+1}-\left(1+\dfrac{1}{1+1}\right)}{t-1}=\frac{t-1}{t-1}+\frac{\dfrac{1}{t+1}-\dfrac{1}{2}}{t-1}$$

$$=1+\frac{2-t-1}{2(t+1)(t-1)}$$

$$=1-\frac{1}{2(t+1)}\ \text{m/sec}$$

As t approaches 1, this approaches the value

$$1-\frac{1}{2(1+1)}=1-\tfrac{1}{4}=\tfrac{3}{4}\,\text{m/sec}$$

The speed of the particle after 1 sec is thus $\frac{3}{4}$ m/sec.

Exercises 9b

1 Plot the graph of the function $A = \mathrm{f}(x)$ obtained in question 1 of Exercise 9a. From your graph determine the value of x which maximises A. What is the maximum value of A?

2 Plot the graphs of the functions $y = 2x^2 - 4x - 3$ and $y = x - 5$ on the same scale. Hence solve the equation $2x^2 - 5x + 2 = 0$. Verify the correctness of your solutions.

3 Plot on the same scale the graphs of the functions $y = \tan x$ and $y = 1/x$. Hence find an acute angle x so that $x \tan x = 1$ (x measured in radians).

4 Find the gradient of the chord joining the points with abscissae 3 and x_2 on the curve $y = x^2 + 5x$. What is the gradient of the tangent to the curve at the point with abscissa 3?

5 A particle moves so that the distance s m travelled after t sec is given by $s = t^2 + 5t$. Find the average speed of the particle during the 4th second and its speed after 3 sec.

6 Find the gradient of the chord joining the points with abscissae 1 and x_2 on the curve $y = (x + 1)^2$. Find the gradient of the tangent to the curve at the point with abscissa 1.

7 Find the gradient of the tangent at the point with abscissa 1 on the curve $y = 3(x + 1)^2 - 2(x + 1)$.

8 Verify by multiplication that $x_2{}^3 - x_1{}^3 = (x_2 - x_1)(x_2{}^2 + x_1x_2 + x_1{}^2)$. Hence show that the gradient of the chord joining the two points with abscissae x_1 and x_2 on the curve $y = x^3$ is $x_2{}^2 + x_1x_2 + x_1{}^2$. What is the gradient of the chord joining the two points with abscissae $x_1 = 1$ and $x_2 =$ (i) 2, (ii) 1·1, (iii) 1·01, (iv) 1·000 01? What is the gradient of the tangent to the curve at the point with abscissa (i) 1, (ii) x?

9 Show that the gradient of the chord joining the points with abscissae x_1 and x_2 on the curve $y = 1/x$ is $-1/x_1x_2$. Deduce the gradient of the tangent at the point with abscissa (i) 1, (ii) x.

10 A particle is dropped from the top of a tall tower. The distance s cm fallen after t sec is given by $s \approx 500t^2$. Find its approximate speed after 1 sec and after it has fallen 20 m.

9.4 Limits and limit notation

We have just seen that the gradient of the tangent at the point with abscissa x, on the curve $y = f(x)$ is the value approached by $(y_2 - y_1)/(x_2 - x_1)$ as x_2 takes values closer and closer to x_1. A notation has been developed which enables us to avoid rather cumbersome expressions like 'x_2 takes values closer and closer to x_1'. This we write as $x_2 \to x_1$. The value approached by $\dfrac{y_2 - y_1}{x_2 - x_1}$ as x_2 tends to x_1 is called the limiting value or the limit of this expression. This we abbreviate to

$$\lim_{x_2 \to x_1} \frac{y_2 - y_1}{x_2 - x_1} \quad \text{or} \quad \lim_{x_2 \to x_1} \frac{f(x_2) - f(x_1)}{x_2 - x_1}$$

which is read as the limit of (or the value approached by) $\dfrac{y_2 - y_1}{x_2 - x_1}$, that is $\dfrac{f(x_2) - f(x_1)}{x_2 - x_1}$, as x_2 gets nearer and nearer to x_1 and this is the gradient of the tangent at the point with abscissa x_1.

It is convenient to use the values x and $x + \delta x$ for the abscissae of the end points of a chord instead of x_1 and x_2. The symbol δx (*delta* x) represents the change or the increment in x and $\delta x \to 0$ is equivalent to $x_2 \to x_1$. (Note: δx is *one* symbol. It is *not* x multiplied by some quantity δ.) In the same way, we use y and $y + \delta y$ instead of y_1 and y_2, where δy represents the change or increment in y. The gradient of the chord joining the points with co-ordinates (x, y), $(x + \delta x, y + \delta y)$, i.e. the average rate of change of the function, is then

$$\frac{y + \delta y - y}{x + \delta x - x} = \frac{f(x + \delta x) - f(x)}{\delta x} = \frac{\delta y}{\delta x} \tag{9.4}$$

and the gradient of the tangent at the point with abscissa x is

$$\lim_{\delta x \to 0} \frac{f(x + \delta x) - f(x)}{\delta x} = \lim_{\delta x \to 0} \frac{\delta y}{\delta x} \tag{9.5}$$

The limiting value of $\delta y/\delta x$ is called the differential coefficient of y with respect to x and is denoted by the one symbol dy/dx. The process of finding this limiting value is called differentiation. The form of (9.5) will depend on the function $f(x)$ and for this reason the differential coefficient of y with respect to x is sometimes called the derived function or the derivative of $f(x)$ and may be written

$$\frac{d}{dx}[f(x)] \quad \text{or} \quad f'(x)$$

Thus we have the following equivalent expressions

$$\frac{dy}{dx} = \frac{d}{dx}[f(x)] = f'(x) = \lim_{\delta x \to 0} \frac{f(x + \delta x) - f(x)}{\delta x}$$

$$= \lim_{\delta x \to 0} \frac{\delta y}{\delta x} \qquad (9.6)$$

$f'(x)$ represents the rate of change of the function $f(x)$ at the value x or the gradient of the tangent to the curve $y = f(x)$ at the point with abscissa x. The gradient of the tangent at the point with abscissa a is $f'(a)$ and is best obtained by substituting $x = a$ in $f'(x)$. It can, of course, be calculated by evaluating

$$\lim_{\delta x \to 0} \frac{f(a + \delta x) - f(a)}{\delta x}$$

Example 1 Find the derivative of the function $y = 3x^2$ and the gradient of the tangent to the curve $y = 3x^2$ at the point with abscissa 3.

$$f'(x) = \lim_{\delta x \to 0} \frac{3(x + \delta x)^2 - 3x^2}{\delta x}$$

$$= \lim_{\delta x \to 0} \frac{6x\delta x + 3(\delta x)^2}{\delta x}$$

$$= \lim_{\delta x \to 0} (6x + 3\delta x)$$

Therefore

$$f'(x) = 6x$$

Thus the gradient of the tangent at the point with abscissa 3 is just

$$f'(3) = 6 \times 3 = 18$$

This could have been calculated by evaluating

$$\lim_{\delta x \to 0} \frac{3(3 + \delta x)^2 - 3.3^2}{\delta x}$$

However, exactly the same algebra would be used to evaluate this limit as in the general case and it is clear that once we have $f'(x) = 6x$, the rate of change of the function or the gradient of the tangent for any value of x is easily obtained without considering the limiting process again.

It can happen that the derivative of a function, although it is defined to the right or to the left of a particular point, is not defined at the point itself. The function $y = f(x) = |x|$, whose graph was shown in *Figure 9.3*, is an example of this. For values of $x < 0$, $f'(x) = -1$; for values of $x > 0$, $f'(x) = 1$; and for $x = 0$, $f'(x)$ does not exist.

The difficulty is fairly obvious geometrically. To the left of the value $x = 0$, the graph has gradient -1, and to the right its gradient is $+1$, but at the point where $x = 0$ the situation is ambiguous. In the case of the definition in terms of limits

$$\lim_{\substack{\delta x \to 0 \\ \text{from the left}}} \frac{f(\delta x) - f(0)}{\delta x} \text{ exists and equals } -1$$

This limit is often written

$$\lim_{\delta x \to 0-} \frac{f(\delta x) - f(0)}{\delta x} = -1$$

The limit from the right

$$\lim_{\delta x \to 0+} \frac{f(\delta x) - f(0)}{\delta x} = +1$$

It is not possible, however, to discuss

$$\lim_{\delta x \to 0} \frac{f(\delta x) - f(0)}{\delta x}$$

since, as δx approaches zero, we obtain one of the numbers -1 or $+1$ according as δx is just less than or just greater than zero. The derivative $f'(x)$ is in this case discontinuous at the point with abscissa $x = 0$, although as we have already mentioned $f(x)$ itself is continuous at this point. The graph of $f'(x)$ is shown in *Figure 9.8*. There is *no* value of $f'(x)$ when $x = 0$, *not* a possible two.

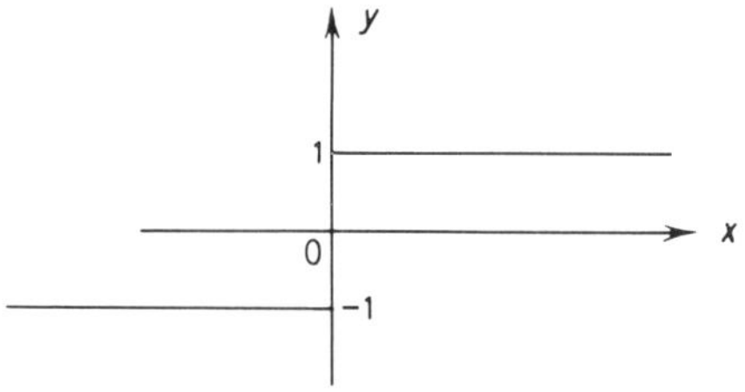

Figure 9.8

9.5 The calculation of the derivative for some common functions

The process of calculating the derivative or differential coefficient is called differentiation. More specifically, the process of calculating the derivative directly from (9.6) is called differentiation from first principles. The need for this phrase may puzzle readers. For the time being, we remark that in the next chapter we shall develop techniques which enable us to differentiate a function without having to deal directly with the expression (9.6). We shall, however, need certain standard results, three of which we shall derive now.

The derivative of ax^n, where a is a constant and n is a positive integer

If $y = ax^n$

$$y + \delta y = a(x + \delta x)^n$$

which by the binomial theorem

$$= a\left[x^n + nx^{n-1}\delta x + \frac{n(n-1)}{2}x^{n-2}(\delta x)^2 + \ldots + (\delta x)^n\right]$$

therefore, by subtraction,

$$\delta y = anx^{n-1}\,\delta x + \frac{an(n-1)}{2}x^{n-2}(\delta x)^2 + \ldots + a(\delta x)^n$$

therefore

$$\frac{\delta y}{\delta x} = anx^{n-1} + \frac{an(n-1)}{2}x^{n-2}\,\delta x + \ldots + a(\delta x)^{n-1}$$

Proceeding to the limit

$$\lim_{\delta x\to 0}\frac{\delta y}{\delta x} = anx^{n-1}$$

since all the other terms on the right contain positive powers of δx which approach zero as δx approaches zero. Thus for $y = \mathrm{f}(x) = ax^n$, where a is a constant and n a positive integer,

$$\frac{\mathrm{d}y}{\mathrm{d}x} = \mathrm{f}'(x) = anx^{n-1} \tag{9.7}$$

We have only proved this result for n a positive integer; the proof for other values of n is deferred to the next chapter.

The derivative of sin *x* (*x* in radians)

If $y = \sin x$

$$y + \delta y = \sin(x + \delta x)$$

$$\delta y = \sin(x + \delta x) - \sin x$$

$$= 2\cos(x + \tfrac{1}{2}\delta x)\sin\tfrac{1}{2}\delta x \quad \text{by (6.50)}$$

$$\frac{\delta y}{\delta x} = \frac{2\cos(x + \frac{1}{2}\delta x)\sin\frac{1}{2}\delta x}{\delta x}$$

$$= \cos(x + \tfrac{1}{2}\delta x)\frac{\sin\frac{1}{2}\delta x}{\frac{1}{2}\delta x}$$

Proceeding to the limit

$$\frac{\mathrm{d}y}{\mathrm{d}x} = \lim_{\delta x\to 0}\frac{\delta y}{\delta x} = \lim_{\delta x\to 0}\cos(x + \tfrac{1}{2}\delta x)\frac{\sin\frac{1}{2}\delta x}{\frac{1}{2}\delta x}$$

$$= \lim_{\delta x\to 0}\cos(x + \tfrac{1}{2}\delta x)\lim_{\delta x\to 0}\frac{\sin\frac{1}{2}\delta x}{\frac{1}{2}\delta x}$$

We have assumed the result that the limit of a product is the product of the two limits. (The proof of this apparently obvious result is beyond the scope of this course.)

Now

$$\lim_{\delta x \to 0} \cos(x + \tfrac{1}{2}\delta x) = \cos x$$

and

$$\lim_{\delta x \to 0} \frac{\sin \frac{1}{2}\delta x}{\frac{1}{2}\delta x} = 1 \quad \text{by (6.60)}$$

therefore

$$\frac{dy}{dx} = \cos x \times 1$$

Thus, if $y = \sin x$

$$\frac{dy}{dx} = \cos x \tag{9.8}$$

The derivative of cos x (x in radians)

If $y = \cos x$

$$\begin{aligned} \delta y &= \cos(x + \delta x) - \cos x \\ &= -2 \sin(x + \tfrac{1}{2}\delta x) \sin \tfrac{1}{2}\delta x \quad \text{by (6.50)} \end{aligned}$$

Therefore

$$\frac{dy}{dx} = \lim_{\delta x \to 0} \frac{\delta y}{\delta x} = \lim_{\delta x \to 0} -\sin(x + \tfrac{1}{2}\delta x) \frac{\sin \frac{1}{2}\delta x}{\frac{1}{2}\delta x}$$

$$\frac{dy}{dx} = -\sin x \tag{9.9}$$

Example 1 Differentiate from first principles $y = x^2 + 3x$.

$$\begin{aligned} y + \delta y &= (x + \delta x)^2 + 3(x + \delta x) \\ \delta y &= (x + \delta x)^2 + 3(x + \delta x) - x^2 - 3x \\ &= 2x\,\delta x + (\delta x)^2 + 3\delta x \end{aligned}$$

Therefore

$$\frac{\delta y}{\delta x} = 2x + 3 + \delta x$$

$$\frac{dy}{dx} = \lim_{\delta x \to 0} \frac{\delta y}{\delta x} = 2x + 3$$

We note that the differential coefficient of the sum of the two functions x^2 and $3x$ is just the sum of their separate differential coefficients, a result which we shall prove to be true quite generally, in the next chapter.

Example 2 Differentiate from first principles $\sin 3x$.

If $y = \sin 3x$

$$\delta y = \sin 3(x + \delta x) - \sin 3x$$

Therefore

$$\frac{\delta y}{\delta x} = 2\cos(3x + \tfrac{3}{2}\delta x)\,\frac{\sin\tfrac{3}{2}\delta x}{\delta x}$$

$$= 3\cos(3x + \tfrac{3}{2}\delta x)\,\frac{\sin\tfrac{3}{2}\delta x}{\tfrac{3}{2}\delta x}$$

(Note: we adjust the final part of the expression, $(\sin\tfrac{3}{2}\delta x)/(\tfrac{3}{2}\delta x)$, to be of the form $\sin\theta/\theta$ which tends to 1 as $\theta \to 0$.)

$$\frac{dy}{dx} = \lim_{\delta x \to 0} 3\cos(3x + \tfrac{3}{2}\delta x)\,\frac{\sin\tfrac{3}{2}\delta x}{\tfrac{3}{2}\delta x}$$

$$= 3\cos 3x \times 1 = 3\cos 3x$$

Exercises 9c

1 By considering $\lim_{\delta x \to 0} \dfrac{(x + \delta x)^3 - x^3}{\delta x}$, show that the derivative of the function $y = f(x) = x^3$ is $f'(x) = 3x^2$. Evaluate $f'(2)$ and check your result by calculating the approximate limit.

2 The function $y = f(x)$ is defined as follows: $y = -x^2$ for $x < 0$, $y = 2x$ for $x \geqslant 0$. Sketch the graph of $y = f(x)$. Is $f(x)$ continuous at $x = 0$?

Evaluate $\lim_{\delta x \to 0} \dfrac{f(x + \delta x) - f(x)}{\delta x}$ for $x < 0$ and for $x > 0$.

Evaluate $\lim_{\delta x \to 0} \dfrac{f(\delta x) - f(0)}{\delta x}$ and $\lim_{\delta x \to 0} \dfrac{f(\delta x) - f(0)}{\delta x}$.

Sketch the graph of $f'(x)$. Is $f'(x)$ continuous at $x = 0$?

3 The function $y = f(x)$ is defined as follows: $y = -x^2$ for $x < 0$, $y = x^3$ for $x \geqslant 0$. Sketch the graph of $y = f(x)$. Is $f(x)$ continuous at $x = 0$?

Evaluate $\lim_{\delta x \to 0} \dfrac{f(x + \delta x) - f(x)}{\delta x}$ for $x < 0$ and for $x > 0$.

Evaluate $\lim_{\delta x \to 0-} \dfrac{f(\delta x) - f(0)}{\delta x}$ and $\lim_{\delta x \to 0+} \dfrac{f(\delta x) - f(0)}{\delta x}$.

Sketch the graph of $f'(x)$. Is $f'(x)$ continuous at $x = 0$?

4 Differentiate from first principles (i) $7x^2$, (ii) $x^4 - 2x^2$, (iii) $\sin 2x$, (iv) $\cos 3x$, (v) $\sin x - x^2$.

5 Show that the derivative of $\dfrac{1}{x^2}$ is $\dfrac{-2}{x^3}$ and the derivative of $\dfrac{1}{x^3}$ is $\dfrac{-3}{x^4}$.

[The two derivatives obtained here show that the result $y = x^n, \dfrac{dy}{dx} = nx^{n-1}$, which we have proved for n a positive integer, is also true for $n = -2$ and $n = -3$.]

Exercises 9

1 An isosceles triangle is circumscribed about a circle of radius r. Express the area A of this triangle as a function of θ one of the equal angles of this triangle.

2 A right circular cone of semi-vertical angle θ is circumscribed about a sphere of radius r. Show that the volume V of the cone is given by

$$V = \tfrac{1}{3}\pi r^3 (1 + \operatorname{cosec}\theta)^3 \tan^2\theta$$

3 $f(x) = x \arctan x$. Evaluate $f(0)$, $f(1)$, $f(-1)$.

4 $\phi(x) = \log_{10} x$. Evaluate $\phi(10)$, $\phi(100)$. For what value of x is $\phi(x) = -3$?

5 Express y explicitly as a function of x if (i) $xy + 4x + y = 3$, (ii) $x^2 + 2xy + y^2 = 0$.

6 Express y explicitly as a function of x if $x^4 y + 3xy - 6x = 0$ and evaluate y when $x = 1$.

7 $y = f(x)$ is a quadratic function of x. If $f(0) = -2$, $f(1) = -2$ and $f(2) = 0$ calculate $f(x)$ explicitly and evaluate $f(3)$.

8 Find the gradient of the chord joining the points with abscissae x_1 and x_2 on the curve $y = x^3 + 3x$. Deduce an expression for the gradient of the chord joining the points with abscissae 1 and x and the gradient of the tangent at the point with abscissa 1.

9 The distance s m travelled by a particle after t sec is $s = t^2 - t$. Find the average speed of the particle in the time interval from 3 to $3\frac{1}{10}$ sec. Find the speed of the particle after 3 sec.

10 For what value of x is the tangent to the curve $y = 2x^2 - 8x + 3$ parallel to the x-axis?

11 Calculate the gradient of the curve $y = x(x-1)(x-2)$ at each of the points where it crosses the axis of x.

12 The distance s m travelled by a particle after t sec is given by $s = ut + \frac{1}{2}at^2$. Show that the speed of the particle after time t is $v = u + at$. What is the initial speed of the particle? What is the acceleration of the particle?

13 Find the value of the constant a such that the tangent at the origin to the curve $y = ax(1-x)$ makes an angle of 60° with the x-axis.

14 Differentiate from first principles (i) $x^2 + x + 1$, (ii) $1/x^4$.

15 Differentiate from first principles (i) $\sin ax$, (ii) $\cos ax$, where a is a constant.

16 Differentiate from first principles (i) $\tan x$, (ii) $\sin^2 x$. [Hint: express $\sin^2 x$ in terms of $\cos 2x$.]

17 Show that the gradient of the chord joining the points with abscissae x and $x + \delta x$ on the curve $y = \sqrt{x}$ can be expressed in the form $\dfrac{1}{\sqrt{(x+\delta x)} + \sqrt{x}}$. Deduce that the derivative of this function is $\dfrac{1}{2\sqrt{x}}$.

18 Show from first principles that if $y = \dfrac{1}{\sqrt{x}}$ then $\dfrac{dy}{dx} = \dfrac{-1}{2x^{3/2}}$. [These last two examples show that the result $y = x^n$, $\dfrac{dy}{dx} = nx^{n-1}$, proved for n a positive integer, is also true for $n = \frac{1}{2}$ and $n = -\frac{1}{2}$.]

19 If $f(x) = \log_a x$, where a is a constant, show that $f(x_1 x_2) = f(x_1) + f(x_2)$.

20 If $f(x) = a^x$, where a is a constant, show that $f(nx) = [f(x)]^n$, where n is a positive integer.

21 The height h and the base radius r of a right circular cone vary in such a way that the volume remains constant. Find the rate of change of h with respect to r at the instant when h and r are equal. [LU]

22 If $f(x) = 10^x$, show that $f(x_1 + x_2) = f(x_1)f(x_2)$.

23 The function $f(x) = x^2 + px + 1$, where p is a constant, is zero when $x = \alpha$ and $x = \beta$; and the function $g(x) = x^2 - 9x + q$, where q is a constant, is zero when $x = \alpha + 2\beta$ and $x = \beta + 2\alpha$. Find p and q, and show that $f(3) = g(3)$. [JMB]

24 If $y = h(\theta) = 7\cos\theta + 24\sin\theta$ and $0 \leqslant \theta \leqslant 2\pi$, show that $-25 \leqslant y \leqslant 25$.

25 A chord PQ of a circle of radius r subtends an angle θ at O, the centre of the circle, and another chord RS, which is parallel to PQ and lies between PQ and O, subtends an angle $\theta + \frac{1}{3}\pi$ at O. Show that the area A of that part of the circle which lies between the two chords can be expressed as a function of θ as $A = r^2[\frac{1}{6}\pi + \frac{1}{2}\cos(\theta - \frac{1}{6}\pi)]$.

10
Some techniques of differentiation

10.1 Introduction

In the previous chapter, we have shown how to calculate the derivative of a function by evaluating $\delta y/\delta x$ and then calculating the limiting value of this quantity, as δx approaches zero. This method was quite satisfactory for the simple functions considered. It could, however, be extremely laborious if we happened to be dealing with a rather complicated function. Fortunately, it is possible to prove several general theorems, which, together with a knowledge of the derivatives of a comparatively few basic functions, enable us to differentiate most functions without having to use the method of differentiation from first principles.

10.2 Differentiation of a constant

If $y = c$, a constant, whatever the value of x, then $y + \delta y = c$, and so δy is identically zero. Hence, $\delta y/\delta x$ is identically zero, and so $\mathrm{d}y/\mathrm{d}x$, its limiting value, is also zero. Therefore,

$$\text{if} \quad y = c \qquad \mathrm{d}y/\mathrm{d}x = 0 \tag{10.1}$$

The differential coefficient of a constant is zero.

Geometrically, the graph of the function $y = c$ is a straight line parallel to the x-axis and so has zero gradient.

10.3 Differentiation of the sum or difference of functions

If $y = u + v$, where u and v are both functions of x, then if x is increased to $x + \delta x$, u and v will change to $u + \delta u$ and $v + \delta v$ respectively. Hence y will change to $y + \delta y$, where

$$y + \delta y = u + \delta u + v + \delta v$$
$$\delta y = \delta u + \delta v$$

therefore

$$\frac{\delta y}{\delta x} = \frac{\delta u}{\delta x} + \frac{\delta v}{\delta x}$$

Thus for the limiting values as $\delta x \to 0$

$$\frac{dy}{dx} = \frac{du}{dx} + \frac{dv}{dx}$$

(We have assumed that the limit of a sum is the sum of the limits.)

The above process could be carried out for the difference of two functions, $y = u - v$ and the result

$$\frac{dy}{dx} = \frac{du}{dx} - \frac{dv}{dx}$$

obtained. Thus, if $y = u \pm v$,

$$\frac{dy}{dx} = \frac{du}{dx} \pm \frac{dv}{dx} \qquad (10.2)$$

The differential coefficient of the sum or difference of two functions is the sum or difference of their differential coefficients.

It is straightforward to extend this result to deal with sums and differences of three or more functions.

If $u, v, w, \ldots, s, t$ are all functions of x and

$$y = u \pm v \pm w \pm \ldots \pm s \pm t$$
$$\delta y = \delta u \pm \delta v \pm \delta w \pm \ \ldots \ \pm \delta s \pm \delta t$$

therefore

$$\frac{\delta y}{\delta x} = \frac{\delta u}{\delta x} \pm \frac{\delta v}{\delta x} \pm \frac{\delta w}{\delta x} \pm \ldots \pm \frac{\delta s}{\delta x} \pm \frac{\delta t}{\delta x}$$

and so

$$\frac{dy}{dx} = \frac{du}{dx} \pm \frac{dv}{dx} \pm \frac{dw}{dx} \pm \ldots \pm \frac{ds}{dx} \pm \frac{dt}{dx} \qquad (10.3)$$

Example 1 Find the differential coefficient of $x^6 - 7x^3 - 6x + 4$.

The differential coefficient of ax^n is nax^{n-1}. Thus (10.3) gives

$$\frac{d}{dx}(x^6 - 7x^3 - 6x + 4) = \frac{d}{dx}(x^6) - \frac{d}{dx}(7x^3) - \frac{d}{dx}(6x) + \frac{d}{dx}(4)$$
$$= 6x^5 - 21x^2 - 6$$

Example 2 Find the derivative of $\cos x - \sin x$.

Expression (10.2) and the basic results for the derivatives of $\cos x$ and $\sin x$ give for the derivative $-\sin x - \cos x$.

10.4 Differentiation of a product

If $y = uv$, where u and v are both functions of x, then if x is given the increment δx, u, v and y will receive the increments δu, δv and δy where

$$\delta y = (u + \delta u)(v + \delta v) - uv$$
$$\delta y = u\delta v + v\delta u + \delta u\delta v$$

therefore

$$\frac{\delta y}{\delta x} = u\frac{\delta v}{\delta x} + (v + \delta v)\frac{\delta u}{\delta x}$$

Now, as δx approaches zero, δv approaches zero, and $\dfrac{\delta y}{\delta x}$, $\dfrac{\delta v}{\delta x}$ and $\dfrac{\delta u}{\delta x}$ tend to their limiting values $\dfrac{dy}{dx}$, $\dfrac{dv}{dx}$ and $\dfrac{du}{dx}$ respectively. Hence

$$\frac{dy}{dx} = u\frac{dv}{dx} + v\frac{du}{dx} \tag{10.4}$$

The differential coefficient of the product of two functions is equal to the first multiplied by the differential coefficient of the second plus the second multiplied by the differential coefficient of the first.

In the particular case when one of the functions, say u, is a constant, we have $u = c$, $\dfrac{du}{dx} = 0$ and $y = cv$, where c is a constant and v a function of x. Therefore

$$\frac{dy}{dx} = c\frac{dv}{dx} \tag{10.5}$$

The differential coefficient of a constant multiple of a function is that same constant multiple of the differential coefficient of the function.

Expression (10.5) can be combined with (10.3) to give the result: if $a, b, \ldots, c$ are constants and $u, v, \ldots, w$ functions of x* and $y = au + bv + \ldots + cw$, then

$$\frac{dy}{dx} = a\frac{du}{dx} + b\frac{dv}{dx} + \ldots + c\frac{dw}{dx} \tag{10.6}$$

The differential coefficient of the sum of constant multiples of a finite number of functions is the sum of the same constant multiples of the differential coefficients of these functions.

The result (10.4) can be extended to deal with the product of three or more functions. If $y = uvw$, we may consider y as the product of the two functions (uv) and w. Thus, by (10.4),

$$\begin{aligned}\frac{dy}{dx} &= uv\frac{dw}{dx} + w\frac{d}{dx}(uv)\\ &= uv\frac{dw}{dx} + w\left(u\frac{dv}{dx} + v\frac{du}{dx}\right)\\ &= uv\frac{dw}{dx} + uw\frac{dv}{dx} + vw\frac{du}{dx}\end{aligned}$$

In general, if $y = uvw \ldots s$

$$\frac{dy}{dx} = uvw \ldots \frac{ds}{dx} + \ldots + uv \ldots s\frac{dw}{dx} + uw \ldots s\frac{dv}{dx} + vw \ldots s\frac{du}{dx} \tag{10.7}$$

* It is conventional to denote constants by the letters at the beginning of the alphabet and variables by letters near the end of the alphabet.

The differential coefficient of the product of any number of functions is obtained by differentiating each function in turn, multiplying by the remaining functions and summing the results.

Example 1 Find the differential coefficient of (i) $6\sin x$, (ii) $8\cos x + 3\sin x$.

(i) $\dfrac{d}{dx}(6\sin x) = 6\dfrac{d}{dx}(\sin x) = 6\cos x$ by (10.5)

(ii) $\dfrac{d}{dx}(8\cos x + 3\sin x) = 8\dfrac{d}{dx}(\cos x) + 3\dfrac{d}{dx}(\sin x)$

$= -8\sin x + 3\cos x$ by (10.6)

Example 2 Find $\dfrac{dy}{dx}$ if (i) $y = x^6\cos x$, (ii) $y = \sin^2 x$.

(i) $y = x^6\cos x$, therefore by (10.4)

$$\begin{aligned}\frac{dy}{dx} &= x^6\frac{d}{dx}(\cos x) + \cos x\frac{d}{dx}(x^6)\\ &= -x^6\sin x + 6x^5\cos x\\ &= x^5(6\cos x - x\sin x)\end{aligned}$$

(ii) $y = \sin^2 x = \sin x\sin x$, therefore by (10.4)

$$\begin{aligned}\frac{dy}{dx} &= \sin x\frac{d}{dx}(\sin x) + \sin x\frac{d}{dx}(\sin x)\\ &= 2\sin x\cos x\end{aligned}$$

Example 3 Find $\dfrac{dy}{dx}$ if $y = 6x^2\sin x\cos x$.

By (10.7)

$$\begin{aligned}\frac{dy}{dx} &= 6x^2\sin x\frac{d}{dx}(\cos x) + 6x^2\cos x\frac{d}{dx}(\sin x) + \sin x\cos x\frac{d}{dx}(6x^2)\\ &= -6x^2\sin^2 x + 6x^2\cos^2 x + 12x\sin x\cos x\\ &= 6x^2(\cos^2 x - \sin^2 x) + 12x\sin x\cos x\\ &= 6x^2\cos 2x + 6x\sin 2x\\ &= 6x(\sin 2x + x\cos 2x)\end{aligned}$$

Exercises 10a

Differentiate with respect to x

1 $7x^5 - 3x^4 + x^2$

2 $8x^3 - \sin x + 6$

3 $6\cos x - 8(x^2 + x)$

4 $(x^3 + 1)(x^4 + 1)$

5 $\sin x(1 - \cos x)$

6 $x(3x + 4\cos x)$

7 $(x^2 + x + 1)(2x^2 + 3x + 1)$

8 $8x^2(1 + \sin x)(1 + \cos x)$

9 $x\cos x + 3(x + 1)(x - 1)$

10 $4x^2\sin x - 3x^2\cos x$

11 $(x^2 + 1)^2$

12 $(x^2 + 1)^2(2x + 1)$

13 $(x^2 - 1)^3$

14 $3x\sin x\cos x$

15 $3x(x^2 + 1)\sin x\cos x$

10.5 Differentiation of a quotient

If $y = u/v$, where u and v are both functions of x, an increment δx in x will result in increments δu, δv and δy in u, v, and y respectively. Therefore

$$\delta y = \frac{u + \delta u}{v + \delta v} - \frac{u}{v}$$

$$= \frac{v(u + \delta u) - u(v + \delta v)}{v(v + \delta u)}$$

$$= \frac{v\delta u - u\delta v}{v(v + \delta v)}$$

$$\frac{\delta y}{\delta x} = \frac{v\dfrac{\delta u}{\delta x} - u\dfrac{\delta v}{\delta x}}{v(v + \delta v)}$$

As $\delta x \to 0$, so $\delta v \to 0$ and $\dfrac{\delta u}{\delta x} \to \dfrac{\mathrm{d}u}{\mathrm{d}x}$, $\dfrac{\delta v}{\delta x} \to \dfrac{\mathrm{d}v}{\mathrm{d}x}$, therefore

$$\frac{\mathrm{d}y}{\mathrm{d}x} = \frac{v\dfrac{\mathrm{d}u}{\mathrm{d}x} - u\dfrac{\mathrm{d}v}{\mathrm{d}x}}{v^2} \tag{10.8}$$

The differential coefficient of the quotient of two functions is equal to the denominator multiplied by the differential coefficient of the numerator minus the numerator multiplied by the differential coefficient of the denominator all divided by the square of the denominator.

In the case where $u = 1$, i.e. $y = 1/v$ is the reciprocal of the function v, $\mathrm{d}u/\mathrm{d}x = 0$ and so we have

$$\frac{\mathrm{d}}{\mathrm{d}x}\left(\frac{1}{v}\right) = \frac{-\dfrac{\mathrm{d}v}{\mathrm{d}x}}{v^2} \tag{10.9}$$

The differential coefficient of the reciprocal of a function is minus the differential coefficient of that function divided by the square of the function.

Example 1 Find the derivative of

(i) $y = \dfrac{x}{x+1}$ (ii) $y = \dfrac{\sin x}{x^2 + \cos x}$ (iii) $y = \dfrac{1}{x^2 + 4}$

(i) $y = \dfrac{x}{x+1}$

$$\frac{\mathrm{d}y}{\mathrm{d}x} = \frac{(x+1)\dfrac{\mathrm{d}}{\mathrm{d}x}(x) - x\dfrac{\mathrm{d}}{\mathrm{d}x}(x+1)}{(x+1)^2} \quad \text{by (10.8)}$$

$$= \frac{x + 1 - x}{(x+1)^2} = \frac{1}{(x+1)^2}$$

(ii) $y = \dfrac{\sin x}{x^2 + \cos x}$

$$\frac{dy}{dx} = \frac{(x^2 + \cos x)\dfrac{d}{dx}(\sin x) - \sin x \dfrac{d}{dx}(x^2 + \cos x)}{(x^2 + \cos x)^2} \quad \text{by (10.8)}$$

$$= \frac{(x^2 + \cos x)\cos x - \sin x(2x - \sin x)}{(x^2 + \cos x)^2}$$

$$= \frac{x^2 \cos x + \cos^2 x - 2x \sin x + \sin^2 x}{(x^2 + \cos x)^2}$$

$$= \frac{1 + x(x \cos x - 2 \sin x)}{(x^2 + \cos x)^2} \quad (\text{since } \sin^2 x + \cos^2 x = 1)$$

(iii) $y = \dfrac{1}{x^2 + 4}$

$$\frac{dy}{dx} = \frac{-2x}{(x^2 + 4)^2} \quad \text{by (10.9)}$$

Example 2 Find $\dfrac{du}{d\theta}$ if (i) $u = \dfrac{\theta \cos \theta}{\theta + 3}$ (ii) $u = \dfrac{\theta \cos \theta}{(\theta + 1) \sin \theta}$

(i) $u = \dfrac{\theta \cos \theta}{\theta + 3}$

$$\frac{du}{d\theta} = \frac{(\theta + 3)\dfrac{d}{d\theta}(\theta \cos \theta) - \theta \cos \theta \dfrac{d}{d\theta}(\theta + 3)}{(\theta + 3)^2} \quad \text{by (10.8)}$$

$$= \frac{(\theta + 3)(\cos \theta - \theta \sin \theta) - \theta \cos \theta(1)}{(\theta + 3)^2}$$

$$= \frac{3 \cos \theta - (\theta + 3)\theta \sin \theta}{(\theta + 3)^2}$$

(ii) $u = \dfrac{\theta \cos \theta}{(\theta + 1) \sin \theta}$

$$\frac{du}{d\theta} = \frac{(\theta + 1) \sin \theta \dfrac{d}{d\theta}(\theta \cos \theta) - \theta \cos \theta \dfrac{d}{d\theta}[(\theta + 1) \sin \theta]}{(\theta + 1)^2 \sin^2 \theta}$$

$$= \frac{(\theta + 1) \sin \theta(\cos \theta - \theta \sin \theta) - \theta \cos \theta[(\theta + 1) \cos \theta + \sin \theta]}{(\theta + 1)^2 \sin^2 \theta}$$

$$= \frac{(\theta + 1) \sin \theta \cos \theta - \theta(\theta + 1) \sin^2 \theta - \theta(\theta + 1) \cos^2 \theta - \theta \sin \theta \cos \theta}{(\theta + 1)^2 \sin^2 \theta}$$

$$= \frac{\sin \theta \cos \theta - \theta(\theta + 1)}{(\theta + 1)^2 \sin^2 \theta}$$

Example 3 Find the gradient of the tangent to the curve $y = \dfrac{x^2}{x^2+1}$ at the point with abscissa 1.

We first find the general expression for $\dfrac{dy}{dx} = f'(x)$ with $y = f(x) = \dfrac{x^2}{x^2+1}$

$$\frac{dy}{dx} = f'(x) = \frac{(x^2+1)2x - x^2.2x}{(x^2+1)^2} = \frac{2x}{(x^2+1)^2}$$

therefore
$$f'(1) = \frac{2}{2^2} = \frac{1}{2}$$

The gradient of the tangent is $\frac{1}{2}$.

Exercises 10b

Differentiate with respect to x

1 $\dfrac{x^2}{x+1}$ **2** $\dfrac{x}{x+\sin x}$

3 $\dfrac{x+\sin x}{1+\cos x}$ **4** $\dfrac{1}{\sin x}$

5 $\dfrac{1}{1+\cos x}$ **6** $\dfrac{3}{(x+1)^2}$

7 $\dfrac{x\sin x}{\cos x+\sin x}$ **8** $\dfrac{x^3+3x}{(x+1)(x+2)}$

9 $\dfrac{x^2\sin x}{(x+1)(x^2-1)}$ **10** $\dfrac{x}{(x+1)(x+2)(x+3)}$

11 $\dfrac{\sin^2 x}{\cos x(\cos x+\sin x)}$ **12** $\dfrac{x(x+1)}{(x+2)(x+3)}$

13 Find the gradient of the tangent to the curve $w = \dfrac{z^3}{z^2+1}$ at the point with abscissa 3.

14 Find the gradient of the tangent to the curve $v = \dfrac{\sin\theta}{\cos\theta+\sin\theta}$ at the point where $\theta = \pi/3$.

15 Find the gradient of the tangent to the curve $y = \dfrac{3x}{x+1}$ at the point with abscissa 2. At what points on the curve is the tangent to the curve parallel to the line $y = 3x + 7$?

10.6 Differentiation of the trigonometric functions

The results of the preceding section enable us to obtain the derivatives of the remaining four basic trigonometric functions: $\tan x$, $\cot x$, $\sec x$ and $\operatorname{cosec} x$. We already have the results that if $y = \sin x$, $\dfrac{dy}{dx} = \cos x$, and if $y = \cos x$, $\dfrac{dy}{dx} = -\sin x$. Thus, for $y = \tan x = \sin x/\cos x$

$$\frac{dy}{dx} = \frac{\cos x \cos x - \sin x(-\sin x)}{\cos^2 x} \quad \text{by (10.8)}$$

$$= \frac{\cos^2 x + \sin^2 x}{\cos^2 x} = \frac{1}{\cos^2 x}$$

Therefore

$$\frac{d}{dx}(\tan x) = \sec^2 x \tag{10.10}$$

$y = \cot x = \cos x/\sin x$ is treated in just the same way:

$$\frac{dy}{dx} = \frac{-\sin^2 x - \cos^2 x}{\sin^2 x} = \frac{-1}{\sin^2 x} \quad \text{by (10.8)}$$

Therefore

$$\frac{d}{dx}(\cot x) = -\operatorname{cosec}^2 x \tag{10.11}$$

The derivatives of sec x and cosec x are obtained by treating them as the reciprocals of cos x and sin x respectively and using (10.9). Thus

$$\frac{d}{dx}(\sec x) = \frac{d}{dx}\left(\frac{1}{\cos x}\right)$$

$$= -\frac{(-\sin x)}{\cos^2 x} = \frac{1}{\cos x} \cdot \frac{\sin x}{\cos x}$$

Therefore

$$\frac{d}{dx}(\sec x) = \sec x \tan x \tag{10.12}$$

We leave it to readers to show that

$$\frac{d}{dx}(\operatorname{cosec} x) = -\operatorname{cosec} x \cot x \tag{10.13}$$

(Note: the basic trigonometric functions which begin with 'co' have a negative sign in their derivatives.)

10.7 Second and higher derivatives

If $y = f(x)$ is a function of x, then in general the derivative $dy/dx = f'(x)$ will be some other function of x. The derivative expresses the rate of change of $f(x)$ with respect to x, as a function of x. We might well enquire what is the rate of change of this derivative with respect to x, i.e. calculate the differential coefficient of dy/dx or $f'(x)$.

An immediate application of this situation arises if we are dealing with a space–time graph, $s = f(t)$. Then ds/dt represents the speed of the body at any time t, and the rate of change of ds/dt, the acceleration of the body.

The derivative of dy/dx is called the second differential coefficient of y with respect to x and is written d^2y/dx^2. In general, d^2y/dx^2 will be a function of x and so may be differentiated to form the third differential coefficient of y with respect to x, or the third derivative of y, which is denoted by $\mathrm{d}^3y/\mathrm{d}x^3$, and so on. The nth differential coefficient of y with respect to x is denoted by $\mathrm{d}^ny/\mathrm{d}x^n$. If the notation $\mathrm{f}(x)$ is used, the first, second, third, ..., nth derivatives are denoted by $\mathrm{f}'(x)$, $\mathrm{f}''(x)$, $\mathrm{f}'''(x)$, ..., $\mathrm{f}^{(n)}(x)$.

In general, the process of calculating the nth derivative of a function is a tedious business and can only be achieved by calculating the successive derivatives in turn. It is generally worth while to consider briefly whether it is possible to simplify the function $\mathrm{d}y/\mathrm{d}x$ before calculating $\mathrm{d}^2y/\mathrm{d}x^2$ and in turn to try to simplify this before calculating $\mathrm{d}^3y/\mathrm{d}x^3$, etc.

Example 1 If $y = \sin x$, show that $\dfrac{\mathrm{d}^2y}{\mathrm{d}x^2} = -y$, $\dfrac{\mathrm{d}^4y}{\mathrm{d}x^4} = y$.

$$\begin{aligned}
y &= \sin x \\
\frac{\mathrm{d}y}{\mathrm{d}x} &= \cos x \\
\frac{\mathrm{d}^2y}{\mathrm{d}x^2} &= -\sin x = -y \quad \text{as required} \\
\frac{\mathrm{d}^3y}{\mathrm{d}x^3} &= -\cos x \\
\frac{\mathrm{d}^4y}{\mathrm{d}x^4} &= -(-\sin x) = \sin x = y \quad \text{as required}
\end{aligned}$$

Example 2 Find the second derivative of $y = \mathrm{f}(\theta) = \dfrac{\sin\theta}{1+\cos\theta}$.

$$\begin{aligned}
\frac{\mathrm{d}y}{\mathrm{d}\theta} = \mathrm{f}'(\theta) &= \frac{(1+\cos\theta)\cos\theta - \sin\theta(-\sin\theta)}{(1+\cos\theta)^2} \\
&= \frac{\cos^2\theta + \cos\theta + \sin^2\theta}{(1+\cos\theta)} \\
&= \frac{1+\cos\theta}{(1+\cos\theta)^2} = \frac{1}{1+\cos\theta}
\end{aligned}$$

Therefore

$$\frac{\mathrm{d}^2y}{\mathrm{d}\theta^2} = \mathrm{f}''(\theta) = -\frac{(-\sin\theta)}{(1+\cos\theta)^2} = \frac{\sin\theta}{(1+\cos\theta)^2} \quad \text{(see (10.9))}$$

Example 3 If $y = \tan\theta$, show that $\dfrac{\mathrm{d}^2y}{\mathrm{d}\theta^2} = 2y(1+y^2)$.

$$\begin{aligned}
y &= \tan\theta \\
\frac{\mathrm{d}y}{\mathrm{d}\theta} &= \sec^2\theta = \sec\theta . \sec\theta
\end{aligned}$$

By (10.4)

$$\frac{d^2y}{d\theta^2} = \sec\theta . \sec\theta\tan\theta + \sec\theta . \sec\theta\tan\theta$$
$$= 2\tan\theta\sec^2\theta$$
$$= 2\tan\theta(1 + \tan^2\theta)$$

Therefore

$$\frac{d^2y}{d\theta^2} = 2y(1 + y^2)$$

Exercises 10c

Differentiate with respect to x

1 $\sec x + \tan x$ **2** $\sec x \tan x$ **3** $\cos x \cot x$

4 $\sec^2 x + \tan x$ **5** $\sec^2 x + \tan^2 x$

6 $(\cos x + \sin x)(\sec x + \tan x)$

7 $\dfrac{\sec x}{\sin x + \cos x}$ **8** $\dfrac{\sec x}{1 + \sec x}$ **9** $\dfrac{1 - \tan x}{1 + \tan x}$

10 A particle moves so that the distance s m travelled after t sec is given by $s = f(t)$. Find expressions for the speed and acceleration of the particle after time t sec, and the speed and acceleration after 1 sec if (i) $s = 4t^2 - 3t$ and (ii) $s = \cos 2\pi t + \sin 2\pi t$.

11 Find $\dfrac{d^2y}{dx^2}$ if (i) $y = \cos^2 x$, (ii) $y = \dfrac{\cos x}{1 - \sin x}$.

12 If $y = \theta^n$, where n is a positive integer, show that (i) $\theta\dfrac{dy}{d\theta} = ny$, (ii) $\theta^2\dfrac{d^2y}{d\theta^2} = n(n-1)y$.

13 If $y = u^n$, where n is a positive integer, show that $\dfrac{d^ny}{du^n} = n!$.

14 If $y = \sec\theta$, show that $\dfrac{d^2y}{d\theta^2} = y(2y^2 - 1)$.

15 If $y = \sin x$, show that

(i) $\dfrac{dy}{dx} = \sin\left(x + \dfrac{\pi}{2}\right)$

(ii) $\dfrac{d^2y}{dx^2} = \sin\left(x + 2\dfrac{\pi}{2}\right)$

(iii) $\dfrac{d^3y}{dx^3} = \sin\left(x + 3\dfrac{\pi}{2}\right)$

What is the value of $\dfrac{d^ny}{dx^n}$?

10.8 Differentiation of a function of a function

The function $y = (2x + 1)^3$ is a function of $2x + 1$, which in turn is a function of x. More specifically, y is the cube of the function $2x + 1$. We say that y is a function of a function in this case. As a second example, consider the function $y = \sin x^2$; y is the sine of the function x^2, and so is another example of a function of a function.

Functions of this type frequently occur in mathematics and in this section we shall develop rules for evaluating their derivatives. In general, a function of a function is expressible as follows: y is a function of some quantity which in turn is a function of x. In symbols

$$y = \mathrm{F}(v) \quad \text{where } v = \mathrm{f}(x)$$

$$y = (2x + 1)^3 \qquad y = v^3 \quad \text{where } v = 2x + 1$$

$$y = \sin x^2 \qquad y = \sin v \quad \text{where } v = x^2$$

$$y = \cos^2 x, \text{ i.e. } (\cos x)^2 \qquad y = v^2 \quad \text{where} \quad v = \cos x$$

The general rule for differentiating a function of a function can be obtained as follows.

If $y = \mathrm{F}(v)$, where $v = \mathrm{f}(x)$, then if x is given the increment δx, v will be given the increment δv (since $v = \mathrm{f}(x)$), which in turn generates the increment δy in y (since $y = \mathrm{F}(v)$). It is convenient to write $\dfrac{\delta y}{\delta x}$ in the form $\dfrac{\delta y}{\delta v} \times \dfrac{\delta v}{\delta x}$, which is possible provided $\delta v \neq 0$. Then

$$\frac{\mathrm{d}y}{\mathrm{d}x} = \lim_{\delta x \to 0} \frac{\delta y}{\delta x} = \lim_{\delta x \to 0} \frac{\delta y}{\delta v} \times \frac{\delta v}{\delta x}$$

$$= \lim_{\delta x \to 0} \frac{\delta y}{\delta v} \times \lim_{\delta x \to 0} \frac{\delta v}{\delta x}$$

(We have again used the result that the limit of a product is the product of the limits.)

Now as $\delta x \to 0$, $\delta v \to 0$, so the first term of the above is equivalent to $\lim\limits_{\delta v \to 0} \dfrac{\delta y}{\delta v}$, i.e. $\dfrac{\mathrm{d}y}{\mathrm{d}v}$. Therefore

$$\frac{\mathrm{d}y}{\mathrm{d}x} = \frac{\mathrm{d}y}{\mathrm{d}v} \cdot \frac{\mathrm{d}v}{\mathrm{d}x} \tag{10.14}$$

We shall apply this rule to the three examples given above. Thus, for $y = (2x + 1)^3 = v^3$, where $v = 2x + 1$,

$$\frac{\mathrm{d}y}{\mathrm{d}v} = 3v^2 \qquad \frac{\mathrm{d}v}{\mathrm{d}x} = 2$$

therefore

$$\frac{\mathrm{d}y}{\mathrm{d}x} = 3v^2 \times 2 = 6(2x + 1)^2$$

We can verify this particular result by writing

$$y = (2x + 1)(2x + 1)(2x + 1)$$

and using (10.7). In this way, we obtain

$$\frac{dy}{dx} = (2x+1)^2 \frac{d}{dx}(2x+1) + (2x+1)^2 \frac{d}{dx}(2x+1) + (2x+1)^2 \frac{d}{dx}(2x+1)$$
$$= 3(2x+1)^2 \times 2 = 6(2x+1)^2 \quad \text{as before}$$

For $y = \sin x^2 = \sin v$, where $v = x^2$

$$\frac{dy}{dv} = \cos v \qquad \frac{dv}{dx} = 2x$$

therefore

$$\frac{dy}{dx} = 2x \cos x^2$$

For $y = \cos^2 x = v^2$, where $v = \cos x$

$$\frac{dy}{dv} = 2v \qquad \frac{dv}{dx} = -\sin x$$

therefore

$$\frac{dy}{dx} = -2 \sin x \cos x$$

Example 1 Find $\frac{dy}{dx}$ when (i) $y = (3x^2 - 4)^4$, (ii) $y = \left(\frac{x-1}{x+1}\right)^2$.

(i) $y = (3x^2 - 4)^4 = v^4$, where $v = 3x^2 - 4$

$$\frac{dy}{dv} = 4v^3 \qquad \frac{dv}{dx} = 6x$$

therefore

$$\frac{dy}{dx} = 4(3x^2 - 4)^3 \times 6x = 24x(3x^2 - 4)^3$$

(ii) $y = \left(\frac{x-1}{x+1}\right)^2 = v^2$, where $v = \frac{x-1}{x+1}$

$$\frac{dy}{dv} = 2v \qquad \frac{dv}{dx} = \frac{(x+1)\frac{d}{dx}(x-1) - (x-1)\frac{d}{dx}(x+1)}{(x+1)^2} \quad \text{by (10.8)}$$
$$= \frac{(x+1) - (x-1)}{(x+1)^2}$$
$$= \frac{2}{(x+1)^2}$$

therefore

$$\frac{dy}{dx} = \frac{2(x-1)}{(x+1)} \times \frac{2}{(x+1)^2} = \frac{4(x-1)}{(x+1)^3}$$

Example 2 Find (i) $\frac{d}{dx}[\sin(4x^2+3x)]$, (ii) $\frac{d}{d\theta}(\sec^2 4\theta)$.

(i) Let $v = 4x^2 + 3x$. Then

$$\frac{d}{dx}[\sin(4x^2+3x)] = \frac{d}{dx}(\sin v) = \frac{d}{dv}(\sin v) \times \frac{dv}{dx}$$

$$= \cos v(8x+3)$$

$$= (8x+3)\cos(4x^2+3x)$$

(ii) With $y = \sec^2 4\theta$, $y = v^2$, where $v = \sec 4\theta$. Then

$$\frac{dy}{d\theta} = \frac{dy}{dv} \times \frac{dv}{d\theta} = 2v\frac{dv}{d\theta}$$

v is still a function of a function: $v = \sec u$, where $u = 4\theta$. Therefore

$$\frac{dv}{d\theta} = \frac{dv}{du}\cdot\frac{du}{d\theta} = \sec u \tan u \times 4$$

$$\frac{dy}{d\theta} = 2\sec 4\theta \times \sec 4\theta \tan 4\theta \times 4$$

$$= 8\sec^2 4\theta \tan 4\theta$$

We have here an example of a function of a function of a function. y is a function of v, which in turn is a function of u, which is a function of θ. The extension of (10.14) which enables us to deal with this situation is

$$\frac{dy}{d\theta} = \frac{dy}{dv} \times \frac{dv}{du} \times \frac{du}{d\theta} \tag{10.15}$$

Readers will find, as they gain experience, that they do not need to introduce the auxiliary variables v or u specifically. However, at first it is probably wise (even at the cost of a little time) to use them.

Exercises 10d

Differentiate with respect to x

1 $(x-1)^5$ **2** $(2x-1)^5$ **3** $(2x^2-3x)^5$
4 $(x^2+2x+1)^4$ **5** $\sec 3x$ **6** $\tan 5x$
7 $x\sin 4x$ **8** $x^2\cos 3x$ **9** $\sin^3 x$
10 $\sin x^3$ **11** $\sec(3x^2+1)$ **12** $\tan^3(3x-4)$
13 $\sin^2(x^2+1)$ **14** $(x+1)(2x-1)^4$ **15** $\sin^3 x \tan 2x$
16 $(1+\sin^2 x)(1-\sin^2 x)$ **17** $\left(\frac{\cos x}{1+\sin x}\right)^3$ **18** $\left(\frac{1-x^2}{1+x^2}\right)^2$
19 $(1-\cos^4 x)(1+\cos^4 x)$ **20** $\left(\frac{\sin 2x}{1+\cos 2x}\right)^2$ **21** $\frac{\sin^2 x}{2+\sin^2 x}$
22 $\sec^3(\tan^2 3x)$
23 Find $du/d\theta$ if $u =$ (i) $\sin^n\theta$, (ii) $\cos^m\theta$, (iii) $\sin^n\theta\cos^m\theta$, where m and n are positive integers.
24 If $y = \sin m\theta$ show that $d^2y/d\theta^2 + m^2y = 0$.
25 $y = (\sec\theta + \tan\theta)^n$, where n is a positive integer. Show that $dy/d\theta = ny\sec\theta$.

10.9 The derivative of x^n, where n is negative or a fraction

In the previous chapter, we saw that if n is a positive integer then the derivative of x^n is nx^{n-1}. We shall show this result to be true for all values of n.

(i) If n is a negative integer, let $n = -m$ so that m is a positive integer. Then if $y = x^n$, we have

$$y = x^{-m} = \frac{1}{x^m}$$

therefore, by (10.9),

$$\frac{dy}{dx} = -\frac{mx^{m-1}}{x^{2m}} = -mx^{-m-1}$$

$$= nx^{n-1} \quad \text{since } -m = n$$

Thus

$$\frac{d}{dx}(x^n) = nx^{n-1} \quad \text{if } n \text{ is any integer}$$

(ii) If n is a fraction, let $n = p/q$, where p and q are integers (not necessarily positive). Then

$$y = x^{p/q} = (x^{1/q})^p$$

With $x^{1/q} = u$, $y = u^p$, where $x = u^q$, so that

$$\frac{dy}{du} = pu^{p-1} \quad \text{and} \quad \frac{dx}{du} = qu^{q-1} \quad \text{by (i)}$$

These two results enable us to obtain dy/dx, for, by the rule for differentiating a function of a function,

$$\frac{dy}{du} = \frac{dy}{dx} \cdot \frac{dx}{du}$$

$$pu^{p-1} = \frac{dy}{dx} \times qu^{q-1}$$

$$\frac{dy}{dx} = \frac{p}{q}u^{p-q} = \frac{p}{q}(x^{1/q})^{p-q}$$

$$\frac{dy}{dx} = \frac{p}{q}x^{p/q-1}$$

$$= nx^{n-1} \quad \text{since } n = p/q$$

Thus, for all values of n,

$$\frac{d}{dx}(x^n) = nx^{n-1} \tag{10.16}$$

Example 1 Find (i) $\frac{d}{dx}(\sqrt{x})$, (ii) $\frac{d}{dx}\left(\frac{1}{x^4}\right)$.

(i) $$\frac{d}{dx}(\sqrt{x}) = \frac{d}{dx}(x^{1/2}) = \frac{1}{2}x^{-1/2} = \frac{1}{2\sqrt{x}}$$

(ii) $$\frac{d}{dx}\left(\frac{1}{x^4}\right) = \frac{d}{dx}(x^{-4}) = -4x^{-5} = \frac{-4}{x^5}$$

Example 2 Find $\frac{dy}{dx}$ when (i) $y = \left(x^2 - \frac{2}{x^2}\right)^2$, (ii) $y = \sqrt{\left(\frac{x}{1+x}\right)}$.

(i) $y = v^2$, where $v = x^2 - \frac{2}{x^2} = x^2 - 2x^{-2}$

$$\frac{dy}{dv} = 2v \quad \text{and} \quad \frac{dv}{dx} = 2x - (2)(-2)x^{-3} = 2x + \frac{4}{x^3}$$

therefore

$$\frac{dy}{dx} = 2\left(x^2 - \frac{2}{x^2}\right)\left(2x + \frac{4}{x^3}\right) = 4\left(x^2 - \frac{2}{x^2}\right)\left(x + \frac{2}{x^3}\right)$$
$$= 4\left(x^3 - \frac{4}{x^5}\right)$$

(ii) $y = \sqrt{v} = v^{1/2}$, where $v = \frac{x}{1+x}$

$$\frac{dy}{dv} = \frac{1}{2}v^{-1/2} = \frac{1}{2\sqrt{v}} \quad \text{and} \quad \frac{dv}{dx} = \frac{(x+1)1 - x.1}{(x+1)^2} = \frac{1}{(x+1)^2}$$

therefore

$$\frac{dy}{dx} = \frac{1}{2}\sqrt{\left(\frac{1+x}{x}\right)}\frac{1}{(1+x)^2} = \frac{1}{2\sqrt{x}(1+x)^{3/2}}$$

Exercises 10e

Find dy/dx when $y =$

1 $x^{3/2}$	**2** $\sqrt[3]{x}$	**3** $(\sqrt[3]{x})^7$
4 $\left(x^3 - \frac{3}{x^3}\right)^2$	**5** $\frac{1}{(2x^2-3)^3}$	**6** $\sqrt{(2x^2 - x)}$
7 $(x^2+1)^{3/2}$	**8** $1/\sqrt{(x^2+1)}$	**9** $(1+x)\sqrt{(1-x^2)}$
10 $\sqrt{x}(1+x)^3$	**11** $\sec\sqrt{x}$	**12** $\sqrt{\sec x}$
13 $\sqrt{(1+\sin x)}$	**14** $\sqrt{\left(\frac{x-1}{x+1}\right)}$	**15** $\sqrt{\left(\frac{1+\cos x}{1-\cos x}\right)}$

10.10 Differentiation of inverse functions

We have already encountered in Chapter 6 the inverse trigonometric functions $y = \arcsin x$, $y = \arccos x$, $y = \arctan x$, etc. (which mean $x = \sin y$, $x = \cos y$,

$x = \tan y$, respectively). These functions do not fall into any of the categories (product, quotient) so far considered in this chapter and we shall develop a new technique in order to calculate their derivatives.

This we shall do for the general inverse function and then apply the technique to the inverse trigonometric functions. It is important to realise that the notion of an inverse function is not confined to the trigonometric functions. In general, if $y = f(x)$, then the value of x will depend on the value of y and so x is a function of y; $x = g(y)$, the inverse function to $f(x)$.

Care needs to be exercised in some cases. If $y = x^2$, then as we have seen (Section 9.1), there are two inverse functions, $x = \sqrt{y}$ and $x = -\sqrt{y}$, and we must be careful to deal with them separately (see Exercises 10f, question 3). If $y = \sin x$, then $x = \arcsin y$, and we must deal with the principal value so that a unique value of x corresponds to each value of y.

In general, if $y = f(x)$, $x = g(y)$ and we shall show how to find the derivative dx/dy of $g(y)$ in terms of the derivative dy/dx of $f(x)$.

We may regard $y = f(x)$, where $x = g(y)$, as being a function of a function. Thus, on differentiating with respect to y, we obtain

$$\frac{d(y)}{dy} = \frac{d}{dx}[f(x)] \cdot \frac{dx}{dy} \quad \text{by (10.14)}$$

$$1 = \frac{dy}{dx} \cdot \frac{dx}{dy}$$

therefore

$$\frac{dy}{dx} = 1 \bigg/ \frac{dx}{dy} \quad \text{or} \quad \frac{dx}{dy} = 1 \bigg/ \frac{dy}{dx} \qquad (10.17)$$

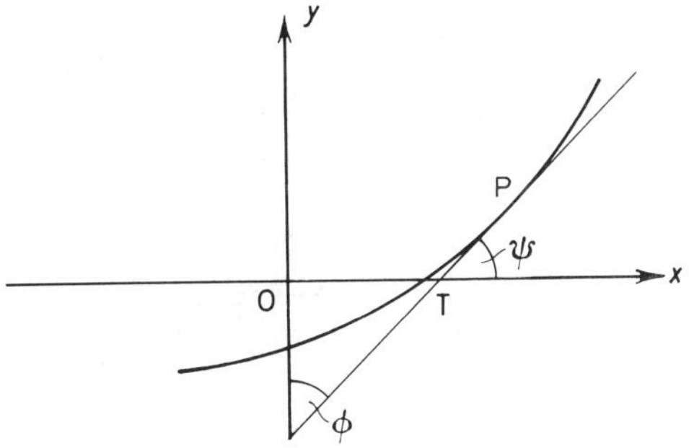

Figure 10.1

Geometrically, if we draw the graph of $y = f(x)$, then at the same time we draw the graph of $x = g(y)$ (*Figure 10.1*).

If PT is the tangent at any point P on this curve and PT makes an angle ψ with the positive direction of the x-axis and an angle ϕ with the positive direction of the y-axis, then

$$\frac{dy}{dx} = \tan \psi \qquad \frac{dx}{dy} = \tan \phi$$

But since $\phi = \frac{1}{2}\pi - \psi$, $\tan\phi = \cot\psi = \dfrac{1}{\tan\psi}$, therefore

$$\frac{dx}{dy} = 1\bigg/\frac{dy}{dx} \quad \text{as before}$$

(Note: this result is only true for the *first* derivative.)

We shall now apply the general result (10.17) to the inverse trigonometric functions.

If $y = \arctan x$,

$$x = \tan y$$

$$\frac{dx}{dy} = \sec^2 y$$

$$= 1 + \tan^2 y$$

$$= 1 + x^2$$

therefore

$$\frac{dy}{dx} = 1\bigg/\frac{dx}{dy} = \frac{1}{1+x^2} \tag{10.18}$$

If $y = \arcsin x$,

$$x = \sin y$$

$$\frac{dx}{dy} = \cos y$$

$$= \sqrt{(1-\sin^2 y)}$$

$$= \sqrt{(1-x^2)}$$

therefore

$$\frac{dy}{dx} = \frac{1}{\sqrt{(1-x^2)}} \tag{10.19}$$

If $y = \arccos x$,

$$x = \cos y$$

$$\frac{dx}{dy} = -\sin y = -\sqrt{(1-\cos^2 y)} = -\sqrt{(1-x^2)}$$

therefore

$$\frac{dy}{dx} = \frac{-1}{\sqrt{(1-x^2)}} \tag{10.20}$$

Note: the results (10.19), (10.20) are in accordance with our convention regarding the square root sign and the principal values of the inverse trigonometric functions. Reference to *Figures 6.20* and *6.21* shows that the gradient of $y = \arcsin x$ is everywhere positive and the gradient of $\arccos x$ is everywhere negative.

Example 1 Find $\dfrac{d}{dx}(\text{arcsec}\, x)$.

If $y = \text{arcsec}\, x$, then

$$x = \sec y$$

$$\frac{dx}{dy} = \sec y \tan y$$

$$= \sec y \sqrt{(\sec^2 y - 1)}$$

$$= x\sqrt{(x^2 - 1)}$$

therefore

$$\frac{d(\text{arcsec}\, x)}{dx} = \frac{dy}{dx} = \frac{1}{x\sqrt{(x^2 - 1)}}$$

Example 2 Find $du/d\theta$ if (i) $u = \arcsin 3\theta$, (ii) $u = \arctan(3\theta^3)$.

(i) $u = \arcsin 3\theta = \arcsin v$, where $v = 3\theta$

$$\frac{du}{d\theta} = \frac{du}{dv} \cdot \frac{dv}{d\theta} = \frac{1}{\sqrt{(1 - v^2)}} \times 3$$

$$= \frac{3}{\sqrt{(1 - 9\theta^2)}}$$

(ii) $u = \arctan(3\theta^3) = \arctan v$, where $v = 3\theta^3$

$$\frac{du}{d\theta} = \frac{du}{dv} \cdot \frac{dv}{d\theta} = \frac{1}{1 + v^2} \times 9\theta^2$$

$$= \frac{9\theta^2}{1 + 9\theta^6}$$

Example 3 Find $\dfrac{d}{dx}\left[\arcsin\left(\dfrac{1 - x^2}{1 + x^2}\right)\right]$.

Let $y = \arcsin\left(\dfrac{1 - x^2}{1 + x^2}\right) = \arcsin u$, where $u = \dfrac{1 - x^2}{1 + x^2}$.

$$\frac{dy}{dx} = \frac{dy}{du} \cdot \frac{du}{dx} = \frac{1}{\sqrt{(1 - u^2)}} \cdot \frac{(1 + x^2)(-2x) - (1 - x^2)(2x)}{(1 + x^2)^2}$$

$$= \frac{1}{\sqrt{[1 - (1 - x^2)^2/(1 + x^2)^2]}} \times \frac{-4x}{(1 + x^2)^2}$$

$$= \frac{-4x}{(1 + x^2)^2} \times \sqrt{\left[\frac{(1 + x^2)^2}{(1 + x^2)^2 - (1 - x^2)^2}\right]}$$

$$= \frac{-4x}{(1 + x^2)^2} \times \frac{(1 + x^2)}{2x} = \frac{-2}{(1 + x^2)}$$

Exercises 10f

1 Show that $\frac{d}{dx}(\operatorname{arccot} x) = \frac{-1}{1+x^2}$.

2 Show that $\frac{d}{dx}(\operatorname{arccosec} x) = \frac{-1}{x\sqrt{(x^2-1)}}$.

3 Use the technique employed in this section to show that if $y = \sqrt{x}$ (i.e. $x = y^2$) then $\frac{dy}{dx} = \frac{1}{2y} = \frac{1}{2\sqrt{x}}$. Show also that if $y = -\sqrt{x}$ (again $x = y^2$) then $\frac{dy}{dx} = \frac{1}{2y} = \frac{-1}{2\sqrt{x}}$. $\left[\text{Note: in both cases } \frac{dy}{dx} = \frac{1}{2y}.\right]$

4 Show that if $y = \sqrt[3]{x}$, then $\frac{dy}{dx} = \frac{1}{3x^{2/3}}$ by the method of this section.

Find dy/dx if $y =$

5 $\arcsin x^2$

6 $\arccos 6x$

7 $\arctan(x+1)$

8 $\operatorname{arcsec} x^2$

9 $\arctan x^2$

10 $x \arcsin x$

11 $\arcsin\left(\frac{1-x}{1+x}\right)$

12 $\arctan\left(\frac{1+x}{1-x}\right)$. Can you explain this result?

13 Show that if a is a constant

(i) $\frac{d}{dx}\left[\arctan\left(\frac{x}{a}\right)\right] = \frac{a}{a^2+x^2}$

(ii) $\frac{d}{dx}\left[\arcsin\left(\frac{x}{a}\right)\right] = \frac{1}{\sqrt{(a^2-x^2)}}$.

14 Show that $\frac{d}{d\theta}(2\arctan\theta) = \frac{d}{d\theta}\left[\arctan\left(\frac{2\theta}{1-\theta^2}\right)\right]$. Why are the two results equal?

15 If $y = \sqrt{(1-\theta^2)}\arcsin\theta$, show that $(1-\theta^2)\frac{dy}{d\theta} = 1-\theta^2-\theta y$.

10.11 Differentiation of implicit functions

The rules which we have established in this chapter have been applied thus far only to explicit functions. In this section, we shall develop the techniques necessary for the differentiation of implicit functions.

Suppose, for example, that y is defined as an implicit function of x by the equation

$$x^2 + y^2 = 4$$

We differentiate each term of the equation above with respect to x and so obtain

$$\frac{d}{dx}(y^2) + \frac{d}{dx}(x^2) = \frac{d}{dx}(4)$$

therefore

$$\frac{d}{dx}(y^2) + 2x = 0$$

Now y^2 is a function of y which is itself a function of x. Thus by (10.14)

$$\frac{d}{dx}(y^2) = \frac{d}{dy}(y^2)\cdot\frac{dy}{dx} = 2y\frac{dy}{dx}$$

therefore

$$2y\frac{dy}{dx} + 2x = 0$$

so that

$$\frac{dy}{dx} = -\frac{x}{y}$$

In this particular example, it is possible to express y explicitly in terms of x; $y = \sqrt{(4 - x^2)}$, i.e. $y = u^{1/2}$, where $u = 4 - x^2$. Therefore

$$\frac{dy}{dx} = \frac{dy}{du}\cdot\frac{du}{dx} = \frac{1}{2}\frac{1}{\sqrt{u}}\cdot(-2x)$$

$$= \frac{-x}{\sqrt{u}} = -\frac{x}{y} \quad \text{as before}$$

However, we could also express y as a function of x as $y = -\sqrt{(4 - x^2)}$, i.e. $y = -u^{1/2}$, where $u = 4 - x^2$. Therefore

$$\frac{dy}{dx} = \frac{dy}{du}\cdot\frac{du}{dx} = -\frac{1}{2}\frac{1}{\sqrt{u}}\cdot(-2x)$$

$$= \frac{x}{\sqrt{u}} = \frac{x}{-y} = -\frac{x}{y}$$

Note, in both cases, the correctness of the result $dy/dx = -x/y$, obtained by the first technique (see *Figure 9.5*).

Generally, however, when it is not possible to express y explicitly in terms of u, we shall have to use the first method.

Example 1 Find dy/dx if $x^2 + y^2 + \sin y = 3$.

We have, on differentiating the equation with respect to x,

$$2x + \frac{d}{dx}(y^2) + \frac{d}{dx}(\sin y) = 0$$

$$2x + \frac{d}{dy}(y^2)\frac{dy}{dx} + \frac{d}{dy}(\sin y)\frac{dy}{dx} = 0$$

$$2x + 2y\frac{dy}{dx} + \cos y\frac{dy}{dx} = 0$$

therefore

$$\frac{dy}{dx} = -\frac{2x}{2y + \cos y}$$

Example 2 Find dy/dx and d^2y/dx^2 if $x + y + \sin y = 3$.

On differentiating the equation above with respect to x, we obtain

$$1 + \frac{dy}{dx} + \frac{d}{dx}(\sin y) = 0$$

$$1 + \frac{dy}{dx} + \cos y \frac{dy}{dx} = 0$$

$$1 + \frac{dy}{dx}(1 + \cos y) = 0$$

therefore

$$\frac{dy}{dx} = \frac{-1}{1 + \cos y}$$

To find d^2y/dx^2, we differentiate this expression with respect to x.

$$\frac{d^2y}{dx^2} = \frac{d}{dx}\left(\frac{dy}{dx}\right) = \frac{d}{dx}\left(\frac{-1}{1 + \cos y}\right) = \frac{d}{dy}\left(\frac{-1}{1 + \cos y}\right)\frac{dy}{dx}$$

$$\frac{d^2y}{dx^2} = \frac{-1[(-1)(-\sin y)]}{(1 + \cos y)^2} \times \frac{-1}{1 + \cos y} = \frac{\sin y}{(1 + \cos y)^3}$$

Alternatively, we may find d^2y/dx^2 by differentiating with respect to x the equation $1 + \frac{dy}{dx} + \frac{dy}{dx}\cos y = 0$ which we obtained earlier. In this way, we obtain

$$0 + \frac{d^2y}{dx^2} + \frac{d}{dx}\left(\frac{dy}{dx}\cos y\right) = 0$$

$$\frac{d^2y}{dx^2} + \frac{dy}{dx}\cdot\frac{d}{dx}(\cos y) + \cos y \frac{d}{dx}\left(\frac{dy}{dx}\right) = 0 \quad \text{by (10.4)}$$

$$\frac{d^2y}{dx^2} + \frac{dy}{dx}\cdot\frac{d}{dy}(\cos y)\cdot\frac{dy}{dx} + \cos y \frac{d^2y}{dx^2} = 0$$

$$\frac{d^2y}{dx^2} - \sin y\left(\frac{dy}{dx}\right)^2 + \cos y \frac{d^2y}{dx^2} = 0$$

therefore

$$\frac{d^2y}{dx^2} = \frac{\sin y\left(\frac{dy}{dx}\right)^2}{1 + \cos y} = \frac{\sin y}{(1 + \cos y)^3} \quad \text{as before}$$

$$\left(\text{since } \frac{dy}{dx} = \frac{-1}{1 + \cos y}\right)$$

Example 3 Find dy/dx and d^2y/dx^2 at the point (1, 1) on the curve $x^3 + y^3 = 2$.

On differentiating the equation with respect to x, we have

$$3x^2 + \frac{d}{dx}(y^3) = 0$$

$$3x^2 + \frac{d}{dy}(y^3)\cdot\frac{dy}{dx} = 0$$

$$3x^2 + 3y^2\frac{dy}{dx} = 0 \qquad \text{(i)}$$

Thus, when $x = 1$, $y = 1$, $\dfrac{dy}{dx}$ is such that $3 + 3\dfrac{dy}{dx} = 0$. Therefore

$$\frac{dy}{dx} = -1 \quad \text{at the point } (1, 1)$$

If we differentiate (i) with respect to x, we obtain

$$6x + \frac{d}{dx}\left(3y^2\frac{dy}{dx}\right) = 0$$

$$6x + 3y^2\frac{d^2y}{dx^2} + \frac{dy}{dx}\cdot\frac{d}{dx}(3y^2) = 0 \quad \text{by (10.4)}$$

$$6x + 3y^2\frac{d^2y}{dx^2} + \frac{dy}{dx}\cdot 6y\frac{dy}{dx} = 0$$

Now when $x = 1$, $y = 1$, $dy/dx = -1$ and so the value of d^2y/dx^2 at this point is such that

$$6 + 3\frac{d^2y}{dx^2} + 6 = 0$$

therefore

$$\frac{d^2y}{dx^2} = -4 \quad \text{at the point } (1, 1)$$

Thus we see that nothing new in the way of differentiation is involved in applying these techniques. Care must be taken, however, to apply the function of a function rule when differentiating a quantity which involves y or its derivatives, with respect to x.

Exercises 10g

Find dy/dx when

1 $xy = 1$
2 $x^2y^2 - x - y = 0$
3 $\sqrt{x} + \sqrt{y} = 1$
4 $x + y + \cos x + \cos y = 2$
5 $xy + \sin y = 1$
6 $x + y + \sin xy = 2$

7 Find dy/dx at the point (1, 1) if $x^2 + y^2 + xy = 3$.
8 Find dy/dx and d^2y/dx^2 if $3xy + x^2 + y^2 = 5$.
9 Find dy/dx and d^2y/dx^2 at the origin on the curve $x^2 + y^2 + x + 3y = 0$.
10 Find dy/dx and d^2y/dx^2 at the point (1, 1) on the curve $3x^2 + 2y^2 + xy + x - 7 = 0$.

10.12 Differentiation from parametric equations

It is often convenient to define y as a function of x by expressing both y and x in terms of a third variable t, known as a parameter. (See Sections 20.5, 20.9, 20.12 and 20.15.)

Thus, in general, $y = f(t)$, $x = \phi(t)$ defines y as a function of x. This follows since y is a function of t which in turn is a function of x. By eliminating t from the two relationships above, we can obtain y as a function of x. (See Section 1.3.)

Thus, if $y = t^2$, $x = 1/t$, we have, since $t = 1/x$, $y = 1/x^2$. Sometimes it will be difficult or even impossible to carry out this elimination although it is always true that y is a function of x since the value of y will depend on x. In such cases, the differential coefficient of y with respect to x can be obtained by regarding y as being a function of a function; y is a function of t and t is a function of x. Thus

$$\frac{dy}{dx} = \frac{dy}{dt} \cdot \frac{dt}{dx}$$

therefore

$$\frac{dy}{dx} = \frac{dy}{dt} \bigg/ \frac{dx}{dt} \qquad (10.21)$$

Example 1 Find dy/dx when (i) $y = t^2$, $x = 1/t$, (ii) $y = \sin\theta$, $x = \cos\theta$.

(i) $$\frac{dy}{dx} = \frac{2t}{-1/t^2} = -2t^3 \quad \text{by (10.21)}$$

(ii) $$\frac{dy}{dx} = \frac{\cos\theta}{-\sin\theta} = -\cot\theta \quad \text{by (10.21)}$$

Example 2 Find dy/dx and d^2y/dx^2 in terms of t when $y = 2t$, $x = t^2$.

$$\frac{dy}{dx} = \frac{dy}{dt} \bigg/ \frac{dx}{dt} = \frac{2}{2t} = \frac{1}{t}$$

$$\frac{d^2y}{dx^2} = \frac{d}{dx}\left(\frac{dy}{dx}\right) = \frac{d}{dt}\left(\frac{dy}{dx}\right) \cdot \frac{dt}{dx}$$

$$\frac{d^2y}{dx^2} = \frac{d}{dt}\left(\frac{dy}{dx}\right) \bigg/ \frac{dx}{dt} \qquad (10.22)$$

(This result is true generally.)

therefore

$$\frac{d^2y}{dx^2} = \frac{d}{dt}\left(\frac{1}{t}\right)\Big/ 2t = \frac{-1}{t^2}\Big/ 2t = -\frac{1}{2t^3}$$

Example 3 Find dy/dx and d^2y/dx^2 at the point with abscissa 1 on the curve $y = 1/t$, $x = 2t$.

The point with abscissa 1 corresponds to $t = \frac{1}{2}$. Now

$$\frac{dy}{dx} = -\frac{1}{t^2}\Big/ 2 = -\frac{1}{2t^2}$$

When $t = \frac{1}{2}$, $\dfrac{dy}{dx} = -1/\frac{1}{2} = -2$

$$\frac{d^2y}{dx^2} = \frac{1}{2}\frac{d}{dt}\left(-\frac{1}{2t^2}\right) = \frac{1}{2t^3} \quad \left(\text{note } \frac{dt}{dx} = \frac{1}{2}\right).$$

When $t = \frac{1}{2}$, $\dfrac{d^2y}{dx^2} = \dfrac{1}{2(\frac{1}{8})} = 4$

Exercises 10h

Find dy/dx in terms of the parameter when

1 $y = t^2$, $x = t^3$

2 $y = 2\sin\theta$, $x = 3\cos\theta$

3 $y = \cos 4t$, $x = \sin 2t$

4 $y = t^2 \cos t$, $x = t \sin t$

5 $y = \dfrac{t^2}{1+t}$, $x = \dfrac{t}{1+t}$

6 $y = \dfrac{t^3}{1+t^3}$, $x = \dfrac{t}{1+t^3}$

Find dy/dx and d^2y/dx^2 in terms of t when

7 $y = \dfrac{2t}{1+t^2}$, $x = \dfrac{1-t^2}{1+t^2}$

8 $y = t$, $x = \dfrac{1}{t^2}$

9 If $x = t^3 + t$ and $y = 2t^2$, find dy/dx in terms of t and show that when $dy/dx = 1$, $x = 2$ or $x = \frac{10}{27}$.

10 The position of a projectile referred to horizontal and vertical axes is given by $x = 8t$, $y = 40t - 16t^2$ after time t sec. Find at what times the projectile is moving (i) horizontally, (ii) at an angle of 45° to the horizontal.

10.13 List of standard forms

The rules of differentiation and the differential coefficients of the important basic functions are listed below. They should be memorised. In the following, $u, v, \ldots, w$ are functions of x; $a, b, \ldots, c$ are constants.

$$\frac{d}{dx}(au + bv + \ldots + cw) = a\frac{du}{dx} + b\frac{dv}{dx} + \ldots + c\frac{dw}{dx}$$

$$\frac{d}{dx}(uv) = u\frac{dv}{dx} + v\frac{du}{dx}$$

$$\frac{d}{dx}(uvw) = uv\frac{dw}{dx} + uw\frac{dv}{dx} + vw\frac{du}{dx}$$

$$\frac{d}{dx}\left(\frac{u}{v}\right) = \frac{v\frac{du}{dx} - u\frac{dv}{dx}}{v^2}$$

$$\frac{dy}{dx} = \frac{dy}{du} \cdot \frac{du}{dx}$$

$$\frac{dx}{dy} = 1 \bigg/ \frac{dy}{dx}$$

$$\frac{d}{dx}(x^n) = nx^{n-1}$$

$$\frac{d}{dx}(\sin x) = \cos x$$

$$\frac{d}{dx}(\sec x) = \sec x \tan x$$

$$\frac{d}{dx}(\cos x) = -\sin x$$

$$\frac{d}{dx}(\operatorname{cosec} x) = -\operatorname{cosec} x \cot x$$

$$\frac{d}{dx}(\tan x) = \sec^2 x$$

$$\frac{d}{dx}(\cot x) = -\operatorname{cosec}^2 x$$

$$\frac{d}{dx}(\arcsin x) = \frac{1}{\sqrt{(1-x^2)}}$$

$$\frac{d}{dx}(\arctan x) = \frac{1}{1+x^2}$$

$$\frac{dy}{dx} = \frac{dy}{dt} \bigg/ \frac{dx}{dt}$$

$$\frac{d}{dx}(\arccos x) = \frac{-1}{\sqrt{(1-x^2)}}$$

Exercises 10

1 Find (i) $\frac{d}{dx}(\frac{3}{4}x^4 - \frac{1}{2}x^2 + 2)$, (ii) $\frac{d}{dt}(\frac{3}{2}t^{3/2} - \frac{1}{2}t^{1/2} + \frac{1}{2}t^{-1/2})$.

2 Find (i) $\frac{d}{dx}[(x^3 + x)(3x^2 + x)]$, (ii) $\frac{d}{d\theta}[\sec\theta(1 + \cot\theta)]$.

3 Find (i) $\frac{d}{dx}\left(\frac{x}{x^2 - 1}\right)$, (ii) $\frac{d}{dx}\left(\frac{x^3 - x}{x^2 + 1}\right)$.

4 Find the derivatives of (i) $\frac{\sqrt{x}}{1 + \sqrt{x}}$, (ii) $\frac{1 + \sqrt{x}}{1 - \sqrt{x}}$.

5 Find $d\theta/dt$ when (i) $\theta = \sin t \sin 3t$, (ii) $\theta = t^2 \sin^{-1} t$. [LU, part]

6 Differentiate with respect to x (i) $\frac{x + 3}{x^2 - 16}$, (ii) $\frac{(x - 1)(x - 2)}{(x + 1)(x + 2)}$.

7 Find dy/dt when (i) $y = \cos t \cos 5t$, (ii) $y = t \arctan t$.

8 Find the derivatives of (i) $\sin^4 x$, (ii) $\sec^3 x$.

9 Find $du/d\theta$ if (i) $u = \sec^3 6\theta$, (ii) $u = \cot^5 \theta^2$.

10 Find (i) $d/d\theta\ (\theta \sec\theta \tan\theta)$, (ii) $d/dx\ (x \sin x \cos 2x)$.

11 Find dy/dx when (i) $y = \sin 1/x$, (ii) $y = \{[\sin(1/x)]/x\}$.

12 Find (i) $\frac{d}{d\theta}\left(\frac{\theta}{1 + \theta^3}\right)$, (ii) $\frac{d}{dt}(\arcsin t^3)$.

13 Differentiate with respect to θ (i) $\arcsin\sqrt{(1 - \theta^2)}$, (ii) $\arctan\left[\frac{1}{\sqrt{(1 - \theta)}}\right]$.

14 Find dy/dx when (i) $y = \arcsin(\tan x)$, (ii) $y = \arctan(\sin x)$.

15 Find (i) $\dfrac{d}{dt}[\arccos(1 - t^2)]$, (ii) $\dfrac{d}{dt}\left[\operatorname{arccot}\left(\dfrac{t}{1+t}\right)\right]$.

16 Differentiate with respect to x (i) $\cos 1/x^2$, (ii) $\arccos\left(\dfrac{1-x^2}{1+x^2}\right)$.

17 If $y = \tan x$, show that $d^2y/dx^2 = 2y + 2y^3$.

18 If $y = \cot^2\theta$, show that $d^2y/d\theta^2 = 2(1 + y)(1 + 3y)$.

19 If $y = \tan x + \frac{1}{3}\tan^3 x$, prove that $dy/dx = (1 + \tan^2 x)^2$.

20 Find dy/dx when (i) $y = \arctan\left(\dfrac{4x^2}{4 - x^2}\right)$, (ii) $y = \arcsin(2x - 5)$.

21 Find dy/dx when (i) $y = x^n \tan nx$, (ii) $y = \arctan(\sin\frac{1}{2}x)$.

22 If y is a function of x find the derivative with respect to x of (i) xy^2, (ii) x/y, (iii) y/x, (iv) $\sin^2 y$.

23 Find dy/dx when (i) $x^3 + y^3 - 3xy + 1 = 0$, (ii) $x^3 - 2x^2y^2 + y^4 = 0$.

24 Find dy/dx when (i) $y^2 + x^2 = 6x + 4y + 1$, (ii) $y^3 + x^3 = 3(x + y)$.

25 (i) If $\sin y = \tan x$, find dy/dx in terms of x. (ii) If $x^3 + y^3 = 3axy$, find dy/dx in terms of x and y, and prove that dy/dx cannot be equal to -1 for finite values of x and y, unless $x = y$. [SUJB]

26 Find the slope of the tangent to the curve $yx^3 - 3x^2y^2 + x^3 - 2x = 0$ at the point where $x = 2$, $y = 1$.

27 Find dy/dx and d^2y/dx^2 at the origin on the curve $x^3y + y^3x = xy + y$.

28 Find dy/dx and d^2y/dx^2 at the point (1, 1) on the curve $2xy - 2x^3 - y^3 + x^2y^2 = 0$.

29 If $x^6 + y^6 = xy$ express d^2y/dx^2 in terms of x and y. [WJC, part]

30 Find dy/dx if $y = b\sin\theta$ and $x = a\cos\theta$.

31 Find dy/dx and d^2y/dx^2 for the curve $x = a\sec\theta$, $y = b\tan\theta$.

32 Find dy/dx and d^2y/dx^2 for the curve $x = a\cos^3\theta$, $y = a\sin^3\theta$.

33 Find an expression for dy/dx for the cycloid given by $x = a(t + \sin t)$, $y = a(1 - \cos t)$.

34 A curve is given by the parametric equations

$$x = a(t\sin t + \cos t - 1) \qquad y = a(\sin t - t\cos t)$$

Find dy/dx and d^2y/dx^2 in terms of t. [JMB, part]

35 The equations of a curve in parametric form are

$$x = 4\cos\theta + 3\sin\theta + 2 \qquad y = 3\cos\theta - 4\sin\theta - 1$$

Find dy/dx at the point where $\theta = \pi/2$. [LU]

36 If $y = (w^2 - 1)^n$, show that $(w^2 - 1)\,dy/dw - 2nwy = 0$.

37 If $y = \sin(m\arcsin x)$, show that $(1 - x^2)\,d^2y/dx^2 - x\,dy/dx + m^2y = 0$.

38 If $y = \sqrt{(4 + 3\sin x)}$, prove that

$$2y\frac{d^2y}{dx^2} + 2\left(\frac{dy}{dx}\right)^2 + y^2 = 4 \qquad \text{[SUJB, part]}$$

39 If $y = \dfrac{\cos\theta}{\theta^2}$, find $dy/d\theta$ and $d^2y/d\theta^2$, and prove that

$$\theta^2\frac{d^2y}{d\theta^2} + 4\theta\frac{dy}{d\theta} + (\theta^2 + 2)y = 0$$

40 If $z = [v + \sqrt{(1 + v^2)}]^p$, show that

$$(1 + v^2)\frac{d^2z}{dv^2} + v\frac{dz}{dv} - p^2z = 0$$

41 The equation of a curve is $x^2 + 4xy + y^2 = 25$. Find the values of dy/dx and d^2y/dx^2 at the point where the curve meets the positive x-axis. [AEB]

42 Given that $y = (\sin x - \cos x)/(\sin x + \cos x)$, show that $dy/dx = 1 + y^2$. Prove that d^2y/dx^2 is zero only when $y = 0$. [JMB]

43 If $y = \sin^2(x^2 + 1)$, find and simplify dy/dx. [AEB]

44 If $y = 1 - \cos 2t$ and $x = \sqrt{1 + t^2}$, show that

$$\frac{dy}{dx} = \frac{2\sqrt{(1 + t^2)}\sin 2t}{t}$$ [AEB]

45 If $x = \cot\theta$ and $y = \sin^2\theta$, find dy/dx in terms of θ and show that $d^2y/dx^2 = 2\sin^3\theta\sin 3\theta$. [JMB]

11
Some applications of differentiation

11.1 The derivative as a rate measurer

In this chapter, we shall apply our knowledge of the derivative of a function to a variety of problems. There are several interpretations of the derivative, but most of the following applications are based on two of these, namely, as measuring the rate of change of the function with respect to the variable, or as measuring the gradient of the tangent to the graph of the function at a particular point. We begin by using the first interpretation.

Example 1 At what rate is the area of a circle changing with respect to its radius when the radius is 1 cm?

If r cm denotes the radius and A cm^2 the area of the circle,

$$A = \pi r^2$$

$$\frac{\mathrm{d}A}{\mathrm{d}r} = 2\pi r$$

Therefore, when $r = 1$ cm,

$$\frac{\mathrm{d}A}{\mathrm{d}r} = 2\pi \text{ cm}^2/\text{cm} \quad \text{(Note the units)}$$

Example 2 The radius of a circle is increasing at the rate of 0·1 cm/sec. At what rate is the area increasing at the instant when $r = 5$ cm?

As before, $A = \pi r^2$ and is given as a function of r, which itself is a function of the time t (since r changes with time). Therefore

$$\frac{\mathrm{d}A}{\mathrm{d}t} = \frac{\mathrm{d}A}{\mathrm{d}r} \cdot \frac{\mathrm{d}r}{\mathrm{d}t}$$

$$= 2\pi r \times 0{\cdot}1$$

Thus, when $r = 5$ cm,

$$\frac{\mathrm{d}A}{\mathrm{d}t} = 2\pi \times 5 \times 0{\cdot}1 \text{ cm}^2/\text{sec}$$

$$= \pi \text{ cm}^2/\text{sec}$$

Therefore, the rate of change of area when $r = 5$ cm is π cm^2/sec.

Example 3 Water is poured into a vessel, in the shape of a right circular cone of vertical angle 90°, with the axis vertical, at the rate of 8 cm^3/sec. At what rate is the water surface rising when the depth of the water is 4 cm?

Let the depth of the water after t sec be x cm, and the volume of water in the vessel at this time V cm^3. Let the radius of the water surface at this time be y cm (*Figure 11.1*).

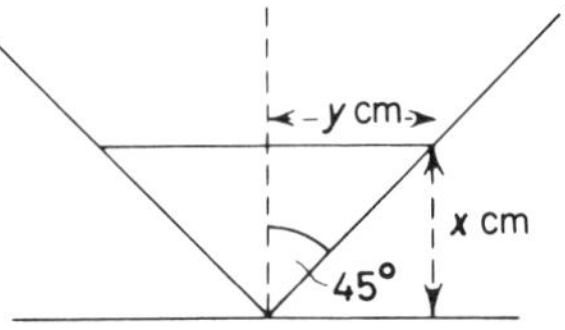

Figure 11.1

The volume of water in the vessel is $V\ \text{cm}^3 = \frac{1}{3}\pi y^2 x$, but since the semi-vertical angle is 45°, $x = y$. Therefore

$$V = \tfrac{1}{3}\pi x^3$$

The rate of increase of V with respect to t is 8 cm^3/sec, that is

$$\frac{dV}{dt} = 8\ \text{cm}^3/\text{sec}$$

But

$$\frac{dV}{dt} = \frac{dV}{dx}\cdot\frac{dx}{dt} = \pi x^2\frac{dx}{dt}$$

Therefore, the value of dx/dt when $x = 4$ cm is given by

$$16\pi\frac{dx}{dt} = 8$$

$$\frac{dx}{dt} = \frac{1}{2\pi}\ \text{cm/sec} = 0{\cdot}159\ \text{cm/sec}$$

Exercises 11a

1 The length l m of a particular rod at temperature t°C is given by $l = 2 + 0{\cdot}000\,0274t + 0{\cdot}000\,000\,0446t^2$. Find the rate at which l is increasing with respect to t when $t = 100$ °C.

2 A spherical balloon is inflated by pumping air into it at the rate of 80 cm^3/min. Find the rate at which the radius is increasing when the radius is 4 cm.

3 The radius of a sphere is increasing at the rate of 0·1 cm/sec. When $r = 5$ cm find the rates at which the surface area and the volume are increasing.

4 A gas expands according to the law pv = constant, where p is the pressure and v the volume of the gas. Initially, $v = 1000\ \text{m}^3$ and $p = 40\ \text{N/m}^2$. If the pressure is decreased at the rate of $5\ \text{N m}^{-2}\,\text{min}^{-1}$, find the rate at which the gas is expanding when its volume is 2000 m^3.

5 The distances u and v of an object and its image from a lens of focal length f are related by the formula $1/v + 1/u = 1/f$. An object 5 cm from a lens whose focal length is 2·5 cm is moved towards the lens at a speed of 10 cm/sec. Find the speed with which the image begins to recede from the lens.

6 Gas is escaping from a spherical balloon at the rate of 30 m^3/min. How fast is the radius decreasing when the radius is 3 m?

7 At what rate is the surface area of the balloon decreasing in question 6?

8 Water is running out of a conical funnel at the rate of 1 cm^3/sec. The radius of the base of the funnel is 5 cm and its height is 10 cm. Find the rate at which the water level is falling when it is 4 cm from the top.

9 A kite, 25 m above the ground, is being carried horizontally by the wind at a speed of 3 m/sec. At what rate is the inclination of the string to the horizontal changing when 50 m of string are out?

10 The radius of a sphere is r cm after t sec. Find the radius when the rate of increase of r and the rate of increase of the surface area are numerically equal.

11.2 Some applications to kinematics

If a body, moving in a straight line, has travelled a distance s after time t, then the rate of change of s with respect to time will be the speed, v, of the body

$$v = \frac{ds}{dt} \tag{11.1}$$

In the same way, the rate of change of the speed with respect to time will be the acceleration, a, of the body

$$a = \frac{dv}{dt} = \frac{d}{dt}\left(\frac{ds}{dt}\right) = \frac{d^2s}{dt^2} \tag{11.2}$$

An alternative expression for a may be obtained as follows:

We may regard v as a function of s, which is a function of t. Thus, by the function of a function rule for differentiation,

$$a = \frac{dv}{dt} = \frac{dv}{ds}\cdot\frac{ds}{dt}$$

$$a = v\frac{dv}{ds} \tag{11.3}$$

It is conventional in dynamics to denote differential coefficients with respect to time by dots placed above the dependent variable. Thus ds/dt is denoted by $\dot{s}$, d^2s/dt^2 by $\ddot{s}$. With this notation,

$$v = \dot{s} \qquad a = \dot{v} = \ddot{s} = v\frac{dv}{ds}$$

Example 1 A body moves in a straight line so that the distance moved s m after time t sec is given by $s = t^3 - 2t^2 + t$. Find an expression for the speed of the body at time t, and find the times at which the body is at rest. What is the acceleration of the body at these times?

$$s = t^3 - 2t^2 + t$$

therefore

$$v = \dot{s} = 3t^2 - 4t + 1$$

When $v = 0$, t satisfies the equation

$$3t^2 - 4t + 1 = 0$$

$$(3t - 1)(t - 1) = 0$$

therefore

$$t = \tfrac{1}{3} \quad \text{or} \quad t = 1$$

The body is at rest after times $\frac{1}{3}$ sec and 1 sec.

The acceleration of the body after time t sec is given by $a = \dot{v} = 6t - 4$. Therefore

$$\text{when } t = \tfrac{1}{3} \qquad a = 2 - 4 = -2 \text{ m/sec}^2$$

$$\text{when } t = 1 \qquad a = 6 - 4 = 2 \text{ m/sec}^2$$

Example 2 The speed of a body varies inversely as the distance it has moved. Show that its acceleration is proportional to the cube of its speed.

With the usual notation

$$v = \frac{k}{s} \quad \text{where } k \text{ is some constant}$$

Therefore

$$a = v\frac{dv}{ds} = v\left(\frac{-k}{s^2}\right) = -\frac{vk}{s^2}$$

$$a = -\frac{vk}{(k/v)^2} \quad \text{since } s = \frac{k}{v}$$

$$a = -v^3/k$$

which proves the result.

So far we have dealt with a body moving along a straight line. The same mathematics can be applied to a body rotating about a fixed axis. If θ is the angle through which the body has turned after time t, the angular speed, ω, of the body is given by

$$\omega = \frac{d\theta}{dt} \tag{11.4}$$

The angular acceleration, Ω, is given by

$$\Omega = \frac{d\omega}{dt} = \frac{d^2\theta}{dt^2} = \omega\frac{d\omega}{d\theta} \tag{11.5}$$

Exercises 11b

1 A body moves a distance s m in t sec along a straight line, where $s = 3t^3$. Find the speed and acceleration of the body after 2 sec and after t sec.
2 The speed v m/sec at time t sec of a body moving along a straight line is proportional to t^3. Find the speed of the body after 2 sec if its acceleration is then 12 m/sec^2.
3 The distance s cm moved by a body after t sec is given by $s = t^3 - 3t$. Find its speed after t sec. After what time(s) is the body at rest?
4 Find the acceleration a of the body in question 3 at any time t sec. When is the acceleration zero?
5 A body moves in a straight line so that the distance s m travelled after time t sec is given by $s = t^3 - 4t^2 + 4t$. Find the two positions of the body when it is momentarily at rest. What is the acceleration of the body at these times?
6 A particle moves along a straight line so that its distance s cm from a fixed point after time t sec is given by $s = \sin nt$, where n is a constant. Show that its acceleration is proportional to s and directed towards the fixed point.
7 Show that the speed of the particle of question 6 is given by $n\sqrt{(1 - s^2)}$.
8 The distance s, moved along a straight line by a particle, after time t is given by $s = \frac{1}{4}t^4$. Show that its speed v and acceleration a satisfy the relation, $a^3 = 27v^2$.
9 If the speed of a body is proportional to the cube of the distance it has travelled, show that its acceleration is proportional to the fifth power of the distance travelled.
10 A body is rotating about a fixed axis so that the angle θ, through which it has rotated after time t sec is given by $\theta = bt + at^2$. If ω denotes the angular speed of the body and Ω the angular acceleration of the body show that

$$\omega^2 - b^2 = 2\Omega\theta$$

11.3 Approximations

The derivative of a function is defined as

$$\lim_{\delta x \to 0} \frac{\delta y}{\delta x} = \frac{\mathrm{d}y}{\mathrm{d}x}$$

It follows that if we write $\delta y/\delta x = \mathrm{d}y/\mathrm{d}x + \alpha$, then the quantity α approaches zero as δx approaches zero, so that if δx is small so also is α. Thus, we may write

$$\delta y = \frac{\mathrm{d}y}{\mathrm{d}x} . \delta x + \alpha\delta x$$

The second term on the right, being the product of two small quantities, will be negligible in comparison with the first term. Thus

$$\delta y \approx \frac{dy}{dx}\delta x \tag{11.6}$$

Figure 11.2 shows the graph of the function $y = f(x)$. P is the point (x, y), Q the point $(x + \delta x, y + \delta y)$. Thus, the change in the function δy as x changes to $x + \delta x$ is given by the length of QN.

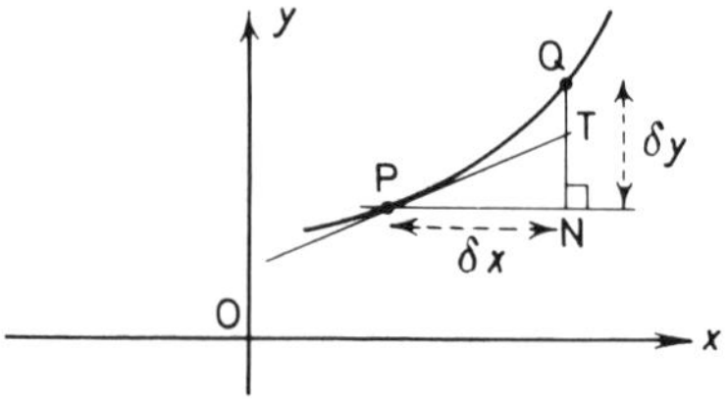

Figure 11.2

If PT is the tangent to the curve at P, then provided δx is small QN $\approx$ TN so that $\delta y \approx$ TN and since

$$\text{TN} = \text{PN} \tan \text{TPN} = \delta x f'(x)$$

$$\delta y \approx f'(x)\,\delta x = \frac{dy}{dx}\delta x \quad \text{as before}$$

Example 1 Given $\cos 45° = 1/\sqrt{2} = 0{\cdot}7071$, calculate the value of $\cos 45°\,1'$.

Consider the function $y = f(x) = \cos x$. Then

$$f'(x) = -\sin x \quad \text{(provided } x \text{ is measured in radians)}$$

Thus, if $x = \pi/4$,

$$\delta x = 1' = \frac{1}{60} \times \frac{\pi}{180} \text{ radian}$$

and so

$$\delta y = -(\sin \tfrac{1}{4}\pi) \times \frac{1}{60} \times \frac{\pi}{180} \quad \text{by (11.6)}$$

$$= -\frac{\pi}{10\,800\sqrt{2}} = -0{\cdot}0002$$

Therefore

$$\cos 45°\,1' \approx y + \delta y$$

$$= 0{\cdot}7071 - 0{\cdot}0002$$

$$\cos 45°\,1' = 0{\cdot}7069 \quad \text{approximately}$$

Example 2 The strength of the magnetic field due to a current I amp in a wire in the form of a circle of radius r cm, at a point x cm from the centre of the circle and on the axis of the circle, is given by

$$H = \frac{\pi I r^2}{5(r^2 + x^2)^{3/2}} \text{ gauss}$$

If $I = 10$ and $r = 4$, find the approximate change in H when x changes from 3 cm to 2·9 cm.

$$H = \frac{\pi I r^2}{5(r^2 + x^2)^{3/2}}$$

$$\frac{\mathrm{d}H}{\mathrm{d}x} = -\frac{3\pi I r^2 x}{5(r^2 + x^2)^{5/2}}$$

therefore

$$\delta H \approx \frac{3\pi I r^2 x}{5(r^2 + x^2)^{5/2}} \times \delta x \quad \text{by (11.6)}$$

When $x = 3$, $\delta x = -0{\cdot}1$, $r = 4$ and $I = 10$. Therefore

$$\delta H \approx -\frac{3\pi \times 10 \times 16 \times 3}{5 \times 25^{5/2}} \times -\frac{1}{10}$$

$$= \frac{1440\pi}{5 \times 31\,250} = 0{\cdot}0289$$

Thus, the magnetic field increases by approximately 0·029 gauss.

Example 3 The value of g, the acceleration due to gravity, is determined by means of a simple pendulum of length l cm. The period T sec of the pendulum is measured and g is calculated from the formula $T = 2\pi\sqrt{(l/g)}$. The experimenter feels that he is able to measure l accurately but realises that his measurement of T is subject to an error of 1 per cent. What is the percentage error in the calculated value of g?

From the above formula for T

$$g = \frac{4\pi^2 l}{T^2}$$

However, since T is subject to the error $\delta T = T \times \frac{1}{100}$, the experimenter will calculate the value $g + \delta g$, where δg is the error in g given by

$$\delta g \approx \frac{\mathrm{d}g}{\mathrm{d}T}\delta T$$

$$\delta g \approx -2 \times \frac{4\pi^2 l}{T^3} \times \frac{T}{100}$$

$$= -\frac{2}{100} \times \frac{4\pi^2 l}{T^2} = -\frac{2}{100}g$$

The percentage error in g is thus approximately 2% and is opposite in sign to the error in T.

Exercises 11c

1 $\tan 45^\circ = 1$, $\cos 45^\circ = \sin 45^\circ = 1/\sqrt{2}$. Evaluate $\tan 45^\circ\,1'$ to four decimal places.

2 Using the values $\sin 45^\circ = \cos 45^\circ = 0{\cdot}7071$, determine the values of $\sin 45^\circ\,1'$ and $\cos 45^\circ\,1'$. Use these two values to determine $\sin 45^\circ\,2'$ and cos 45° 2′ and then calculate sin 45° 3′ and cos 45° 3′. [In this way, we could construct a complete table for $\sin\theta$ and $\cos\theta$ at 1′ intervals for θ. In fact the same calculation is carried out each time with slightly different numbers. This is just the type of calculation which an electronic computer can handle very rapidly and efficiently.]

3 The radius of a circle increases from 2 cm to 2·03 cm. Find the approximate increase in its area. Find the actual increase.

4 The volume of water in a hemispherical bowl of radius 12 cm is given by $V = \frac{1}{3}\pi(36x^2 - x^3)$, where x cm is the greatest depth of the water. Find the approximate volume of water necessary to raise the depth from 2 cm to 2·1 cm. If the water is poured in at the constant rate of 3 cm^3/sec, at what rate is the level rising when the depth is 3 cm?

5 Find the approximate percentage change in the volume of a cube of side x cm caused by increasing the sides by 1 per cent.

6 The radius of a spherical balloon is decreased from 10 cm to 9·9 cm. Find the approximate change in its volume.

7 Find the approximate change in the surface area of the balloon of question 6.

8 The volume of a gas expanding adiabatically is related to its pressure p by the law pv^γ = constant (where γ is a constant). If δp and δv denote corresponding small changes in p and v respectively, show that $\dfrac{\delta p}{p} = -\gamma\dfrac{\delta v}{v}$.

9 The area of a triangle is calculated from the formula $\Delta = \frac{1}{2}ab\sin C$ with the usual notation. The sides a and b are measured accurately as 10 cm and 12 cm, but C is subject to an error of anything up to $\pm\frac{1}{2}°$ about the measured value of 40°. Find approximately the maximum error in Δ.

10 $y = x^2$. If x is decreased by 0·2 per cent, find the approximate percentage decrease in y. Hence find an approximate value for $(99{\cdot}8)^2$.

11.4 The tangent and normal to a curve

$f'(x)$ measures the gradient of the tangent to the curve $y = f(x)$ at the point with abscissa x. The gradient of the tangent at the point with abscissa x_1 is then $f'(x_1)$. Thus the equation of the tangent to the curve at the point (x_1, y_1) on the curve is by (18.6)

$$y - y_1 = f'(x_1)(x - x_1) \tag{11.7}$$

The normal to the curve at the point (x_1, y_1) on the curve is the line through this point perpendicular to the tangent to the curve. The gradient of the normal is thus $-1/f'(x_1)$ and the equation of the normal is, by (18.6),

$$y - y_1 = \frac{-1}{f'(x_1)}(x - x_1) \tag{11.8}$$

Example 1 Find the equation of the tangent and normal to the curve $y = 3x^2 - 5x$ at the point $(1, -2)$

$$y = f(x) = 3x^2 - 5x$$
$$f'(x) = 6x - 5$$
$$f'(1) = 6.1 - 5 = 1$$

Therefore, the tangent at (1, −2) has the equation

$$y-(-2) = 1(x-1)$$
$$y+2 = x-1$$
$$y = x-3$$

The normal at (1, −2) has the equation

$$y-(-2) = -1(x-1)$$
$$y = -x-1$$

Example 2 Find the equation of the tangent to the circle $x^2 + y^2 = 2a^2$ at the point (a, a).

$$x^2 + y^2 = 2a^2$$

therefore

$$\frac{\mathrm{d}}{\mathrm{d}x}(x^2) + \frac{\mathrm{d}}{\mathrm{d}x}(y^2) = 0$$
$$2x + 2y\frac{\mathrm{d}y}{\mathrm{d}x} = 0$$
$$\frac{\mathrm{d}y}{\mathrm{d}x} = -\frac{x}{y}$$

Thus at the point (a, a),

$$\frac{\mathrm{d}y}{\mathrm{d}x} = -1$$

Therefore, the tangent to the circle at (a, a) has the equation

$$y-a = -(x-a)$$
$$y = -x+2a$$

Example 3 Find the equation of the tangent to the curve $y = 2x^2 - x + 3$ which is parallel to the line $y = 3x - 2$.

$$y = \mathrm{f}(x) = 2x^2 - x + 3$$

therefore

$$\frac{\mathrm{d}y}{\mathrm{d}x} = \mathrm{f}'(x) = 4x - 1$$

The line $y = 3x - 2$ has the gradient 3

$$\mathrm{f}'(x) = 3 \quad \text{when} \quad 4x - 1 = 3 \quad \text{i.e.} \quad x = 1$$

Thus the point of contact of the required tangent is (1, 4). Therefore the equation of this tangent is

$$y-4 = 3(x-1)$$
$$y = 3x+1$$

Example 4 Find the equation of the tangent to the curve $27y^2 = 4x^3$ at the point $(3p^2, 2p^3)$.

We see that $x = 3p^2$, $y = 2p^3$ are parametric equations for the curve $27y^2 = 4x^3$ since

$$27(2p^3)^2 = 108p^6 \quad \text{and} \quad 4(3p^2)^3 = 108p^6$$

The tangent at the point with parameter p has gradient

$$\frac{dy}{dx} = \frac{dy}{dp} \bigg/ \frac{dx}{dp}$$
$$= 6p^2/6p = p$$

Thus the tangent has the equation

$$y - 2p^3 = p(x - 3p^2)$$
$$y = px - p^3$$

Exercises 11d

1 Find the equation of the tangent and normal to the curve $y = 3x^2 - 6x + 1$ at the point (2, 1).

2 Find the equation of the tangent and normal to the curve $y = x^3 - 3x^2 + 2$ at the point (1, 0).

3 Find the equations of the tangents to the curve $y = x(x - 1)(x - 2)$ at the points where it crosses the x-axis.

4 Find the equation of the tangent to the hyperbola $x^2 - y^2 = 16$ at the point (5, 3).

5 Find the equation of the tangent and normal to the curve $x^2 + xy + y^2 = 3$ at the point (1, 1).

6 Find the equations of the tangent and normal to the curve $y^2 = 4ax$ at the point $(a, 2a)$.

7 Find the equations of the tangents to the curve $y = x^3 - 6x^2 + 9x + 4$ which are parallel to the x-axis.

8 Find the equations of the tangents to the curve $y = x^3 - 5x^2 + 8x + 1$ which are parallel to the line $y = 5x - 7$.

9 Find the equation of the tangent to the curve $y = \dfrac{x - 1}{x + 1}$ which is perpendicular to the tangent at the point $(1, -1)$ to the curve $y = x^2 - 4x + 2$.

10 Find the equations of the tangents to the curve $y = \dfrac{x}{x + 1}$ which are parallel to the line $y = x$.

11 For the curve with parametric equations $x = a\cos\theta$, $y = a\sin\theta$, show that $dy/dx = -\cot\theta$. Hence find the equation of the tangent to the curve at the point where $\theta = \pi/4$.

12 If $x = at^2$, $y = 2at$, find the equation of the tangent to this curve at the point where $t = 1$ [cf. question 6].

13 Show that the points of intersection of the line $y = mx + c$ and the curve $y^2 = 4x$ are given by the solutions of the equation $m^2x^2 + 2(mc - 2)x + c^2 = 0$. Hence show that if $y = mx + c$ is a tangent to the curve then $c = 1/m$. [See also Section 20.4.]

14 Show that the x-axis is a tangent to the curve $y = x\sin x$.

15 Show that $x = ct$, $y = c/t$ are parametric equations for the curve $xy = c^2$. Deduce that the normal to the curve at the point with parameter t has the equation $ty - t^3x = c(1 - t^4)$.

11.5 The maximum and minimum values of a function

Figure 11.3 shows the graph of a function $y = f(x)$. The point A is called a local maximum of this function. The value of the function at A exceeds its values in a certain neighbourhood of A. Similarly, C is a local maximum and B a local minimum. As can be seen, B is not the absolute minimum value of the function, the value at D for example being less than the value at B. This is the reason for the term 'local' minimum, although in much of the literature this word is omitted although generally implied.

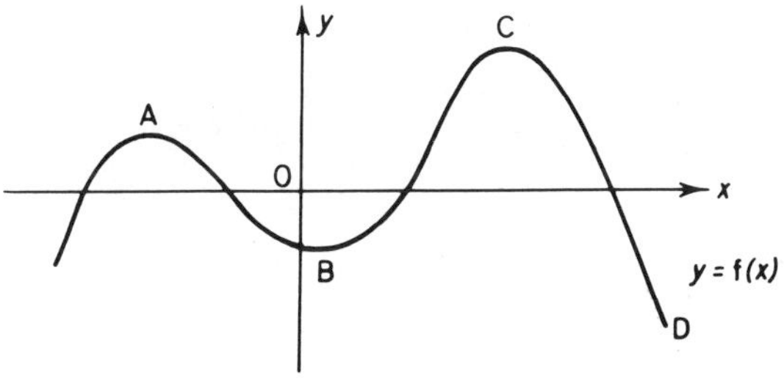

Figure 11.3

The positions of the points A, B, C may be determined by using the property that the derivative is zero (the tangent is parallel to the x-axis) at local maxima or minima.

To distinguish between local maxima and local minima, we shall examine the derivative in the neighbourhood of A and B respectively. Near A, the derivative is positive to the left of A, zero at A and negative to the right of A, i.e. it changes sign from positive to negative as x increases. Near B, the derivative is negative to the left of B, zero at B, and positive to the right of B, i.e. it changes sign from negative to positive as x increases. *Figure 11.4* shows the graph $y = f(x)$ together with the sign (or zeros) of its derivative marked on it.

In those regions in which $f'(x)$ is positive, we say that $f(x)$ is an increasing function of x; in those regions in which $f'(x)$ is negative, we say that $f(x)$ is

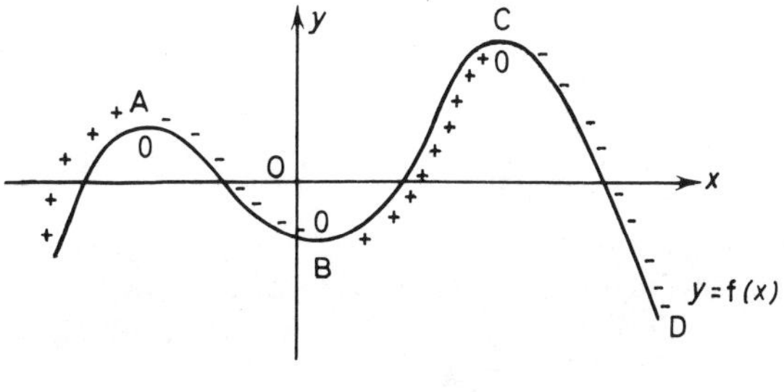

Figure 11.4

a decreasing function of x; and at points where $f'(x)$ is zero, we say that $f(x)$ is stationary. In the case of local maxima and minima (see Section 11.6 for other cases) the stationary points are sometimes called turning points. We can state the following rules for determining the stationary points which are also (local) maxima and minima, i.e. turning points, and for distinguishing between them.

RULE I At a turning point, $f'(x) = 0$.
RULE II At a local maximum, $f'(x)$ changes from positive to negative as x increases.
RULE III At a local minimum, $f'(x)$ changes from negative to positive as x increases.

In practice we evaluate (or at least examine the sign of) $f'(x)$ for values of x just less than and just greater than its value at the turning point.

Example 1 Find the nature of the turning points of the function $y = x^3 - 2x^2 + x + 4$.

$$\frac{dy}{dx} = 3x^2 - 4x + 1$$

At the turning points $dy/dx = 0$, that is

$$3x^2 - 4x + 1 = 0$$
$$(3x - 1)(x - 1) = 0$$
$$x = \tfrac{1}{3} \quad \text{or} \quad x = 1$$

Consider the value $x = \frac{1}{3}$. When $x = \frac{1}{4}$ (a convenient value just less than $\frac{1}{3}$)

$$\frac{dy}{dx} = 3 \times \tfrac{1}{16} - 4 \times \tfrac{1}{4} + 1 = \tfrac{3}{16}$$

When $x = \frac{1}{2}$ (a convenient value just greater than $\frac{1}{3}$)

$$\frac{dy}{dx} = 3 \times \tfrac{1}{4} - 4 \times \tfrac{1}{2} + 1 = -\tfrac{1}{4}$$

Thus, since dy/dx changes sign from positive to negative, when $x = \frac{1}{3}$, y is a maximum, the value of y being

$$\tfrac{1}{27} - \tfrac{2}{9} + \tfrac{1}{3} + 4 = \tfrac{112}{27}$$

For the value $x = 1$. When $x = 0{\cdot}9$

$$\frac{dy}{dx} = 3 \times 0{\cdot}81 - 4 \times 0{\cdot}9 + 1 = -0{\cdot}17 \quad \text{is negative}$$

When $x = 1{\cdot}1$

$$\frac{dy}{dx} = 3 \times 1{\cdot}21 - 4 \times 1{\cdot}1 + 1 = 0{\cdot}23 \quad \text{is positive}$$

Thus when $x = 1$, y is a minimum, the value of y being $1 - 2 + 1 + 4 = 4$.

A second procedure for distinguishing between maximum and minimum values may be obtained as follows. In the region of a maximum, $f'(x)$ changes sign from positive to negative as x increases. Thus $f'(x)$ is a decreasing function of x in the region so that $f''(x)$ is negative. Near a local minimum, $f'(x)$ is an increasing function so that $f''(x)$ is positive.

Hence at stationary points giving maximum values, $f''(x) < 0$; and at turning points giving minimum values, $f''(x) > 0$.

If at a stationary point $f''(x) = 0$, *no conclusions* can be drawn using the above argument and we have to resort to our original criterion for distinguishing between maximum and minimum values.

For the example just considered.

$$f''(x) = \frac{d^2y}{dx^2} = 6x - 4$$

When $x = \frac{1}{3}$

$$\frac{d^2y}{dx^2} = 2 - 4 = -2 \quad \text{is negative}$$

When $x = 1$

$$\frac{d^2y}{dx^2} = 6 - 4 = 2 \quad \text{is positive}$$

Thus $x = \frac{1}{3}$ gives a maximum and $x = 1$ gives a minimum for y.

Example 2 Find the maximum and minimum values of $y = x^3 - 6x^2 + 9x$.

$$\frac{dy}{dx} = 3x^2 - 12x + 9$$
$$= 3(x^2 - 4x + 3) = 3(x - 1)(x - 3)$$

Therefore, $dy/dx = 0$ when $x = 1$ or $x = 3$ and these could give turning values.

$$\frac{d^2y}{dx^2} = 6x - 12$$

When $x = 1$

$$\frac{d^2y}{dx^2} = -6 < 0 \quad \text{giving a maximum value of 4 for } y$$

When $x = 3$

$$\frac{d^2y}{dx^2} = 18 - 12 = 6 > 0 \quad \text{giving a minimum value of 0 or } y$$

Example 3 Rectangles are inscribed in a circle of radius r. Find the dimensions of the rectangle which has maximum area.

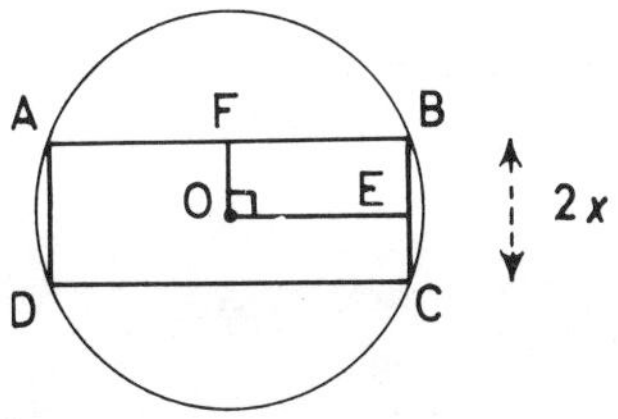

Figure 11.5

Figure 11.5 shows the circle with one such rectangle ABCD inscribed in it. O is the centre of the circle.

Let BC = $2x$. Let E and F be the mid-points of BC and AB. Then since $FB^2 = OB^2 - BE^2 = r^2 - x^2$, $FB = \sqrt{(r^2 - x^2)}$. Therefore $AB = 2\sqrt{(r^2 - x^2)}$ and the area of ABCD is

$$A = 4x\sqrt{(r^2 - x^2)}$$

For the maximum value of A, $dA/dx = 0$, therefore

$$4\sqrt{(r^2 - x^2)} - \frac{4x^2}{\sqrt{(r^2 - x^2)}} = 0$$

$$\frac{4(r^2 - x^2) - 4x^2}{\sqrt{(r^2 - x^2)}} = 0$$

$$\frac{4(r^2 - 2x^2)}{\sqrt{(r^2 - x^2)}} = 0$$

so that

$$r^2 - 2x^2 = 0 \quad \text{i.e.} \quad x = \frac{r}{\sqrt{2}} \quad \text{(only the positive value is valid)}$$

When $x < r/\sqrt{2}$, dA/dx is positive since then $r^2/2 > x^2$. When $x > r/\sqrt{2}$, dA/dx is negative. Thus A is a maximum when $x = r/\sqrt{2}$. [This method for distinguishing between maximum and minimum values is more convenient in this example than calculating d^2A/dx^2.]

Thus A has a maximum when the rectangle has dimensions $\sqrt{2}r$ by $\sqrt{2}r$; i.e. it is a square.

Example 4 The point X is 21 km south of the point Y. At noon a boy starts from X and jogs due east at 9 km/h. At the same time a second boy starts from Y and jogs south at 12 km/h. Find their least distance apart.

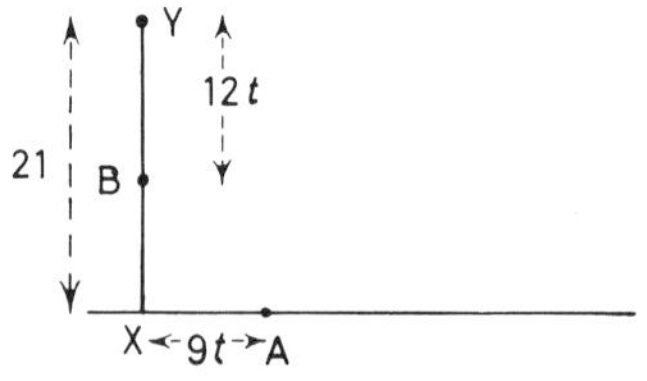

Figure 11.6

At time t hours after noon, the first boy is at A, $9t$ km east of X, and the second boy is at B, $12t$ km south of Y (*Figure 11.6*).

If d is their distance apart,

$$\begin{aligned} d^2 &= BX^2 + AX^2 \\ &= (21 - 12t)^2 + (9t)^2 \\ &= 225t^2 - 504t + 441 \end{aligned}$$

Since d is positive, d is a minimum when d^2 is a minimum, i.e. when

$$\frac{\mathrm{d}}{\mathrm{d}t}(d^2) = 450t - 504 = 0$$

$$t = \tfrac{504}{450} = \tfrac{56}{50}$$

$$\frac{\mathrm{d}^2}{\mathrm{d}t^2}(d^2) = 450 \quad \text{which is always positive}$$

Therefore, $t = \frac{56}{50}$ gives a minimum for d^2 of

$$441 - \frac{504 \times 56}{50} + 225 \times \frac{56^2}{50^2} = 441 - \frac{1008 \times 56}{100} + \frac{56 \times 504}{100}$$

$$= 441 - \frac{56 \times 504}{100} = 158{\cdot}76$$

Thus the minimum value of $d \approx 12{\cdot}6$ km

A further note on maxima and minima

Although in most cases $\mathrm{f}'(x)$ will be zero at a maximum or minimum value, the essential property of such points is that $\mathrm{f}'(x)$ should change sign. *Figure 11.7* shows the graph of the function $y = 1 - x^{2/3}$; the sign of $\mathrm{d}y/\mathrm{d}x$ is marked on the graph. This function has a maximum at $x = 0$ but $\mathrm{d}y/\mathrm{d}x = \frac{2}{3}x^{-1/3}$, although it changes sign as x passes through 0, is not defined for $x = 0$! This is also true for the function $y = |x|$, whose graph is shown in *Figure 9.3*.

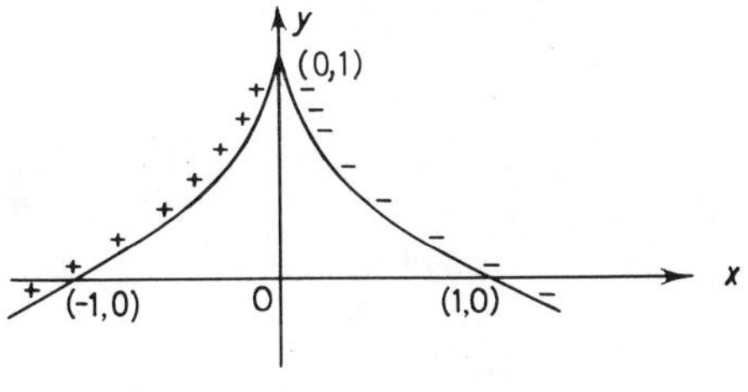

Figure 11.7

Exercises 11e

1 If $y = (x - 1)(x + 2)^2$, find the maximum and minimum values of y.
2 Find the maximum and minimum values of $y = x(x - 1)^2$.
3 Find the maximum and minimum values of $y = \dfrac{x}{x^2 + 1}$.
4 Find the maximum and minimum values of $\sin t + \frac{1}{2}\cos 2t$.
5 Show that the maximum value of $a\cos\theta + b\sin\theta$ is $\sqrt{(a^2 + b^2)}$. Can you show this without using the calculus? What is the minimum value?
6 Find the dimensions of the largest right circular cylinder which can be cut from a sphere of radius r.
7 An isosceles triangle of vertical angle 2θ is inscribed in a circle of radius r. Find an expression for the area of the triangle as a function of θ, and show that this is a maximum when the triangle is equilateral.

8 A right circular cone is constructed to have a total surface area A. Show that its volume $V = \frac{1}{3}r\sqrt{(A^2 - 2\pi A r^2)}$, where r is the radius of its base. Hence show that the largest such cone has semi-vertical angle $\arctan(1/2\sqrt{2})$.

9 The force exerted on a small magnet placed at a distance x from the centre of a plane circular coil of radius a, and along the axis of the coil is proportional to $x/(x^2 + a^2)^{5/2}$ when an electric current flows in the coil. Show that the force is a maximum when $x = \frac{1}{2}a$.

10 A man is situated at A, a km from a road XY (*Figure 11.8*). He wishes to reach the point Y where XY = b km. His speed on the road is u km/h, and his speed across country is v km/h ($u > v$). If he wishes to reach Y as quickly as possible, find the position of the point P where he joins the road.

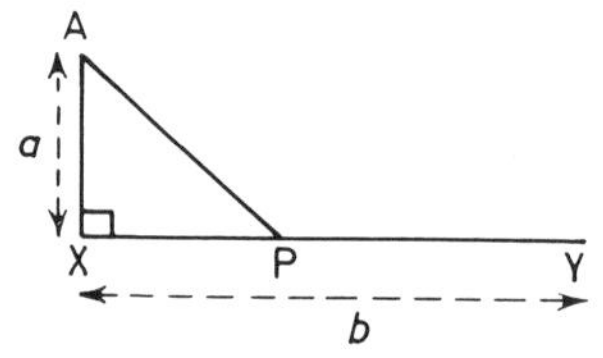

Figure 11.8

11.6 Points of inflexion

Consider the function $y = x^3(x - 4) = x^4 - 4x^3$.

$$\frac{dy}{dx} = 4x^3 - 12x^2 = 4x^2(x - 3)$$

$$\frac{dy}{dx} = 0 \quad \text{when} \quad x = 0 \quad \text{or} \quad x = 3$$

Near $x = 3$, dy/dx changes sign from negative to positive as x increases through the value 3. Thus $x = 3$ gives a minimum value of -27 for y. Near $x = 0$, dy/dx is negative for x just below zero, is zero when x is zero, and is negative again for x just greater than zero. Thus, although dy/dx is zero, since dy/dx does not change sign as x passes through this value, this point gives neither a maximum nor a minimum value for y.

Figure 11.9 shows the graph of the function $y = x^3(x - 4)$. The sign of dy/dx is indicated on the graph.

At B the function has a minimum value and at the origin there is a point called a point of inflexion. At such a point, the graph of the function changes from being concave up to concave down or vice versa. (In our particular case, it is the former.) As the value of x increases through zero, the *derivative* changes from negative to zero, and then to negative again, i.e. at O the derivative has a maximum. This is true quite generally; at a point of inflexion the derivative has a maximum value or a minimum value. This latter condition enables us to give a criterion for finding points of inflexion, for at points at which dy/dx is a maximum or a minimum

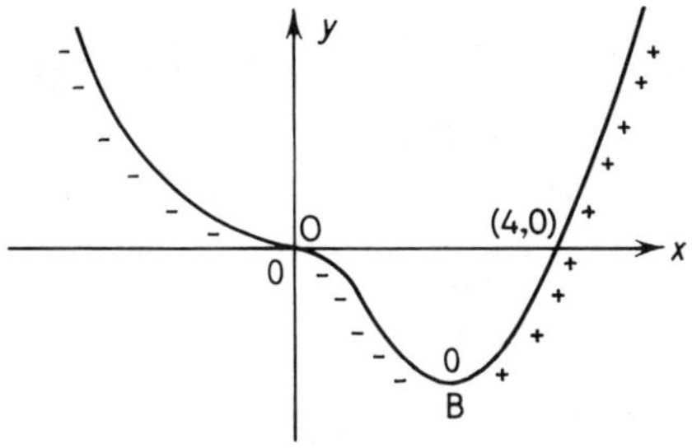

Figure 11.9

$$\frac{d^2y}{dx^2} = 0 \quad \text{and changes sign}$$

This is the case in our present example where

$$\frac{d^2y}{dx^2} = 12x^2 - 24x = 12x(x-2)$$

which vanishes and changes sign at the origin from positive to negative.

Although it is so in the example chosen, it is not necessary that dy/dx be zero at a point of inflexion (see *Figure 11.10*). In *Figure 11.10a*, the curve

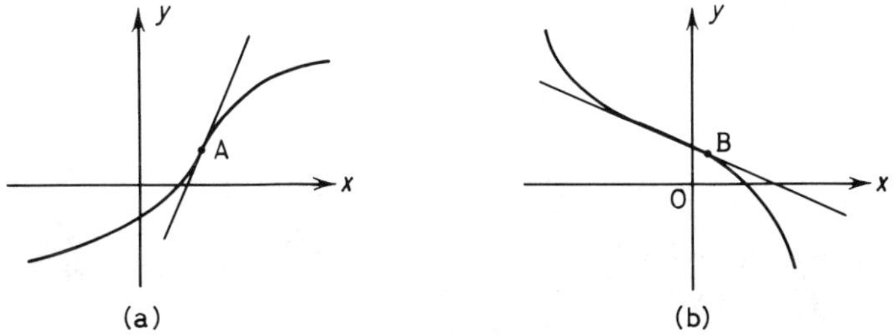

Figure 11.10

has a point of inflexion at A; dy/dx is a maximum. In *Figure 11.10b*, the curve has a point of inflexion at B; dy/dx is a minimum. In both cases, $d^2y/dx^2 = 0$ and changes sign but $dy/dx \neq 0$.

To sum up the results of this and the previous section:

$$\left.\begin{array}{l} \text{If } \dfrac{dy}{dx} = 0 \text{ and } \dfrac{d^2y}{dx^2} < 0\text{, there is a maximum value.} \\[2ex] \text{If } \dfrac{dy}{dx} = 0 \text{ and } \dfrac{d^2y}{dx^2} > 0\text{, there is a minimum value.} \\[2ex] \text{If } \dfrac{d^2y}{dx^2} = 0 \textit{ and changes sign}\text{, there is a point of inflexion.} \end{array}\right\} \quad (11.9)$$

Example 1 Find the maximum and minimum values and the points of inflexion of $y = x^3 - 6x^2 + 9x + 1$.

$$\frac{dy}{dx} = 3x^2 - 12x + 9 = 3(x-1)(x-3)$$

therefore, dy/dx is zero when $x = 1$ or when $x = 3$

$$\frac{d^2y}{dx^2} = 6x - 12$$

When $x = 1$

$$\frac{d^2y}{dx^2} = -6 < 0$$

When $x = 3$

$$\frac{d^2y}{dx^2} = 6 > 0$$

$d^2y/dx^2 = 6x - 12$ is zero when $x = 2$ and changes sign. Thus

when $x = 1$ y has a maximum value of 5
when $x = 3$ y has a minimum value of 1

and there is just one point of inflexion at the point (2, 3).

Exercises 11f

1 Find the position of the point of inflexion of the curve $y = 2x^3 - 5x^2 - 4x + 1$.
2 Find the positions of the points of inflexion of the curve $y = 3x^4 - 4x^3 + 2$.
3 Find the positions of the turning values and the point of inflexion of the curve $y = x^3 - 2x^2 + x + 3$.
4 Show that for the curve $y = x^3$, $d^2y/dx^2 = 0$ and changes sign at the origin. Plot the graph.
5 Show that for the curve $y = x^4$, $d^2y/dx^2 = 0$ but does not change sign at the origin. Plot the graph.

11.7 Curve sketching

On many occasions it is useful to make a rough sketch of a curve without plotting a large number of its points. Outlined below is a systematic procedure which should enable the shape of the curve to be obtained. It is not always necessary to consider every point detailed below.

(i) Determine if the curve is symmetrical about either of the co-ordinate axes. If its equation involves only even powers of x, it will be symmetrical about the y-axis; if only even powers of y are involved, it will be symmetrical about the x-axis.
(ii) Examine the behaviour of the function for large positive and large negative values of x, i.e. examine y as $x \to \pm\infty$.
(iii) Seek values of x for which y is not defined. Some common examples

will be values of x which make the denominator of a rational function zero or which make y^2 negative.

(iv) Find the value of y when $x = 0$, and if convenient the value(s) of x when $y = 0$. This will give the points where the curve crosses the axes.

(v) Calculate dy/dx and examine its sign. Where dy/dx is positive, the graph will slope up from left to right; where dy/dx is negative, the graph will slope down from left to right.

(vi) Find the turning points and points of inflexion.

Example 1 Sketch the curve $y = x^4 - 24x^2 + 64x + 10$.

(i) There is no symmetry about either axis.

(ii) When $x \to \pm\infty$, $y \to \infty$, the dominant term being x^4.

(iii) y is defined for all x.

(iv) $y = 10$ when $x = 0$. $y = 0$ when x satisfies $x^4 - 24x^2 + 64x + 10 = 0$ and this equation is not easily solved.

(v) $dy/dx = 4x^3 - 48x + 64 = 4(x^3 - 12x + 16)$
$$= 4(x-2)^2(x+4)$$
Therefore, dy/dx is positive for $x > -4$ and negative for $x < -4$ since $(x-2)^2$ is always positive.

(vi) dy/dx is zero when $x = 2$ or $x = -4$

$$\frac{d^2y}{dx^2} = 12x^2 - 48 = 12(x-2)(x+2)$$

When $x = -4$, $d^2y/dx^2 = 144 > 0$ so $x = -4$ gives a minimum value of -374 for y.

When $x = 2$, d^2y/dx^2 is zero and changes sign. Thus $x = 2$ gives a point of inflexion with the tangent parallel to the x-axis. $x = -2$ gives a second point of inflexion. The curve is sketched in *Figure 11.11*.

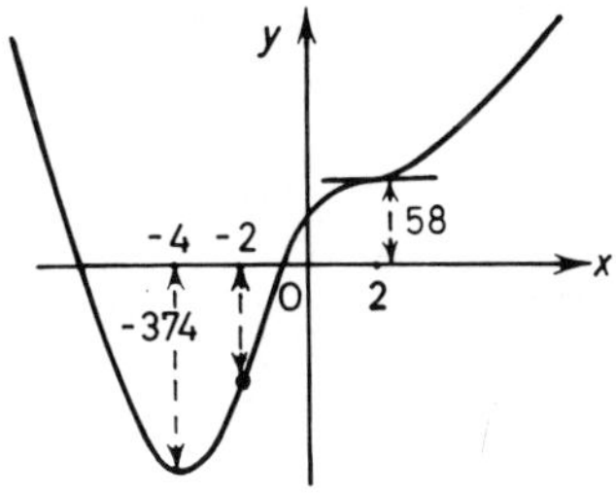

Figure 11.11

Example 2 Sketch the curve $y^2 = x$.

(i) The curve is symmetrical about the x-axis.

(ii) When $x \to \infty$, $y \to \infty$; when $x \to -\infty$, y is not defined.

(iii) y is defined only if x is not negative.

(iv) $y = 0$ when $x = 0$. The curve passes through the origin.

(v) With $y^2 = x$, $2y\,dy/dx = 1$, i.e. $dy/dx = 1/2y$. Thus when y is positive (negative), dy/dx is positive (negative).

(vi) dy/dx is never zero so there are no turning values, but at the origin $dy/dx \to \infty$ as $y \to 0$, i.e. the curve is vertical. It is shown in *Figure 11.12.*

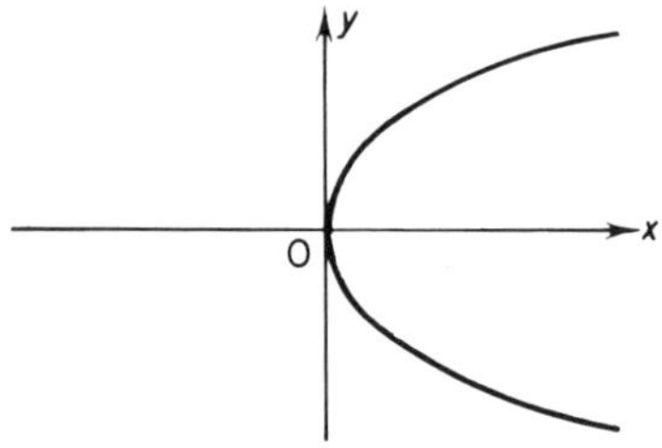

Figure 11.12

Example 3 Sketch the curve $y = \dfrac{2x+1}{x-1}$.

(i) There is no symmetry about either axis.

(ii) When $x \to \pm\infty$, $y \to 2$. For large (positive or negative) values of x the graph approaches the line $y = 2$. (A horizontal asymptote.)

(iii) y is defined for all x except $x = 1$. When x is just less than one, y is large and negative; for x just greater than one, y is large and positive. $x = 1$ is a vertical asymptote.

(iv) When $x = 0$, $y = -1$ and when $y = 0$, $x = -\frac{1}{2}$.

(v) $$\frac{dy}{dx} = \frac{(2x-1)-2x-1}{(x-1)^2} = -\frac{3}{(x-1)^2}$$

Therefore, dy/dx is always negative, and since it is never zero there are no turning values. The curve is shown in *Figure 11.13.*

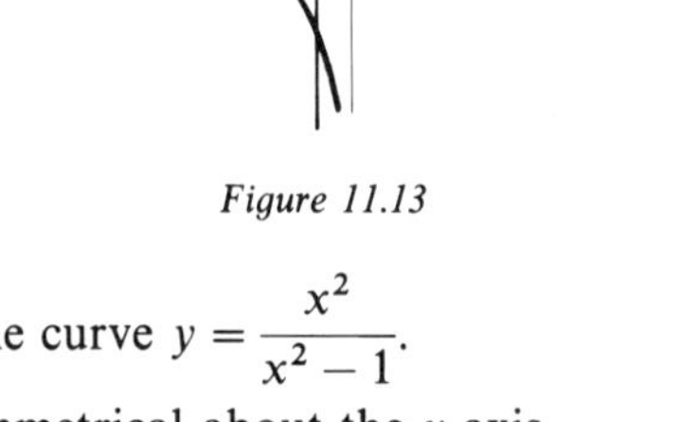

Figure 11.13

Example 4 Sketch the curve $y = \dfrac{x^2}{x^2-1}$.

(i) The curve is symmetrical about the y-axis.

(ii) When $x \to \pm\infty$, $y \to 1$. Indeed, $y = 1 + \dfrac{1}{x^2-1}$, so for large x the curve approaches the horizontal asymptote $y = 1$ from above the asymptote whether x is positive or negative. The last point is also clear from the symmetry.

(iii) $y = \dfrac{x^2}{(x-1)(x+1)}$ is defined for all x except $x = 1$ or $x = -1$ when the denominator is zero. When x is just less than -1, y is large and positive. When x is just greater than -1, y is large and negative. When x is just less than 1, y is large and negative, and when x is just greater than 1, y is large and positive. The lines $x = -1$ and $x = 1$ are vertical asymptotes.

(iv) When $x = 0$, $y = 0$.

(v) $\dfrac{dy}{dx} = \dfrac{-2x}{(x^2-1)^2}$. This is zero when $x = 0$ and changes sign from positive to negative as x increases through the value zero. Hence there is a maximum value when $x = 0$. The curve is sketched in *Figure 11.14*.

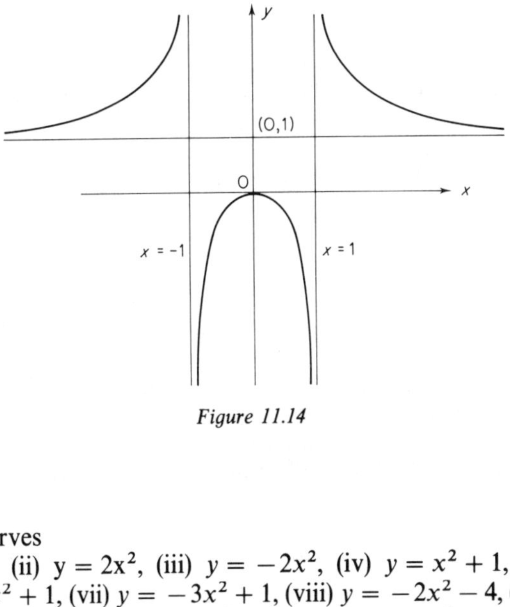

Figure 11.14

Exercises 11g

Sketch the curves

1 (i) $y = x^2$, (ii) $y = 2x^2$, (iii) $y = -2x^2$, (iv) $y = x^2 + 1$, (v) $y = x^2 - 3$, (vi) $y = 6x^2 + 1$, (vii) $y = -3x^2 + 1$, (viii) $y = -2x^2 - 4$, (ix) $y = ax^2 + b$.

2 (i) $y = (x-1)^2$, (ii) $y = 2(x-1)^2$, (iii) $y = -2(x-1)^2$, (iv) $y = (x-1)^2 + 1$, (v) $y = (x-2)^2$, (vi) $y = (x+1)^2$, (vii) $y = 6(x+1)^2$, (viii) $y = -2(x-2)^2$, (ix) $y = 2(x-2)^2 + 3$, (x) $y = 3(x-2)^2 + 8$, (xi) $y = -(x-5)^2 + 4$, (xii) $y = a(x-b)^2 + c$.

3 (i) $y = x^3$, (ii) $y = x^4$, (iii) $y = x^5$, (iv) $y = x^6$.

4 (i) $y = 2x^3$, (ii) $y = -3x^3$, (iii) $y = x^3 + 2$, (iv) $y = x^3 - 6$, (v) $y = x^3 + b$, (vi) $y = ax^3 + b$, (vii) $y = 3x^4 - 4$, (viii) $y = 3(x-1)^3 + 4$, (ix) $y = 6(x-2)^4 + 3$, (x) $y = (x-1)^3 - 7$.

5 (i) $y = x(x-1)$, (ii) $y^2 = x(x-1)$, (iii) $y = x^2(x-1)$, (iv) $y^2 = x^2(x-1)$, (v) $y = x(x-1)(x-2)$, (vi) $y^2 = x(x-1)(x-2)$, (vii) $y = x(x-2)^2$, (viii) $y^2 = x(x-2)^2$.

6 (i) $y = x^2 - 6x - 7$, (ii) $y = 3 - 7x + 4x^2$.

7 (i) $y = x^3 - 6x + 4$, (ii) $y = x^5 - 5x^4 + 5x^3 - 2$.

8 $y^2 = x^3$ (pay particular attention to the form of the curve near the origin).

9 (i) $y = \dfrac{1}{x-1}$ (ii) $y = \dfrac{1}{2x-1}$ (iii) $y = \dfrac{x}{x-1}$

(iv) $y = \dfrac{x}{2x-1}$ (v) $y = \dfrac{x+1}{x-1}$ (vi) $y = \dfrac{x-2}{x+3}$

(vii) $y = \dfrac{2x+1}{3x-1}$ (viii) $y = \dfrac{x-4}{x+5}$ (ix) $y = \dfrac{x^2-12x+27}{x^2-4x+5}$

(x) $y = \dfrac{2x^2+4x+7}{x^2+2x+5}$

10 Sketch the curve $y = x^n$. Consider four cases: (i) n is a positive even integer, (ii) n is a positive odd integer, (iii) n is a negative even integer and (iv) n is a negative odd integer.

Exercises 11

1 Water is poured into a hemispherical bowl of radius 6 cm at a rate of 5 cm^3/sec. At what rate is the water rising in the bowl when the depth of water is 2 cm? (The volume of a cap, of a sphere of radius R, whose height is h is $\pi h^2(R - h/3)$.)

2 The distances u and v of a point and its image from a lens of focal length f are connected by the relation $1/u + 1/v = 1/f$. If $f = 10$ cm and the object is moved towards the lens at 2 cm/sec find the speed of its image when this is 25 cm from the lens.

3 The efficiency of an engine is given by $E = 100(1 - r^{-1/4})$ where r is the compression ratio. Find the rate at which E is changing with respect to r when $r = 7$.

4 A pipe delivers V m^3 of water in t sec, where $V = 12t - t^2/10$. At what rate is the water delivered after 10 sec?

5 Sand falling from a chute forms a conical pile whose height is always $\frac{5}{4}$ times the radius of the base. How fast is the radius of the base increasing when it is 3 m if the sand falls at the rate of 24 m^3/min?

6 A body moves along a line according to the law $s = t^3 - 9t^2 + 24t$. Find the positions of the body when its speed is zero and when its acceleration is zero.

7 A particle moves along a straight line Ox in the time interval $0 \leqslant t \leqslant \pi$; after t sec its distance from O is x m, where $x = t + \sin 2t$. Calculate the values of t between 0 and π when the direction of motion changes, and show that the particle always remains on the same side of O. Find also the times at which the acceleration is zero. Sketch the graph of x for $0 \leqslant t \leqslant \pi$, and state the largest value of x in this interval. [JMB]

8 A vehicle moves from rest on level ground in such a way that its speed is v m/sec; when it has covered a distance x m, x is given by the relation $x = \dfrac{2v^2}{60 - v}$. Sketch a graph showing v as a function of x, and show that the acceleration of the vehicle is $\dfrac{(60-v)^2}{2(120-v)}$ m/sec^2. [JMB]

9 The period of oscillation of a pendulum is calculated from the formula

$T = 2\pi\sqrt{(l/g)}$, where l is the length of the pendulum and g the acceleration due to gravity. Find the percentage error in the calculated value of T if g is taken to be 1000 cm/sec^2 instead of 980 cm/sec^2.

10 The side of a triangle is calculated by means of the formula $a^2 = b^2 + c^2 - 2bc \cos A$. If an error of 1° is made in the measurement of A, find the approximate error in the calculated value of a when $b = 10$ cm, $c = 15$ cm and $A = 60°$.

11 The pressure p units and the volume v units of an expanding gas are related by the law $pv^{1\cdot 4} = k$, where k is a constant. If the volume increases by 0·3 per cent, estimate the percentage change in the pressure. [JMB, part]

12 Prove that the gradient of the curve $y = x^3 + 6x^2 + 15x + 36$ is positive for all values of x. Show that the curve has a point of inflexion when $x = -2$, and state the gradient of the curve at this point. Write down the equation of the tangent to the curve at the point where $x = 0$, and find the co-ordinates of the point where this tangent meets the curve again. [JMB]

13 Find the abscissae of the points on the curve $y = x^3 - 3x^2 - 2x + 1$ at which the tangent is equally inclined to the co-ordinate axes.

14 Find the equation of the tangent to the curve $y^2 = \frac{1}{3}x^2(x+1)$ at the point (2, 2). Show that this tangent intersects the curve again at a point R and that it is the normal to the curve at R.

15 The curves (i) $x^2 - y^2 = 15$, and (ii) $xy = 4$ intersect at a point in the first quadrant. Find the equations of the tangents to both curves at the point, and show that they are at right angles to one another. [WJC]

16 If y is such that $dy/dx = x^3(x-1)^2(x^2+1)$, find the values of x for which y has stationary values and state the nature of the stationary values. [WJC, part]

17 The curve whose equation is $y = ax^3 + bx^2 + cx + d$ has a point of inflexion at $(-1, 4)$, has a turning point when $x = 2$ and passes through the point $(3, -7)$. Find the values of a, b, c, d and the position of the other turning point.

18 Show that $\dfrac{(x+4)}{(x^2+12)^{2/3}}$ has a maximum value of 0·945 approximately, and find its minimum value. [WJC]

19 (i) Find the maximum and minimum values of the function $\tan 2x \cot^2 x$, and the values of x, in the range $0 \leqslant x \leqslant \pi$, at which they occur.

(ii) Find the maximum and minimum values of y, and the corresponding values of x if $9y^2 + 6xy + 4x^2 - 24y - 8x + 4 = 0$. [WJC]

20 A right circular cone is inscribed in a sphere. Prove that the volume of the cone cannot exceed $\frac{8}{27}$ of the volume of the sphere.

21 Show that the function $f(x) = x^{1/\gamma}(1 - x^{\gamma - 1/\gamma})^{1/2}$, where $\gamma(>1)$ is a constant, has a maximum when $x = \left(\dfrac{2}{\gamma+1}\right)^{\gamma/\gamma - 1}$.

22 Find the maximum and minimum values and the points of inflexion of the function $\dfrac{x}{x^2+1}$, and show that the points of inflexion lie on the line $4y = x$.

23 Find the stationary values of the function $f(x) = 1 - \dfrac{9}{x^2} + \dfrac{18}{x^4}$ and determine their nature. Sketch the curve $y = f(x)$. [LU]

24 A tree trunk is in the form of a frustrum of a right circular cone, the radii of the end faces being a and b respectively ($a > b$) and the distance between these faces being l. A log in the form of a right circular cylinder is cut out of the trunk, the axis of the cylinder being perpendicular to the end faces of the frustrum. Show that, if $b < 2a/3$, the volume of the log is a maximum when its length is $al/3(a-b)$. If $b > 2a/3$, what is the length of the log when its volume is as great as possible. [LU]

25 A right circular cone of semi-vertical angle θ is circumscribed about a sphere of radius R. Show that the volume of the cone is given by $V = \frac{1}{3}\pi R^3\ (1 + \operatorname{cosec}\theta)^3 \tan^2\theta$ and find the value of θ when V is a minimum.

26 If $y = 2\sin x + \tan x$, prove that $d^2y/dx^2 = 2\sin x(\sec^2 x - 1)$. Show that for $0 < x < \pi/2$ the gradient of the function is greater than 3, and that for $\pi/2 < x < \pi$ the function has a turning point and is zero for a value of x other than π. Sketch these two branches of the graph of the function. [SUJB]

27 Find the maximum and minimum values of the function $x/(x^2 + 3x + 1)$. Sketch the graph of the function. [SUJB]

28 Find the abscissa of the point of inflexion on the curve $y = ax^3 + bx^2 + cx + d$ where a, b, c and d are constants.

29 Sketch the graphs of (i) $y = \dfrac{4x+1}{2x-3}$, (ii) $y = \dfrac{x+3}{x-1}$, (iii) $y = \dfrac{x^2 - 12x + 27}{x^2 - 4x + 15}$,

(iv) $y = \dfrac{2x^2 + 4x + 7}{x^2 + 2x + 5}$.

30 Find the equation of the tangent to the curve $y = 1/x$ at the point $(1, 1)$ and the equation of the tangent to the curve $y = \cos x$ at the point $(\pi/2, 0)$. Deduce that $1/x > \cos x$ for $0 \leqslant x \leqslant \pi/2$. [JMB, part]

31 Sketch with the same pair of axes the graphs of the functions $y = (x-2)^2$, $y = (x-2)^2 + 4$, $y = (x-2)^2 - 4$. Indicate on each graph the co-ordinates of the turning point and of the points where the graph crosses the axes. [JMB]

32 A straight line of variable slope passes through the fixed point (a, b) in the positive quadrant. Its intercepts on the co-ordinate axes are p and q (p, q both positive). Show that the maximum value of $p + q$ is $(\sqrt{a} + \sqrt{b})^2$.

33 Show that there is just one tangent to the curve $y = x^3 - x + 2$ which passes through the origin. Find its equation and point of contact with the curve.

34 A right circular cone is inscribed in a sphere of radius a. If its volume is a maximum, show that its altitude is $4a/3$. In the cone of maximum volume a right circular cylinder is inscribed. Show that the maximum volume of this cylinder is 32/243 of the volume of the sphere.

35 The efficiency of a jack is given by $E = \dfrac{\tan\theta}{\tan(\theta + A)}$ where θ is acute and A is constant. Show the maximum value of E is $(1 - \sin A)/(1 + \sin A)$ and find the value of θ for which this occurs.

36 The area of the sheet metal used in the manufacture of a closed cylindrical can is 400 cm^2. Find to the nearest cm^3, the largest possible volume of the can. [AEB]

37 A rectangle ABCD is inscribed in an equilateral triangle PQR of side $2a$, the points A and B lying on QR, C lying on RP and D on PQ. Find the length of AB for which the area of the rectangle is a maximum. [LU]

38 The equation of a curve is $y = \cot x - 8\cos x\,(0 < x < \pi)$. Find the co-ordinates of the points on the curve where $\mathrm{d}y/\mathrm{d}x = 0$. Sketch the curve. [C]

39 Find the co-ordinates of the points of intersection of the curves whose equations are $y = (x-1)(x-2)$, $y = \dfrac{3(x-1)}{x}$. State or obtain (i) the co-ordinates of the turning point of the first curve; (ii) the equations of the asymptotes of the second curve. Sketch the two curves on the same diagram. [JMB]

40 Sketch the curve $y = x/(1+x^2)$, finding its turning points, showing that the origin is a point of inflexion and indicating the behaviour of y when x is large. [LU]

41 Find the turning points on the curve $y = \cos x + 2\cos\frac{1}{2}x$ for $0 \leqslant x \leqslant 2\pi$. Hence sketch the curve over this range. [AEB]

42 Find the equations of the tangents to the curve $27y^2 = 4x^3$ at the points $\mathrm{P}(3p^2, 2p^3)$ and $\mathrm{Q}(3q^2, 2q^3)$. Show that these tangents intersect at the point $\mathrm{R}(\alpha, \beta)$, where $\alpha = p^2 + pq + q^2$, $\beta = pq(p+q)$. The points P and Q move along the curve in such a way that the tangents at P and Q are always perpendicular. Prove that R moves on the parabola $y^2 = x - 1$. [JMB]

43 A right pyramid having a square base is inscribed in a sphere of radius R, all five vertices of the pyramid lying on the sphere. The height of the pyramid is x; show that the four vertices forming the base of the pyramid lie on a circle of radius r, where $r^2 = 2Rx - x^2$. Hence, or otherwise, show that the volume, V, of the pyramid is given by the formula $V = \frac{2}{3}x^2(2R - x)$. If R is fixed but x may vary, find the greatest possible value of V. [C]

44 The fixed points A, O, B and C are on a straight line such that AO = OB = BC = 1 unit. The points A and B are also joined by a semicircle of radius 1 unit, and P is a variable point on this semicircle such that the angle POC is θ. Calculate the value of θ for which the area of the region R, bounded by the arc AP of the semicircle and the straight lines PC and AC is a maximum. Show that the perimeter of R is of length $L = 3 + \pi - \theta + (5 - 4\cos\theta)^{1/2}$. Prove that L has just one stationary point and that this occurs at the same value of θ for which the area of R is a maximum. Find the greatest value and the least value of L in the interval $0 \leqslant \theta \leqslant \pi$. [JMB]

45 If y is defined in terms of x by the equation $x^3 + y^3 - 3xy = 0$, show that the graph of y has a turning point at $(2^{1/3}, 2^{2/3})$. Determine whether this gives a maximum or minimum value of y. [O]

12

The logarithmic and exponential functions

12.1 The logarithmic function $y = \log_a x$

Figure 12.1 shows the graph of $y = \log_{10} x$. y is only defined for positive values of x. As x approaches zero, y becomes large and negative, changing very rapidly with respect to x. Thus when $x = 0{\cdot}01$, $y = -2$; when $x = 0{\cdot}000\,001$, $y = -6$, etc. As x takes on large positive values, y increases but only very slowly with respect to x. Thus when $x = 100$, $y = 2$; when $x = 1000$, $y = 3$; when $x = 10\,000$, $y = 4$, etc. When $x = 1$, $y = 0$. The general characteristics described above and the sketch of *Figure 12.1* are true for the function $y = \log_a x$, where a is any constant.

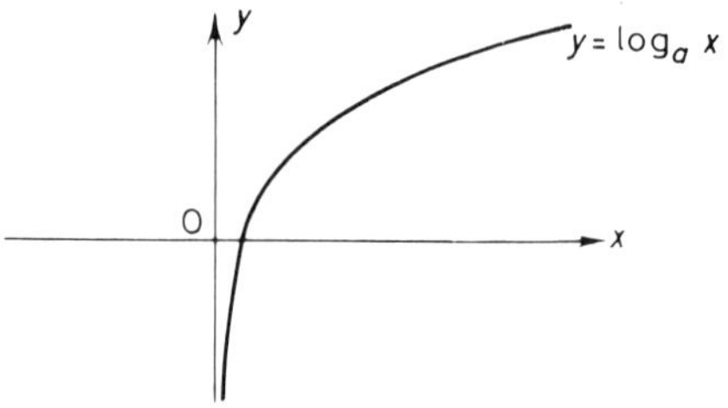

Figure 12.1

We shall now evaluate precisely the derivative of the logarithmic function as opposed to the rather vague statements above. None of the methods of Chapter 10 is applicable to the function $y = \log_a x$ and we have to resort to the method of differentiation from first principles.

If $y = \log_a x$, $y + \delta y = \log_a (x + \delta x)$, therefore

$$\delta y = \log_a (x + \delta x) - \log_a x$$

$$= \log_a \left(\frac{x + \delta x}{x} \right) = \log_a \left(1 + \frac{\delta x}{x} \right)$$

$$\frac{\delta y}{\delta x} = \frac{1}{\delta x} \log_a \left(1 + \frac{\delta x}{x} \right)$$

$$\frac{\delta y}{\delta x} = \frac{1}{x}\cdot\frac{x}{\delta x}\log_a\left(1 + \frac{\delta x}{x}\right)$$
$$= \frac{1}{x}\log_a\left(1 + \frac{\delta x}{x}\right)^{x/\delta x}$$

Therefore

$$\frac{dy}{dx} = \lim_{\delta x \to 0}\frac{\delta y}{\delta x} = \frac{1}{x}\lim_{\delta x \to 0}\log_a\left(1 + \frac{\delta x}{x}\right)^{x/\delta x} \tag{A}$$

To evaluate $\lim_{\delta x \to 0}\log_a(1 + \delta x/x)^{x/\delta x}$, we replace $x/\delta x$ by n. Then $\delta x \to 0$ is equivalent to n approaching infinity ($n \to \infty$) and the required limit may be written*

$$\log_a\left[\lim_{n \to \infty}\left(1 + \frac{1}{n}\right)^n\right]$$

A full investigation of $\lim_{n \to \infty}\left(1 + \frac{1}{n}\right)^n$ is beyond the scope of this course. For the present, we shall content ourselves with the evaluation of $\left(1 + \frac{1}{n}\right)^n$ for $n = 10, 100, 1000, 10\,000$. With the aid of six-figure logarithms or a calculator, we have

$$(1 + \tfrac{1}{10})^{10} = 2{\cdot}5936$$
$$(1 + \tfrac{1}{100})^{100} = 2{\cdot}7046$$
$$(1 + \tfrac{1}{1000})^{1000} = 2{\cdot}7164$$
$$(1 + \tfrac{1}{10\,000})^{10\,000} = 2{\cdot}7182$$

A brief study of these results should convince readers that as n increases $\left(1 + \frac{1}{n}\right)^n$ is going to approach a limiting value between 2·5 and 3·0. A fuller study of the problem reveals that $\lim_{n \to \infty}\left(1 + \frac{1}{n}\right)^n = 2{\cdot}718\,28$ to five decimal places. This number holds a very important place in all higher mathematics and is denoted by the symbol e. Although the above argument is at best tentative, the conclusion is correct and we have

$$e = \lim_{n \to \infty}\left(1 + \frac{1}{n}\right)^n \approx 2{\cdot}718\,28 \tag{12.1}$$

Thus, if we return to (A), we see that

$$\frac{d}{dx}(\log_a x) = \frac{1}{x}\log_a e \tag{12.2}$$

* We have taken lim(log...) = log(lim...). To prove this 'apparently obvious' result is beyond the scope of this course.

For a given value of a, $\log_a e$ is, of course, just a constant. If $a = e$ so that $\log_e a = 1$, we have the result

$$\frac{d}{dx}(\log_e x) = \frac{1}{x} \qquad (12.3)$$

$y = \log_e x$ is called the natural logarithm of x. (The notation $\ln x$ is also used to denote the logarithm of x to base e and is becoming increasingly popular. We shall, however, use $\log_e x$ in all our workings, but ln does occur in some of the chapter-end exercises derived from the examining boards.)

We may, of course, calculate the logarithm of x to any base—for example, logarithms to the base 10 are very convenient for calculations—but natural logarithms are very important in the theoretical aspects of mathematics. So it is important to remember that the derivative of the logarithmic function is $1/x$ *only when* the base is e, otherwise it is $(1/x)\log_a e$.

Example 1 Find dy/dx if (i) $y = \log_e 1/x$, (ii) $y = \log_e \sec x$.

(i) $y = \log_e 1/x = \log_e u$, where $u = 1/x$

$$\frac{dy}{dx} = \frac{dy}{du}\cdot\frac{du}{dx} = \frac{1}{u}\left(-\frac{1}{x^2}\right) = x\left(\frac{-1}{x^2}\right) = \frac{-1}{x}$$

$$\left[\frac{d}{dx}\left(\log_e\frac{1}{x}\right) = -\frac{d}{dx}(\log_e x) \quad \text{because} \quad \log_e\frac{1}{x} = -\log_e x\right]$$

(ii) $y = \log_e \sec x = \log_e u$, where $u = \sec x$

$$\frac{dy}{dx} = \frac{dy}{du}\cdot\frac{du}{dx} = \frac{1}{u}.\sec x \tan x$$

$$\frac{dy}{dx} = \tan x$$

Example 2 Find dy/dx if $y = \log_{10} x^2$.

By (1.30)

$$y = \log_{10} x^2 = \log_{10} e \times \log_e x^2 = 0{\cdot}4343 \log_e x^2$$
$$= 0{\cdot}8686 \log_e x$$

therefore

$$\frac{dy}{dx} = \frac{0{\cdot}8686}{x}$$

Example 3 Find dy/dx if $y = \log_e f(x)$.

$$y = \log_e f(x) = \log_e u \quad \text{where} \quad u = f(x)$$

$$\frac{dy}{dx} = \frac{dy}{du}\cdot\frac{du}{dx} = \frac{1}{u}f'(x)$$

therefore

$$\frac{d}{dx}[\log_e f(x)] = \frac{f'(x)}{f(x)} \qquad (12.4)$$

Exercises 12a

Find dy/dx when

1 $y = \log_e 2x$
2 $y = \log_e 1/x^2$
3 $y = \log_e \sin x$
4 $y = \log_e(ax + b)$
5 $y = \log_e \sqrt{(x - 1)}$
6 $y = \log_e \tan x$
7 $y = \log_e(\sec x + \tan x)$
8 $y = \log_e \sin^2 x$
9 $y = x \log_e x - x$
10 $y = \dfrac{\log_e x}{x}$
11 $y = \log_e \dfrac{1 - x}{1 + x}$
12 $y = \cos x \log_e \sin x$

13 Find dy/dx if $x + y + \log_e xy = 2$.
14 If $y = (\log_e x)/x$, show that dy/dx is zero when $x = e$.
15 If $y = x \log_e x$, find d^2y/dx^2.

12.2 The exponential function

The exponential function is the inverse of the logarithmic function. Thus if $y = e^x$, $x = \log_e y$. *Figure 12.2* shows a sketch of the function $y = e^x$. The function $y = a^x$ possesses the same general characteristics. *Figure 12.2* may be obtained directly from *Figure 12.1*.

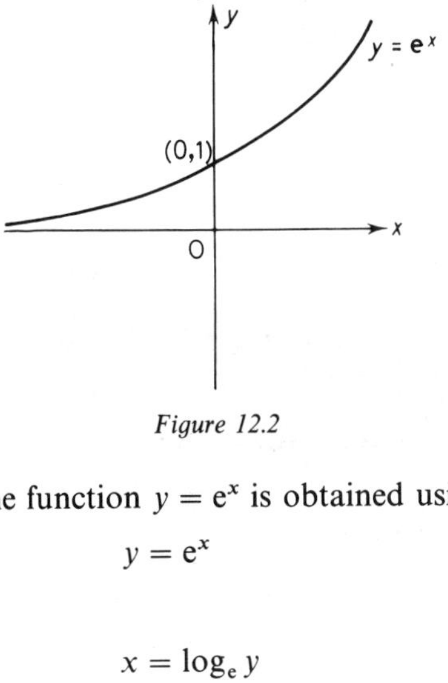

Figure 12.2

The derivative of the function $y = e^x$ is obtained using (10.17). Thus if

$$y = e^x$$

then

$$x = \log_e y$$

therefore

$$\frac{dx}{dy} = \frac{1}{y} \quad \text{by (12.3)}$$

$$\frac{dy}{dx} = y$$

that is

$$\frac{d(e^x)}{dx} = e^x \tag{12.5}$$

For the function $y = a^x$ it is convenient to write the constant a in the form $a = e^{\log_e a}$. Then

$$y = a^x = (e^{\log_e a})^x = e^{x \log_e a}$$

To differentiate y with respect to x, we need the rule for differentiating a function of a function

$$y = e^v \quad \text{where} \quad v = x \log_e a$$

$$\frac{dy}{dx} = \frac{dy}{dv} \cdot \frac{dv}{dx} = \log_e a . e^{x \log_e a}$$

that is

$$\frac{dy}{dx} = \log_e a . a^x \tag{12.6}$$

Example 1 Find dy/dx when (i) $y = e^{\cos x}$, (ii) $y = (\sin 2x) e^{2x}$.

(i) $y = e^{\cos x} = e^u$, where $u = \cos x$

$$\frac{dy}{dx} = \frac{dy}{du} \cdot \frac{du}{dx} = e^u . (-\sin x) = -e^{\cos x} \sin x$$

(ii) $y = (\sin 2x) e^{2x}$

Therefore, by (10.4)

$$\frac{dy}{dx} = \sin 2x \frac{d}{dx}(e^{2x}) + e^{2x} \frac{d}{dx}(\sin 2x)$$

$$\frac{d}{dx}(e^{2x}) = \frac{d}{dx}(e^u) \quad \text{where} \quad u = 2x$$

$$= \frac{d}{du}(e^u) \cdot \frac{du}{dx} = e^u \times 2 = 2e^{2x}$$

$$\frac{d}{dx}(\sin 2x) = \frac{d}{dx}(\sin u) \quad \text{where} \quad u = 2x$$

$$= \frac{d}{du}(\sin u) \cdot \frac{du}{dx} = \cos u \times 2 = 2 \cos 2x$$

Therefore

$$\frac{dy}{dx} = 2e^{2x} \sin 2x + 2e^{2x} \cos 2x = 2e^{2x}(\sin 2x + \cos 2x)$$

Example 2 If $y = e^{3x}$, show that $d^2y/dx^2 - 5dy/dx + 6y = 0$.

With $y = e^{3x}$

$$\frac{dy}{dx} = \frac{d}{du}(e^u) \cdot \frac{du}{dx} \quad \text{where} \quad u = 3x$$

$$= 3e^{3x}$$

$$\frac{d^2y}{dx^2} = 3 \frac{d}{dx}(e^{3x}) = 9e^{3x}$$

Therefore

$$\frac{d^2y}{dx^2} - 5\frac{dy}{dx} + 6y = 9e^{3x} - 5 \times 3e^{3x} + 6e^{3x} = 0$$

Example 3 Find dy/dx if $y = e^{f(x)}$.

$$y = e^{f(x)} = e^u \quad \text{where} \quad u = f(x)$$

$$\frac{dy}{dx} = \frac{dy}{du} \cdot \frac{du}{dx} = e^u . f'(x)$$

$$\frac{d}{dx}[e^{f(x)}] = f'(x)e^{f(x)} \qquad (12.7)$$

Exercises 12b

Find dy/dx when

1 $y = e^{3x}$ **2** $y = e^{-x^2}$ **3** $y = e^{\sin x}$
4 $y = e^{-x}$ **5** $y = e^{ax+b}$ **6** $y = 2^x$
7 $y = 3^{x^2}$ **8** $y = \cos xe^x$ **9** $y = xe^{-x^2}$
10 $y = \dfrac{e^x}{e^x + e^{-x}}$ **11** $y = e^x \log_e x$ **12** $y = e^{-x^2} \sin x^2$
13 If $y = ae^{px} + be^{-px}$ show that $d^2y/dx^2 = p^2y$.
14 If $y = e^{2x}$ show that $d^2y/dx^2 - 3\,dy/dx + 2y = 0$.
15 Find m if $y = e^{mx}$ is such that $d^2y/dx^2 - 3\,dy/dx - 4y = 0$.

12.3 Logarithmic differentiation

Logarithmic differentiation is the name given to a particular technique which can be very useful in helping us to obtain the derivatives of certain functions. The technique is illustrated by the following examples.

Example 1 Find dy/dx if $y = e^{x^3}$.
With $y = e^{x^3}$, we first take logarithms on each side to give

$$\log_e y = \log_e (e^{x^3}) = x^3$$

On differentiating this with respect to x, we have

$$\frac{d}{dx}(\log_e y) = \frac{d}{dx}(x^3) = 3x^2$$

$$\frac{d}{dy}(\log_e y) \cdot \frac{dy}{dx} = 3x^2$$

Therefore

$$\frac{1}{y}\frac{dy}{dx}^* = 3x^2$$

* The result $d/dx(\log_e y) = (1/y)dy/dx$ should be memorised once it is thoroughly understood how to obtain it.

so that

$$\frac{dy}{dx} = y \times 3x^2 = 3x^2e^{x^3}$$

This same result can be obtained by (12.7). Consider, however, the next example.

Example 2 Find dy/dx if $y = x^x$, where (12.7) cannot be used directly.

We have $\log_e y = \log_e x^x = x \log_e x$ by (1.29), therefore

$$\frac{d}{dx}(\log_e y) = \frac{d}{dx}(x \log_e x)$$

$$\frac{1}{y}\frac{dy}{dx} = x\frac{d}{dx}(\log_e x) + \log_e x \frac{d}{dx}(x) \quad \text{by (10.4)}$$

$$\frac{1}{y}\frac{dy}{dx} = x\left(\frac{1}{x}\right) + \log_e x\,(1) = 1 + \log_e x$$

$$\frac{dy}{dx} = y(1 + \log_e x) = x^x(1 + \log_e x)$$

In order to apply (12.7), we would first have to write x in the form $x = e^{\log_e x}$. Then $(e^{x \log_e x})^x = e^{x \log_e x}$. Expression (12.7) then yields the result above.

These examples indicate one class of function for which the technique of logarithmic differentiation is appropriate, namely, those functions which contain a power that involves the variable, although if powers of e are involved (12.7) is more direct. The properties of logarithms sometimes make the method of value in dealing with algebraic functions.

Example 3 Find dy/dx if $y = \left(\frac{x-1}{x+1}\right)^{1/3}$

The complications involved in differentiating this function arise from the cube root and the quotient. The technique of logarithmic differentiation removes them, as follows:

$$y = \left(\frac{x-1}{x+1}\right)^{1/3}$$

$$\log_e y = \log_e\left(\frac{x-1}{x+1}\right)^{1/3} = \frac{1}{3}\log\frac{x-1}{x+1} \quad \text{by (1.29)}$$

$$\log_e y = \frac{1}{3}[\log_e(x-1) - \log_e(x+1)] \quad \text{by (1.28)}$$

Therefore

$$\frac{1}{y}\frac{dy}{dx} = \frac{1}{3}\left(\frac{1}{x-1} - \frac{1}{x+1}\right) = \frac{2}{3(x-1)(x+1)}$$

$$\frac{dy}{dx} = \frac{2}{3(x^2-1)}\left(\frac{x-1}{x+1}\right)^{1/3}$$

Readers should verify this result by other methods.

Exercises 12c

Find dy/dx if

1 $y = e^{\tan x}$ **2** $y = e^{x^4}$ **3** $y = x^{\sin x}$

4 $y = (\sin x)^x$ **5** $y = (\log_e x)^x$ **6** $y = x^{x-1}$

7 $y = e^x + x^x$ **8** $y = e^{x^2} + x^{x^2}$ **9** $y = \left(\dfrac{x+1}{x-1}\right)^{1/3}$

10 $y = \left(\dfrac{x^2-1}{x^2+1}\right)^{1/4}$

12.4 Polynomial approximations for a function and Maclaurin's series

Consider the function $f(x) = \dfrac{1}{1-x}$. Provided $x \neq 1$,

$$\frac{1}{1-x} = 1 + \frac{x}{1-x}$$

as is readily verified by simple algebra. Multiplication of this identity by x gives

$$\frac{x}{1-x} = x + \frac{x^2}{1-x}$$

so that, on substitution, we obtain

$$\frac{1}{1-x} = 1 + x + \frac{x^2}{1-x}$$

that is

$$\frac{1}{1-x} = 1 + x \quad \text{with error} \quad \frac{x^2}{1-x}$$

Multiplication by x shows that

$$\frac{x}{1-x} = x + x^2 + \frac{x^3}{1-x}$$

so that

$$\frac{1}{1-x} = 1 + x + x^2 + \frac{x^3}{1-x}$$

that is

$$\frac{1}{1-x} = 1 + x + x^2 \quad \text{with error} \quad \frac{x^3}{1-x}$$

In this way, we may obtain the successive polynomial approximations 1, $1+x$, $1+x+x^2$, $1+x+x^2+x^3$, etc. to the function $\dfrac{1}{1-x}$ with respective errors $\dfrac{x}{1-x}, \dfrac{x^2}{1-x}, \dfrac{x^3}{1-x}, \dfrac{x^4}{1-x}$, etc.

For $|x| < 1$, the errors involved by using the approximations will be small. Thus if $x = 0{\cdot}1$, the error involved in using the approximation $1 + x + x^2 + x^3$ for the function $1/(1 - x)$ is $(0{\cdot}1)^4/0{\cdot}9$, which is a percentage error of $100\,(0{\cdot}1)^4$, i.e. 0·01 per cent. This is not so if $x > 1$ and the approximations are no longer so useful.

The first polynomial above (1) is equal to $f(x)$ when $x = 0$. The second polynomial $(1 + x)$ and its first derivative are equal to $f(x)$ and its first derivative respectively when $x = 0$. The third polynomial and its first and second derivatives are equal to $f(x)$ and its first and second derivatives respectively when $x = 0$, etc.

This suggests that, in general, we may be able to find a polynomial approximation to any function $f(x)$ by setting

$$f(x) \approx a_0 + a_1 x + a_2 x^2 + \dots + a_n x^n \tag{12.8}$$

and choosing the a's so that the function and its first n derivatives when $x = 0$ equal the polynomial and its first n derivatives respectively when $x = 0$.

The first n derivatives of the right-hand side of (12.8) are

$$\begin{gathered} a_1 + 2a_2 x + 3a_3 x^2 + \dots + na_n x^{n-1} \\ 2!a_2 + 3!a_3 x + 4.3a_4 x^2 + \dots + n(n-1)a_n x^{n-2} \\ 3!a_3 + 4!a_4 x + \dots + n(n-1)(n-2)a_n x^{n-3} \\ \dots \quad \dots \quad \dots \quad \dots \quad \dots \\ n!a_n \end{gathered}$$

When $x = 0$, these derivatives have the values

$$a_1,\ 2!a_2,\ 3!a_3,\ 4!a_4,\ \dots,\ n!a_n \quad \text{respectively}$$

On equating these to $f'(0), f''(0), f'''(0), \dots, f^{(n)}(0)$, we have

$$a_1 = f'(0),\ a_2 = \frac{f''(0)}{2!},\ a_3 = \frac{f'''(0)}{3!},\ \dots,\ a_r = \frac{f^{(r)}(0)}{r!},\ \dots,\ a_n = \frac{f^{(n)}(0)}{n!}$$

a_0 is determined by making the polynomial of (12.8) when $x = 0$ equal to $f(0)$, i.e. $a_0 = f(0)$.

Thus we obtain

$$f(x) \approx f(0) + xf'(0) + x^2\frac{f''(0)}{2!} + \dots + \frac{x^n f^{(n)}(0)}{n!} \tag{12.9}$$

We can now see immediately that for this procedure to be possible $f(x)$ and its first n derivatives must exist and be continuous at $x = 0$. The method will fail, for example, for the function $f(x) = \log_e x$, since neither this function nor its derivatives are defined at $x = 0$.

In order to obtain a satisfactory approximation, the difference between $f(x)$ and the polynomial (12.9) must decrease with increasing n. For the function $1/(1 - x)$, we saw that this was the case provided $|x| < 1$. In general, it is not easy to evaluate the difference between $f(x)$ and the polynomial (12.9). Further consideration of this difficult problem is beyond the scope of this course and we shall only record the following very important results: there exist many functions $f(x)$ which together with all their derivatives are defined and continuous at $x = 0$, and for which (for some values of x) the infinite series

$$f(0) + xf'(0) + \frac{x^2f''(0)}{2!} + \frac{x^3f'''(0)}{3!} + \dots + \frac{x^rf^{(r)}(0)}{r!} + \dots$$

is convergent.

For such functions f(x), the limit of the sum of this series is f(x), provided x lies within the interval for which the series is convergent, and we may write

$$f(x) = f(0) + xf'(0) + \frac{x^2f''(0)}{2!} + \frac{x^3f'''(0)}{3!} + \dots + \frac{x^rf^{(r)}(0)}{r!} + \dots \quad (12.10)$$

The series on the right of (12.10) is known as the Maclaurin series for f(x).

Example 1 Show that the Maclaurin series for $\dfrac{1}{1-x}$ is

$$\frac{1}{1-x} = 1 + x + x^2 + x^3 + \dots$$

With

$$\begin{aligned}
f(x) &= \frac{1}{1-x} = (1-x)^{-1} && \text{therefore} \quad f(0) = 1 \\
f'(x) &= (1-x)^{-2} && \text{therefore} \quad f'(0) = 1! \\
f''(x) &= 1.2(1-x)^{-3} && \text{therefore} \quad f''(0) = 2! \\
f'''(x) &= 1.2.3(1-x)^{-4} && \text{therefore} \quad f'''(0) = 3! \\
&\dots \quad \dots \quad \dots && \dots \quad \dots \quad \dots \\
f^{(r)}(x) &= 1.2.3 \dots r(1-x)^{-(r+1)} && \text{therefore} \quad f^{(r)}(0) = r!
\end{aligned}$$

By (12.10)

$$\frac{1}{1-x} = 1 + x.1 + x^2.\frac{2!}{2!} + x^3.\frac{3!}{3!} + \dots + x^r\frac{r!}{r!} + \dots$$

therefore

$$\frac{1}{1-x} = 1 + x + x^2 + \dots + x^r + \dots$$

This is an infinite geometric series and is convergent for $-1 < x < 1$ (see Section 2.4).

Example 2 Assuming the series below is convergent, show that

$$\sin x = x - \frac{x^3}{3!} + \frac{x^5}{5!} - \frac{x^7}{7!} + \dots$$

With

$$\begin{aligned}
f(x) &= \sin x & f(0) &= 0 \\
f'(x) &= \cos x & f'(0) &= 1 \\
f''(x) &= -\sin x & f''(0) &= 0 \\
f'''(x) &= -\cos x & f'''(0) &= -1 \\
f''''(x) &= \sin x & f''''(0) &= 0 \quad \text{etc.}
\end{aligned}$$

We can now see that the values of the successive derivatives when $x = 0$ are going to repeat the sequence 0, 1, 0, -1, 0, 1, 0, -1, etc. Thus by (12.10)

$$\sin x = 0 + x.\frac{1}{1!} + x^2\frac{0}{2!} + x^3\frac{(-1)}{3!} + x^4\frac{(0)}{4!} + x^5\frac{(1)}{5!} + \ldots$$

$$\sin x = x - \frac{x^3}{3!} + \frac{x^5}{5!} - \frac{x^7}{7!} + \frac{x^9}{9!} + \ldots$$

It can be shown that this result is true for all x (in radians).

Exercises 12d

1 Show that

$$\frac{1}{1+x} = 1 - \frac{x}{1+x} = 1 - x + \frac{x^2}{1+x} = 1 - x + x^2 - \frac{x^3}{1+x}$$
$$= 1 - x + x^2 - x^3 + \frac{x^4}{1+x}$$

2 Show that the Maclaurin series for $\dfrac{1}{1+x}$ is

$$\frac{1}{1+x} = 1 - x + x^2 - x^3 + \ldots + (-1)^r x^r + \ldots$$

For what values of x will this result be valid?

3 Evaluate x, $x - \dfrac{x^3}{3!}$, $x - \dfrac{x^3}{3!} + \dfrac{x^5}{5!}$, $x - \dfrac{x^3}{3!} + \dfrac{x^5}{5!} - \dfrac{x^7}{7!}$ for $x = \pi/6$. Compare the values with $\sin \pi/6 = 0{\cdot}5$. [Compare with Example 2.]

4 Assuming the convergence of the series below, show that

$$\cos x = 1 - \frac{x^2}{2!} + \frac{x^4}{4!} - \frac{x^6}{6!} + \ldots$$

5 Using the previous question and question 3, compare these polynomial approximations for $\sin x$ and $\cos x$ with those obtained in (6.60) and (6.61).

6 Starting with the inequality $\sin x < x$ [see (6.59)], show by integration* from 0 to x that $\cos x > 1 - \dfrac{x^2}{2!}$, and then by integration from 0 to x that $\sin x > x - \dfrac{x^3}{3!}$, and then by integration from 0 to x that $\cos x < 1 - \dfrac{x^2}{2!} + \dfrac{x^4}{4!}$, etc. Compare with the previous examples.

7 Show that if n is a positive integer, (12.9) gives the polynomials 1, $1 + nx$, $1 + nx + \dfrac{n(n-1)}{2!}x^2, \ldots,$ $1 + nx + \dfrac{n(n-1)}{2!}x^2 + \ldots + nx^{n-1} + x^n$ as successive approximations to $(1 + x)^n$. The last approximation is, of course, exact. What is the Maclaurin series for $(1 + x)^n$ when (i) n is a positive integer, (ii) n is not a positive integer?

* See next chapter.

8 For $f(x) = \tan x$ show that $f^{i}(x) = \sec^2 x$, $f^{ii}(x) = 2\sec^2 x \tan x$, $f^{iii}(x) = 4\sec^2 x \tan^2 x + 2\sec^4 x$, $f^{iv}(x) = 8 \tan x(\sec^2 x \tan^2 x + 2\sec^4 x)$, $f^{v}(x) = 8 \sec^2 x(\sec^2 x \tan^2 x + 2\sec^4 x) + 8 \tan x\, d/dx(\sec^2 x \tan^2 x + 2\sec^4 x)$. Hence show that (12.9) gives as successive approximations to the function $\tan x$: x, $x + \frac{1}{3}x^3$, $x + \frac{1}{3}x^3 + \frac{2}{15}x^5$.

9 Evaluate the polynomials of question 8 for $x = \pi/6$. Compare the values obtained with $\tan \pi/6$.

10 Show that, for the polynomial $f(x) = a + bx + cx^2 + dx^3 + ex^4$, (12.9) gives as successive approximations ($n = 1, 2, 3, 4$), $a + bx$, $a + bx + cx^2$, $a + bx + cx^2 + dx^3$, $a + bx + cx^2 + dx^3 + ex^4$. What is the Maclaurin series for $f(x)$?

12.5 The series for e^x and $\log_e(1+x)$

In this section, we shall use (12.10) to obtain the Maclaurin series for the functions e^x and $\log_e(1+x)$. (We have already mentioned that the method of (12.9) will fail for the function $\log_e x$ but, as we shall see, it is satisfactory for $\log_e(1+x)$.)

If $f(x) = e^x$, then

$$f'(x) = f''(x) = f'''(x) = \ldots = f^{(r)}(x) = \ldots = e^x$$

therefore

$$f'(0) = f''(0) = f'''(0) = \ldots = f^{(r)}(0) = \ldots = 1$$

By (12.10)

$$e^x = 1 + x.1 + x^2.\frac{1}{2!} + x^3.\frac{1}{3!} + \ldots + x^r.\frac{1}{r!} + \ldots$$

$$e^x = 1 + x + \frac{x^2}{2!} + \frac{x^3}{3!} + \frac{x^4}{4!} + \ldots + \frac{x^r}{r!} + \ldots \qquad (12.11)$$

It may be shown (although the methods required are beyond the scope of this book) that this series is valid for all values of x.

If we replace x by $(-x)$, we obtain

$$e^{-x} = 1 + (-x) + \frac{(-x)^2}{2!} + \frac{(-x)^3}{3!} + \frac{(-x)^4}{4!} + \ldots$$

that is

$$e^{-x} = 1 - x + \frac{x^2}{2!} - \frac{x^3}{3!} + \frac{x^4}{4!} - \frac{x^5}{5!} + \ldots + \frac{(-1)^r x^r}{r!} + \ldots \qquad (12.12)$$

For the function $\log_e(1+x)$, we have

$$\begin{aligned} f(x) &= \log_e(1+x) & f(0) &= \log_e 1 = 0 \\ f'(x) &= (1+x)^{-1} & f'(0) &= 1 \\ f''(x) &= -1(1+x)^{-2} & f''(0) &= -1 \\ f'''(x) &= (-1)(-2)(1+x)^{-3} & f'''(0) &= 2! \\ f''''(x) &= (-1)(-2)(-3)(1+x)^{-4} & f''''(0) &= -3! \end{aligned}$$

$$f^{(n)}(x) = (-1)(-2)\ldots[-(n-1)](1+x)^{-n}$$
$$= (-1)^{n-1}(n-1)!(1+x)^{-n} \qquad f^{(n)}(0) = (-1)^{n-1}(n-1)!$$

Thus the Maclaurin series for $\log_e(1+x)$ is

$$\log_e(1+x) = 0 + x.1 + x^2.\frac{(-1)}{2!} + x^3.\frac{2!}{3!} + x^4\frac{(-3!)}{4!} + \ldots$$

that is

$$\log_e(1+x) = x - \frac{x^2}{2} + \frac{x^3}{3} - \frac{x^4}{4} + \frac{x^5}{5} - \frac{x^6}{6} + \ldots$$

It may be shown that this series is convergent only if $-1 < x \leqslant 1$. Thus, more precisely, we have

$$\log_e(1+x) = x - \frac{x^2}{2} + \frac{x^3}{3} - \frac{x^4}{4} + \ldots + \frac{(-1)^{r+1}x^r}{r} + \ldots$$
$$\text{for } -1 < x \leqslant 1 \qquad (12.13)$$

If we replace x by $(-x)$ in (12.13), we obtain

$$\log_e(1-x) = -x - \frac{(-x)^2}{2} + \frac{(-x)^3}{3} - \frac{(-x)^4}{4} \quad \text{for } -1 \leqslant x < 1$$

that is

$$\log_e(1-x) = -\left(x + \frac{x^2}{2} + \frac{x^3}{3} + \frac{x^4}{4} + \frac{x^5}{5} + \ldots + \frac{x^r}{r} + \ldots\right)$$
$$\text{for } -1 \leqslant x < 1 \qquad (12.14)$$

Example 1 Calculate e to three decimal places.
From (12.11)

$$e^x = 1 + x + \frac{x^2}{2!} + \frac{x^3}{3!} + \frac{x^4}{4!} + \ldots$$

With $x = 1$

$$e = 1 + 1 + \frac{1}{2!} + \frac{1}{3!} + \frac{1}{4!} + \frac{1}{5!} + \frac{1}{6!} + \ldots$$
$$= 1 + 1 + 0{\cdot}5 + 0{\cdot}1667 + 0{\cdot}0417$$
$$+ 0{\cdot}0083 + 0{\cdot}0014 + 0{\cdot}0002 + \ldots$$
$$= 2{\cdot}718$$

This result is accurate to three decimal places since the remaining terms of the series do not affect these places. (Compare this value with the value obtained in Section 12.1.) For values of $x < 1$ (or > 1), it will require fewer (or more) terms of the series (12.11) to obtain the same accuracy for e^x.

Example 2 Find the Maclaurin series for e^{3x}.
From (12.11)

$$e^X = 1 + X + \frac{X^2}{2!} + \frac{X^3}{3!} + \frac{X^4}{4!} + \ldots$$

Therefore, with $X = 3x$,

$$e^{3x} = 1 + 3x + \frac{(3x)^2}{2!} + \frac{(3x)^3}{3!} + \ldots$$

$$= 1 + 3x + \frac{9x^2}{2!} + \frac{27x^3}{3!} + \frac{81x^4}{4!} + \ldots + \frac{3^r x^r}{r!} + \ldots$$

Example 3 Evaluate $\log_e(1{\cdot}1)$ to four decimal places.

$$\log_e(1+x) = x - \frac{x^2}{2} + \frac{x^3}{3} - \frac{x^4}{4} + \frac{x^5}{5} - \ldots \quad \text{for } -1 < x \leqslant 1$$

Therefore, with $x = \frac{1}{10}$, which is within the range of convergence,

$$\log_e(1{\cdot}1) = 0{\cdot}1 - \frac{0{\cdot}01}{2} + \frac{0{\cdot}001}{3} - \frac{0{\cdot}0001}{4} + \frac{0{\cdot}000\,01}{5} - \ldots$$

$$= 0{\cdot}1 - 0{\cdot}005 + 0{\cdot}000\,33 - 0{\cdot}000\,025 + 0{\cdot}000\,002 - \ldots$$
$$= (0{\cdot}100\,335) - (0{\cdot}005\,025) + \ldots$$
$$= 0{\cdot}0953 \quad \text{to four decimal places}$$

Example 4 Show that for small values of x

$$(1+2x)e^{-x} + \log_e(1+2x) \approx 1 + 3x - \frac{7x^2}{2} + \frac{7x^3}{2}$$

By (12.12) and (12.13)

$$(1+2x)e^{-x} + \log_e(1+2x) = (1+2x)\left(1 - x + \frac{x^2}{2} - \frac{x^3}{6} + \ldots\right)$$
$$+ \left[2x - \frac{(2x)^2}{2} + \frac{(2x)^3}{3} - \frac{(2x)^4}{4} + \ldots\right]$$
$$= 1 - x + \frac{x^2}{2} - \frac{x^3}{6} + 2x - 2x^2 + x^3$$
$$+ 2x - 2x^2 + \frac{8x^3}{3} + \ldots$$
$$= 1 + 3x - \frac{7}{2}x^2 + \frac{7}{2}x^3 \quad \text{as far as terms in } x^3$$

The calculation of logarithms

The series (12.13) and (12.14) for $\log_e(1+x)$ and $\log_e(1-x)$ are convergent for $-1 < x \leqslant 1$ and $-1 \leqslant x < 1$ respectively. They will therefore only allow us to calculate the natural logarithms of numbers from just above zero to two.

Some simple manipulations of these series enable us to extend this range. Provided $-1 < x < 1$

$$\log_e\left(\frac{1+x}{1-x}\right) = \log_e(1+x) - \log_e(1-x)$$
$$= x - \frac{x^2}{2} + \frac{x^3}{3} - \frac{x^4}{4} + \frac{x^5}{5} - \frac{x^6}{6} + \ldots$$
$$-\left(-x - \frac{x^2}{2} - \frac{x^3}{3} - \frac{x^4}{4} - \frac{x^5}{5} - \frac{x^6}{6}\right)$$

Therefore

$$\log_e\left(\frac{1+x}{1-x}\right) = 2\left(x + \frac{1}{3}x^3 + \frac{1}{5}x^5 + \frac{1}{7}x^7 + \frac{1}{9}x^9 + \ldots\right) \quad (12.15)$$

With $x = \frac{1}{2}$

$$\log_e\left(\frac{\frac{3}{2}}{\frac{1}{2}}\right) = 2\left(\frac{1}{2} + \frac{1}{24} + \frac{1}{160} + \frac{1}{896} + \frac{1}{4608} + \ldots\right)$$

therefore

$$\log_e 3 \approx 2(0{\cdot}5 + 0{\cdot}041\,667 + 0{\cdot}006\,25 + 0{\cdot}001\,1161$$
$$+ 0{\cdot}000\,217\,02)$$
$$\approx 2(0{\cdot}549\,29)$$
$$\approx 1{\cdot}0986$$

Example 5 Show that if n is positive

$$\log_e n = 2\left[\frac{n-1}{n+1} + \frac{1}{3}\left(\frac{n-1}{n+1}\right)^3 + \frac{1}{5}\left(\frac{n-1}{n+1}\right)^5 + \ldots\right]$$

From (12.15)

$$\log_e\left(\frac{1+x}{1-x}\right) = 2\left(x + \frac{1}{3}x^3 + \frac{1}{5}x^5 + \frac{1}{7}x^7 + \ldots\right)$$

With $\dfrac{1+x}{1-x} = n$, $x = \dfrac{n-1}{n+1}$, which is less than 1 if n is positive. Therefore

$$\log_e n = 2\left[\frac{n-1}{n+1} + \frac{1}{3}\left(\frac{n-1}{n+1}\right)^3 + \frac{1}{5}\left(\frac{n-1}{n+1}\right)^5 + \ldots\right] \quad (12.16)$$

This series is quite useful for calculating logarithms. However, for large values of n, $\dfrac{n-1}{n+1}$ is near to one and a large number of terms of the series (12.16) will be needed to obtain a satisfactory approximation to $\log_e n$.

With $n = 2$, we have

$$\log_e 2 = 2\left(\frac{1}{3} + \frac{1}{3}\cdot\frac{1}{3^3} + \frac{1}{5}\cdot\frac{1}{3^5} + \ldots\right)$$

This series is clearly a better way of calculating $\log_e 2$ than the series (12.13) with $x = 1$, namely,

$$\log_e 2 = 1 - \tfrac{1}{2} + \tfrac{1}{3} - \tfrac{1}{4} + \tfrac{1}{5} - \tfrac{1}{6} + \ldots$$

Exercises 12e

1 Use the series (12.11) to evaluate $\sqrt{e}$ and $1/e$ correct to four places of decimals. Check your results from tables or by calculator.

2 Use the series (12.13) to evaluate $\log_e(1{\cdot}2)$ and $\log_e 0{\cdot}9$ correct to four places of decimals. Check your results using tables or calculator.

3 Use (12.11) to write down the first few terms of the Maclaurin series for (i) e^{2x}, (ii) e^{-3x}, (iii) e^{x^2}.

4 Use (12.13) to write down the first few terms of the Maclaurin series for (i) $\log_e(1+2x)$, (ii) $\log_e(1-3x)$, (iii) $\log_e(1+x^2)$. For what values of x are these series valid?

5 Show that $\log_e(3+4x) = \log_e 3 + \frac{4}{3}x - \frac{8}{9}x^2 + \frac{64}{81}x^3 - \ldots$ and state the limits between which x must lie for the expansion to be valid.

6 Show that

$$\frac{1}{2}\left(e - \frac{1}{e}\right) = 1 + \frac{1}{3!} + \frac{1}{5!} + \frac{1}{7!} + \ldots + \frac{1}{(2r-1)!} + \ldots$$

and write down similar series for $\dfrac{1}{2}\left(e + \dfrac{1}{e}\right)$.

7 If x is so small that x^4 and higher powers of x can be neglected, show that $e^x + \log_e(1-x) = 1 - \frac{1}{6}x^3$ approximately.

8 Show that

$$\log_e \sec^2 x = \tan^2 x - \frac{\tan^4 x}{2} + \frac{\tan^6 x}{3} - \frac{\tan^8 x}{4} + \ldots$$

For what values of x is the expansion valid?

9 If $x > 1$ show that

$$\log_e\left(\frac{x+1}{x-1}\right) = 2\left(\frac{1}{x} + \frac{1}{3x^3} + \frac{1}{5x^5} + \ldots\right)$$

By putting $x = 3$, evaluate $\log_e 2$.

10 Show that if x is so small that x^4 and higher powers of x may be neglected $\log_e(1-2x-3x^2) = -2x - 5x^2 - \frac{26}{3}x^3$.

[Hint: $\log_e(1-2x-3x^2) = \log_e(1-3x)(1+x)$
$= \log_e(1-3x) + \log_e(1+x)$]

Exercises 12

1 Differentiate the following functions with respect to x (i) $(3x^2-1)\log_e x$, (ii) $e^x \log_e 2x$, (iii) $e^{-2x}\cos 4x$.

2 Differentiate with respect to t (i) $\log_e(1+e^{2t})$, (ii) $\log_e(\tan\frac{1}{2}t)$, (iii) $\log_e(\cot t + \operatorname{cosec} t)$.

3 If $y = x^n \log_e x$, show that $x\,dy/dx = x^n + ny$.

4 If $y\log_e y = x$, find dy/dx in terms of x and y. [JMB, part]

5 If $y = e^{3x}\cos 4x$, find dy/dx and express it in the form $Re^{3x}\cos(4x+\alpha)$, where R is a positive constant; state the cosine and sine of the constant angle α. Hence write down d^2y/dx^2 in a similar form. [JMB, part]

6 Find (i) $\dfrac{d}{dt}\left(\dfrac{e^t}{1-e^{2t}}\right)$, (ii) $\dfrac{d}{d\theta}\left(\dfrac{\log_e \theta}{\theta + \log_e \theta}\right)$.

7 Find (i) $\dfrac{d}{dx}(e^{x \sin x})$, (ii) $\dfrac{d}{dx}[\sin(\log_e x)]$.

8 If $y = \sin(\log_e x)$, show that $x^2\dfrac{d^2y}{dx^2} + x\dfrac{dy}{dx} + y = 0$.

9 Show that

$$\text{(i)}\quad \frac{d}{dx}\{\log_e[x + \sqrt{(x^2+1)}]\} = \frac{1}{\sqrt{(x^2+1)}}$$

$$\text{(ii)}\quad \frac{d}{dx}\{\log_e[x + \sqrt{(x^2-1)}]\} = \frac{1}{\sqrt{(x^2-1)}}$$

10 Differentiate with respect to x (i) $\log_e \sqrt{(x+1)}$, (ii) $\log_e[\sqrt{(x+1)} + \sqrt{(x-1)}]$.
11 Show that if $y = xe^{-x}$, $d^2y/dx^2 + 2\,dy/dx + y = 0$.
12 Show that if $y = e^{-2x} \sin 5x$, $d^2y/dx^2 + 4\,dy/dx + 29y = 0$.
13 If $y = 2e^{-4x} - e^{3x}$, show that $d^2y/dx^2 + 2dy/dx - 8y = ke^{3x}$, where k is a constant, and state the value of k. [JMB, part]
14 If $y = \log_e(1 + \cos x)$, show that $d^3y/dx^3 + d^2y/dx^2 . dy/dx = 0$.
15 If $y = e^{\arctan x}$, show that $(1 + x^2)\,d^2y/dx^2 - (1 - 2x)\,dy/dx = 0$.
16 If $y = e^{-2x} \cos 4x$, show that $d^2y/dx^2 + 4\,dy/dx + 20y = 0$.
17 If $y = \tan^{-1}(e^x)$, show that $d^2y/dx^2 = 2(dy/dx)^2 \cot 2y$.
18 If $x = t^2$, $y = \log_e t$, find dy/dx and d^2y/dx^2 in terms of t.
19 If $x = e^t$, $y = \log_e t$, find dy/dx and d^2y/dx^2 in terms of t.
20 Find the stationary value of $(\log_e x)/x$.
21 Show that $y = xe^{-x}$ has one maximum value which occurs when $x = 1$.
22 Find the positions of the points of inflexion of the curve $y = e^{-x^2/2}$.
23 Find the equation of the tangent to the curve $y = e^{2x}$ at the point (0, 1).
24 Find the position of the point of inflexion on the curve $y = xe^{-x}$. Sketch the graph.
25 The speed of signalling in a submarine cable is given by $Kx^2 \log_e(1/x)$, where K is a constant and x is the ratio of the radius of the core to the thickness of the insulating material. Show that the speed of signalling is a maximum when $x = 1/\sqrt{e}$.
26 Show that if $v = 100p(1 + \log_e r) - 10Gqr$, where p and q are constants, then v is a maximum when $r = p/q$.
27 Find the position of the point of inflexion on the curve $y = \log_e x + (2/x)$. Sketch the graph.
28 For what values of x is the derivative of xe^{-x^2} zero?
29 Differentiate with respect to x (i) $(\log_e x)^{\log_e x}$, (ii) $x^x + e^{\tan x}$.
30 If $y = x^x[1 + x/(\alpha + 2)]$, find dy/dx.
31 Find the gradient of the tangent from the origin to the curve $y = \log_e x$. Hence, by considering a sketch of the curve, find the range of values of the constant K for which the equation $\log_e x = Kx$ has two unequal roots. Draw a graph of $y = \log_e x$ from $x = 1$ to $x = 1{\cdot}9$ and use it to find to two decimal places the smaller root of $4 \log_e x = x$. [JMB]

32 Show that $1 < e^x < e^c$ for $0 < x < c$. By integrating* this inequality from 0 to x show that $x < e^x - 1 < e^c x$ and then by the same method that

$$\frac{x^2}{2!} < e^x - x - 1 < e^c \frac{x^2}{2!}$$

and that

$$\frac{x^3}{3!} < e^x - \frac{x^2}{2!} - x - 1 < e^c \frac{x^3}{3!}$$

etc. to

$$\frac{x^{n+1}}{(n+1)!} < e^x - \frac{x^n}{n!} - \frac{x^{n-1}}{(n-1)!} \cdots - \frac{x^2}{2} - x - 1 < e^c \frac{x^{n+1}}{(n+1)!}$$

Deduce that the difference between e^x and $1 + x + \frac{x^2}{2!} + \ldots + \frac{x^2}{n!}$ tends to zero with increasing n.

33 Show that $e^{x^2} \log_e (1 + x^2) \approx x^2 + \frac{1}{2}x^4 + \frac{1}{3}x^6$ for $x < 1$. Find the approximate value of $\int_0^{0\cdot1} e^{x^2} \log_e (1 + x^2)\, dx$*.

34 Assuming the convergence of the series below show that
(i) $\sec x = 1 + \frac{1}{2}x^2 + \frac{5}{24}x^4 + \ldots$
(ii) $\log_e \cos x = -\frac{1}{2}x^2 - \frac{1}{12}x^4 - \frac{1}{45}x^6 + \ldots$
(iii) $\log_e (1 + e^x) = \log_e 2 + \frac{1}{2}x + \frac{1}{8}x^2 - \frac{1}{192}x^4 + \ldots$

35 Write down the expansions in powers of x, as far as the term in x^3, of (i) e^{-2x}, (ii) $(1 - 4x)^{1/2}$. Use your series to find to five decimal places the difference in the values of these two functions when $x = 0\cdot01$. [SUJB]

36 Show that if $-\frac{1}{2} < x \leqslant \frac{1}{2}$, then

$$\log_e (1 + x - 2x^2) = x - \frac{5}{2}x^2 + \frac{7}{3}x^3 - \frac{17}{4}x^4 + \ldots$$ [LU, part]

37 Obtain the expansion of $\log_e (1 + x + x^2)$ (if $x < 1$) in powers of x. State the coefficients of x^{3n-1}, x^{3n}, x^{3n+1}.

$$\left[\text{Hint: } 1 + x + x^2 = \frac{1 - x^3}{1 - x}.\right]$$

38 (a) Expand $\log_e \cos \theta$ in a series of ascending powers of $\sin^2 \theta$, giving the terms up to $\sin^6 \theta$, and the general term. For what values of θ, in the interval $0 \leqslant \theta \leqslant \pi$, is the expansion valid?
(b) Given that $y = (2 + x)^2 e^{-x}$, find the expansion of y in ascending powers of x as far as the term in x^3. Find also the expansion of $\log_e y$ in ascending powers of x as far as the term in x^3, and state the coefficient of x^n. [JMB]

39 (a) Expand the function $e^{-2x}/(1 - x)^2$ in a series of ascending powers of x up to and including the term in x^3.
(b) Prove that $\dfrac{1 + \cos \theta}{1 - \cos \theta} = \cot^2 \frac{1}{2}\theta$. Write down the first three terms in the expansion of $\log_e (1 + x)$ in ascending powers of x. Express $\log_e \cot^2 \frac{1}{2}\theta$ as a series of powers of $\cos \theta$, giving the first three terms and the nth term. [JMB]

* See next chapter.

40 Use (12.15) to show that

$$\log_e\left(\frac{n}{n-1}\right) = 2\left[\frac{1}{2n-1} + \frac{1}{3}\left(\frac{1}{2n-1}\right)^3 + \frac{1}{5}\left(\frac{1}{2n-1}\right)^5 + \ldots\right]$$

provided $n > 1$. Hence evaluate in succession $\log_e 2$, $\log_e 3$, $\log_e 4$, $\log_e 5$, $\log_e 6$. (Some of these may be checked using the first two values.)

41 State the first five terms of the expansions of $\log_e(1+x)$ and $\log_e(1-x)$. Deduce that, when $n > 1$,

$$\log_e\left(\frac{n+1}{n-1}\right) = 2\left[\frac{1}{n} + \frac{1}{3n^3} + \frac{1}{5n^5} + \ldots\right]$$

Hence evaluate $\log_e 1{\cdot}25$ correct to four places of decimals. [AEB]

42 Find the coefficient of x^3 in the expansion, in ascending powers of x, of the function $f(x) = (1-x)^2 e^{2x}$. Find also the coefficients of x^n in this expansion of $f(x)$ and deduce that the only values of n for which the coefficient of x^n is zero are 1 and 4. [JMB]

43 Given that $y = \dfrac{e^x - e^{-x}}{e^x + e^{-x}}$, show that $\dfrac{dy}{dx} = 1 - y^2$. [C]

44 Show that $e^{-x}\sin x$ has turning values when $x = (r + \frac{1}{4})\pi$, where r is an integer. Determine which are maxima and which are minima. [O]

45 A curve has the equation $y = \dfrac{a}{b - e^{-cx}}$ $(a \neq 0,\ b > 0,\ c > 0)$. Show that it has one point of inflexion, and that the value of y at the point of inflexion is half the limiting value of y as $x \to \infty$. [O]

46 Given that $y = e^{-2x}(3\sin 3x - 2\cos 3x)$, find dy/dx in its simplest form. [C]

47 Simplify $\log_e\left[\dfrac{(1+x)e^{-2x}}{1-x}\right]^{1/2}$ and show that its derivative is $\dfrac{x^2}{1-x^2}$. Hence, or otherwise, evaluate dy/dx at $x = 0$ for the function

$$y = \left[\frac{(1+x)e^{-2x}}{1-x}\right]^{1/2}$$

[JMB]

48 Given that $y = (1+x)^2 e^{-2x}$, find dy/dx. Find also the co-ordinates of the turning points of the graph of y. [LU]

49 Expand $\log_e(1 + 3x + 2x^2)$ and $\log_e \dfrac{1+3x+2x^2}{1-3x+2x^2}$ in series of ascending powers of x up to, and including, the term in x^3. Use the series for $\log_e \dfrac{1+3x+2x^2}{1-3x+2x^2}$ to show that $\log_e 5$ is approximately 1·6. [AEB]

50 Find the minimum value of $x\log_e x$, where x is positive. Given that $x\log_e x$ tends to 0 as x tends to 0, sketch the curve $y = x\log_e x$.

13
The basic ideas of integration

13.1 Introduction

In the preceding chapters, we have been considering the problem of finding the differential coefficient or rate of change of a given function. The integral calculus to which we now turn our attention is concerned with the inverse problem, namely, given the rate of change of a function, to find the function. In symbols, we require to find $f(x)$ where

$$\frac{df(x)}{dx} = g(x) \tag{13.1}$$

and $g(x)$ is given. It is more usual to write

$$f(x) = \int g(x)\,dx \tag{13.2}$$

and we define integration as follows.

The integral of a function $g(x)$ with respect to x is the function whose differential coefficient with respect to x is $g(x)$ and it is written $\int g(x)\,dx$.

The reason for this notation will be explained later (see Section 13.6); meanwhile (13.2) is to be regarded as an alternative way of writing (13.1).

If we are required to find $\int 3x^2\,dx$, then $\int 3x^2\,dx = x^3$ because $\dfrac{d(x^3)}{dx} = 3x^2$.

Similarly

$$\int \sin x\,dx = -\cos x \qquad \text{because} \qquad \frac{d(-\cos x)}{dx} = \sin x$$

$$\int \frac{dx}{x} = \log_e x \qquad \text{because} \qquad \frac{d}{dx}(\log_e x) = \frac{1}{x}$$

13.2 Arbitrary constant

We recall that the differential coefficient of a constant is zero; hence there is not a perfectly definite value for the integral. In the previous three cases, we have the more general results

$$\int 3x^2 \mathrm{d}x = x^3 + C \qquad \text{because} \quad \frac{\mathrm{d}}{\mathrm{d}x}(x^3 + C) = 3x^2$$

$$\int \sin x \,\mathrm{d}x = -\cos x + C \qquad \text{because} \quad \frac{\mathrm{d}}{\mathrm{d}x}(-\cos x + C) = \sin x$$

$$\int \frac{\mathrm{d}x}{x} = \log_e x + C \qquad \text{because} \quad \frac{\mathrm{d}}{\mathrm{d}x}(\log_e x + C) = \frac{1}{x}$$

An arbitrary constant can always be added to the result and hence

$$\int \mathrm{g}(x)\mathrm{d}x = \mathrm{f}(x) + C$$

This is known as the indefinite integral of $\mathrm{g}(x)$.

13.3 Standard forms

To find $\mathrm{f}(x)$ given $\mathrm{g}(x)$ means that we have to retrace the steps we made in the process of differentiation and then add an arbitrary constant. Unfortunately, there is no general method for doing this, but a few of the more common integrals can be stated from our knowledge of differential coefficients. These results are known as standard forms.

$$\int \sin x \,\mathrm{d}x = -\cos x + C$$

$$\int \cos x \,\mathrm{d}x = \sin x + C$$

$$\int \mathrm{e}^x \,\mathrm{d}x = \mathrm{e}^x + C$$

$$\int x^n \,\mathrm{d}x = \frac{x^{n+1}}{n+1} + C \qquad \text{provided } n \neq -1$$

$$\int \frac{\mathrm{d}x}{x} = \log_e x + C$$

$$\left.\begin{aligned} \int \frac{\mathrm{d}x}{a^2 + x^2} &= \frac{1}{a}\arctan\frac{x}{a} + C \\ \int \frac{\mathrm{d}x}{\sqrt{(a^2 - x^2)}} &= \arcsin\frac{x}{a} + C \end{aligned}\right\} \quad a \text{ is a constant}$$

The '$\mathrm{d}x$' which appears in all these integrals indicates that the integration is with respect to x. Thus, while we have $\int \cos x \,\mathrm{d}x = \sin x + C$, $\int \cos x \,\mathrm{d}y$ cannot be evaluated unless more information is available to enable us to change the integral with respect to y into one with respect to x.

It cannot be emphasised too much that the 'x' in the above list stands for any variable quantity and could just as well be written as y, z, u, v, etc. Thus

$$\int e^y \, dy = e^y + C$$

$$\int \frac{du}{u^2+9} = \tfrac{1}{3}\arctan \tfrac{1}{3}u + C$$

$$\int \frac{dz}{z} = \log_e z + C$$

Example 1 Integrate the following functions with respect to x: (i) x^8, (ii) $\sqrt[3]{x}$, (iii) $1/x^6$, (iv) $1/\sqrt{x^3}$, (v) $1/\sqrt{(9-x^2)}$, (vi) $1/(25+x^2)$, (vii) $\sin^2 \tfrac{1}{2}x$.

(i) $$\int x^8 \, dx = \tfrac{1}{9}x^9 + C$$

(ii) $$\int \sqrt[3]{x}\, dx = \int x^{1/3} \, dx = \frac{x^{4/3}}{\frac{4}{3}} + C = \tfrac{3}{4}x^{4/3} + C$$

(iii) $$\int \frac{1}{x^6} \, dx = \int x^{-6} \, dx = \frac{x^{-5}}{-5} + C = -\frac{1}{5x^5} + C$$

(iv) $$\int \frac{1}{\sqrt{x^3}} \, dx = \int x^{-3/2} \, dx = \frac{x^{-1/2}}{-\frac{1}{2}} + C = \frac{-2}{\sqrt{x}} + C$$

(v) $$\int \frac{1}{\sqrt{(9-x^2)}} \, dx = \arcsin \tfrac{1}{3}x + C$$

(vi) $$\int \frac{1}{25+x^2} \, dx = \tfrac{1}{5}\arctan \tfrac{1}{5}x + C$$

(vii) For $\int \sin^2 \tfrac{1}{2}x \, dx$ we notice that this integral is not included in our list of standard forms. However, and this is not an uncommon device, it is possible to rewrite $\sin^2 \tfrac{1}{2}x$ by means of a trigonometric identity in a form which is immediately integrable. Thus since

$$\cos x = 1 - 2\sin^2 \tfrac{1}{2}x$$
$$\sin^2 \tfrac{1}{2}x = \tfrac{1}{2}(1 - \cos x)$$

therefore

$$\int \sin^2 \tfrac{1}{2}x \, dx = \int (\tfrac{1}{2} - \tfrac{1}{2}\cos x)\, dx$$
$$= \tfrac{1}{2}x - \tfrac{1}{2}\sin x + C$$

Example 2 Find y in terms of x if $d^2y/dx^2 = 6x - 4$ and further $y = 0$ when $x = 0$ and $dy/dx = 3$ when $x = 0$.

Since $d^2y/dx^2 = 6x - 4$, therefore

$$\frac{dy}{dx} = \int (6x-4)\, dx = 3x^2 - 4x + C$$

Since $dy/dx = 3$ when $x = 0$, then $3 = C$ and

$$\frac{dy}{dx} = 3x^2 - 4x + 3$$

$$y = \int (3x^2 - 4x + 3)dx = x^3 - 2x^2 + 3x + D$$

But since $y = 0$ when $x = 0$, $D = 0$, so that $y = x^3 - 2x^2 + 3x$.

Exercises 13a

Integrate the following functions with respect to x:

1 $x^{5/3}, x^{11}, x^{-2/3}, \sqrt[3]{x^2}, x^{-1}, \sqrt[4]{x^5}, x^{21}$

2 $\frac{1}{x^3}, \frac{1}{x^2}, \frac{1}{\sqrt{x}}, \frac{1}{x}, \frac{1}{x^{11}}, \frac{1}{\sqrt[3]{x^4}}, \frac{1}{\sqrt[5]{x^2}}$

3 $\frac{1}{\sqrt{(16 - x^2)}}, \frac{1}{\sqrt{(1 - x^2)}}, \frac{1}{\sqrt{(\frac{1}{4} - x^2)}}, \frac{1}{\sqrt{(\frac{1}{9} - x^2)}}, \frac{1}{\sqrt{(36 - x^2)}}$

4 $\frac{1}{x^2 + 4}, \frac{1}{x^2 + 1}, \frac{1}{x^2 + 9}, \frac{1}{x^2 + \frac{1}{9}}, \frac{1}{x^2 + \frac{1}{25}}$

5 $\sec^2 x$, $\operatorname{cosec}^2 x$, $\cos^2 \frac{1}{2}x$ (see Example 1 (vii) above), $\tan^2 x$ [Hint: use $\sec^2 x = 1 + \tan^2 x$], $\cot^2 x$

6 $\frac{5 + x - 2x^2}{x^5}, \frac{1 + 3x - 5x^5}{\sqrt{x}}$

7 $\frac{ax^3 + bx + c}{x^7}, \frac{2x^n + 3x^p}{x^5}, \frac{1 + 3x + 5x^2}{x^n}$

8 $(1 - x^3)^2, (1 + 5x)^2, \frac{(ax + b)^2}{x^2}$

9 $\frac{6}{\operatorname{cosec} x}, \frac{7}{\sec x}, 8 \cos x - 6 \sin x$

10 Find y in terms of x if $dy/dx = 3x^2 - 6x + 2$ and $y = 7$ when $x = 0$.

11 Find v in terms of t given that $dv/dt = 5 - 2kt$, where k is a constant. If $v = 0$ when $t = 0$ and $v = 1$ when $t = 1$ find the value of k.

12 Given $d^2x/dt^2 = 3 \sin t$ and that when $t = 0$, $dx/dt = -3$ and $x = 0$, find x in terms of t. Hence show that $\frac{d^2x}{dt^2} + x = 0$.

13 The slope of a curve at any point (x, y) is equal to $\sin x$ and the curve passes through the point $(0, 2)$. Find its equation.

14 What curve passing through the origin has its slope given by the equation $dy/dx = (x^2 - x)^2$?

15 A particle starts from rest at the origin and moves along the x-axis. The acceleration of the particle after time t is given by

$$d^2x/dt^2 = 12t^2 - 60t + 32$$

Find an expression for x at time t. Hence find the times at which the particle again passes through the origin.

13.4 Five important rules

All these rules follow from the definition of integration as the reverse of differentiation. The first two rules will just be stated.

RULE I The integral of a sum of a finite number of functions is the sum of their separate integrals. ('Sum' includes the addition of negative quantities, i.e. 'difference.')

Example 1
$$\int (x^2 + \sin x + \sqrt{x})\mathrm{d}x = \int x^2\,\mathrm{d}x + \int \sin x\,\mathrm{d}x + \int \sqrt{x}\,\mathrm{d}x$$
$$= \tfrac{1}{3}x^3 - \cos x + \tfrac{2}{3}x^{3/2} + C$$

Example 2
$$\int (\mathrm{e}^x - \cos x)\mathrm{d}x = \int \mathrm{e}^x\,\mathrm{d}x - \int \cos x\,\mathrm{d}x$$
$$= \mathrm{e}^x - \sin x + C$$

RULE II A constant factor may be brought outside the integral sign.

Example 3
$$\int 6x^4\,\mathrm{d}x = 6\int x^4\,\mathrm{d}x = \tfrac{6}{5}x^5 + C$$

Example 4
$$\int \frac{5}{u}\,\mathrm{d}u = 5\int \frac{\mathrm{d}u}{u} = 5\log_e u + C$$

Example 5
$$\int (6\cos x - 4x^2)\mathrm{d}x = 6\int \cos x\,\mathrm{d}x - 4\int x^2\,\mathrm{d}x$$
$$= 6\sin x - \tfrac{4}{3}x^3 + C$$

The third and fourth rules extend the applications of our standard forms.

Consider $\int \cos(x+3)\mathrm{d}x$. From our standard forms, the result is possibly $\sin(x+3)$. On differentiating $\sin(x+3)$ with respect to x, we do, in fact, obtain $\cos(x+3)$. Therefore, $\int \cos(x+3)\,\mathrm{d}x = \sin(x+3) + C$. Similarly

$$\int \mathrm{e}^{x-2}\,\mathrm{d}x = \mathrm{e}^{x-2} + C$$
$$\int \frac{\mathrm{d}x}{x+a} = \log_e(x+a) + C \quad (a \text{ is a constant})$$
$$\int \frac{\mathrm{d}x}{(x-2)^2+9} = \tfrac{1}{3}\tan^{-1}\tfrac{1}{3}(x-2) + C$$

hence the rule:

RULE III The addition of a constant to the variable makes no difference to the form of the result.

Now consider $\int \cos 5x\,\mathrm{d}x$. From our standard forms, the result $\sin 5x$ is suggested but on differentiating this latter function we obtain $5\cos 5x$. Since this only differs by a constant factor 5 and not a variable factor from the required $\sin 5x$, we find that $\tfrac{1}{5}\sin 5x$ when differentiated gives the required result. Therefore

$$\int \cos 5x \, dx = \tfrac{1}{5}\sin 5x + C$$

Similarly

$$\int e^{-2x}\, dx = -\tfrac{1}{2}e^{-2x} + C$$

$$\int \cos \tfrac{1}{3}x \, dx = (\sin \tfrac{1}{3}x)/\tfrac{1}{3} = 3\sin \tfrac{1}{3}x + C$$

$$\int \frac{dx}{\sqrt{(9-4x^2)}} = \int \frac{dx}{\sqrt{[9-(2x)^2]}} = \tfrac{1}{2}\arcsin \tfrac{2}{3}x + C$$

hence the rule:

RULE IV Multiplying the variable by a constant makes no difference to the form of the result but we have to divide by the constant.

Rules III and IV may be applied together.

Example 6 $\displaystyle\int e^{3x-4}\, dx = \tfrac{1}{3}e^{3x-4} + C$

Example 7 $\displaystyle\int \frac{dx}{2-x} = -\log_e(2-x) + C$

Example 8

$$\int \sqrt{(5x+3)}\, dx = \int (5x+3)^{1/2}\, dx = \frac{1}{5}\frac{(5x+3)^{3/2}}{(3/2)} + C$$

Example 9

$$\int \frac{dx}{(4-3x)^5} = \int (4-3x)^{-5}\, dx = \frac{(4-3x)^{-4}}{-4}\left(\frac{1}{-3}\right) + C = \frac{1}{12(4-3x)^4} + C$$

Example 10 $\displaystyle\int \frac{dx}{x^2+4x+13}$. We express the denominator as a sum of squares.

$$\int \frac{dx}{x^2+4x+13} = \int \frac{dx}{(x+2)^2+3^2} = \tfrac{1}{3}\arctan \tfrac{1}{3}(x+2) + C$$

Example 11 $\displaystyle\int \frac{dx}{\sqrt{(12x-9x^2)}}$. We render the expression under the root sign as the difference of two squares.

$$\int \frac{dx}{\sqrt{(12x-9x^2)}} = \int \frac{dx}{\sqrt{[2^2-(3x-2)^2]}} = \tfrac{1}{3}\arcsin \tfrac{1}{2}(3x-2) + C$$

Another useful rule is obtained by considering the derivative of $\log_e[f(x)]$:

$$\frac{d}{dx}[\log_e f(x)] = \frac{1}{f(x)}\frac{d}{dx}f(x) = \frac{f'(x)}{f(x)}$$

hence

$$\int \frac{f'(x)}{f(x)}\,dx = \log_e[f(x)] + C$$

RULE V The integral of a fraction whose numerator is the derivative of its denominator is the logarithm of the denominator.

Example 12 $\displaystyle\int \frac{4x^3-1}{x^4-x+2}\,dx = \log_e(x^4-x+2) + C$

Example 13 $\displaystyle\int \frac{3x^2}{x^3+1}\,dx = \log_e(x^3+1) + C$

In some cases, a constant factor has to be inserted to make the numerator exactly equal to the derivative of the denominator.

Example 14
$$\int \frac{x+1}{x^2+2x+5}\,dx = \frac{1}{2}\int \frac{2x+2}{x^2+2x+5}\,dx = \frac{1}{2}\log_e(x^2+2x+5) + C$$

Example 15 $\displaystyle\int \frac{e^{3x}}{e^{3x}-1}\,dx = \frac{1}{3}\int \frac{3e^{3x}}{e^{3x}-1}\,dx = \frac{1}{3}\log_e(e^{3x}-1) + C$

The separation into $f'(x)/f(x)$ may not always be obvious.

Example 16 $\displaystyle\int \frac{dx}{x\log_e x} = \int \frac{\frac{1}{x}}{\log_e x}\,dx = \log_e(\log_e x) + C$

Example 17 $\displaystyle\int \tan x\,dx = \int \frac{\sin x\,dx}{\cos x} = -\int \frac{-\sin x}{\cos x}\,dx = -\log_e \cos x + C$

Example 18 $\displaystyle\int \frac{dx}{(1+x^2)\arctan x} = \int \frac{\frac{1}{1+x^2}}{\arctan x}\,dx = \log_e(\arctan x) + C$

Exercises 13b

Evaluate

1 (i) $\displaystyle\int (2x+3)^{10}\,dx$ (ii) $\displaystyle\int (5-x)^{11}\,dx$ (iii) $\displaystyle\int \sqrt{(7t+5)}\,dt$

(iv) $\displaystyle\int (3u-5)^{5/2}\,du$

2 (i) $\int \frac{dx}{(3x+1)^{15}}$ (ii) $\int \frac{dx}{\sqrt{(2x+1)}}$ (iii) $\int \frac{dx}{(1-x)^{3/2}}$

(iv) $\int \frac{dy}{\sqrt[3]{(1-3y)}}$

3 (i) $\int \sin(3x+3)\,dx$ (ii) $\int \cos(5u-1)\,du$ (iii) $\int \sin(1-y)\,dy$

(iv) $\int \sin^2 x\,dx$ (v) $\int \cos^2 x\,dx$ [Hint: Express $\sin^2 x$ and $\cos^2 x$ in terms of $\cos 2x$.]

4 (i) $\int e^{2-x}\,dx$ (ii) $\int e^{5(t+2)}\,dt$ (iii) $\int e^{1-6u}\,du$

5 (i) $\int \frac{dx}{2x+1}$ (ii) $\int \frac{dx}{1-2x}$ (iii) $\int \cot x\,dx$

(iv) $\int \frac{5x}{x^2+1}\,dx$ (v) $\int \frac{2x+1}{x^2+x-1}\,dx$ (vi) $\int \frac{e^{-t}}{2-e^{-t}}\,dt$

(vii) $\int \frac{xe^{x^2}}{e^{x^2}+3}\,dx$ (viii) $\int \frac{du}{u\log_e u^2}$ (ix) $\int \frac{dx}{x\log_e 3x}$

(x) $\int \frac{\cos x+\sin x}{\cos x-\sin x}\,dx$

6 (i) $\int \frac{dx}{x^2-2x+5}$ (ii) $\int \frac{dt}{16t^2+1}$ (iii) $\int \frac{dx}{x^2+x+\frac{5}{4}}$

7 (i) $\int \frac{d\theta}{\sqrt{(1-16\theta^2)}}$ (ii) $\int \frac{dx}{\sqrt{(15-4x-4x^2)}}$ (iii) $\int \frac{du}{\sqrt{(8+2u-u^2)}}$

8 (i) $\int \frac{x^2\,dx}{x^3+1}$ (ii) $\int (2-3t)^{1/2}\,dt$ (iii) $\int \frac{e^x\,dx}{1+e^x}$

9 (i) $\int \frac{du}{1+4u^2}$ (ii) $\int \frac{u\,du}{1+4u^2}$ (iii) $\int \frac{du}{\sqrt{(1-4u^2)}}$

10 (i) $\int \frac{2\tan x\sec^2 x}{1+\tan^2 x}\,dx$ (ii) $\int \frac{\sin 2u}{1-\sin^2 u}\,du$

Note on $\log_e x$

Consider

$$\int \frac{dx}{ax-b} = \frac{1}{a}\log_e(ax-b) + C$$

Alternatively,

$$\int \frac{dx}{ax-b} = -\int \frac{dx}{b-ax} = \frac{1}{a}\log_e(b-ax) + C$$

Either of these forms may be valid and the correct result is $\frac{1}{a}\log_e|ax-b|$ where $|ax-b|$ is the positive numerical value of $ax-b$ (the modulus of $ax-b$). While the modulus sign will not always be used, it must be remembered especially for definite integration (see Section 13.6).

13.5 Applications to geometry and mechanics

The problem of finding a function when its differential coefficient is given has many applications in geometry and mechanics. Generally, the arbitrary constant which arises can be evaluated by referring to the initial conditions or to some specific value the function must possess.

Example 1 A curve passes through the point (1, 6) and is such that its slope at any point equals twice the abscissa of that point; find its equation.

Here we have $dy/dx = 2x$. Therefore, on integrating, we have $y = x^2 + C$. But the point (1, 6) lies on the curve, hence

$$6 = 1^2 + C$$
$$5 = C$$

Therefore, the required equation is $y = x^2 + 5$.

Example 2 A particle starts from rest with an acceleration $(10-2t)$ m/sec^2 at any time t. When and where will it come to rest again?

Since acceleration is the rate of change of speed, v, with respect to time

$$\frac{dv}{dt} = 10 - 2t$$

Hence

$$v = \int (10-2t)dt$$
$$v = 10t - t^2 + C$$

But the particle starts from rest, so that $v = 0$ when $t = 0$; hence $C = 0$ and

$$v = 10t - t^2$$
$$v = t(10-t)$$

The body is at rest when $v = 0$, that is when $t(10-t) = 0$ or when $t = 0$ or 10 sec. If s is the distance travelled in t sec

$$\frac{ds}{dt} = v = 10t - t^2$$

therefore

$$s = \int (10t - t^2)\mathrm{d}t = 5t^2 - \tfrac{1}{3}t^3 + D$$

where D is an arbitrary constant. If s is the distance measured from the starting point, $s = 0$ when $t = 0$; hence $D = 0$. Thus

$$s = 5t^2 - \tfrac{1}{3}t^3$$

and when $t = 10$

$$s = 5 \times 10^2 - \tfrac{1}{3} \times 10^3$$

that is, $s = 166\tfrac{2}{3}$ m is the distance travelled before the particle comes to rest again.

Exercises 13c

1 Find the equation of the curve whose gradient is $1 - 2x^2$ and which passes through the point $x = 0$, $y = 1$. [LU]

2 At a point on a curve the product of the slope of the curve and the square of the abscissa of the point is 2. If the curve passes through the point $x = 1$, $y = -1$, find its equation. [LU]

3 A particle starts with an initial speed of 20 m/sec. Its acceleration at any time t is $18 - 2t$ m/sec^2. Find the speed at the end of 6 sec and the distance travelled in that time.

4 A particle starts with an initial speed u. It moves in a straight line with an acceleration which varies as the square of the time the particle has been in motion. Find the speed at any time t, and the distance travelled.

5 A particle is projected upwards with a velocity of 3000 cm/sec. In addition to being subject to gravity* it is acted on by a retardation of $500t$ where t is the time from the commencement of the motion. What is the greatest height the particle will reach?

13.6 Integration as a summation

We shall now show that an alternative way of regarding integration is as the limiting value of a summation. This method of approach is of great value in applying integration to physical problems. Incidentally, it also explains the use of the symbol $\int$ which is an elongated 'S' for 'sum'.

Consider *Figure 13.1* where K and B are the points $(a, 0)$ and $(b, 0)$ respectively.

DC is an arc of the curve $y = \mathrm{f}(x)$; P is a variable point on the curve with co-ordinates (x, y); Q is a neighbouring point on the curve whose co-ordinates are $(x + \delta x, y + \delta y)$. We denote the area DPLK by A. Since D and K are fixed, A depends on the position of $\mathrm{P}(x, y)$ and is therefore a function of x, $A(x)$. The area PQML can be denoted by δA, the increase in A due to an

* The acceleration due to gravity can be taken as 1000 cm/sec^2 in the downwards direction.

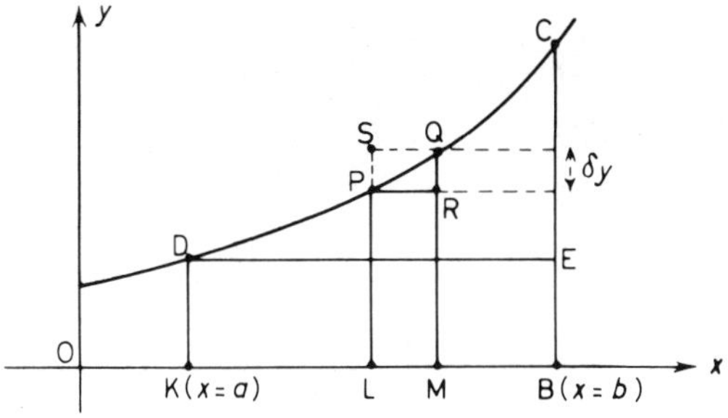

Figure 13.1

increase δx in x. Referring to the diagram, we have

$$\text{area rectangle PRML} < \delta A < \text{area rectangle SQML}$$

(If the slope of the curve is negative, both inequality signs are reversed.)
Therefore

$$\text{PL.LM} < \delta A < \text{QM.LM}$$
$$y\delta x < \delta A < (y + \delta y)\delta x \tag{13.3}$$

Now let KB be subdivided into n equal parts each of length δx (such as LM). Then by drawing ordinates at all the points of subdivision, n strips like SQML are obtained. Summing over all such strips, we have

$$\sum_{x=a}^{x=b} y\delta x < \text{area DKBC} < \sum_{x=a}^{x=b} (y + \delta y)\delta x \tag{13.4}$$

Now consider the difference between the two extreme quantities in the inequality (13.4)

$$\sum_{x=a}^{x=b} (y + \delta y)\delta x - \sum_{x=a}^{x=b} y\delta x = \sum_{x=a}^{x=b} \delta y\delta x$$
$$= \delta x \times \sum_{x=a}^{x=b} \delta y$$

Since δx is the same for all points of subdivision (see *Figure 13.1*),

$$\delta x \times \sum_{x=a}^{x=b} \delta y = \delta x.\text{CE}$$

Now since

$$\delta x = \frac{\text{BK}}{n} = \frac{b - a}{n} \tag{13.5}$$

δx may be made arbitrarily small by increasing n sufficiently. Hence the difference between $\Sigma_{x=a}^{x=b}(y + \delta y)\delta x$ and $\Sigma_{x=a}^{x=b} y\delta x$ can be made arbitrarily small. Since the area DKBC lies between these two, it follows that

$$\text{area DKBC} = \lim_{\delta x \to 0} \sum_{x=a}^{x=b} y\delta x \tag{13.6}$$

Returning to the inequality (13.3), since we are dealing with small but finite quantities, we may divide throughout by δx and hence

$$y < \frac{\delta A}{\delta x} < y + \delta y$$

Now as $\delta x \to 0$, $\delta A \to 0$, and $\delta y \to 0$, so we have

$$\lim_{\delta x \to 0} \frac{\delta A}{\delta x} = y$$

therefore

$$\frac{dA}{dx} = y$$

hence $A = \int y\,dx$ from our definition of integration as the reverse of differentiation.

We note that, as yet, there is no definite value for the area because $\int y\,dx$ involves an arbitrary constant. This is because A, as we remarked earlier, is a function of x. Thus $\int y\,dx$ gives the area measured from an arbitrary origin to the point x. Referring to *Figure 13.1*,

$$\begin{aligned}\text{area DKBC} &= \text{area up to CB}(x = b) - \text{area up to DK}(x = a)\\ &= A(b) - A(a)\end{aligned}$$

Thus, to find the area between the curve $y = f(x)$ and the x-axis, we first find the indefinite integral $\int y\,dx$ or $\int f(x)\,dx$. We then substitute $x = b$ and $x = a$ respectively in the indefinite integral and subtract the two results. The notation adopted for this definite integral is $\int_{x=a}^{x=b} f(x)\,dx$ or the shorter $\int_a^b f(x)\,dx$ or $\int_a^b y\,dx$. Thus

$$\text{area DKBC} = \int_a^b y\,dx \tag{13.7}$$

Finally, we note from (13.6) and (13.7) that

$$\text{area DKBC} = \lim_{\delta x \to 0} \sum_{x=a}^{x=b} y\delta x = \int_a^b y\,dx \tag{13.8}$$

Example 1 Find the area between the curve $y = x^3$, the x-axis and the ordinates $x = 2$ and $x = 6$.

$$\begin{aligned}\text{Area} &= \int_2^6 x^3\,dx\\ &= \left[\frac{x^4}{4}\right]_2^6\\ &= \frac{6^4}{4} - \frac{2^4}{4}\\ &= 320 \text{ square units}\end{aligned}$$

Example 2 Evaluate $\int_2^{2\sqrt{3}} \frac{dx}{x^2+4}$

$$\int_2^{2\sqrt{3}} \frac{dx}{x^2+4} = \left[\frac{1}{2}\arctan\frac{x}{2}\right]_2^{2\sqrt{3}}$$

$$= \frac{1}{2}\arctan\left(\frac{2\sqrt{3}}{2}\right) - \frac{1}{2}\arctan\left(\frac{2}{2}\right)$$

$$= \frac{1}{2}\cdot\frac{\pi}{3} - \frac{1}{2}\cdot\frac{\pi}{4} = \frac{\pi}{24}$$

Example 3 Find the area between the curve $y = x(6 - x)$ and the line $y = 5$.

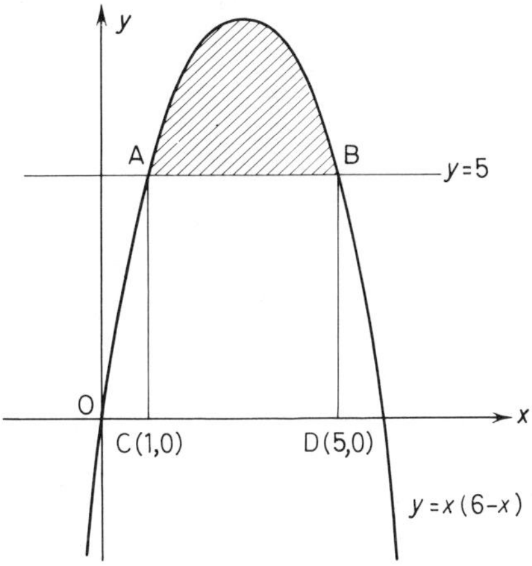

Figure 13.2

Figure 13.2 shows the curve and the line. The area required is the shaded region. The abscissae of A and B are given by the solutions of the equation

$$x(6-x) = 5$$
$$x^2 - 6x + 5 = 0$$
$$(x-1)(x-5) = 0$$
$$x = 1 \quad \text{or} \quad x = 5$$

$$\text{Required area} = \int_1^5 x(6-x)dx - \text{area ABCD}$$

$$= \left[3x^2 - \frac{x^3}{3}\right]_1^5 - 4 \times 5$$

$$= \left(75 - \frac{125}{3}\right) - \left(3 - \frac{1}{3}\right) - 20$$

$$= 72 - \frac{124}{3} - 20 = \frac{32}{3} \text{ square units}$$

Exercises 13d

Evaluate the following definite integrals:

1 (i) $\int_0^1 x^2\,dx$ (ii) $\int_2^3 x^3\,dx$ (iii) $\int_{-1}^1 x^5\,dx$

2 (i) $\int_0^1 y^2\,dy$ (ii) $\int_2^3 y^3\,dy$ (iii) $\int_{-1}^1 y^5\,dy$

3 (i) $\int_1^4 \sqrt{x}\,dx$ (ii) $\int_0^1 \sqrt[3]{x}\,dx$ (iii) $\int_1^8 x^{2/3}\,dx$

4 (i) $\int_0^3 \frac{dx}{x^2+9}$ (ii) $\int_{-1}^1 \frac{dy}{y^2+1}$ (iii) $\int_0^3 \frac{du}{u^2+3}$

5 (i) $\int_{-3}^{-2} \frac{dx}{x^3}$ (ii) $\int_{-3}^{-2} \frac{dy}{y^3}$ (iii) $\int_{-3}^{-2} \frac{dw}{w^3}$

6 (i) $\int_{-1}^4 \sqrt{(x+5)}\,dx$ (ii) $\int_4^{-1} \sqrt{(x+5)}\,dx$

7 (i) $\int_{\pi/6}^{\pi/3} \cos 3x\,dx$ (ii) $\int_{-\pi/2}^0 \sin(5x+\frac{1}{2}\pi)\,dx$

8 (i) $\int_2^3 e^{3x}\,dx$ (ii) $\int_{-1}^0 e^{1-5x}\,dx$ (iii) $\int_0^1 e^{-2x}\,dx$

9 (i) $\int_1^3 \frac{dx}{x}$ (ii) $\int_3^5 \frac{dx}{x}$ (iii) $\int_1^5 \frac{dx}{x}$

10 (i) $\int_{-3}^3 \frac{x\,dx}{x^2+5}$ (ii) $\int_0^2 \frac{dx}{\sqrt{(4-x^2)}}$ (iii) $\int_1^4 \frac{dx}{\sqrt{(8+2x-x^2)}}$

11 Find the area between the curve $y = x^3 + 9x$, the x-axis, and the ordinates at $x = 0$ and $x = 3$.

12 Find the area between the curve $y = x^4 + x^2$, the x-axis and the ordinates at $x = 0$ and $x = 1$.

13 Find the area between the curve $y = \sin x$ and the x-axis between $x = 0$ and $x = \pi$.

14 Find the area between the curve $y = \frac{1}{1+x^2}$, the x-axis and the ordinates at $x = -1$ and $x = +1$.

15 Find the area between the curve $y = 3 + 7x - x^2$ and the line $y = 9$.

Exercises 13

1 Evaluate (i) $\int (2x-1)^3\,dx$, (ii) $\int (2x-1)^{16}\,dx$.

2 Evaluate (i) $\int (2x^2-1)^3\,dx$, (*ii*) $\int (2x^4-1)^3\,dx$.

3 Find the indefinite integrals with respect to x of (i) $\sqrt{x}(x-1)^2$, (ii) $(x-a) \times (b-x)$, where a, b are constants.

4 Evaluate

(i) $\int \frac{(6 + x - x^4)}{x^2} \mathrm{d}x$ (ii) $\int \frac{1 + 2x + 2x^2}{\sqrt{x}} \mathrm{d}x$

(iii) $\int \frac{5 + x + 3x^2}{\sqrt[3]{x}} \mathrm{d}x$

5 Evaluate

(i) $\int_{\pi/2}^{\pi} (\sin 2x + \cos \tfrac{1}{3}x) \mathrm{d}x$ (ii) $\int_{\pi/6}^{\pi/3} (\sec^2 x + \operatorname{cosec}^2 x) \mathrm{d}x$

6 Find the equation of the curve whose slope at any point is $2x - 3x^2$ and which passes through the point (1, 1).

7 The gradient of a curve at the point with abscissa x is given by $\mathrm{d}y/\mathrm{d}x = a + bx$. If the curve passes through the origin and has slope 1 at this point, find the value of a. If the curve also passes through the point (1,3), find its equation.

8 From any point P on a curve, PA is drawn perpendicular to the y-axis. The tangent at P meets the y-axis at B. If PA.AB $= k^2$ find the equation of the curve.

9 If $\frac{\mathrm{d}^2 y}{dx^2} = 4 + \frac{1}{x^3}$ and $\frac{\mathrm{d}y}{\mathrm{d}x} = 0$ when $x = \frac{1}{2}$, find y as a function of x, given $y = \frac{3}{2}$ when $x = 1$.

10 Find the area between the curve $y = \cos x$, the y-axis, the x-axis and the ordinate at $x = \pi/4$.

11 * The gradient of a curve at the point (x, y) is $\left(x - \frac{1}{x}\right)$ and the curve passes through the point (1, 2). Find the equation of the curve. Show that the area enclosed by the curve, the x-axis and the ordinates $x = 1$, $x = 2$ is $\frac{11}{3} - 2 \log_e 2$. [LU]

12 Find the area between the curve $y = (\sin x + \cos x)^2$, the x-axis and the ordinates at $x = 0$ and $x = \pi/2$.

13 Find the area between the curve $y = 1 + 9x - x^2$ and the line $y = 9$.

14 Find the area between the curve $y = 5x - 2x^2$ and the line $y = x$.

15 A body moves under a constant acceleration f. If its initial velocity is u and it starts from some origin at time $t = 0$, show that its velocity and displacement s from the origin are given by $v = u + ft$, $s = ut + \frac{1}{2}ft^2$. By writing its acceleration as $\frac{1}{2}\frac{\mathrm{d}}{\mathrm{d}s}(v^2)$ show that $v^2 = u^2 + 2fs$.

16 Show that the expression for acceleration $\mathrm{d}v/\mathrm{d}t$ can be rewritten $\frac{1}{2}\frac{\mathrm{d}}{\mathrm{d}s}(v^2)$. Hence, if the acceleration of a particle is equal to $16s$ and $v = 4$ m/sec when $s = 1$ m, find the velocity of the particle in terms of s.

17 The equation of a curve is of the form $y = ax^2 + bx + c$. It meets the x-axis where $x = -1$ and $x = 3$; also $y = 12$ when $x = 1$. Find the equation of the curve and the area between it and the x-axis. [LU]

*Note that $\int \log_e x \,\mathrm{d}x = x \log_e x - x$.

18 A particle is subject to retardation equal to $32 + 16t$ at any time t. Initially its velocity is 40 m/sec. Find how long it takes to come to rest and how far it is then from its starting point.

19 Verify the following results:

(i) $\displaystyle\int_0^1 \frac{dx}{\sqrt{(3-2x)}} = \sqrt{3} - 1$ (ii) $\displaystyle\int_0^{\sqrt{(1/3)}} \frac{dx}{\sqrt{(2-3x^2)}} = \frac{\pi}{4\sqrt{3}}$

20 Show that $\sin 3x = 3\sin x - 4\sin^3 x$ and hence evaluate $\int_0^{\pi/2} \sin^3 x\, dx$. In a similar manner evaluate $\int_0^{\pi/2} 2\cos^3 x\, dx$.

21 Given that

$$f(x) = \frac{1+x}{(1-x)(1+x^2)}$$

show that

$$f(x) = \frac{1}{1-x} + \frac{1}{2}\cdot\frac{2x}{1+x^2}$$

Hence evaluate

$$\int_0^{3/4} f(x)\, dx$$

22 Express $\dfrac{1}{(2-x)(1+x)}$ in partial fractions and hence, or otherwise,

evaluate $\displaystyle\int_0^1 \frac{dx}{(2-x)(1+x)}$. [O]

23 Sketch the curve $y = \dfrac{x^2 - 7x + 10}{x - 6}$, indicating clearly any vertical asymptotes and turning values. Determine the finite area bounded by the x-axis and that part of the curve between the points where it crosses the x-axis. [O]

24 Find the x co-ordinates of the stationary points of the curve whose equation is $y = x(x-3)^4$, and determine whether these points are maxima, minima or points of inflexion. Sketch the curve. Find the area of the region bounded by the curve and that part of the x-axis lying between $x = 0$ and $x = 3$. [JMB]

25 Show that $\cos 5x \cos 3x = \frac{1}{2}(\cos 8x + \cos 2x)$. Hence show that $\int_0^{2\pi} \cos 5x \cos 3x\, dx = 0$.

14
Some methods of integration

14.1 Introduction

In the previous chapter, we introduced the basic ideas of integration. The examples used involved only simple integrals, which were obtained from the inverses of differential coefficients (Section 13.3) or the simple extensions made possible by the five rules (Section 13.4) The object of this chapter is to examine several ways in which more involved integrals can be resolved into simpler forms which can then be recognised as standard integrals. While several important methods of integration will be examined, it must be realised that not all available methods are covered here.

Ability to integrate readily only comes with experience and the student is well advised to work through as many exercises as possible.

14.2 Integration of rational algebraic fractions

We now consider the integration of rational algebraic functions, by which we mean fractions whose numerator and denominator each contain only positive integral powers of x with constant coefficients. In all cases, if the numerator is of the same or higher degree than the denominator, we first divide out. Thus we shall have one or more terms (in x, x^2, etc. or a constant) which can be immediately integrated and a fraction whose numerator is of a lesser degree than the denominator. It is with such fractions that we shall now be concerned.

Denominator of the first degree

In this case, after any necessary division, the integral can be immediately evaluated.

Example 1 $$\int \frac{2x^3 - x^2 - x}{2x - 3}\,\mathrm{d}x = \int\left(x^2 + x + 1 + \frac{3}{2x-3}\right)\mathrm{d}x$$
$$= \frac{x^2}{3} + \frac{x^2}{2} + x + \frac{3}{2}\log_e(2x - 3) + C$$

Example 2 $$\int \frac{7 + x - 2x^2}{2 - x}\,\mathrm{d}x = \int\left(2x + 3 + \frac{1}{2-x}\right)\mathrm{d}x$$
$$= x^2 + 3x - \log_e(2 - x) + C$$

Exercises 14a

Evaluate

1 $\displaystyle\int\frac{x^2\,dx}{x-1}$ **2** $\displaystyle\int\frac{x^3\,dx}{x-1}$ **3** $\displaystyle\int\frac{x\,dx}{x-1}$

4 $\displaystyle\int\frac{t}{1-3t}\,dt$ **5** $\displaystyle\int\frac{t^2}{1-3t}\,dt$ **6** $\displaystyle\int\frac{2-x}{1-x}\,dx$

7 $\displaystyle\int_0^{1/2}\frac{2\theta-3\theta^2}{1-\theta}\,d\theta$ **8** $\displaystyle\int\frac{t}{1+4t}\,dt$ **9** $\displaystyle\int_0^1\frac{2x-8x^2}{1+4x}\,dx$

10 $\displaystyle\int\frac{x}{a+bx}\,dx$

Denominator of the second degree which does not resolve into rational factors

We shall discuss two cases here: (i) the numerator is a constant, (ii) the numerator is a linear expression in x.

Case 1 Consider

$$\int\frac{k}{ax^2+bx+c}\,dx$$

This can always be put in the form

$$\frac{k}{a}\int\frac{dx}{(x+\alpha)^2\pm\beta^2}$$

We shall restrict ourselves to the case of $+\beta^2$. In this case

$$\frac{k}{a}\int\frac{dx}{(x+\alpha)^2+\beta^2}=\frac{k}{a}\cdot\frac{1}{\beta}\arctan\frac{(x+\alpha)}{\beta}+C$$

by our standard form for $\arctan x$ and its extension (rule III, Section 13.4).

Example 1

$$\int\frac{2x^2+11}{x^2+4}\,dx=\int\left(2+\frac{3}{x^2+4}\right)dx=2x+\tfrac{3}{2}\arctan\tfrac{1}{2}x+C$$

Example 2

$$\int_1^4\frac{5}{x^2-2x+10}\,dx=5\int_1^4\frac{dx}{(x-1)^2+9}=\frac{5}{3}\left[\arctan\left(\frac{x-1}{3}\right)\right]_1^4$$
$$=\tfrac{5}{3}(\arctan 1)-\tfrac{5}{3}(\arctan 0)=\tfrac{5}{12}\pi$$

Example 3

$$\int\frac{7}{2x^2+2x+5}\,dx=\frac{7}{2}\int\frac{dx}{x^2+x+\frac{5}{2}}=\frac{7}{2}\int\frac{dx}{(x+\frac{1}{2})^2+\frac{9}{4}}$$
$$=\frac{7}{2}\cdot\frac{1}{(3/2)}\arctan\frac{(x+\frac{1}{2})}{(3/2)}=\tfrac{7}{3}\arctan\tfrac{1}{3}(2x+1)+C$$

Exercises 14b

Evaluate

1 $\displaystyle\int \frac{5}{x^2+2x+2}\,dx$

2 $\displaystyle\int \frac{7}{x^2-6x+13}\,dx$

3 $\displaystyle\int_0^4 \frac{dx}{x^2-4x+8}$

4 $\displaystyle\int \frac{5x^2+6}{x^2+1}\,dx$

5 $\displaystyle\int \frac{dx}{9x^2-6x+37}$

6 $\displaystyle\int_{-2}^{3} \frac{5}{2x^2-2x+13}\,dx$

7 $\displaystyle\int_{-1}^{1} \frac{x^2+2x}{x^2+2x+2}\,dx$

8 $\displaystyle\int \frac{x^4}{x^2+9}\,dx$

9 $\displaystyle\int \frac{x^2+x+2}{x^2+x+1}\,dx$

10 $\displaystyle\int_{1/4}^{1} \frac{16x^3+6x+8}{8x^2-4x+5}\,dx$

Case 2 If the numerator is a linear expression in x, we put it equal to $k \times$ (the derivative of the denominator) $+ l$, where k, l can be determined by inspection. The integral now splits into two parts, in the first of which the numerator is the derivative of the denominator, and hence this integral is the logarithm of the denominator; the second is of the type considered above.

Example 1

$$\int \frac{x+3}{x^2+25}\,dx = \int \frac{\frac{1}{2}(2x)+3}{x^2+25}\,dx$$

$$= \frac{1}{2}\int \frac{2x}{x^2+25}\,dx + 3\int \frac{dx}{x^2+25}$$

$$= \tfrac{1}{2}\log_e(x^2+25) + \tfrac{3}{5}\arctan \tfrac{1}{5}x + C$$

Example 2

$$\int \frac{3x-2}{2x^2+2x+5}\,dx = \int \frac{\frac{3}{4}(4x+2)-\frac{7}{2}}{2x^2+2x+5}\,dx$$

$$= \frac{3}{4}\int \frac{(4x+2)\,dx}{2x^2+2x+5} - \frac{7}{2}\int \frac{dx}{2x^2+2x+5}$$

$$= \frac{3}{4}\log_e(2x^2+2x+5) - \frac{7}{4}\int \frac{dx}{(x+\frac{1}{2})^2+\frac{9}{4}}$$

$$= \tfrac{3}{4}\log_e(2x^2+2x+5) - \tfrac{7}{6}\arctan \tfrac{1}{3}(2x+1) + C$$

Exercises 14c

Integrate the following functions with respect to x:

1 $\dfrac{x+7}{x^2+16}$ **2** $\dfrac{3x-5}{x^2+36}$ **3** $\dfrac{2x+3}{x^2+2x+10}$

4 $\dfrac{3x+5}{x^2-6x+10}$ **5** $\dfrac{1-3x}{x^2-8x+25}$ **6** $\dfrac{x}{x^2-x+1}$

7 $\dfrac{x^3}{x^2-6x+10}$ **8** $\dfrac{5x+1}{3x^2-12x+13}$ **9** $\dfrac{x^2}{x^2+2x+5}$

10 $\dfrac{2x^3+4x^2}{2x^2+2x+5}$

Denominator which resolves into rational factors of the first and second degree

If the denominator factorises, we use the technique of partial fractions to express the integrand in a form suitable for integration. The three possible types of fraction are

$$\frac{1}{ax+b} \qquad \frac{px+q}{ax^2+bx+c} \qquad \frac{1}{(ax+b)^2}$$

We have just considered the first two types and the third type integrates to

$$\frac{-1}{a(ax+b)}$$

Example 1 $\displaystyle\int \frac{7x-4}{2x^2-3x-2}\,dx$

We note that $2x^2-3x-2=(x-2)(2x+1)$. Put

$$\frac{7x-4}{2x^2-3x-2} \equiv \frac{A}{x-2}+\frac{B}{2x+1}$$

On multiplying throughout by the common denominator, we obtain

$$7x-4 \equiv A(2x+1)+B(x-2)$$

With $x=2$

$$10 = 5A+0$$
$$2 = A$$

With $x=-\frac{1}{2}$

$$-7\tfrac{1}{2} = 0-2\tfrac{1}{2}B$$
$$3 = B$$

Rewriting the integrand in partial fractions, we have

$$\int \frac{7x-4}{2x^2-3x-2}\,dx = \int \frac{2}{x-2}\,dx + \int \frac{3}{2x+1}\,dx$$
$$= 2\log_e(x-2)+\tfrac{3}{2}\log_e(2x+1)+C$$

Example 2 $\displaystyle\int_3^6 \frac{22-5x^3-5x^4}{(x-1)(x+2)}\,dx$

We must first divide out because the numerator is of higher degree than the denominator. This division can be carried out using the technique of partial fractions. By inspection, we see that the highest power of x obtained by division is x^2 and we allow for this and all lower powers of x, including the constant in the partial fractions. Thus we set

$$\frac{22-5x^3-5x^4}{(x-1)(x+2)} \equiv Ax^2+Bx+C+\frac{D}{x-1}+\frac{E}{x+2}$$

therefore

$$22-5x^3-5x^4 \equiv (Ax^2+Bx+C)(x-1)(x+2)+D(x+2)+E(x-1)$$

With $x = 1$

$$12 = 0+3D+0 \qquad \text{therefore} \quad D = 4$$

With $x = -2$

$$-18 = 0+0-3E \qquad \text{therefore} \quad E = 6$$

The other constants are found by equating the coefficients of various powers of x.

For coefficient of x^4

$$-5 = A$$

For the constant term

$$\begin{aligned} 22 &= -2C+2D-E \\ &= -2C+8-6 \\ C &= -10 \end{aligned}$$

Then with $x = -1$

$$\begin{aligned} 22 &= (A-B+C)(-2)(1)+D(1)+E(-2) \\ 22 &= (-B-15)(-2)+4-12 \\ 30 &= 2B+30 \\ B &= 0 \end{aligned}$$

Hence

$$\int_3^6 \frac{(22-5x^3-5x^4)}{(x-1)(x+2)}\,dx$$

$$\begin{aligned} &= \int_3^6 (-5x^2-10)\,dx + \int_3^6 \frac{4}{(x-1)}\,dx + \int_3^6 \frac{6}{(x+2)}\,dx \\ &= [-\tfrac{5}{3}x^3 - 10x + 4\log_e(x-1) + 6\log_e(x+2)]_3^6 \\ &= (-420+4\log_e 5+6\log_e 8)-(-75+4\log_e 2+6\log_e 5) \end{aligned}$$

(Since $\log_e 8 = \log_e 2^3 = 3\log_e 2$)

$$\begin{aligned} &= (-345-2\log_e 5+14\log_e 2) \\ &= -338{\cdot}5 \quad \text{approximately} \end{aligned}$$

Example 3 $\int \frac{2x^2 - 10x}{(x+3)(x-1)^2}\,dx$

Set

$$\frac{2x^2 - 10x}{(x+3)(x-1)^2} \equiv \frac{A}{(x+3)} + \frac{B}{(x-1)} + \frac{C}{(x-1)^2}$$

(Note the partial fractions for the repeated linear factor $(x-1)^2$.)

Therefore

$$2x^2 - 10x \equiv A(x-1)^2 + B(x+3)(x-1) + C(x+3)$$

With $x = 1$

$$-8 = 0 + 0 + 4C \qquad C = -2$$

With $x = -3$

$$48 = 16A + 0 + 0 \qquad A = 3$$

Equating the coefficients of x^2 gives $2 = A + B$, therefore $B = -1$ and so

$$\int \frac{2x^2 - 10x}{(x+3)(x-1)^2}\,dx$$

$$= \int \frac{3}{(x+3)}\,dx - \int \frac{dx}{(x-1)} - \int \frac{2\,dx}{(x-1)^2}$$

$$= 3\log_e(x+3) - \log_e(x-1) - 2\int (x-1)^{-2}\,dx$$

$$= 3\log_e(x+3) - \log_e(x-1) + \frac{2}{x-1} + C$$

Example 4 $\int \frac{(3x+1)}{x^3 + 2x^2 + x + 2}\,dx = \int \frac{(3x+1)}{(x^2+1)(x+2)}\,dx$

Set

$$\frac{3x+1}{(x^2+1)(x+2)} \equiv \frac{Ax+B}{x^2+1} + \frac{C}{x+2}$$

therefore

$$3x + 1 \equiv (Ax+B)(x+2) + C(x^2+1)$$

With $x = -2$

$$-5 = 0 + 5C \qquad \text{therefore} \quad C = -1$$

Equating coefficients of x^2 gives $0 = A + C$, therefore $A = 1$.

Equating constant terms gives $1 = 2B + C$, therefore $B = 1$.

Substituting these values of A, B and C gives

$$\int \frac{3x+1}{(x^2+1)(x+2)}\,dx = \int \frac{(x+1)}{(x^2+1)}\,dx - \int \frac{1}{(x+2)}\,dx$$

$$= \frac{1}{2}\int \frac{2x\,dx}{(x^2+1)} + \int \frac{dx}{(x^2+1)} - \int \frac{dx}{(x+2)}$$

$$= \tfrac{1}{2}\log_e(x^2+1) + \arctan x - \log_e(x+2) + C$$

Exercises 14d

Integrate the following functions with respect to x:

1 $\dfrac{x+1}{x^2+5x+6}$

2 $\dfrac{1}{6x^2-5x+1}$

3 $\dfrac{x+5}{x-x^2}$

4 $\dfrac{4x^2-2x-7}{2x^2-3x-2}$

5 $\dfrac{2x^3+7x^2+2}{2x^2+x}$

6 $\dfrac{x+62}{(3x-1)^2(2x+3)}$

7 $\dfrac{4x^2-3x+5}{(x+2)(x-1)^2}$

8 $\dfrac{6x^2+5x-2}{(2x+1)^2(x-1)}$

9 $\dfrac{6x^3+10x^2-13x-6}{3x^3+x^2}$

10 $\dfrac{4x}{(x^2+4)(x^2+8)}$

11 $\dfrac{1}{(x^2+4)(x^2+8)}$

12 $\dfrac{8x^2+3x-3}{(2x^2-1)(2x+3)}$

13 $\dfrac{(x-2)^2}{x^3+1}$

14 $\dfrac{1}{x^4+5x^2+4}$

15 $\dfrac{10}{(x-1)(x^2+9)}$

16 $\dfrac{5}{(x+1)(x^2+4)}$

17 $\dfrac{1}{x^2-(p+q)x+pq}$

18 $\dfrac{x^4}{x^2-9}$

19 $\dfrac{x}{(x-a)(x-b)(x-c)}$

20 $\dfrac{x}{(x^2+a^2)(x^2+b^2)}$

14.3 Change of variable

Another widely used device in integration is to change the independent variable, say x, to another one, u, where the relation between x and u is known. Suppose

$$I = \int 2x\cos x^2\,dx$$

Let $u = x^2$, so that $du/dx = 2x$. Then

$$I = \int \frac{du}{dx} \cos u \, dx$$

and since integration is the inverse of differentiation

$$\frac{dI}{dx} = \frac{du}{dx} \cos u$$

$$\frac{dI}{dx} \cdot \frac{dx}{du} = \cos u$$

$$\frac{dI}{du} = \cos u \quad \text{by (10.14)}$$

$$I = \int \cos u \, du$$

which is recognisable as a standard form, therefore

$$I = \sin u + C = \sin x^2 + C \quad \text{(substituting for } u\text{)}$$

In general, we have

$$\int f(u) \frac{du}{dx} dx = \int f(u) \, du \tag{14.1}$$

For with

$$I = \int f(u) \frac{du}{dx} dx$$

$$\frac{dI}{dx} = f(u) \frac{du}{dx}$$

$$\frac{dI}{dx} \cdot \frac{dx}{du} = f(u)$$

$$\frac{dI}{du} = f(u)$$

Therefore

$$I = \int f(u) \, du$$

The difficulty of the method lies in finding the relation $u = \phi(x)$ which simplifies the integral. It must be remembered that:

(i) One part of the integrand supplies the du/dx which has to be introduced.
(ii) The rest of the integrand must be easily expressible in terms of u.

Example 1 Evaluate $I = \int \sqrt{(x^3 - 5)} 3x^2 \, dx$.

Let $u = x^3 - 5$, therefore $du/dx = 3x^2$, which we note is part of the integrand. In fact, it is because we foresaw this that we chose the substitution. Therefore

$$\begin{aligned} I &= \int \sqrt{u}\,\frac{du}{dx}\,dx \\ &= \int u^{1/2}\,du \\ &= \tfrac{2}{3}u^{3/2} + C \\ &= \tfrac{2}{3}(x^3 - 5)^{3/2} + C \end{aligned}$$

Occasionally, $k\,du/dx$ (*k a constant*) may be included in the integrand instead of just du/dx.

Example 2 Evaluate $I = \int e^{-x^3}x^2\,dx$.

Let $u = -x^3$, therefore $du/dx = -3x^2$ and so, after substitution,

$$\begin{aligned} I &= \int e^u\left(-\frac{1}{3}\frac{du}{dx}\right)dx \\ &= -\frac{1}{3}\int e^u\frac{du}{dx}\,dx \\ &= -\frac{1}{3}\int e^u\,du \\ &= -\tfrac{1}{3}e^u + C \\ &= -\tfrac{1}{3}e^{-x^3} + C \end{aligned}$$

Example 3 $I = \displaystyle\int \frac{x^3\,dx}{(3x^4 - 5)^6}$

Let $u = 3x^4 - 5$, therefore $du/dx = 12x^3$ and so, after substitution,

$$\begin{aligned} I &= \int \frac{1}{u^6}\,\frac{1}{12}\,\frac{du}{dx}\,dx \\ &= \frac{1}{12}\int u^{-6}\,du \\ &= \frac{1}{12}\cdot\frac{u^{-5}}{-5} + C \\ &= \frac{-1}{60}\cdot\frac{1}{u^5} + C \\ &= \frac{-1}{60(3x^4 - 5)^5} + C \end{aligned}$$

It is the more usual practice to make the substitution $u = \phi(x)$, differentiate it to obtain $du/dx = \phi'(x)$ and then replace dx by $du/\phi'(x)$ $\left(dx = du\Big/\frac{du}{dx} = \frac{dx}{du}\,du\right)$.

Example 4 $I = \displaystyle\int (x - 1)\sqrt[3]{(x^2 - 2x + 3)}\,dx$

Let $u = x^2 - 2x + 3$, therefore $du/dx = 2x - 2$ and so

$$\begin{aligned} I &= \int (x-1)\sqrt[3]{u}\,\frac{du}{2x-2} \\ &= \int (x-1)\sqrt[3]{u}\,\frac{du}{2(x-1)} \\ &= \frac{1}{2}\int u^{1/3}\,du \\ &= \tfrac{1}{2}\cdot\tfrac{3}{4}u^{4/3} + C \\ &= \tfrac{3}{8}(x^2 - 2x + 3)^{4/3} + C \end{aligned}$$

It will be noted that, of the five rules given earlier, III, IV and V are special cases of integration by substitution.

Example 5 $\int e^{x-2}\,dx$

Let $u = (x - 2)$, therefore $du/dx = 1$ and so

$$\int e^{x-2}\,dx = \int e^u\,du = e^u + C = e^{x-2} + C$$

Example 6 $\int \sin 5x\,dx$

Let $u = 5x$, therefore $du/dx = 5$ and so

$$\begin{aligned} \int \sin 5x\,dx &= \int \sin u\,\frac{du}{5} = \frac{1}{5}\int \sin u\,du \\ &= -\tfrac{1}{5}\cos u + C \\ &= -\tfrac{1}{5}\cos 5x + C \end{aligned}$$

Example 7 $\int \frac{dx}{(4-3x)^3}$

Let $u = 4 - 3x$, therefore $du/dx = -3$ and so

$$\begin{aligned} \int \frac{dx}{(4-3x)^3} &= \int \frac{1}{u^3}\left(-\frac{1}{3}\right)du = -\frac{1}{3}\int u^{-3}\,du \\ &= -\frac{1}{3}\cdot\frac{u^{-2}}{-2} + C = \frac{1}{6u^2} + C = \frac{1}{6(4-3x)^2} + C \end{aligned}$$

Example 8 $\int_{\pi/6}^{\pi/4} \cot x\,dx$

Consider the indefinite integral $I = \int \cot x\,dx$:

$$I = \int \frac{\cos x}{\sin x}\,dx$$

Let $u = \sin x$, therefore $du/dx = \cos x$ and substitution gives

$$\begin{aligned} I &= \int \frac{\cos x}{u} \frac{du}{\cos x} \\ &= \int \frac{du}{u} \\ &= \log_e u + C \\ &= \log_e (\sin x) + C \end{aligned}$$

Hence

$$\begin{aligned} \int_{\pi/6}^{\pi/4} \cot x \, dx &= \Big[\log_e (\sin x) \Big]_{\pi/6}^{\pi/4} \\ &= \log_e (\sin \tfrac{1}{4}\pi) - \log_e (\sin \tfrac{1}{6}\pi) \\ &= \log_e \frac{1}{\sqrt{2}} - \log_e \frac{1}{2} \\ &= \log_e \left(\frac{1}{\sqrt{2}} \Big/ \frac{1}{2} \right) \\ &= \log_e \sqrt{2} \end{aligned}$$

An alternative approach when dealing with definite integrals is to change the limits of integration for the variable x into corresponding limits for the variable u. Thus for the above example, with $u = \sin x$, when

$$x = \frac{\pi}{6} \qquad u = \sin \frac{\pi}{6} = \frac{1}{2}$$

and when

$$x = \frac{\pi}{4} \qquad u = \sin \frac{\pi}{4} = \frac{1}{\sqrt{2}}$$

Therefore

$$\begin{aligned} \int_{\pi/6}^{\pi/4} \cot x dx &= \int_{1/2}^{1/\sqrt{2}} \frac{du}{u} = \Big[\log_e u \Big]_{1/2}^{1/\sqrt{2}} \\ &= \log_e \sqrt{2} \quad \text{as before} \end{aligned}$$

Two devices may sometimes be necessary.

Example 9 $\displaystyle\int_2^3 \frac{2x \, dx}{(x^4 - 1)}$

Let $u = x^2$, therefore $du/dx = 2x$.

When $x = 2$, $u = 4$; when $x = 3$, $u = 9$. Therefore

$$\int_2^3 \frac{2x \, dx}{x^4 - 1} = \int_4^9 \frac{1}{u^2 - 1} 2x \frac{du}{2x} = \int_4^9 \frac{du}{u^2 - 1}$$

Now we use partial fractions.

$$= \frac{1}{2}\int_4^9 \left[\frac{1}{(u-1)} - \frac{1}{(u+1)}\right] du$$

$$= \left[\frac{1}{2}\log_e(u-1) - \frac{1}{2}\log_e(u+1)\right]_4^9$$

$$= \frac{1}{2}[(\log_e 8 - \log_e 10) - (\log_e 3 - \log_e 5)]$$

$$= \frac{1}{2}(\log_e 8 - \log_e 10 - \log_e 3 + \log_e 5)$$

$$= \tfrac{1}{2}\log_e \frac{8 \times 5}{3 \times 10} = \tfrac{1}{2}\log_e \tfrac{4}{3}$$

Exercises 14e

Integrate the following functions with respect to x:

1 xe^{-x^2} **2** $x\sqrt{(a^2 - x^2)}$ **3** $\dfrac{x}{\sqrt{(9 - x^2)}}$

4 $\dfrac{1}{x}(\log_e x)^3$ **5** $\dfrac{\cos x}{(1 - \sin x)}$ **6** $\dfrac{\sin x \cos x}{(\cos^2 x + 2\sin^2 x)}$

7 $\dfrac{x^4}{(x^5 + 6)^7}$ **8** $x\sqrt{(x + 1)}$ **9** $\dfrac{\log_e x}{x}$

10 $\dfrac{1}{x \log_e x}$ **11** $\tan^5 x \sec^2 x$ **12** $\dfrac{e^{1/x}}{x^2}$

13 $\dfrac{x^2}{(1 + x^3)(2 + x^3)}$ **14** $x^2(x^3 - 2)^{14}$ **15** $\dfrac{\arcsin x}{\sqrt{(1 - x^2)}}$

Evaluate the following definite integrals:

16 $\displaystyle\int_0^{\pi/4} \tan^3 x \sec^2 x \, dx$ **17** $\displaystyle\int_0^{\pi/4} \frac{\sec^2 \theta}{(1 + \tan \theta)} d\theta$

18 $\displaystyle\int_1^2 \frac{(\log_e x)^n}{x} dx \quad (n > 1)$ **19** $\displaystyle\int_0^1 \frac{e^{2u}}{(3e^{2u} + 2)} du$

20 $\displaystyle\int_1^2 \frac{ye^{y^2}}{(e^{y^2} - 1)} dy$ **21** $\displaystyle\int_1^2 \frac{x \, dx}{(x^4 - 2x^2 + 10)}$

22 $\displaystyle\int_1^2 \frac{e^x}{(1 - e^{2x})} dx$ **23** $\displaystyle\int_{1/2}^{1/\sqrt{2}} \frac{\arcsin \theta}{\sqrt{(1 - \theta^2)}} d\theta$

24 $\displaystyle\int_{-1}^1 \frac{t \, dt}{(t^4 - 4)}$ **25** $\displaystyle\int_{e^4}^{e^2} \frac{1}{x} \sin(\log_e x) \, dx$

14.4 Trigonometric substitutions

If the integrand involves

(i) $\sqrt{(a^2 - x^2)}$, we try $x = a\sin\theta$ because then

$$\sqrt{(a^2 - x^2)} = \sqrt{(a^2 - a^2\sin\theta)} = \sqrt{(a^2\cos^2\theta)} = a\cos\theta$$

(ii) $\sqrt{(a^2 + x^2)}$, we try $x = a\tan\theta$ because then

$$\sqrt{(a^2 + x^2)} = \sqrt{(a^2 + a^2\tan^2\theta)} = \sqrt{(a^2\sec^2\theta)} = a\sec\theta$$

(iii) $\sqrt{(x^2 - a^2)}$, we try $x = a\sec\theta$ because then

$$\sqrt{(x^2 - a^2)} = \sqrt{(a^2\sec^2\theta - a^2)} = \sqrt{(a^2\tan^2\theta)} = a\tan\theta$$

Example 1 $\displaystyle\int_2^{2\sqrt{3}} \frac{\mathrm{d}x}{x^2\sqrt{(4 + x^2)}}$

Let $x = 2\tan\theta$, then

$$\frac{\mathrm{d}x}{\mathrm{d}\theta} = 2\sec^2\theta$$

When $x = 2$, $\tan\theta = 1$, therefore $\theta = \pi/4$.

When $x = 2\sqrt{3}$, $\tan\theta = \sqrt{3}$, therefore $\theta = \pi/3$.

Hence

$$\begin{aligned}\int_2^{2\sqrt{3}} \frac{\mathrm{d}x}{x^2\sqrt{(4 + x^2)}} &= \int_{\pi/4}^{\pi/3} \frac{2\sec^2\theta}{4\tan^2\theta\, 2\sec\theta}\,\mathrm{d}\theta \\ &= \frac{1}{4}\int_{\pi/4}^{\pi/3} \frac{\sec\theta}{\tan^2\theta}\,\mathrm{d}\theta \\ &= \frac{1}{4}\int_{\pi/4}^{\pi/3} \frac{\cos\theta}{\sin^2\theta}\,\mathrm{d}\theta \\ &= \frac{1}{4}\int_{\pi/4}^{\pi/3} \operatorname{cosec}\theta\cot\theta\,\mathrm{d}\theta \\ &= \frac{1}{4}\Big[-\operatorname{cosec}\theta\Big]_{\pi/4}^{\pi/3} \\ &= \frac{1}{4}\left(-\frac{2}{\sqrt{3}} + \sqrt{2}\right) \\ &\approx 0{\cdot}065\end{aligned}$$

Example 2 $\displaystyle\int_0^a \sqrt{(a^2 - x^2)}\,\mathrm{d}x$

Let $x = a\sin\theta$, therefore

$$\frac{\mathrm{d}x}{\mathrm{d}\theta} = a\cos\theta$$

When $x = 0$, $\sin\theta = 0$, therefore $\theta = 0$.

When $x = a$, $\sin\theta = 1$, therefore $\theta = \pi/2$.

Hence

$$\begin{aligned}\int_0^a \sqrt{(a^2 - x^2)}\,\mathrm{d}x &= \int_0^{\pi/2} a\cos\theta . a\cos\theta\,\mathrm{d}\theta \\ &= a^2 \int_0^{\pi/2} \cos^2\theta\,\mathrm{d}\theta \\ &= \frac{a^2}{2}\int_0^{\pi/2} (1 + \cos 2\theta)\,\mathrm{d}\theta \\ &= \frac{a^2}{2}\left[\theta + \frac{\sin 2\theta}{2}\right]_0^{\pi/2} \\ &= \frac{a^2}{2}\left(\frac{\pi}{2} + 0\right) - (0) \\ &= \frac{\pi a^2}{4}\end{aligned}$$

Some times other trigonometric substitutions can be used.

Example 3 $\int_0^2 \sqrt{\left(\frac{x}{4 - x}\right)}\,\mathrm{d}x$

Let $x = 4\sin^2\theta$, then

$$\frac{\mathrm{d}x}{\mathrm{d}\theta} = 8\sin\theta\cos\theta$$

When $x = 2$, $\sin^2\theta = \frac{1}{2}$, therefore $\theta = \pi/4$.

When $x = 0$, $\sin^2\theta = 0$, therefore $\theta = 0$.

Hence

$$\begin{aligned}\int_0^2 \sqrt{\left(\frac{x}{4 - x}\right)}\,\mathrm{d}x &= \int_0^{\pi/4} \sqrt{\left(\frac{4\sin^2\theta}{4 - 4\sin^2\theta}\right)} 8\sin\theta\cos\theta\,\mathrm{d}\theta \\ &= \int_0^{\pi/4} \frac{2\sin\theta}{2\cos\theta} . 8\sin\theta\cos\theta\,\mathrm{d}\theta \\ &= 8\int_0^{\pi/4} \sin^2\theta\,\mathrm{d}\theta \\ &= 4\int_0^{\pi/4} (1 - \cos 2\theta)\,\mathrm{d}\theta \\ &= 4\left[\theta - \frac{\sin 2\theta}{2}\right]_0^{\pi/4} \\ &= 4\left(\tfrac{1}{4}\pi - \tfrac{1}{2}\right) \\ &= \pi - 2\end{aligned}$$

Integration by substitution is a fairly straightforward technique. The question of what is the correct substitution for any particular integral is not quite so straightforward. We have mentioned, at the beginning of this section, three substitutions which often prove helpful. Some other useful substitutions were mentioned in the preceding section. We mention two others which are generally very satisfactory. If the integrand involves $e^{f(x)}$, put $u = f(x)$. If the integrand involves $\sqrt{(a + x)}$, put $u = \sqrt{(a + x)}$.

Exercises 14f

Integrate with respect to x

1 $x\sqrt{(x^2 - a)}$ **2** $x\sqrt{(x^2 + 4)}$ **3** $\dfrac{\sqrt{(16 - x^2)}}{x^2}$

4 $\dfrac{x}{\sqrt{(1 - x^2)}}$ **5** $\sqrt{\left(\dfrac{1 + x}{1 - x}\right)}$ [Hint: put $x = \cos 2\theta$]

6 $x\sqrt{(x + 1)}$ **7** $x^2\sqrt{(x - 1)}$

Evaluate the following definite integrals:

8 $\displaystyle\int_0^1 x\sqrt{(1 - x^2)}\,dx$ **9** $\displaystyle\int_0^1 x\sqrt{(4 + x^2)}\,dx$

10 $\displaystyle\int_0^{1/\sqrt{2}} \frac{x^2\,dx}{\sqrt{(1 - x^2)}}$

14.5 Integration of trigonometric functions

$$\int \sin ax\,dx = -\frac{1}{a}\cos ax + C$$

$$\int \cos ax\,dx = \frac{1}{a}\sin ax + C$$

$$\int \sec^2 ax\,dx = \frac{1}{a}\tan ax + C$$

$$\int \operatorname{cosec}^2 ax\,dx = -\frac{1}{a}\cot ax + C \quad (a \text{ is a constant})$$

Certain trigonometric functions may be integrated after we have used the identities of Chapter 6 to express the integrand in terms of the standard forms given above. Of some importance are the two results

$$\int \sin^2 x\,dx = \int \tfrac{1}{2}(1 - \cos 2x)\,dx = \tfrac{1}{2}(x - \tfrac{1}{2}\sin 2x) + C$$
$$= \tfrac{1}{2}x - \tfrac{1}{4}\sin 2x + C$$
$$\int \cos^2 x\,dx = \int \tfrac{1}{2}(1 + \cos 2x)\,dx = \tfrac{1}{2}(x + \sin 2x) + C$$
$$= \tfrac{1}{2}x + \tfrac{1}{4}\sin 2x + C$$

If the integrand is a product of a sine and/or a cosine of a multiple angle, it may be expressed as a sum by means of the identities (see (6.46) to (6.49)).

$$\sin mx \cos nx = \tfrac{1}{2}[\sin(m+n)x + \sin(m-n)x]$$
$$\cos mx \cos nx = \tfrac{1}{2}[\cos(m+n)x + \cos(m-n)x]$$
$$\sin mx \sin nx = \tfrac{1}{2}[\cos(m-n)x - \cos(m+n)x]$$

Example 1 $\int \sin 3x \cos x \, dx = \int \tfrac{1}{2}(\sin 4x + \sin 2x)\, dx$

$$= \frac{-\cos 4x}{8} - \frac{\cos 2x}{4} + C$$

Example 2 $\int \sin 5x \sin 2x \, dx = \int \tfrac{1}{2}(\cos 3x - \cos 7x)\, dx$

$$= \frac{\sin 3x}{6} - \frac{\sin 7x}{14} + C$$

The integral $\int \sin^m x \cos^n x \, dx$ can be evaluated quite easily if m or n is an odd integer. If m is odd, the substitution $u = \cos x$ is used; if n is odd, the substitution $u = \sin x$ is used.

Example 3 $\int \sin^3 x \cos^2 x \, dx$

Put $u = \cos x$, therefore $du/dx = -\sin x$ and so

$$\int \sin^3 x \cos^2 x \, dx = -\int \sin x \sin^2 x . u^2 \frac{du}{\sin x}$$
$$= -\int \sin^2 x . u^2 \, du$$
$$= -\int (1-u^2)u^2 \, du \quad (\sin^2 x = 1 - \cos^2 x)$$
$$= -\tfrac{1}{3}u^3 + \tfrac{1}{5}u^5 + C$$
$$= -\tfrac{1}{3}\cos^3 x + \tfrac{1}{5}\cos^5 x + C$$

Example 4 $\displaystyle\int_{\pi/4}^{\pi/2} \frac{\cos^3 dx}{\sin^6 x} dx$

Put $u = \sin x$, therefore $du/dx = \cos x$.

When $x = \tfrac{1}{2}\pi$, $u = \sin \tfrac{1}{2}\pi = 1$.

When $x = \tfrac{1}{4}\pi$, $u = \sin \tfrac{1}{4}\pi = 1/\sqrt{2}$.

Therefore

$$\int_{\pi/4}^{\pi/2} \frac{\cos^3 x}{\sin^6 x} dx = \int_{1/\sqrt{2}}^{1} \frac{\cos^3 x}{u^6} \cdot \frac{du}{\cos x}$$
$$= \int_{1/\sqrt{2}}^{1} \frac{1-u^2}{u^6} du \quad (\text{Note: } \cos^2 x = 1 - \sin^2 x = 1 - u^2)$$
$$= \int_{1/\sqrt{2}}^{1} (u^{-6} - u^{-4})\, du$$

$$= \left[\frac{-1}{5} \cdot \frac{1}{u^5} + \frac{1}{3} \cdot \frac{1}{u^3} \right]_{1/\sqrt{2}}^{1}$$
$$= [-\tfrac{1}{5} + \tfrac{1}{3}] - [-\tfrac{1}{5}(\sqrt{2})^5 + \tfrac{1}{3}(\sqrt{2})^3]$$
$$= \tfrac{2}{15} + \tfrac{2}{15}\sqrt{2}$$

Example 5 $\int \sqrt[3]{(\cos x)} \sin^3 x \, dx$

Put $u = \cos x$, therefore $du/dx = -\sin x$ and so

$$\int \sqrt[3]{(\cos x)} \sin x^3 \, dx = \int \sqrt[3]{u} \sin^3 x \left(-\frac{du}{\sin x} \right)$$
$$= -\int u^{1/3}(1 - u^2) \, du$$
$$\text{(Note: } \sin^2 x = 1 - \cos^2 x = 1 - u^2\text{)}$$
$$= -\int (u^{1/3} - u^{7/3}) \, du$$
$$= -\tfrac{3}{4}u^{4/3} + \tfrac{3}{10}u^{10/3} + C$$
$$= -\tfrac{3}{4}(\cos x)^{4/3} + \tfrac{3}{10}(\cos x)^{10/3} + C$$

Exercises 14g

Integrate with respect to x

1 $\sin 7x \cos 2x$ **2** $\sin 3x \cos 8x$ **3** $\cos 5x \cos 6x$
4 $\sin 7x \sin 5x$ **5** $\cos^3 x \sin^4 x$ **6** $\cos^6 x \sin^5 x$
7 $\sqrt{(\sin x)} \cos x$ **8** $\dfrac{\sin^5 x}{\cos^2 x}$

Evaluate the following definite integrals:

9 $\int_0^{2\pi} \sin 5x \cos 3x \, dx$ **10** $\int_{\pi/3}^{\pi/2} \dfrac{\cos^5 x}{\sin^7 x} \, dx$

14.6 Integration by parts

This is a method for integrating a product of two functions. From

$$\frac{d}{dx}(uv) = v\frac{du}{dx} + u\frac{dv}{dx}$$

on integrating both sides with respect to x, we have

$$uv = \int v\frac{du}{dx} \, dx + \int u\frac{dv}{dx} \, dx$$

whence

$$\int u\frac{dv}{dx} \, dx = uv - \int v\frac{du}{dx} \, dx \qquad (14.2)$$

The product to be integrated is $u \times dv/dx$ and to obtain a slightly different version of the above result we consider u and dv/dx as being the 'first' and 'second' parts of the product, noting that, if dv/dx is '2nd function', $v = \int$ '2nd function.' The formula is then

$$\int 1\text{st} \times 2\text{nd} = 1\text{st} \times \int 2\text{nd} - \int (\text{derivative of } 1\text{st} \times \int 2\text{nd}) \qquad (14.3)$$

Note that

(i) one function, '2nd', *must* be integrable;
(ii) the other function, '1st', is never integrated.

Thus if this method is used to integrate the product of two functions, we first look for a function which can be integrated immediately. If there is only one, this is taken to be the '2nd' function; if both functions are integrable, we generally choose as the '1st' function the one which simplifies most on differentiation.

Example 1 $\int xe^{3x}\,dx$

Both x and e^{3x} are easily integrable, but as x becomes simpler on differentiation, we use them in the given order.

$$\int xe^{3x}\,dx = x\int e^{3x}\,dx - \int\left[\frac{d}{dx}(x)\int e^{3x}\,dx\right]dx$$

$$= \frac{xe^{3x}}{3} - \int 1 \times \frac{e^{3x}}{3}\,dx$$

$$= \tfrac{1}{3}xe^{3x} - \tfrac{1}{9}e^{3x} + C$$

Example 2 $\int x^2 \sin x\,dx$

By the same reasoning as before, we treat this product in the order given.

$$\int x^2 \sin x\,dx = x^2\int \sin x\,dx - \int\left[\frac{d}{dx}(x^2)\int \sin x\,dx\right]dx$$

$$= -x^2\cos x + \int 2x\cos x\,dx$$

We now apply the rule to the second integral, taking care to keep the trigonometric function as the '2nd' function, therefore

$$\int x^2 \sin x\,dx = -x^2\cos x + 2x\int \cos x\,dx - \int\left[\frac{d}{dx}(2x)\int \cos x\,dx\right]dx$$

$$= -x^2\cos x + 2x\sin x - 2\int \sin x\,dx$$

$$= -x^2\cos x + 2x\sin x + 2\cos x + C$$

Example 3 $\int x^3 \log_e x\,dx$

Of the two functions involved, we see that x^3 is the only one which is immediately integrable, therefore

$$\begin{aligned}\int x^3 \log_e x \, dx &= \int \log_e x . x^3 \, dx \\ &= \log_e x \int x^3 \, dx - \int \left[\frac{d}{dx} (\log_e x) \int x^3 \, dx \right] dx \\ &= \log_e x . \frac{x^4}{4} - \int \frac{1}{x} \cdot \frac{x^4}{4} \, dx \\ &= \tfrac{1}{4} x^4 \log_e x - \tfrac{1}{16} x^4 + C\end{aligned}$$

Example 4 $\int x \arctan x \, dx = \int \arctan x . x \, dx$

$$\begin{aligned} &= \arctan x \int x \, dx - \int \left[\frac{d}{dx} (\arctan x) \int x \, dx \right] dx \\ &= \tfrac{1}{2} x^2 \arctan x - \int \frac{1}{1 + x^2} \, \frac{x^2}{2} \, dx \\ &= \tfrac{1}{2} x^2 \arctan x - \int \left[\frac{1}{2} - \frac{\frac{1}{2}}{1 - x^2} \right] dx \\ &= \tfrac{1}{2} x^2 \arctan x - \tfrac{1}{2} x + \tfrac{1}{2} \arctan x + C\end{aligned}$$

Example 5 $\int \log_e x \, dx = \int \log_e x \times 1 \, dx$

$$\begin{aligned}\int \log_e x \times 1 \, dx &= \log_e x \int 1 \, dx - \int \left[\frac{d}{dx} (\log_e x) \times \int 1 \, dx \right] dx \\ &= x \log_e x - \int \frac{1}{x} x \, dx \\ &= x \log_e x - x + C\end{aligned}$$

Example 6 $\int \arcsin x \, dx = \int \arcsin x \times 1 \, dx$

$$\begin{aligned}\int \arcsin x \times 1 \, dx &= \arcsin x \int 1 \, dx - \int \left[\frac{d}{dx} (\arcsin x) \times \int 1 \, dx \right] dx \\ &= x \arcsin x - \int \frac{x}{\sqrt{(1 - x^2)}} \, dx\end{aligned}$$

For the latter integral let $u = 1 - x^2$, therefore $du/dx = -2x$ and

$$\begin{aligned}\int \arcsin x \, dx &= x \arcsin x - \int \frac{x}{\sqrt{u}} \left(-\frac{du}{2x} \right) \\ &= x \arcsin x + \frac{1}{2} \int \frac{du}{\sqrt{u}} \\ &= x \arcsin x + \sqrt{u} + C \\ &= x \arcsin x + \sqrt{(1 - x^2)} + C\end{aligned}$$

Exercises 14h

Find the following integrals

1 $\int x e^{-x}\, dx$ **2** $\int_0^{\pi/2} x\cos 2x\, dx$ **3** $\int \theta^2 \cos 2\theta\, d\theta$

4 $\int x^3 e^x$ **5** $\int x^3 \sin x\, dx$ **6** $\int_1^2 t^4 \log_e t\, dt$

7 $\int x \arcsin x\, dx$ **8** $\int x^5 \log_e 3x\, dx$ **9** $\int_1^{\sqrt{3}} \arctan\theta\, d\theta$

10 $\int \theta \sin m\theta\, d\theta$ **11** $\int_0^{\pi/4} x\sec^2 x\, dx$ **12** $\int (\log_e x)^2\, dx$

13 $\int \frac{x+\sin x}{1+\cos x}\, dx$ **14** $\int \frac{x\cos x}{\sin^2 x}\, dx$ **15** $\int \sec^3 x\, dx$

14.7 Further integration by parts

The following examples indicate some useful ways of proceeding.

Example 1 $$\int e^x \sin x\, dx = e^x \int \sin x - \int \left[e^x \int \sin x\, dx \right] dx$$
$$= -e^x \cos x + \int e^x \cos x\, dx$$

Therefore

$$\int e^x \sin x\, dx = -e^x \cos x + e^x \int \cos x\, dx - \int \left[e^x \int \cos x\, dx \right] dx$$
$$= -e^x \cos x + e^x \sin x - \int e^x \sin x\, dx$$

We note that on the RHS we have the original integral, therefore

$$2\int e^x \sin x\, dx = e^x(\sin x - \cos x)$$
$$\int e^x \sin x\, dx = \tfrac{1}{2} e^x(\sin x - \cos x) + C$$

Example 2 $\int \sqrt{(9-x^2)}\, dx$

We have previously used a trigonometric substitution for this type integral (see Section 14.14, Example 2).

$$\begin{aligned}\int\sqrt{(9-x^2)}\,dx &= \int\sqrt{(9-x^2)}\times 1\,dx\\ &= \sqrt{(9-x^2)}\int 1\,dx-\int\left[\frac{d}{dx}\sqrt{(9-x^2)}\int 1\,dx\right]dx\\ &= x\sqrt{(9-x^2)}-\int\frac{-x^2\,dx}{\sqrt{(9-x^2)}}\\ &= x\sqrt{(9-x^2)}-\int\frac{9-x^2-9}{\sqrt{(9-x^2)}}\,dx\\ &= x\sqrt{(9-x^2)}-\int\frac{9-x^2}{\sqrt{(9-x^2)}}\,dx+9\int\frac{dx}{\sqrt{(9-x^2)}}\end{aligned}$$

Therefore

$$\begin{aligned}\int\sqrt{(9-x^2)}\,dx &= x\sqrt{(9-x^2)}-\int\sqrt{(9-x^2)}\,dx+9\arcsin\tfrac{1}{3}x\\ 2\int\sqrt{(9-x^2)}\,dx &= x\sqrt{(9-x^2)}+9\arcsin\tfrac{1}{3}x\\ \int\sqrt{(9-x^2)}\,dx &= \frac{x\sqrt{(9-x^2)}}{2}+\frac{9}{2}\arcsin\frac{x}{3}+C\end{aligned}$$

Exercises 14i

Integrate the following functions with respect to x:

1 $e^x\cos x$ **2** $e^{-2x}\sin 3x$ **3** $e^{5x}\cos\frac{1}{2}x$
4 $\sqrt{(16+x^2)}$ **5** $\operatorname{cosec}^3 x$

Exercises 14

1 Show that
(i) $\int\cos x\sin^2 x\,dx=\frac{1}{3}\sin^3 x+C$
(ii) $\int\tan^5 x\sec^2 x\,dx=\frac{1}{6}\tan^6 x+C$
(iii) $\int 2x(x^2+1)^3\,dx=\frac{1}{4}(x^2+1)^4+C$

2 Find the following indefinite integrals:

(i) $\displaystyle\int(2x+1)^{5/2}\,dx$ (ii) $\displaystyle\int\frac{dx}{1-3x}$ (iii) $\displaystyle\int\sec^4\theta\,d\theta$

(iv) $\displaystyle\int\tan^3 x\sec x\,dx$ (v) $\displaystyle\int\frac{dx}{\sqrt{(3-2x-x^2)}}$ (vi) $\displaystyle\int\frac{dx}{x^2+6x+10}$

(vii) $\displaystyle\int\frac{\cos x\,dx}{3+\sin x}$ (viii) $\displaystyle\int\frac{6e^{2x}}{1+e^{2x}}\,dx$ (ix) $\displaystyle\int\frac{dx}{\sqrt{(1-x^2)}\sin^{-1}x}$

(x) $\displaystyle\int\cos 5x\sin 2x\,dx$ (xi) $\displaystyle\int\frac{\sin x}{\sqrt{\cos x}}\,dx$ (xii) $\displaystyle\int\sin^3 x\cos^3 x\,dx$

3 Find the following integrals:

(i) $\displaystyle\int \frac{x+1}{x^2-3x+2}\,dx$ (ii) $\displaystyle\int \frac{(x-1)(x+2)}{x(x+1)}\,dx$

(iii) $\displaystyle\int \frac{dx}{(x+1)^2(x^2+4)}$ (iv) $\displaystyle\int_{-2}^{2} \frac{dx}{x^2-16}$

(v) $\displaystyle\int \frac{(3x+2)\,dx}{x(x-1)(x^2+4)}$ (vi) $\displaystyle\int \frac{x\,dx}{(x+1)(x^2+9)}$

(vii) $\displaystyle\int \frac{(5-3x+4x^2)\,dx}{(1-x)(2-x-x^2)}$

4 At any point P(x, y) on a curve the product of x^2 and the slope of the curve is 2. If the curve passes through the point (1, 4), find its equation.

5 Show that

$$\int_1^2 x\log_e(x^2+1)\,dx = \tfrac{5}{2}\log_e 5 - \log_e 2 - \tfrac{3}{2}$$ [WJC, part]

6 Evaluate

(i) $\displaystyle\int_0^{\pi/2} \frac{dx}{1+\cos x}$ (ii) $\displaystyle\int_0^{\pi/4} x\sec^2 x\,dx$

(iii) $\displaystyle\int_0^1 x^2\arctan x\,dx$ (iv) $\displaystyle\int_0^1 \frac{\arctan x}{1+x^2}\,dx$

7 Evaluate

(i) $\displaystyle\int_0^2 \frac{x\,dx}{(x+2)(x^2+8)}$ (ii) $\displaystyle\int_2^3 \frac{3x^2-1}{x(x^2-1)}\,dx$

(iii) $\displaystyle\int_0^{\pi/2} \sin^5 x\,dx$ (iv) $\displaystyle\int_0^{\pi} x^2\sin x\,dx$

8 (a) Evaluate

(i) $\displaystyle\int_0^{\pi/4} \sin 5x\cos 3x\,dx$ (ii) $\displaystyle\int_0^1 xe^{-3x}\,dx$

(b) Show that

$$\int_0^{\pi/4} (\tan^3 x + \tan x)\,dx = \tfrac{1}{2}$$

Hence evaluate

$$\int_0^{\pi/4} \tan^3 x\,dx$$ [JMB]

9 (a) Use the substitution $y = \sin x$ to evaluate

$$\int_0^{\alpha} \frac{\cos x\,dx}{2\cos 2x - 1} \quad \text{where } \alpha = \sin^{-1}(\tfrac{1}{3})$$

(b) Evaluate $\int_0^1 \frac{\sin^{-1} x}{\sqrt{(1+x)}}\,dx$ [JMB]

10 (i) Evaluate

(a) $\int_2^4 \frac{(x^2-1)^2}{x}\,dx$ (b) $\int_0^{\pi/4} \sin 3x \sin 2x\,dx$

(ii) Find $\int \frac{x^3}{x^2+1}\,dx$ [LU]

11 Evaluate, to three significant figures, the integrals

$$\int_1^2 \frac{dx}{x(x+1)} \qquad \int_0^1 \frac{dx}{x^2+x+1} \qquad \int_1^2 x^2 \log x\,dx$$

[SUJB]

12 Find

(i) $\int x\sqrt{(6-x)}\,dx$ (ii) $\int \frac{x^2+x+1}{x^2+4}\,dx$ (iii) $\int \sin^2 x(2-\cos x)\,dx$

[WJC, part]

13 Show that $1+\sin 2\theta = (\sin\theta + \cos\theta)^2$. Hence evaluate

$$\int_{\pi/2}^{\pi} \sqrt{(1+\sin 2\theta)}\,d\theta$$

14 Prove that $\cos mx \cos nx = \frac{1}{2}[\cos(m+n)x + \cos(m-n)x]$. Hence show that

$$\int_0^{2\pi} \cos mx \cos nx\,dx = 0 \quad m \neq n$$

Find the value of the integral when $m = n$.

15 Use the substitution $x = a\cos^2\theta + b\sin^2\theta$ to evaluate

$$\int \frac{dx}{(x-a)(x-b)}$$

Also evaluate

$$\int \frac{dx}{2\sqrt{(x-a)(b-x)}} \quad (a < b)$$

16 Use the substitution $u = t - 1/t$ to show that

$$\int \frac{t^2+1}{t^4+1}\,dt = \int \frac{du}{u^2+2}$$

and use the substitution $v = t + 1/t$ to show that

$$\int \frac{t^2-1}{t^4+1}\,dt = \int \frac{dv}{v^2-2}$$

Hence evaluate $\int \frac{dt}{t^4+1}$ in terms of u and v.

17 Evaluate $\int \frac{\log_e x}{x^n}\,dx$, considering separately the three cases $n = 0$, $n = 1$, and $n \neq 0$ or 1.

18 Prove by two different methods that

$$\int \sqrt{(x^2 - a^2)}\,dx = \tfrac{1}{2}x\sqrt{(x^2 - a^2)} - \tfrac{1}{2}a^2 \log\{x + \sqrt{(x^2 - a^2)}\}$$

19 Evaluate

(i) $\displaystyle\int_0^1 \frac{\arctan x}{1 + x^2}\,dx$ (ii) $\displaystyle\int_0^\pi \sin x \cos^n x\,dx$

20 Show that $\displaystyle\int_0^1 2x^3 e^{-x^2}\,dx = 1 - \frac{2}{e}$

21 Show that

$$\int e^{ax} \cos bx\,dx = \frac{e^{ax}}{a^2 + b^2}(b \sin bx + a \cos bx)$$

Hence evaluate $\displaystyle\int_0^{\pi/2} e^{2x} \cos 3x\,dx$

22 Prove that $\displaystyle\int_0^{\pi/2} x^2 \cos x\,dx = \tfrac{1}{4}\pi^2 - 2$

23 Use the substitution $x = \cos 2\theta$ to prove that

$$\int_{-1}^{1} \sqrt{\left(\frac{1 - x}{1 + x}\right)}\,dx = \pi$$

24 Evaluate $\displaystyle\int \sqrt{\left(\frac{x}{1 - x}\right)}\,dx$

25 Evaluate

(i) $\displaystyle\int \frac{3x + 4}{\sqrt{(5 - 2x)}}\,dx$ (ii) $\displaystyle\int \frac{x^2 + 5}{(x - 1)(x - 2)}\,dx$

26 Evaluate

(i) $\displaystyle\int_0^{\pi/2} \cos^4 x \sin x\,dx$ (ii) $\displaystyle\int_0^{\pi/2} (\sqrt{\cos x}) \sin^2 x\,dx$ (iii) $\displaystyle\int_{3/2}^{3\sqrt{3}/2} \frac{dx}{x^2\sqrt{(9 - x^2)}}$

27 Prove that $\displaystyle\frac{d}{dx}\{\log(\tan \tfrac{1}{2}x)\} = \frac{1}{\sin x}$

Hence show that

$$\int \frac{1 - \cos x}{\sin x}\,dx = 2 \log(\sec \tfrac{1}{2}x)$$

Verify this result by using the formulae $1 - \cos x = 2 \sin^2 \tfrac{1}{2}x$ and $\sin x = 2 \sin \tfrac{1}{2}x \cos \tfrac{1}{2}x$.

28 Evaluate $\int \sqrt{(1-x^2)}\,\mathrm{d}x$

29 Integrate $x^3/(x^2+1)^3$ with respect to x, by the substitutions (a) $x = \tan\theta$, (b) $x^2+1=u$. Verify that the two results agree.

30 Prove that

$$\int_0^{\pi/2} x^2 \sin x \cos x\,\mathrm{d}x = \tfrac{1}{16}\pi^2 - \tfrac{1}{4}$$

31 (a) Evaluate $\int_0^{\pi/4} (\sin x + \cos x)^2\,\mathrm{d}x$

(b) Find $\int \frac{\mathrm{d}x}{\mathrm{e}^x + \mathrm{e}^{-x}}$ [AEB]

32 (i) Use the substitution $t = \tan\frac{1}{2}x$ to show that

$$\int_0^{\pi/2} \frac{\mathrm{d}x}{1+\sin x + \cos x} = \ln 2$$

(ii) Obtain $\int x \sec^2 x\,\mathrm{d}x$ [LU]

33 Evaluate

(a) $\int_0^{\pi/2} \sin 2\theta \sin\theta\,\mathrm{d}\theta$

(b) $\int_{-1}^{1} x\mathrm{e}^x\,\mathrm{d}x$

(c) $\int_0^1 x\sqrt{(1-x)}\,\mathrm{d}x$, by putting $x = 1 - t^2$ or otherwise. [LU]

34 (a) Find constants a and b such that, for all values of x,

$$2\sin x + 3\cos x = a(3\sin x + 2\cos x) + b(-2\sin x + 3\cos x)$$

Using this result, find

$$\int \frac{2\sin x + 3\cos x}{3\sin x + 2\cos x}\,\mathrm{d}x$$

(b) By using the substitution $x = u - 1$, or otherwise, evaluate

$$\int_0^3 \frac{2x+1}{\sqrt{(x+1)}}\,\mathrm{d}x$$ [C]

35 Show that

$$\int_0^1 \frac{1}{(x+1)(2x+1)}\,\mathrm{d}x = \log_e \alpha$$

where α is a rational number to be determined. [JMB]

36 Evaluate

(i) $\displaystyle\int_0^{1/2} \frac{x\,\mathrm{d}x}{\sqrt{(1-x^2)}}$ (ii) $\displaystyle\int_0^1 x\mathrm{e}^{-x}\,\mathrm{d}x$ [O]

37 Find the constants A, B, C in the identity

$$\frac{3x^2 - ax}{(x-2a)(x^2+a^2)} \equiv \frac{A}{(x-2a)} + \frac{Bx + Ca}{x^2 + a^2}$$

where a is a constant. Hence prove that

$$\int_0^a \frac{3x^2 - ax}{(x-2a)(x^2+a^2)}\,\mathrm{d}x = \tfrac{1}{4}\pi - \tfrac{3}{2}\log_e 2$$ [JMB]

38 Find

(i) $\displaystyle\int (\mathrm{e}^x - 1)^2\,\mathrm{d}x$ (ii) $\displaystyle\int \cos\theta\sqrt{(1+\sin\theta)}\,\mathrm{d}x$ [AEB]

39 By substituting $x = \sin\theta$, or otherwise, evaluate

$$\int_0^{1/\sqrt{2}} \frac{x^2}{(1-x^2)^{3/2}}\,\mathrm{d}x$$

leaving your answer in terms of π. [JMB]

40 Prove that

$$\int_a^b \mathrm{f}(x)\,\mathrm{d}x = \int_a^b \mathrm{f}(a+b-t)\,\mathrm{d}t$$

Hence prove that, if $0 < \beta < \frac{1}{2}\pi$,

$$\int_\beta^{\pi-\beta} \frac{\theta\,\mathrm{d}\theta}{\sin\theta} = \pi \ln \cot \tfrac{1}{2}\beta$$ [O]

15
Some applications of the integral calculus

15.1 Further examples on area

Example 1 Find the area under the curve $y = \sin(3x + \pi/3)$ between $x = -\pi/18$ and $x = \pi/9$.

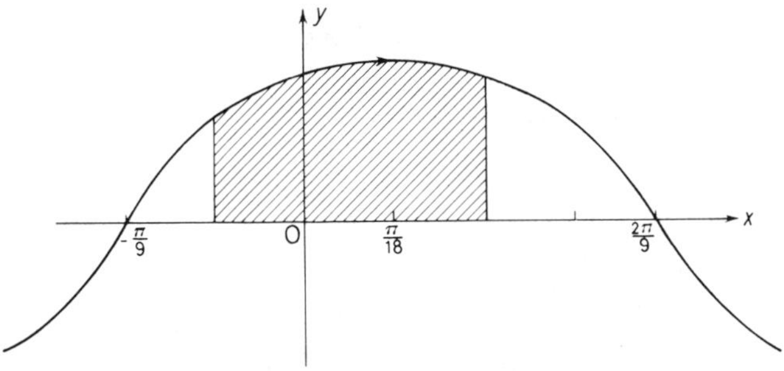

Figure 15.1

$$\text{Area} = \int_{-\pi/18}^{\pi/9} y\,\mathrm{d}x \quad \text{(see } \textit{Figure 15.1}\text{)}$$

$$= \int_{-\pi/18}^{\pi/9} \sin(3x + \pi/3)\,\mathrm{d}x$$

$$= \Big[-\tfrac{1}{3}\cos(3x + \pi/3)\Big]_{-\pi/18}^{\pi/9}$$

$$= -\tfrac{1}{3}\cos\tfrac{2}{3}\pi + \tfrac{1}{3}\cos\tfrac{1}{6}\pi$$

$$= \frac{1 + \sqrt{3}}{6}$$

Example 2 Find the area contained between the two parabolas $4y = x^2$ and $4x = y^2$.

From *Figure 15.2* the area required is OABC. We first find the point of intersection of the two curves, i.e. B. At B, $4y = x^2$ and $4x = y^2$ and we solve these two simultaneous equations. From the first equation, $y = x^2/4$; substituting in the second equation, we have

$$4x = \frac{x^4}{16}$$

$$x^4 - 64x = 0$$

$$x(x^3 - 64) = 0$$

$$x = 0 \quad \text{or} \quad x^3 = 64 \quad \text{giving } x = 4$$

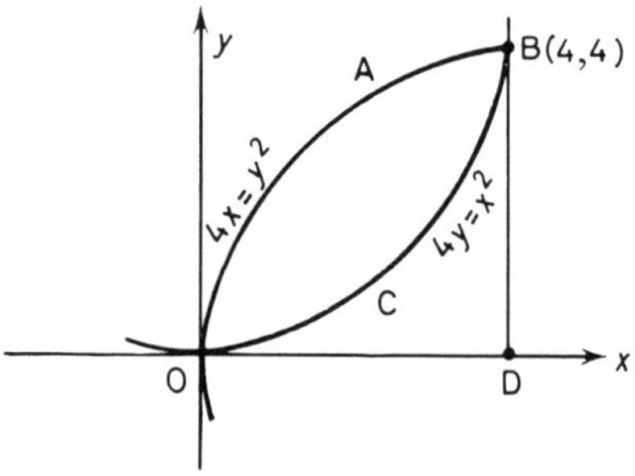

Figure 15.2

The points of intersection of the two curves are thus the origin (0, 0) and B (4, 4). If BD is the ordinate at B,

$$\text{area OABC} = \text{area OABD} - \text{area OCBD}$$

$$= \int_0^4 \sqrt{4x}\,dx - \int_0^4 \frac{x^2}{4}\,dx$$

(Note: the positive square root is taken because we are dealing with the top half of the curve $4x = y^2$.)

Therefore

$$\text{area OABC} = 2\int_0^4 x^{1/2}\,dx - \frac{1}{4}\int_0^4 x^2\,dx$$

$$= 2\left[\frac{2}{3}x^{3/2}\right]_0^4 - \frac{1}{4}\left[\frac{x^3}{3}\right]_0^4$$

$$= 2\left[\frac{16}{3} - 0\right] - \frac{1}{4}\left[\frac{64}{3} - 0\right]$$

$$= 16/3$$

When the area between the y-axis and a curve is required, the integral is, by symmetry,

$$\int_{y=c}^{y=d} x\,dy \qquad (15.1)$$

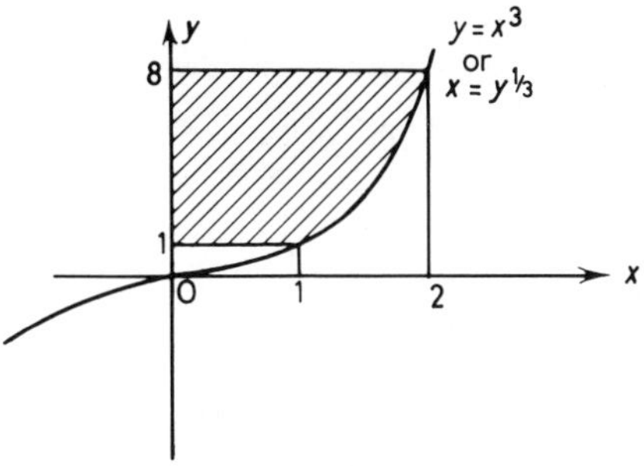

Figure 15.3

Example 3 Find the area between the curve $y = x^3$, the axis of y and the lines $y = 1$, $y = 8$ (*Figure 15.3*).

$$\text{Area} = \int_1^8 x\,\mathrm{d}y$$

$$= \int_1^8 y^{1/3}\,\mathrm{d}y$$

$$= [\tfrac{3}{4}y^{4/3}]_1^8$$

$$= \tfrac{3}{4}[8^{4/3}] - \tfrac{3}{4}[1^{4/3}]$$

$$= \tfrac{3}{4} \times 16 - \tfrac{3}{4}$$

$$= 11\tfrac{1}{4}$$

We note that, in representing the area under a curve between the ordinates $x = a$ and $x = b$ by $\int_a^b y\,\mathrm{d}x$, we have assumed that $b > a$ and that the ordinates are positive throughout the range of integration. If this is not so, it is clear from *Figure 15.4* that the integral $\int_a^b y\,\mathrm{d}x$ gives the numerical value of the area but with a positive or negative sign, according as the area is to the right or left of the curve, which is supposed described in the direction from P to Q.

If the curve cuts the axis in the range, the integral gives the difference (positive or negative) between the area to the right and that to the left.

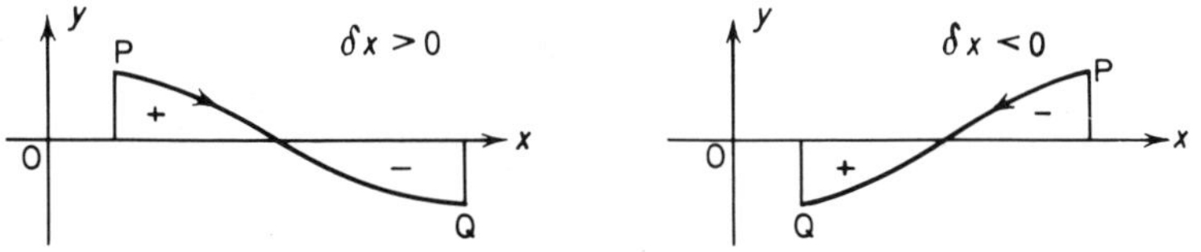

Figure 15.4

Example 4 Find the area included between the curve $y = x^3 - 4x^2 + 3x$ and the x-axis.

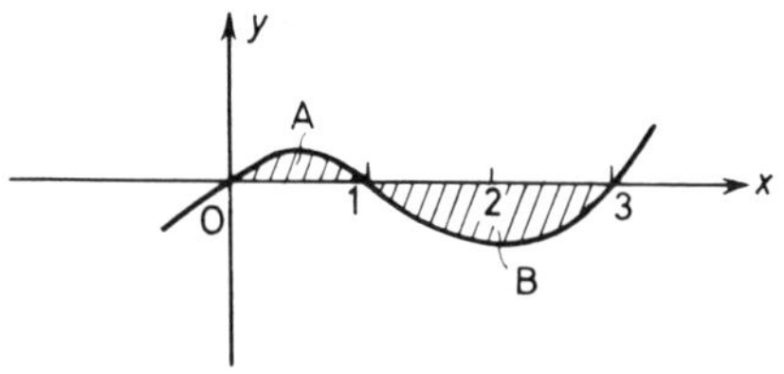

Figure 15.5

From the sketch of the curve (*Figure 15.5*), we see that we have to find the areas of the two parts A and B respectively.

The curve cuts the x-axis where $y = 0$, therefore

$$x^3 - 4x^2 + 3x = 0$$
$$x(x^2 - 4x + 3) = 0$$
$$x(x - 1)(x - 3) = 0$$
$$x = 0 \quad 1 \quad \text{or} \quad 3$$

The areas A and B are found as follows:

$$\begin{aligned}
\text{area } A &= \int_0^1 y\,\mathrm{d}x \\
&= \int_0^1 (x^3 - 4x^2 + 3x)\,\mathrm{d}x \\
&= [\tfrac{1}{4}x^4 - \tfrac{4}{3}x^3 + \tfrac{3}{2}x^2]_0^1 \\
&= [\tfrac{1}{4} - \tfrac{4}{3} + \tfrac{3}{2}] - 0 = \tfrac{5}{12} \\
\text{area } B &= \int_1^3 y\,\mathrm{d}x = \int_1^3 (x^3 - 4x^2 + 3x)\,\mathrm{d}x \\
&= [\tfrac{1}{4}x^4 - \tfrac{4}{3}x^3 + \tfrac{3}{2}x^2]_1^3 \\
&= [\tfrac{81}{4} - 36 + \tfrac{27}{2}] - [\tfrac{1}{4} - \tfrac{4}{9} + \tfrac{3}{2}] \\
&= -2\tfrac{1}{4} - \tfrac{5}{12} \\
&= -2\tfrac{2}{3}
\end{aligned}$$

Therefore, the total area $= \frac{5}{12} + 2\frac{2}{3} = 3\frac{1}{12}$.

Example 5 Find the area of the curve $x^2 + 3xy + 3y^2 = 1$.

The area is the limit of the sum of the areas of strips like PQ of width δx (see *Figure 15.6*). Now if P has ordinate y_1 and Q ordinate y_2, noting that neither upward nor downward movement of Ox alters the length PQ, we have

$$\begin{aligned}
\text{area} &= \lim_{\delta x \to 0} \sum PQ\,\delta x \\
&= \lim_{\delta x \to 0} \sum (y_1 - y_2)\,\delta x \\
&= \int (y_1 - y_2)\,\mathrm{d}x \quad \text{with the appropriate limits}
\end{aligned}$$

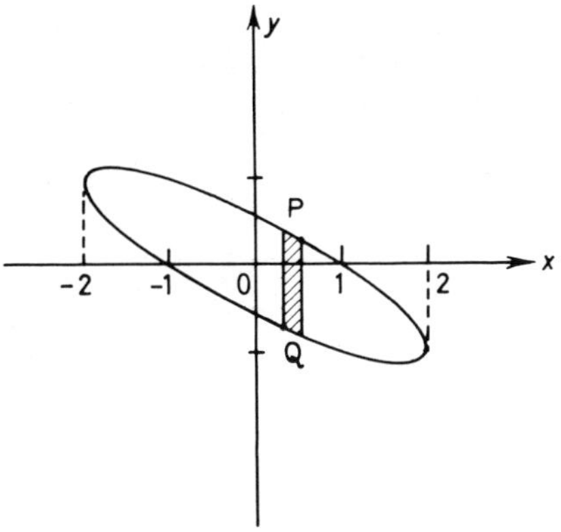

Figure 15.6

The two values of y are found, in general, by solving the equation as a quadratic in y, namely, $3y^2 + 3xy + x^2 - 1 = 0$:

$$y = \frac{-3x \pm \sqrt{[9x^2 - 12(x^2 - 1)]}}{6}$$

$$= \frac{-3x \pm \sqrt{(12 - 3x^2)}}{6}$$

The two values of y are

$$\frac{-3x + \sqrt{(12 - 3x^2)}}{6} \quad \text{and} \quad \frac{-3x - \sqrt{(12 - 3x^2)}}{6}$$

Hence

$$y_1 - y_2 = \frac{2\sqrt{(12 - 3x^2)}}{6} = \frac{\sqrt{[3(4 - x^2)]}}{3}$$

Now the limits of x are the values for which $y_1 - y_2 = 0$, that is

$$4 - x^2 = 0 \qquad x = \pm 2$$

therefore

$$\text{area} = \int_{-2}^{2} \frac{\sqrt{[3(4 - x^2)]}}{3} \, dx$$

$$= \frac{1}{\sqrt{3}} \int_{-2}^{2} \sqrt{(4 - x^2)} \, dx$$

To evaluate this integral, let $x = 2 \sin \theta$, then

$$\text{area} = \frac{1}{\sqrt{3}} \int_{-\pi/2}^{\pi/2} 2 \cos \theta . 2 \cos \theta \, d\theta$$

$$= \frac{2}{\sqrt{3}} \int_{-\pi/2}^{\pi/2} 2 \cos^2 \theta \, d\theta$$

$$= \frac{2}{\sqrt{3}} \int_{-\pi/2}^{\pi/2} (1 + \cos 2\theta)\,d\theta$$

$$= \frac{2}{\sqrt{3}} \left[\theta + \frac{\sin 2\theta}{2} \right]_{-\pi/2}^{\pi/2}$$

$$= \frac{2}{\sqrt{3}}\left(\frac{\pi}{2} + 0\right) - \frac{2}{\sqrt{3}}\left(\frac{-\pi}{2} + 0\right) = \frac{2\pi}{\sqrt{3}}$$

Exercises 15a

1 Find the area between the curve $y = \tan x$, the x-axis and the ordinates $x = 0$ and $x = \pi/4$.
2 Find the area between the curve $y = x^2 + 1/x$, the x-axis and the ordinates $x = 1$ and $x = 3$.
3 Find the area between the curve $y = -x^2 + 5x$ and the line $y = 6$.
4 A (2, 8) is a point on the curve $y = x^3$, O is the origin. Lines AB, AC are drawn from A perpendicular to Ox and Oy and meet these lines at B and C respectively. Find the areas OCA and OBA and verify that their sum is the same as the area of the rectangle OBAC.
5 Find the area contained between the parabola $9y = x^2$ and the line $3y = x + 6$.
6 Find the area enclosed between the curve $9y = x^2$, the y-axis, the line $y = 4$ and the line $x = -1$.
7 Find the area enclosed between the parabolas $y^2 = x$ and $x^2 = y$.
8 Find the area of the two segments bounded by the x-axis and each of the curves (i) $y = x^3 - x$, (ii) $y = x^3 + 2x^2 - 3x$.
9 Find the area of each of the curves (i) $2x^2 + 6xy + 6y^2 = 1$, (ii) $5x^2 - 12xy + 12y^2 = 2$, (iii) $3x^2 + 10xy + 10y^2 = 2$.
10 Find the area of the loop of the curve $y^2 = x^2(x - 1)^2$.

15.2 Mean values

Suppose that the function $\phi(x)$ is continuous and finite in the range $x = a$ to $x = b$. Divide the range $b - a$ into n equal parts each of length δx. Hence

$$n\,\delta x = (b - a) \tag{15.2}$$

Let $\phi(x_1), \phi(x_2), \ldots, \phi(x_n)$ be the values of the function at some convenient point in each interval, say the middle (sometimes the beginning is taken), then

$$\lim_{n\to\infty} \frac{1}{n}[\phi(x_1) + \phi(x_2) + \ldots + \phi(x_n)] \tag{15.3}$$

is known as the mean value of the function $\phi(x)$ over the range a to b with respect to x. It is a natural extension of the usual average. Since from (15.2)

$$\frac{1}{n} = \frac{\delta x}{b-a}$$

the mean value of $\phi(x)$, over the range a to b with respect to x,

$$= \lim_{n\to\infty} \frac{\delta x}{b-a}[\phi(x_1) + \dots + \phi(x_n)]$$

$$= \frac{1}{b-a} \lim_{\delta x \to 0} \sum_a^b \phi(x)\,\delta x \quad ((b-a) \text{ is a constant})$$

$$= \frac{1}{b-a}\int_a^b \phi(x)\,\mathrm{d}x \quad \text{by (13.8)} \qquad (15.4)$$

Geometrically, $\int_a^b \phi(x)\,\mathrm{d}x$ is the area BCNK under the curve (see *Figure 15.7*) and $(b-a)$ is the distance KN. If LMNK is a rectangle whose area is equal to the area under the curve, then the mean value is represented by the height MN of this rectangle.

If it is possible to express $\phi(x)$ as a function of another variable, say u, the mean value with respect to u will, in general, differ from the mean value with respect to x, and it is important to notice which mean value is required.

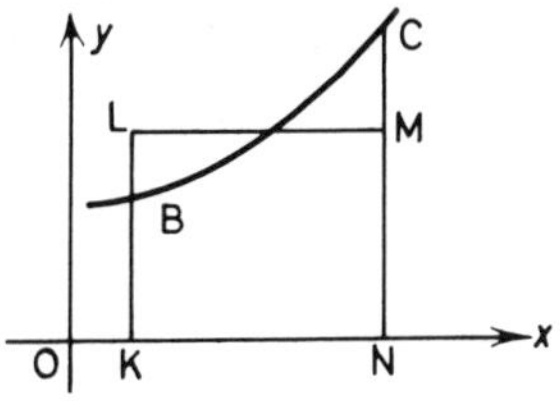

Figure 15.7

Example 1 When a particle falls freely from rest, its speed at any time t sec from the commencement of its motion is given by $v = 980t$ or $v^2 = 1960s$, where s is the distance fallen. If its speed on impact is 2450 cm/sec, find its mean speed (i) with respect to time, (ii) with respect to distance.

(i) Since $v = 980t$ on impact, $2450 = 980T$, where T is the time taken to fall. Therefore, the time of falling is $2\frac{1}{2}$ sec. Thus

$$\text{mean speed} = \frac{1}{2{\cdot}5}\int_0^{2\cdot5} v\,\mathrm{d}t$$

$$= \frac{1}{2{\cdot}5}\int_0^{2\cdot5} 980t\,\mathrm{d}t$$

$$= \frac{1}{2{\cdot}5}[490t^2]_0^{2\cdot5}$$

$$= 1225 \text{ cm/sec}$$

(ii) Since $v^2 = 1960s$ (given) on impact, $2450^2 = 1960s$, where s is the distance fallen. Therefore $s = 3062{\cdot}5$ cm. Thus

$$\text{mean speed} = \frac{1}{3062{\cdot}5}\int_0^{3062\cdot5} v\,\mathrm{d}s$$

$$= \frac{1}{3062{\cdot}5}\int_0^{3062\cdot 5} \sqrt{1960s}\,\mathrm{d}s$$

$$= \frac{1}{3062{\cdot}5}\int_0^{3062\cdot 5} 14\sqrt{10}\,s^{1/2}\,\mathrm{d}s$$

$$= \frac{1}{3062{\cdot}5} \times 14\sqrt{10}\left[\frac{2}{3}s^{3/2}\right]_0^{3062\cdot 5}$$

$$= \frac{1}{3062{\cdot}5} \times 14\sqrt{10} \times \frac{2}{3}(169478{\cdot}3 - 0)$$

$$= 1633{\cdot}3 \text{ cm/sec}$$

Another mean value which is used, particularly in electrical engineering, is the root mean square or RMS value of a function over a given interval:

$$\text{RMS} = \sqrt{\left\{\frac{1}{b-a}\int_a^b [\phi(x)]^2\,\mathrm{d}x\right\}} \qquad (15.5)$$

and is the square root of the mean value of the square of the function.

Example 2 An alternating current is given by $i = I\sin(5t + \pi/3)$. Find the RMS value for i taken over the interval 0 to $2\pi/5$ sec.

$$(\text{RMS value})^2 = \frac{5}{2\pi}\int_0^{2\pi/5} I^2\sin^2\left(5t + \frac{\pi}{3}\right)\mathrm{d}t$$

$$= \frac{5I^2}{2\pi}\int_0^{2\pi/5} \frac{1}{2}\left[1 - \cos\left(10t + \frac{2\pi}{3}\right)\right]\mathrm{d}t$$

$$= \frac{5I^2}{4\pi}\left[t - \frac{1}{10}\sin\left(10\mathrm{t} + \frac{2\pi}{3}\right)\right]_0^{2\pi/5}$$

$$= \frac{5I^2}{4\pi}\left[\frac{2\pi}{5} - \frac{1}{10}\sin\left(4\pi + \frac{2\pi}{3}\right)\right]$$
$$- \frac{5I^2}{4\pi}\left[0 - \frac{1}{10}\sin\left(0 + \frac{2\pi}{3}\right)\right]$$

$$= \frac{I^2}{2} - \frac{5I^2}{40\pi}\sin\frac{2\pi}{3} - 0 + \frac{5I^2}{40\pi}\sin\frac{2\pi}{3}$$

$$= I^2/2$$

Hence RMS value is $I/\sqrt{2}$.

In practical problems, the values of the function $\phi(x)$ can often be found only at isolated points. If the intervals between these points are all equal, then an approximate mean value is found from

$$\frac{1}{n}[\phi(x_1) + \phi(x_2) + \dots + \phi(x_n)]$$

If the intervals are not all equal, the values may be plotted on a graph. Then a smooth curve is drawn through the points and a set of values of $\phi(x)$ at equal intervals, read off from the graph. These latter values may then be used to find an approximate mean value.

Exercises 15b

1 A quantity of steam follows the law $pv^{0\cdot75} = 10\,000$, p being measured in N/m^2. Find the mean pressure as v increases from 1 m^3 to 16 m^3.

2 Find the mean value and the RMS value in each of the following cases:

(i) $\sin\theta$ in the range 0 to π.

(ii) $\sin\theta$ in the range 0 to 2π.

(iii) $\sin\theta + \cos\theta$ in the range 0 to 2π.

(iv) $I\sin(10t + \pi/4)$, the values of t being taken over one period $t = 0$ to $t = 2\pi/10$.

3 A body is dropped from a height of 1960 cm. Show that the mean value of its velocity until just before it hits the ground is (a) 960 cm/sec with respect to time and (b) 1306·66 cm/sec with respect to distance.

4 Show that the mean value of the ordinates of a semicircle of radius a drawn through equidistant points on the diameter is $\frac{1}{4}\pi a$.

5 The following table gives the values of a current, i amps, in a circuit at various times, t sec. Find the mean value of the current.

t	0	6	8	11	17	20	23	28	33	37	40
i	0	510	640	800	975	1000	975	840	580	275	0

15.3 Volume of a solid of revolution

Consider the volume swept out when the area enclosed by the curve $y = \phi(x)$, the x-axis, and the ordinates $x = a$, $x = b$ is rotated through 2π radians about Ox (see *Figure 15.8*).

Let KB and NC be the ordinates at the points $x = a$ and $x = b$ respectively. Divide KN into n parts each of width δx. Let L, M be two consecutive points

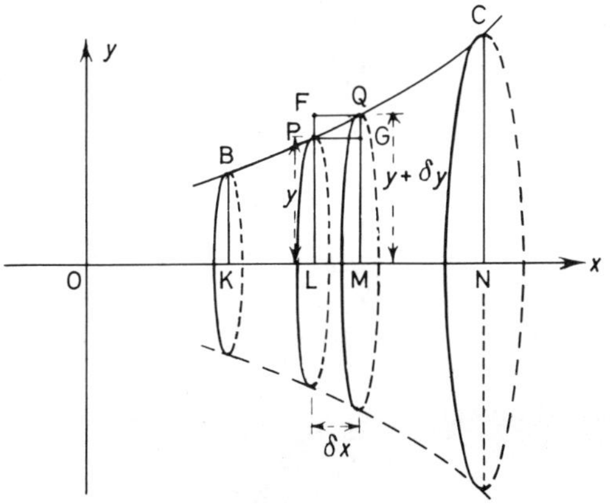

Figure 15.8

of subdivision, LP, MQ the respective ordinates. Complete the rectangles PGML, FQML.

Then all such rectangles as PGML will sweep out thin circular discs of area πy^2 and thickness δx, i.e. of volume $\pi y^2\,\delta x$. The sum of all such discs will be less than the volume required. Similarly, all such rectangles as FQML will sweep out thin circular discs of volume $\pi(y+\delta y)^2\,\delta x$ and the sum of all such discs will be greater than the volume required (note the curve has been taken as increasing from $x = a$ to $x = b$). Thus

$$\sum_a^b \pi y^2\,\delta x \leqslant \text{required volume of revolution} \leqslant \sum_a^b \pi(y+\delta y)^2\,\delta x$$

Now in the limit as $\delta x \to 0$, $\delta y \to 0$, and we have

$$\begin{aligned}\text{required volume of revolution} &= \lim_{\delta x \to 0} \sum_a^b \pi y^2\,\delta x \\ &= \int_a^b \pi y^2\,\mathrm{d}x \qquad (15.6)\end{aligned}$$

Example 1 Find the volume cut from a sphere of radius a by two parallel planes distances h_1, h_2 from the centre ($h_2 > h_1$) and both measured in the same direction.

A sphere is swept out by the rotation through 2π radians of a semicircle about its bounding diameter. Take the centre of the circle as origin and the bounding diameter as the x-axis. Then by (15.6)

$$\text{required volume} = \int_{h_1}^{h_2} \pi y^2\,\mathrm{d}x$$

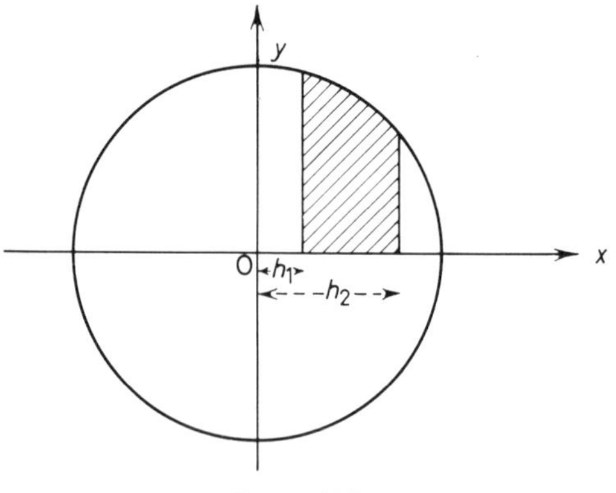

Figure 15.9

The equation of the semicircle is $x^2 + y^2 = a^2$, therefore

$$\text{required volume} = \pi \int_{h_1}^{h_2} (a^2 - x^2)\,\mathrm{d}x$$

$$= \pi\left[a^2x - \frac{x^3}{3}\right]_{h_1}^{h_2}$$

$$= \pi[a^2(h_2 - h_1) - \tfrac{1}{3}(h_2{}^3 - h_1{}^3)]$$

$$= \tfrac{1}{3}\pi(h_2 - h_1)[3a^2 - (h_2{}^2 + h_1h_2 + h_1{}^2)]$$

Note that if we put $h_2 = a$ and $h_1 = a - k$, we obtain the volume of a spherical cap of height k, that is

$$\begin{aligned}\text{volume} &= \tfrac{1}{3}\pi[a - (a - k)]\{3a^2 - [a^2 + a(a - k) + (a - k)^2]\} \\ &= \tfrac{1}{3}\pi k(3a^2 - a^2 - a^2 + ak - a^2 + 2ak - k^2) \\ &= \tfrac{1}{3}\pi k(3ak - k^2) \\ &= \pi k^2(a - \tfrac{1}{3}k) \qquad (15.7)\end{aligned}$$

When any portion of the area contained between a curve and the y-axis is rotated about the y-axis, the volume swept out will, by symmetry, be $\int_{y=c}^{y=d} \pi x^2\,dy$ but in other cases we may not be able to quote these formulae but have to return to first principles. Consider Example 2.

Example 2 Find the volume swept out by revolving the area between the curve $y = e^{2x}$, the x-axis and the ordinates $x = 1$, $x = 2$, through 2π radians about Oy.

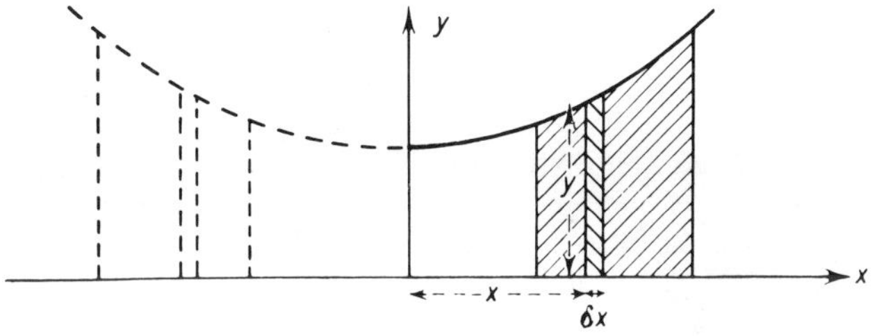

Figure 15.10

Referring to *Figure 15.10*, divide the volume into the shells by cylinders whose axes are the y-axis and whose radii are at equal intervals δx from $x = 1$ to $x = 2$. The volume of a typical shell contained between two cylinders of radii x and $x + \delta x$ can be obtained if we consider the shell cut and flattened into a plate which will be approximately of length y, width, $2\pi x$ and thickness δx. That is, whose volume is $2\pi xy\,\delta x$. Therefore

$$\text{required volume} = \lim_{\delta x \to 0} \sum_{x=1}^{x=2} 2\pi xy\,\delta x$$

$$= \int_1^2 2\pi xy\,dx \quad \text{(see (13.8))}$$

But $y = e^{2x}$, thus

$$\begin{aligned}\text{required volume} &= 2\pi\int_1^2 xe^{2x}\,dx\\ &= 2\pi\left[\frac{xe^{2x}}{2}-\frac{e^{2x}}{4}\right]_1^2 \quad \text{(by integration by parts)}\\ &= 2\pi(e^4-\tfrac{1}{4}e^4)-2\pi(\tfrac{1}{2}e^2-\tfrac{1}{4}e^2)\\ &= 2\pi(\tfrac{3}{4}e^4-\tfrac{1}{4}e^2)\\ &\approx 245{\cdot}7\end{aligned}$$

Exercises 15c

1 Find the volume swept out when the area between the parabola $y = x^2 + 1$, the x-axis and the ordinates at $x = 2$ and $x = 3$ is rotated through 2π radians about the x-axis.
2 Show that the volume of a sphere of radius a is $\frac{4}{3}\pi a^3$.
3 Find the volume generated by rotating the area bounded by the axes and the curve $y = \cos x$ between $x = 0$ and $x = \pi/2$, through 2π radians about the x-axis.
4 The portion of the curve $y = x^2 + 2$ between the points (0, 2) and (1, 3) is rotated through 2π radians about the y-axis to form the surface of a bowl. Find the volume of the bowl.
5 Find the volume swept out when the area between the parabola $y^2 = 4ax$, the x-axis and the ordinate $x = h$ rotates through 2π radians about the x-axis.
6 Find the volume of a cylinder of height h, radius of base a.
7 The ellipse $x^2/a^2 + y^2/b^2 = 1 (a > b)$ is rotated through π radians about its major axis. Find the volume swept out. What would the volume be if the ellipse were rotated about the minor axis?
8 Show that in the solid generated by the revolution of the rectangular hyperbola $x^2 - y^2 = a^2$ about the x-axis, the volume of a segment of height a measured from the vertex is $\frac{4}{3}\pi a^3$.
9 Find the volume generated by rotating a loop of the curve $y^2 = x^2(x - 1)^2$ about the x-axis.
10 Find the volume swept out when the area between the curve $y = e^{3x}$, the x-axis, the y-axis, and the ordinate $x = 3$ is rotated through 2π radians about Oy.

15.4 Centres of gravity

Consider a number of particles of masses, $m_1, m_2, \ldots$ situated at points whose co-ordinates are $(x_1, y_1), (x_2, y_2)\ldots$. Then the point G $(\bar{x}, \bar{y})$ whose co-ordinates are given by the equations

$$\left.\begin{aligned}\bar{x} &= \frac{m_1x_1+m_2x_2+\ldots}{m_1+m_2+\ldots} = \frac{\Sigma mx}{\Sigma m}\\ \bar{y} &= \frac{m_1y_1+m_2y_2+\ldots}{m_1+m_2+\ldots} = \frac{\Sigma my}{\Sigma m}\end{aligned}\right\} \quad (15.8)$$

is defined as the centre of gravity or centre of mass of the system.

We have assumed that the masses all lie in a plane. If they are not coplanar, then each point will have a third co-ordinate $z_1, z_2, \ldots$ and G will also have a third co-ordinate defined by $\bar{z} = \Sigma mz/\Sigma m$.

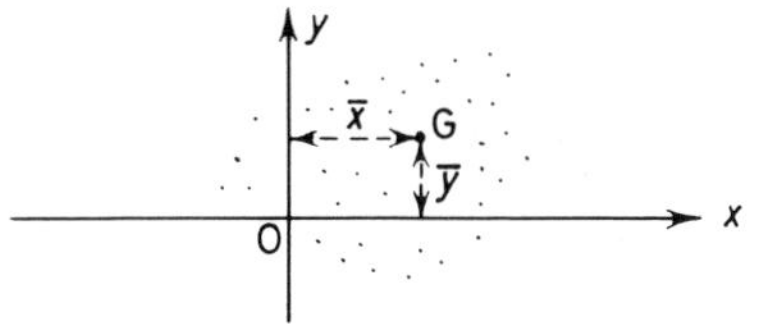

Figure 15.11

In the case of a solid body (*Figure 15.11*), we consider it split up into a large number of very small elements each of mass δm and then the centre of gravity is defined as the point G whose co-ordinates $(\bar{x}, \bar{y}, \bar{z})$ are given by

$$\left.\begin{aligned} \bar{x} &= \frac{\lim\limits_{\delta m \to 0} \sum x\,\delta m}{\lim\limits_{\delta m \to 0} \sum \delta m} = \frac{\int x\,\mathrm{d}m}{\int \mathrm{d}m} \quad \text{by (13.8)} \\ \bar{y} &= \frac{\int y\,\mathrm{d}m}{\int \mathrm{d}m} \qquad \bar{z} = \frac{\int z\,\mathrm{d}m}{\int \mathrm{d}m} \end{aligned}\right\} \tag{15.9}$$

The summations and integrals are taken throughout the whole body.

Example 1 A circular arc of radius a, subtends an angle 2α at its centre. Find the centre of gravity of the arc.

Take the centre of the circle as the origin and the x-axis along the medial line. Let ρ be the mass of the arc per unit length, then the length of a small element of the arc is $a\delta\theta$ and its mass is $\rho a\delta\theta$ (*Figure 15.12*). Therefore

$$\bar{x} = \frac{\int_{\theta=-\alpha}^{\theta=\alpha} x\rho a\,\mathrm{d}\theta}{\int_{\theta=-\alpha}^{\theta=\alpha} \rho a\,\mathrm{d}\theta} \qquad (\bar{y} = 0 \text{ by symmetry})$$

$$= \frac{\int_{-\alpha}^{\alpha} a\cos\theta\,\rho a\,\mathrm{d}\theta}{\int_{-\alpha}^{\alpha} \rho a\,\mathrm{d}\theta}$$

$$= \frac{a[\sin\theta]_{-\alpha}^{\alpha}}{[\theta]_{-\alpha}^{\alpha}}$$

$$= \frac{a\sin\alpha}{\alpha}$$

Note that for a semicircle $\alpha = \pi/2$ and $\bar{x} = 2a/\pi$.

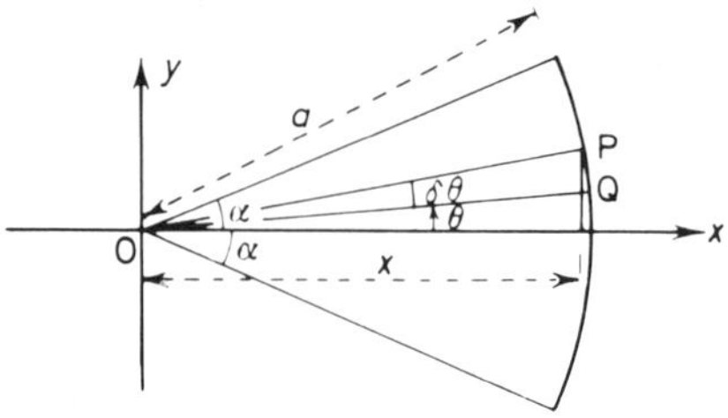

Figure 15.12

Our small elements of mass δm have been taken as small arcs of length $a\delta\theta$ and mass $a\rho\delta\theta$. Their moment about the y-axis ($x\delta m$) has been expressed in the form $a\cos\theta\ \rho a\delta\theta$, which is a suitable expression for integration with respect to θ.

In general, when we are dealing with a lamina or a solid, the small elemental masses δm, and their moments $x\delta m$ or $y\delta m$ about the axes, are expressed in the form $f(\theta)\,\delta\theta$, where θ is some convenient variable for which the integrations can be carried out. We shall find it convenient to make use of the symmetry of the body (if it exists) in choosing the variable θ and the elemental masses. It is important to notice that the 'x' in (15.9) is now the distance of the centre of gravity of the elemental mass from the y-axis. Similarly, the 'y' is the distance of the centre of gravity of the elemental mass from the x-axis.

Example 2 A sector of a circle of radius a subtends an angle 2α at the centre of the circle. Find its centre of gravity.

We take the centre of the circle as origin and the x-axis along the medial line (*Figure 15.13*). By symmetry, $\bar{y} = 0$.

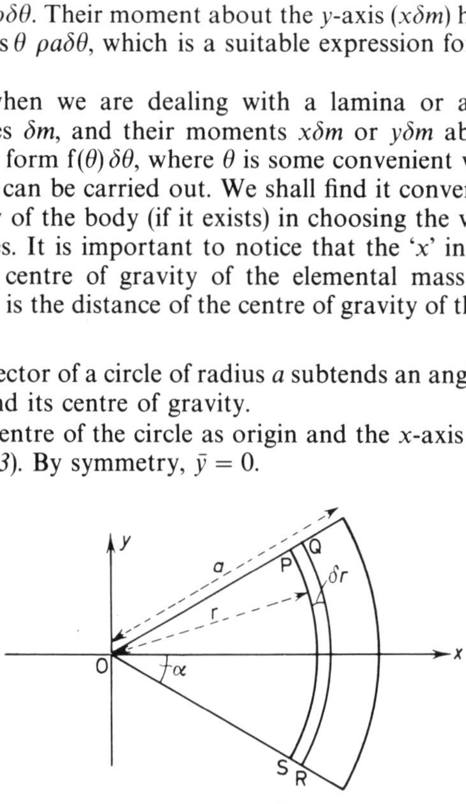

Figure 15.13

We divide the sector by a large number of concentric arcs into sections such as PQRS. Let OP $= r$, OQ $= r + \delta r$. Then

$$\text{area PQRS} \approx 2r\alpha\delta r$$
$$\text{mass PQRS} \approx 2r\alpha\rho\delta r$$

where ρ is the surface density of the sector.

Now by the result of Example 1, the centre of gravity of PQRS is at a distance $(r\sin\alpha)/\alpha$ from O along Ox. Its moment about Oy is thus $2r\alpha\rho\delta r(r\sin\alpha)/\alpha$. Therefore

$$\bar{x} = \frac{\int_0^a 2r\alpha\rho \frac{r \sin \alpha}{\alpha} \,\mathrm{d}r}{\int_0^a 2r\alpha\rho \,\mathrm{d}r} \quad \text{(since } \rho \text{ is assumed to be constant)}$$

$$= \frac{\sin \alpha}{\alpha} \frac{\int_0^a r^2 \,\mathrm{d}r}{\int_0^a r \,\mathrm{d}r}$$

$$= \frac{\sin \alpha}{\alpha} \frac{\frac{1}{3}a^3}{\frac{1}{2}a^2}$$

$$= \frac{2}{3}a\frac{\sin \alpha}{\alpha}$$

For a semicircle, $\alpha = \pi/2$ and so

$$\bar{x} = \frac{4a}{3\pi} \tag{15.10}$$

Example 3 Find the centre of gravity of a uniform solid hemisphere.

Take the centre of the hemisphere as origin and its axis of symmetry as Ox. Let ρ be its density.

Divide the hemisphere into elemental discs of width δx by planes parallel to its plane end (*Figure 15.14*). The mass of a disc is approximately $\pi y^2 \rho \,\delta x$ and the distance from the origin to the centre of gravity of a disc is approximately x. Since by symmetry $\bar{y} = 0 = \bar{z}$, we have

$$\bar{x} = \frac{\int_0^a x\pi y^2 \rho \,\mathrm{d}x}{\int_0^a \pi y^2 \rho \,\mathrm{d}x}$$

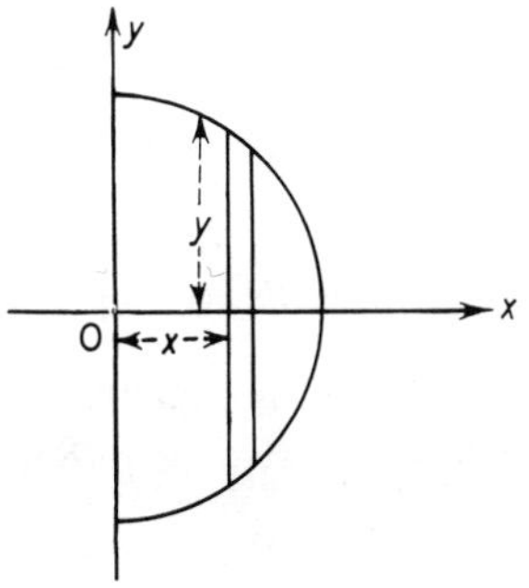

Figure 15.14

The equation of the bounding circle is $x^2 + y^2 = a^2$. Thus

$$\bar{x} = \frac{\pi\rho \int_0^a x(a^2 - x^2)\,dx}{\pi\rho \int_0^a (a^2 - x^2)\,dx}$$

$$= \frac{\left[\dfrac{a^2x^2}{2} - \dfrac{x^4}{4}\right]_0^a}{\left[a^2x - \dfrac{x^3}{3}\right]_0^a}$$

$$= \frac{\frac{1}{4}a^4}{\frac{2}{3}a^3}$$

$$= \tfrac{3}{8}a$$

Exercises 15d

1 Find the centre of gravity of the following (in all cases the mass per unit area is assumed to be constant):
(a) the area enclosed by the parabola $y^2 = 4x$ and by the line $x = 1$.
(b) The area between the curve $y = \sin x$ and the x-axis from $x = 0$ to $x = \pi$.
(c) The area formed by one loop of the curve $y^2 = x(x - 1)^2$.

2 If the loops of the curves in question 1(a), (b) and (c) are each rotated through 2π radians about Ox, find the centres of gravity of the solids so formed.

3 Find the centres of gravity of the following: (a) a plane isosceles triangle, (b) a quadrant of a circle, (c) a solid cone, (d) a frustrum of a cone height h, radii of its ends a and b $(a > b)$.

4 By dividing the sector of a circle of angle 2α into elemental sectors of angle $\delta\theta$, find the position of its mean centre. (Use the result of question 3(a).) Compare your result with that of Example 2, Section 15.4.

5 Find the position of the centre of gravity of the area contained between the positive co-ordinate axes and the astroid $x^{2/3} + y^{2/3} = a^{2/3}$.

15.5 Area enclosed by a curve in polar form

Section 17.1 introduces polar co-ordinates and in Section 17.8 the polar equation of a curve is discussed.

Figure 15.15 shows the curve $r = f(\theta)$. We shall find the area enclosed by the curve and the two lines OA and OB joining the origin to two points A and B on the curve.

Let AOx $= \theta_A$ and BOx $= \theta_B$. We divide the arc AB into n parts. Let P$(r, \angle\theta)$ and Q$(r + \delta r, \angle\theta + \delta\theta)$ be two neighbouring points on the arc. $\angle$POx $= \theta$, $\angle$QOP $= \delta\theta$.

$$\text{Area of sector POQ} \approx \text{area of } \triangle\text{POQ} = \tfrac{1}{2}r(r + \delta r)\sin\delta\theta \approx \tfrac{1}{2}r^2\delta\theta$$

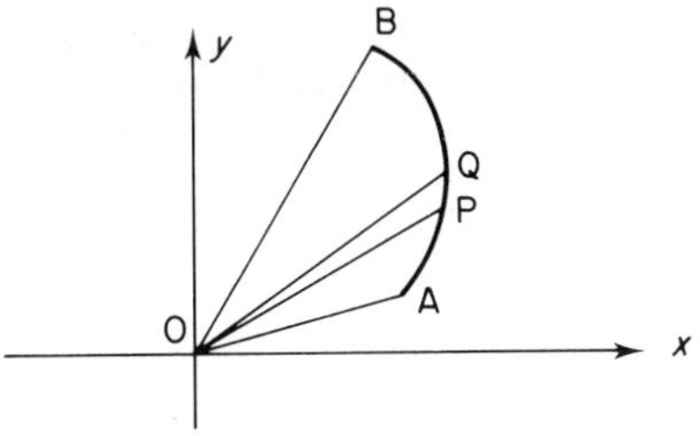

Figure 15.15

Therefore

$$\text{area of sector AOB} = \lim_{\delta\theta \to 0} \sum \tfrac{1}{2}r^2 \delta\theta$$

$$= \int_{\theta_A}^{\theta_B} \tfrac{1}{2}r^2 \, d\theta \qquad (15.11)$$

where θ_A and θ_B are the polar angles of A and B.

Example 1 A has cartesian co-ordinates (2, 0). The line AB has the polar equation $r\cos\theta = 2$ and $\angle AOB = \pi/3$. Use (15.11) to find the area of the triangle AOB and verify that the result is correct.

The point with cartesian co-ordinates (x, y) has polar co-ordinates $(r, \angle\theta)$, where $r\cos\theta = x$ and $r\sin\theta = y$ (see (17.1) and (17.2)). Thus the line $r\cos\theta = 2$ is equivalent to $x = 2$ and lies parallel to the y-axis at a distance of two units from it. The situation is shown in *Figure 15.16*.

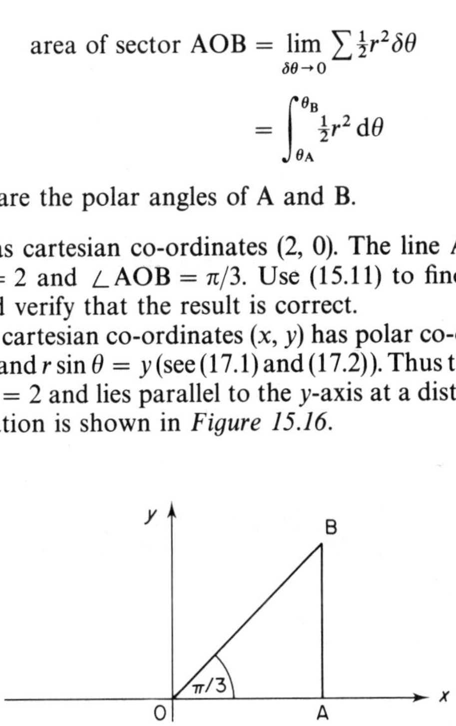

Figure 15.16

AB has equation $r\cos\theta = 2$, i.e. $r = 2\sec\theta$. Therefore

$$\text{area AOB} = \int_0^{\pi/3} \tfrac{1}{2}r^2 \, d\theta = \int_0^{\pi/3} 2\sec^2\theta \, d\theta$$

$$= [2\tan\theta]_0^{\pi/3} = 2\tan\tfrac{1}{3}\pi = 2\sqrt{3}$$

This is clearly correct since OA = 2, $AB = OA\tan\tfrac{1}{3}\pi = 2\sqrt{3}$ and area $AOB = \tfrac{1}{2}OA.AB$.

Example 2 Find the area enclosed by the cardioid $r = a(1 + \cos\theta)$.

From *Figure 15.17* (see also *Figure 17.17* for further details), the required area is, by symmetry,

$$A = 2\int_0^\pi \tfrac{1}{2}r^2\,\mathrm{d}\theta$$

$$= a^2\int_0^\pi (1+\cos\theta)^2\,\mathrm{d}\theta$$

$$= a^2\int_0^\pi (1+2\cos\theta+\cos^2\theta)\,\mathrm{d}\theta$$

$$= a^2\int_0^\pi \left(1+2\cos\theta+\frac{1+\cos 2\theta}{2}\right)\mathrm{d}\theta \quad (\cos 2\theta = 2\cos^2\theta - 1)$$

$$= a^2\left[\theta + 2\sin\theta + \tfrac{1}{2}\theta + \tfrac{1}{4}\sin 2\theta\right]_0^\pi = \tfrac{3}{2}\pi a^2$$

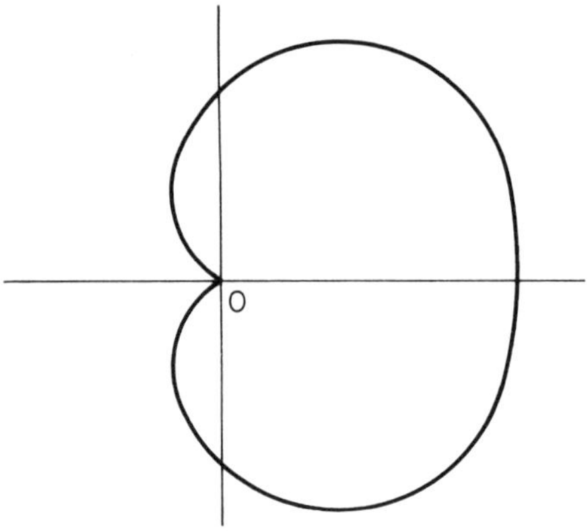

Figure 15.17

Exercises 15e

1 Verify using polar co-ordinates that the area of the circle $r = 3$ is 9π.
2 Verify using polar co-ordinates that the area of the circle $r = a\cos\theta$ is $\pi a^2/4$.
3 Find the area bounded by the spiral $r = a\theta$ and the lines $\theta = 0$ and $\theta = \pi/4$.
4 Find the area bounded by the curve $r = a(1+\cos\theta)$ and the lines $\theta = 0$ and $\theta = \pi/2$.
5 Find the area bounded by the spiral $r = \mathrm{e}^\theta$ and the lines $\theta = 0$, $\theta = 3\pi/2$.

Exercises 15

1 Show that the curve $y^2 = x^4(4 - x)$ possesses a loop and find the area of the loop. [WJC]

2 Find the area of one loop of the curve $4y^2 = x^2(4 - x^2)$. Also find the position of the mean centre of this area.

3 Sketch the curves $y^2 = 2x$, $x^3 = 4y$ giving the co-ordinates of the points of intersection. Find the area they enclose and the volume this area sweeps out when revolved through 2π radians about Ox.

4 The area enclosed by the parabola $y^2 = 4ax$, the x-axis, and the ordinate $x = h$ is rotated through 2π radians about the x-axis. Show that the volume swept out is $2\pi ah^2$.

5 PAQ is an arc of the curve $y = \sin x$ from $x = 2\pi$ to $x = 3\pi$, A being the mid-point of the arc. Show that, if P and Q are the points on the curve where $x = 2\pi$ and $x = 3\pi$ respectively, the area between the arc and the x-axis is divided by the line PA approximately in the ratio 0·12 : 1.

6 Determine the mean value of the function $x(4 - x)$ between $x = 0$ and $x = 4$.

7 Find the RMS value of $I = \sin(wt + \pi/3)$, the value of t being taken over one period from $t = 0$ to $t = 2\pi/w$.

8 Sketch the curve $y^2 = (x - 1)^2(x + 1)$. If the curve is rotated about the x-axis through an angle $\pi/2$, find the volume enclosed by the surface swept out by the loop of the curve.

9 Find the area of the portion of the plane enclosed by the curve $y = 1 + \sin x$, the axis of y, and the axis of x from 0 to $\frac{3}{2}\pi$. Find also the volume of the solid obtained by rotating this area about the axis of x. [SUJB]

10 Sketch the graph of $y = \dfrac{x}{2 + x}$ for $x \geqslant 0$. Find the area enclosed by the curve, the lines $x = 0$, $x = 1$ and the line $y = 1$. Also find the volume generated when this area revolves through 2π radians about the line $y = 1$.

11 Find the mean centre of the area between the curve $y = (x - 1)(4 - x)$ and the axis of x.

12 Find the area enclosed by the two parabolas $ay = 2x^2$, $y^2 = 4ax$. Also find the position of the mean centre of this area.

13 Find (i) the area bounded by the axes and the part of the curve $y = \cos 2x$ between $x = 0$ and $x = \pi/4$; (ii) the volume described when that area is rotated through four right angles about the x-axis; (iii) the centre of gravity of that area. [SUJB]

14 Prove that the area common to the two parabolas $y^2 = 4ax$ and $x^2 = 4ay$ is $16a^2/3$. Find the centroid of the common area. Show that if the area is rotated, through four right angles about the x-axis, the volume generated is $96\pi a^3/5$. [JMB]

15 Sketch the curve whose equation is $a^2y^2 = 4x^2(a^2 - x^2)$. Prove that the area contained by one loop of the curve is $4a^2/3$. Find the volume swept out when one loop is rotated through two right angles about the x-axis. [JMB]

16 The curves $y = 7 - x^2$ and $xy = 6$ intersect at the points A and B in the first quadrant. Find the co-ordinates of A and B. Find the area contained between the two curves. This area is rotated through four right angles about the y-axis. Prove that the volume swept out is $3\pi/2$. [JMB]

17 ABC is a triangular lamina in which AB = AC and the perpendicular distance of A from BC is h. The density of a thin strip of the lamina which is parallel to BC and at a distance x from A is kx, where k is a constant. Prove that the centre of gravity of the lamina is at a distance $\frac{3}{4}h$ from A.

18 Prove that the area bounded by the two parabolas $3y = 2x^2$, $y^2 = 12x$ is 6 square units. Find the co-ordinates of the centroid of this area.

19 The co-ordinates of a variable point P are given by the equations $x = 4 - t^2$, $y = 1 + 3t$, where t is a parameter. Find the value of t for which the tangent to the locus of P is parallel to the y-axis. Find also the x co-ordinate of the centroid of the area bounded by the curve and the y-axis.

20 A curve whose equation has the form $y = x(x-2)(ax+b)$ touches the x-axis at the point where $x = 2$ and the line $y = 2x$ at the origin. Find the values of a and b, sketch the curve and prove that the area enclosed by an arc of the curve and a segment of the line $y = 2x$ is $\frac{32}{3}$. [LU]

21 Sketch the graph of $y = x^2 \sin 2x$ (x being measured in radians) from $x = 0$ to $x = \pi$, and prove that the ratio of the two areas bounded by the curve and the axis of x is $\dfrac{\pi^2 - 4}{5\pi^2 - 4}$. [SUJB]

22 In a triangle ABC the angle $C = 2\pi/3$. Express c^2 in terms of a and b. The triangle is rotated about A in its own plane, through an angle $\theta (< \pi)$. Find, in terms of a, b and θ, the area swept out by (i) AC, (ii) BC. [JMB]

23 Draw a rough sketch of the curve defined by the equations $x = 2(\theta - \sin\theta)$, $y = 2(1 - \cos\theta)$ as θ increases from 0 to 2π. Evaluate for this curve the integrals

$$\text{(i)} \int_0^{2\pi} y\frac{dx}{d\theta}\,d\theta \qquad \text{(ii)} \int_0^{2\pi} \sqrt{\left[\left(\frac{dx}{d\theta}\right)^2 + \left(\frac{dy}{d\theta}\right)^2\right]}\,d\theta$$

24 Sketch the curve given by the parametric equations $x = 2 - t^2$, $y = t^3$. The area enclosed by this curve and the axis of y is rotated about the axis through four right angles. Find the volume of the solid so described. [SUJB]

25 The points A(c, a), B(0, b) and C($-c$, a), where ($b > a > 0$), lie on a curve of the type $y = Px^2 + Qx + R$. Determine the constants P, Q and R and hence show that the equation of the curve is $c^2y + (b-a)x^2 = bc^2$. AM and CN are ordinates. Find the volume of the solid formed by revolving the area bounded by the curve, the x-axis and these ordinates through four right angles about the x-axis. [LU]

26 Given that $y = \dfrac{6x - 10}{(x-3)^2(x+1)}$, express y in partial fractions. Determine the mean value of y with respect to x in the range $x = 4$ to $x = 6$. Find the expansion of y in ascending powers of x up to and including the term in x^2. [AEB]

27 Find the co-ordinates of the centroid of the finite region R bounded by the curve $y = e^x$, the co-ordinates axes and the line $x = 1$. This region R is rotated about the x-axis to form a solid of revolution. Find the co-ordinates of the centroid of this solid. (Leave answers in terms of e.) [LU]

28 Given that $y = \sqrt{(4 - x)}$ find, for values of x in the interval $0 \leqslant x \leqslant 4$,

(i) the mean value of y with respect to x, (ii) the mean value of xy with respect to x. [JMB]

29 Using the same axes, sketch the curves which are given by $y = e^x$ and $y = 3 - 4x - x^2$. If t is the positive root of the equation $e^x = 3 - 4x - x^2$, prove that the area of the finite region in the first quadrant enclosed by the curves and the y-axis is given by $\frac{1}{3}(21t - 3t^2 - t^3 - 6)$. [AEB]

30 The area enclosed by the ellipse defined by the parametric equations $x = a\cos\theta$, $y = b\sin\theta$ is rotated about the y-axis through an angle of π. Show that the volume swept out is represented by

$$\pi a^2 b \int_{-\pi/2}^{\pi/2} \cos^3\theta \, d\theta$$

Evaluate the integral. [JMB]

16
Differential equations

16.1 Introduction

We have seen in Section 13.1 that an equation of the type

$$\frac{dv}{dt} = 6 \tag{16.1}$$

can be solved by integration (integration is the inverse of differentiation) to give as its solution

$$v = 6t + C \tag{16.2}$$

This is also known as the general solution of the equation (16.1) because it contains the arbitrary constant of integration.

Such equations which involve differential coefficients are known as differential equations and they occur very often when practical problems are expressed in mathematical symbols. The equation (16.1) arose from the question: if the acceleration of a particle is constant and equal to 6 cm/sec^2, what is its speed? We now see the practical importance of the constant of integration: two objects may have the same acceleration but different speeds, depending on the initial conditions. For example, if we know that initially ($t = 0$) the speed is 11 cm/sec, then substituting in (16.2)

$$11 = 6.0 + C$$

and we obtain

$$C = 11$$

therefore

$$v = 6t + 11 \tag{16.3}$$

This is known as a particular solution of the differential equation (16.1).

As another example of how differential equations arise, consider one of the laws of chemical reaction.

Example 1 In a certain chemical reaction, the amount x of one substance at any time t is related to the speed of the reaction, dx/dt, by the equation

$$\frac{dx}{dt} = k(a - x)(b - x) \quad (a, b, k \text{ constants})$$

Find a relation between x, a, b, k and t.

To solve the equation, we rewrite it in the form

$$\frac{dt}{dx} = \frac{1}{k(a-x)(b-x)}$$

therefore

$$t = \frac{1}{k}\int \frac{dx}{(a-x)(b-x)}$$

$$kt = \frac{1}{(b-a)}\int \left[\frac{1}{(a-x)} - \frac{1}{(b-x)}\right] dx$$

$$= \frac{1}{(b-a)}[-\log_e (a-x) + \log_e (b-x)] + C$$

$$= \frac{1}{(b-a)} \log_e \left(\frac{b-x}{a-x}\right) + C$$

Although we now have a relation between t and x, the three original unknown constants and the arbitrary constant of integration are also involved, and more information is needed before the result is of practical value.

Example 2 A beaker containing water at 100 °C is placed in a room which has a constant temperature of 20 °C. The rate of cooling at any moment is proportional to the difference between the temperature of the room and the liquid. If after 5 min the temperature of the water is 60 °C, what will it be after 10 min?

The law stated here is the physical reality known as Newton's law of cooling.

Let the temperature of the water at any time t min be θ °C. Then the rate of change of temperature is $d\theta/dt$ (see Section 11.1), thus the rate of cooling is $-d\theta/dt$. Hence

$$-\frac{d\theta}{dt} \propto \theta - 20$$

therefore

$$-\frac{d\theta}{dt} = k(\theta - 20) \qquad (16.4)$$

which is the mathematical expression of Newton's law.

By considering the reciprocal of both sides, we have

$$-\frac{dt}{d\theta} = \frac{1}{k(\theta - 20)}$$

$$-t = \frac{1}{k}\int \frac{d\theta}{(\theta - 20)}$$

$$-kt = \log_e (\theta - 20) + C$$

$$-C - kt = \log_e (\theta - 20)$$

$$e^{-C-kt} = \theta - 20$$

$$e^{-C}e^{-kt} = \theta - 20$$

$$Ae^{-kt} = \theta - 20 \qquad \text{where } A = e^{-C}$$

$$20 + Ae^{-kt} = \theta$$

Now initially $t = 0$ and $\theta = 100$, therefore

$$20 + A = 100$$
$$A = 80$$

which gives

$$\theta = 20 + 80e^{-kt} \tag{16.5}$$

Also when $t = 5$, $\theta = 60$, therefore

$$60 = 20 + 80e^{-5k}$$
$$e^{-5k} = 0{\cdot}5 \tag{16.6}$$

It is possible to find k exactly from this equation but it is not necessary in this example for when $t = 10$, we have, by substitution in (16.5),

$$\theta = 20 + 80e^{-10k}$$
$$= 20 + 80(e^{-5k})^2$$

which from (16.6) gives

$$\theta = 20 + 80(0{\cdot}5)^2$$
$$\theta = 40\,^\circ\text{C} \quad \text{after 10 min}$$

Applications to mechanics

Problems in mechanics often involve acceleration, which may be expressed in any of the following three forms:

(i) $\dfrac{dv}{dt}$

(ii) $\dfrac{d^2s}{dt^2}$ $\left(\text{since } v = \dfrac{ds}{dt}\right)$

(iii) $v\dfrac{dv}{ds}$ $\left(\text{since } \dfrac{dv}{dt} = \dfrac{ds}{dt}\cdot\dfrac{dv}{ds} = v\dfrac{dv}{ds}\right)$

Example 3 A particle moves in a straight line with constant acceleration a. If at any time t sec its speed is v cm/sec and the distance travelled is s cm find expressions for (i) v in terms of a and t, (ii) s in terms of a and t, (iii) v in terms of a and s. Assume that when $t = 0$, $s = 0$ and $v = u$.

(i) In this case, since v and t are to be linked, we use $dv/dt = a$. Therefore

$$v = at + C$$

But when $t = 0$, $v = u$, therefore

$$u = 0 + C$$
$$v = u + at$$

(ii) From the preceding result

$$\frac{ds}{dt} = u + at$$

hence

$$s = ut + \tfrac{1}{2}at^2 + A$$

Now when $t = 0$, $s = 0$, hence $A = 0$ and so

$$s = ut + \tfrac{1}{2}at^2$$

(iii) In this case, since v, s and a are to be linked, we use

$$v\frac{\mathrm{d}v}{\mathrm{d}s} = a$$

that is

$$\frac{\mathrm{d}(\frac{1}{2}v^2)}{\mathrm{d}s} = a$$

hence

$$\tfrac{1}{2}v^2 = as + B$$

But when $s = 0$, $v = u$, therefore

$$\tfrac{1}{2}u^2 = 0 + B$$

and so

$$\tfrac{1}{2}v^2 = as + \tfrac{1}{2}u^2$$
$$v^2 = u^2 + 2as$$

A differential equation sometimes expresses a physical relation better than its general solution. To illustrate this point, and also to show that, given a general solution, the differential equation can be formed by differentiation and elimination of the arbitrary constants, consider the following example.

Example 4 Given that a particle moves so that its distance, s, from a fixed point at time t is given by $s = A\sin(3t + \varepsilon)$, where A, ε are constants, find the original differential equation governing the motion of the particle.

Since there are two constants, we shall find that the original differential equation involves *second-order* differential coefficients.

$$s = A\sin(3t + \varepsilon) \tag{16.7}$$

$$\frac{\mathrm{d}s}{\mathrm{d}t} = 3A\cos(3t + \varepsilon)$$

$$\frac{\mathrm{d}^2s}{\mathrm{d}t^2} = -9A\sin(3t + \varepsilon) \tag{16.8}$$

From (16.7) and (16.8)

$$\frac{\mathrm{d}^2s}{\mathrm{d}t^2} = -9s$$

that is, the acceleration is $-9s$.

Thus the acceleration is proportional to the distance s from the point and (since it is negative) directed towards that point. This kind of motion is known as simple harmonic motion.

In subsequent sections of this chapter, a systematic approach will be made to the solution of differential equations generally. It is, however, always worth while considering if some of the usual mathematical processes will help.

Example 5 Solve the differential equation

$$\left(\frac{dy}{dx}\right)^2 - \left(\frac{1}{x} - x\right)\frac{dy}{dx} - 1 = 0$$

and interpret the result geometrically.

In this case, the equation will factorise giving

$$\left(\frac{dy}{dx} + x\right)\left(\frac{dy}{dx} - \frac{1}{x}\right) = 0$$

Hence either

$$\frac{dy}{dx} = -x \quad \text{or} \quad \frac{dy}{dx} = \frac{1}{x} \tag{16.9}$$

Therefore

$$y = -\tfrac{1}{2}x^2 + A \quad \text{or} \quad y = \log_e x + B$$

Note that although the two constants are both arbitrary, they may have different values.

To interpret the result geometrically, we note that as A is given different values (say 0, ± 1, $\pm 2, \ldots$), $y = -\frac{1}{2}x^2 + A$ gives rise to a family of parabolas all with their vertices on the y-axis and with their axes coincident with the y-axis (see *Figure 16.1*).

Similarly, as the value of B varies, $y = \log_e x + B$ gives a family of logarithmic curves (see *Figure 16.1*).

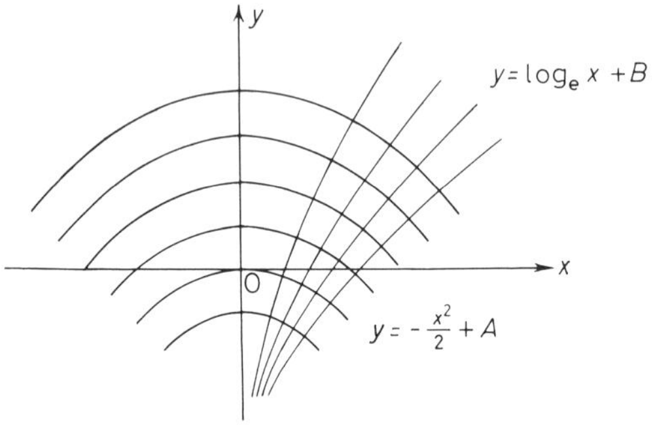

Figure 16.1

In this example, it is interesting to note that from (16.9) the slopes of the two families of curves are $-x$ and $1/x$, and that the product of these slopes is -1, so that the curves cut at right angles.

Two such families of curves in which every member of one family cuts every member of the other family at right angles are said to be orthogonal trajectories.

Exercises 16a

1 Solve the following differential equations:

(i) $\dfrac{dy}{dx} = \sin x$ (ii) $\dfrac{dx}{dt} - t^2 + 2 = 0$

(iii) $\dfrac{dx}{dt} = 3x$ (iv) $\dfrac{d^2y}{dt^2} = 5$

2 Form the differential equations whose complete solutions are

(i) $y = Ax$ (ii) $y = Ae^{2x}$

(iii) $y = Ax^2 + \dfrac{B}{x}$ (iv) $y = Ae^{3x} + Be^{-2x}$

3 Prove that for any straight line through the origin, $dy/dx = y/x$ and interpret this result geometrically.

4 The rate of decay of a radioactive substance at any time is proportional to the amount remaining at that time, the constant of proportionality being k. If initially the amount of substance is 10 g find an expression for the amount remaining after t sec.

5 A particle moves in a straight line so that its acceleration is always towards a fixed point and varies inversely as the square of its distance x from that point. Show that its speed v is given by $\frac{1}{2}v^2 = \mu/x + C$ (μ and C being constants).

This law is true for the case of the earth attracting a meteorite. If x is measured in km and v in km/sec, then $\mu = 0{\cdot}4 \times 10^6$. Neglecting the effect of the earth's atmosphere and assuming its radius is 6400 km, find with what speed a meteorite would reach the earth after moving from a very great distance under its attraction.

6 A particle falls from rest in air. If the resistance of the air is assumed to vary as its speed, v, then it can be shown that its acceleration a is given by $a = 980 - kv$ (where k is a constant). Show that $v = 980(1 - e^{-kt})/k$ and hence find the limiting value to which v tends as t increases (known as the *terminal velocity*).

7 Find the two general solutions of $(dy/dx)^2 + (x + y)\,dy/dx + xy = 0$ and illustrate the results geometrically.

8 The current i flowing in a circuit at any time t is given by

$$L\frac{di}{dt} + Ri = E \quad (L, R, E \text{ constant})$$

where L is the self inductance of the circuit, R its resistance and E the external electromotive force. Find i in terms of E, R, L and t, and given that initially the current is zero, find the value of the current as t becomes very large.

9 Find V if $d[r^2(dV/dr)]/dr = 0$ and $V = V_1$ at $r = a$, $V = 0$ at $r = b$. (V is the potential at a distance r from the common centre of two spherical conductors radii a, b at potentials V_1 and 0 respectively.)

10 The ordinate and normal through a point P on a curve meet the x-axis in N and G respectively, and $NG = kNP^2$, where k is a constant. Find the equation of the curve if it passes through the point (1, 1) with gradient 2.

16.2 First-order differential equations with variables separable

The order of a differential equation is the order of the highest differential coefficient contained in the equation. In this section, we shall consider only first-order differential equations, i.e. equations which contain only a first-order differential coefficient dy/dx, dy/dt, etc. and no higher derivatives. Functions of x and y or x and t will also figure in the equations. First-order equations with variables separable are equations which may be put into the form

$$\frac{dy}{dx} = f(x)g(y) \qquad (16.10)$$

i.e. dy/dx is equal to an expression which can be resolved into two factors, one containing x only, the other y only. Expression (16.10) can be rewritten

$$\frac{1}{g(y)}\frac{dy}{dx} = f(x)$$

Integration with respect to x gives

$$\int \frac{1}{g(y)}\,dy = \int f(x)\,dx \quad \text{(see Section 14.3)}$$

We have separated the variables whence the name of this type of equation.

Example 1 Find the general solution of the differential equation $x^2y(dy/dx) = x + 1$.

$$x^2y\frac{dy}{dx} = x + 1$$

$$y\frac{dy}{dx} = \frac{1}{x} + \frac{1}{x^2}$$

Therefore

$$\int y\frac{dy}{dx}\,dx = \int\left(\frac{1}{x} + \frac{1}{x^2}\right)dx$$

$$\int y\,dy = \int (x^{-1} + x^{-2})\,dx$$

$$\frac{1}{2}y^2 = \log_e x - \frac{1}{x} + C$$

It will be noted that, although there are two separate integrations, only one arbitrary constant is necessary, because the sum or difference of two arbitrary constants is another arbitrary constant.

Example 2 During a fermentation process, the rate of decomposition of a substance at any time t varies directly as the amount of substance y and also as the amount of active ferment x. If the constant of proportionality is 0·5, the value of x at any time t is $4/(1 + t)^2$, and initially $y = 10$, find y as a function of t. Also deduce the amount of substance remaining as t becomes very large.

The rate of change of the substance is $\mathrm{d}y/\mathrm{d}t$, therefore the rate of decomposition is $-(\mathrm{d}y/\mathrm{d}t)$. Hence

$$-\frac{\mathrm{d}y}{\mathrm{d}t} \propto xy$$

But the constant of proportionality is given as 0·5, thus

$$\frac{\mathrm{d}y}{\mathrm{d}t} = -0{\cdot}5xy$$

However

$$x = \frac{4}{(1+t)^2}$$

therefore

$$\frac{\mathrm{d}y}{\mathrm{d}t} = -0{\cdot}5 \times \frac{4}{(1+t)^2}y$$

$$\frac{1}{y}\frac{\mathrm{d}y}{\mathrm{d}t} = -\frac{2}{(1+t)^2}$$

$$\int \frac{1}{y}\frac{\mathrm{d}y}{\mathrm{d}t}\,\mathrm{d}t = -2\int \frac{\mathrm{d}t}{(1+t)^2}$$

$$\int \frac{\mathrm{d}y}{y} = \frac{2}{1+t} + C$$

$$\log_e y = \frac{2}{1+t} + C$$

$$y = \mathrm{e}^{C+2/(1+t)}$$

$$= \mathrm{e}^C\mathrm{e}^{2/(1+t)}$$

which may be written as

$$y = A\mathrm{e}^{2/(1+t)}$$

where $A = \mathrm{e}^C$ is a convenient form for the arbitrary constant.

Now initially $t = 0$ and $y = 10$, therefore

$$10 = A\mathrm{e}^2$$

$$10\,\mathrm{e}^{-2} = A$$

Thus

$$y = 10\,\mathrm{e}^{-2}\mathrm{e}^{2/(1+t)}$$

$$= 10\,\mathrm{e}^{2/(1+t)-2} = 10\,\mathrm{e}^{-2t/(1+t)}$$

As $t \to \infty$, $-2t/(1+t) \to -2$ and so $y \to 10\,\mathrm{e}^{-2} = 1{\cdot}36$.

If we now reconsider the examples in Section 16.1, it will be seen that they could nearly all be treated by this method.

Exercises 16b

1 Solve the differential equations

(i) $\dfrac{dy}{dx} = \dfrac{y^2}{1+x^2}$ (ii) $\dfrac{dy}{dx} = xy(y-2)$

(iii) $(1+x)^2 \dfrac{dy}{dx} + y^2 = 1$ (iv) $\dfrac{dy}{dx} = e^{x+y}$

2 A particle moves in a straight line in a resisting medium so that its acceleration a is given by $a = 10(v^3 + 9v)$. If the particle passes through the origin with a speed u, find an expression for its distance s from the origin in terms of u and v.

3 Solve the equation $\dfrac{d}{dr}\left(r\dfrac{d\theta}{dr}\right) = 0$.

4 Show that the general solution of the equation $(1+y^2) + (1+x^2)(dy/dx) = 0$ can be written in the form $y = (k-x)/(1+kx)$, where k is the arbitrary constant. Hence find the particular solution for which $y = \frac{2}{3}$ when $x = 1$.

5 The normal and the ordinate at any point P on a curve meet the x-axis at G and N respectively. The difference between the length of GN and the x co-ordinate of P is one unit. Find the general equation of the curve.

6 Show that the equation $y = 2x(dy/dx)$ represents a family of parabolas with a common axis and a common tangent at the vertex.

7 Find a function whose rate of change is proportional to the square of its value and whose value is 1 when $x = 0$ and 3 when $x = 1$.

8 In a suspension bridge with a uniform horizontal load, the form of the cables is determined by the equation $2y = x(dy/dx)$, where the lowest point is taken as the origin of co-ordinates and the tangent at this point as the x-axis. Show that the form of the cables is a parabola with its axis vertical.

9 Find the curve such that the normals all pass through the origin.

10 In a reservoir which is discharging over a weir, it is known that $\dfrac{dt}{dH} = \dfrac{270}{81 - H^2}$, where H m is the height of the surface above the sill of the weir at any time t min. If initially $H = 1$ m find an expression for H.

16.3 The differential equation $d^2x/dt^2 = kx$

The equation $d^2x/dt^2 = kx$ is a very simple example of a second-order differential equation but it is of intrinsic importance in kinematics. It arises when the acceleration of a particle is proportional to its distance from a fixed point, that is acceleration $\propto$ distance

$$\frac{d^2x}{dt^2} \propto x$$

$$\frac{d^2x}{dt^2} = kx \tag{16.11}$$

If the acceleration is directed away from the point, k is positive and can be written as n^2. Hence

$$\frac{d^2x}{dt^2} = n^2x \tag{16.12}$$

If the acceleration is directed towards the point, k is negative and can be written as $-n^2$. Hence

$$\frac{d^2x}{dt^2} = -n^2x \tag{16.13}$$

The method of solution is to multiply both sides by $2(dx/dt)$. Thus (16.11) becomes

$$2\frac{dx}{dt}\cdot\frac{d^2x}{dt^2} = 2kx\frac{dx}{dt}$$

which can be written

$$\frac{d\left(\frac{dx}{dt}\right)^2}{dt} = 2kx\frac{dx}{dt}$$

Hence

$$\left(\frac{dx}{dt}\right)^2 = \int 2kx\frac{dx}{dt}\,dt$$

$$\left(\frac{dx}{dt}\right)^2 = \int 2kx\,dx \quad \text{(see Section 14.3)}$$

$$\left(\frac{dx}{dt}\right)^2 = kx^2 + c$$

Therefore

$$\frac{dx}{dt} = \pm\sqrt{(kx^2 + C)}$$

This is a separable differential equation and by Section 16.2

$$t = \pm\int\frac{dx}{\sqrt{(kx^2 + C)}} \tag{16.14}$$

General solutions

If k is positive and equal to n^2 (see (16.12)), the integral (16.14) becomes

$$t = \pm \int \frac{dx}{\sqrt{(n^2x^2 + C)}}$$

the solution of which is beyond the scope of this volume. However, if we return to (16.12), the solution of the differential equation can be obtained as follows:

$$\frac{d^2x}{dt^2} = n^2x$$

Subtract $n(dx/dt)$ from both sides of the equation to give

$$\frac{d^2x}{dt^2} - n\frac{dx}{dt} = -n\frac{dx}{dt} + n^2x$$

$$\frac{d}{dt}\left(\frac{dx}{dt} - nx\right) = -n\left(\frac{dx}{dt} - nx\right)$$

If we now set $z = dx/dt - nx$, we have

$$\frac{dz}{dt} = -nz$$

This is a separable differential equation (see Section 16.2) and we have

$$\int \frac{dz}{z} = \int -n\,dt$$

therefore

$$\log_e z = -nt + A$$

from which $z = Ce^{-nt}$. Resubstituting for z

$$\frac{dx}{dt} - nx = Ce^{-nt} \qquad (16.15)$$

Reconsider the original equation

$$\frac{d^2x}{dt^2} = n^2x$$

and this time *add* $n(dx/dt)$ to both sides of the equation; then

$$\frac{d^2x}{dt^2} + n\frac{dx}{dt} = n\frac{dx}{dt} + n^2x$$

$$\frac{d}{dt}\left(\frac{dx}{dt} + nx\right) = n\left(\frac{dx}{dt} + nx\right)$$

If we now set $z = dx/dt + nx$, we have

$$\frac{dz}{dt} = nz$$

This is again a separable differential equation and we have

$$\int \frac{dz}{z} = \int n\,dt$$

therefore

$$\log_e z = nt + B$$

from which $z = De^{nt}$. Resubstituting for z

$$\frac{dx}{dt} + nx = De^{nt} \tag{16.16}$$

We now subtract (16.15) from (16.16) to eliminate dx/dt and we have

$$2nx = De^{nt} - Ce^{-nt}$$

$$x = \frac{D}{2n}e^{nt} - \frac{C}{2n}e^{-nt}$$

therefore

$$x = Ae^{nt} + Be^{-nt} \tag{16.17}$$

where $A = D/2n$, $B = -C/2n$ are convenient forms for the arbitrary constants.
If k is negative and equal to $-n^2$ (see (16.14)), then the solution is

$$t = \pm \int \frac{dx}{\sqrt{(C - n^2x^2)}}$$

$$t = \pm \frac{1}{n} \int \frac{dx}{\sqrt{\left(\frac{C}{n^2} - x^2\right)}}$$

If we let $p^2 = C/n^2$, the solution to this is

$$t = \pm \frac{1}{n} \arcsin \frac{x}{p} + K$$

or

$$nt - nK = \pm \arcsin \frac{x}{p}$$

that is

$$nt + \varepsilon = \pm \arcsin \frac{x}{p} \quad (\text{putting } \varepsilon = -nK)$$

Hence $x = \pm|p| \sin(nt + \varepsilon)$.

The ambiguous sign can be absorbed by changing the sign of p, since p is arbitrary, or by replacing ε by $\varepsilon + \pi$, since ε is arbitrary; hence the solution is

$$x = p \sin(nt + \varepsilon) \tag{16.18}$$

This may be rewritten

$$x = p \sin nt \cos \varepsilon + p \cos nt \sin \varepsilon$$

that is

$$x = A \sin nt + B \cos nt \tag{16.19}$$

where $A = p \cos \varepsilon$, $B = p \sin \varepsilon$ (see also Sections 6.8 and 16.1, Example 4).

Note that x is a periodic function, period $2\pi/n$.

The solutions (16.17), (16.18), (16.19) should be remembered.

Example 1 Solve the equation $d^2x/dt^2 = 4x$, given that when $t = 0$, $x = 5$ and $dx/dt = 2$.

From (16.17) the solution of $d^2x/dt^2 = 4x$ is

$$x = Ae^{-2t} + Be^{2t} \tag{i}$$

Hence

$$\frac{dx}{dt} = -2Ae^{-2t} + 2Be^{2t} \tag{ii}$$

We are given that when $t = 0$, $x = 5$ and $dx/dt = 2$. Substituting in (i) and (ii), we have

$$5 = A + B$$
$$2 = -2A + 2B$$

Hence

$$A = 2 \qquad B = 3$$

and

$$x = 2e^{-2t} + 3e^{2t}$$

Example 2 The speed of a particle is given by $v = \sqrt{(16 - 9x^2)}$. Find an expression for its acceleration in terms of x. Given that $x = 0$ when $t = 0$, find an expression for x.

$$v = \sqrt{(16 - 9x^2)}$$

$$\frac{dv}{dx} = \frac{-9x}{\sqrt{(16 - 9x^2)}}$$

$$\text{acceleration} = v\frac{dv}{dx}$$

$$= \sqrt{(16 - 9x^2)}\frac{-9x}{\sqrt{(16 - 9x^2)}}$$

$$= -9x$$

therefore

$$\frac{d^2x}{dt^2} = -9x$$

From (16.19)

$$x = A\sin 3t + B\cos 3t$$

Initially $t = 0$, $x = 0$, therefore

$$x = A\sin 3t$$

Now $v = dx/dt = 3A\cos 3t$ and from the given expression when $x = 0\,(t = 0)$, $v = 4$. Therefore

$$4 = 3A$$

$$A = \tfrac{4}{3}$$

giving

$$x = \tfrac{4}{3}\sin 3t$$

Example 3 The acceleration of a particle is proportional to its distance from a fixed point O and is directed towards that point.

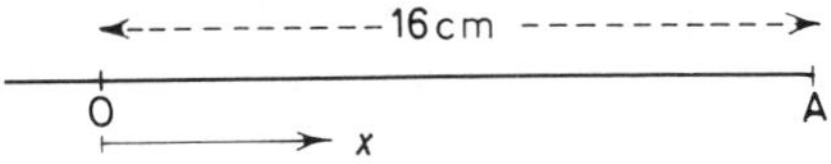

Figure 16.2

The particle starts from rest at a point A distance 16 cm from O. If after 3 sec the particle reaches O, find when it was at a distance of 8 cm from O.

Measure the distance x from O in the direction OA. Because the acceleration is directed towards O, we have

$$-\frac{d^2x}{dt^2} \propto x$$

$$\frac{d^2x}{dt^2} = -n^2x$$

From (16.18) the general solution of this is

$$x = p\sin(nt + \varepsilon) \qquad \text{(i)}$$

Thus the speed $v\,(= dx/dt)$ is given by

$$v = pn\cos(nt + \varepsilon) \qquad \text{(ii)}$$

From the initial conditions, when $t = 0$, $x = 16$ and $v = 0$. Substituting in (i) and (ii), we have

$$16 = p\sin\varepsilon \qquad \text{(iii)}$$

$$0 = pn\cos\varepsilon \qquad \text{(iv)}$$

From (iii) and (iv)

$$p = 16 \quad \text{and} \quad \varepsilon = \tfrac{1}{2}\pi$$

thus

$$x = 16\sin(nt + \tfrac{1}{2}\pi)$$

Now when $t = 3$, $x = 0$, hence

$$0 = 16\sin(3n + \tfrac{1}{2}\pi)$$

that is

$$3n + \tfrac{1}{2}\pi = 0, \quad \pm\pi, \quad \pm 2\pi, \quad \ldots$$

The only acceptable solution is

$$3n + \tfrac{1}{2}\pi = \pi$$

thus $n = \pi/6$, giving as the complete solution

$$x = 16\sin(\tfrac{1}{6}\pi t + \tfrac{1}{2}\pi) \tag{v}$$

To find when the particle was at a distance of 8 cm from O, substitute $x = 8$ in (v), then

$$8 = 16\sin(\tfrac{1}{6}\pi t + \tfrac{1}{2}\pi)$$

$$\sin(\tfrac{1}{6}\pi t + \tfrac{1}{2}\pi) = \tfrac{1}{2}$$

Therefore

$$\tfrac{1}{6}\pi t + \tfrac{1}{2}\pi = n\pi + (-)^n\tfrac{1}{6}\pi \quad \text{(see 7.4)}$$

The value $n = 1$ gives the smallest value for t for which $x = 8$, thus

$$\tfrac{1}{6}\pi t + \tfrac{1}{2}\pi = \pi - \tfrac{1}{6}\pi$$

that is

$$t = 2\text{ sec}$$

Example 4 A particle of mass m is suspended from a light elastic string. Its acceleration is given by $g - (\lambda x/ma)$, where g, λ and a are constants and x is the distance of the particle from a fixed point A. Find an expression for x at any time t.

$$\text{Acceleration} = g - \frac{\lambda x}{ma}$$

$$\frac{d^2x}{dt^2} = g - \frac{\lambda x}{ma}$$

This can be written

$$\frac{d^2x}{dt^2} = -\frac{\lambda}{ma}\left(x - \frac{mag}{\lambda}\right) \tag{i}$$

Now let $z = x - (mag/\lambda)$. Then $d^2z/dt^2 = d^2x/dt^2$ and equation (i) becomes

$$\frac{d^2z}{dt^2} = -\frac{\lambda}{ma}z$$

This is the same as (16.13) with

$$\frac{\lambda}{ma} = n^2$$

Thus its solution is

$$z = p \sin\left[\sqrt{\left(\frac{\lambda}{ma}\right)}t + \varepsilon\right]$$

that is

$$x = \frac{mag}{\lambda} + p \sin\left[\sqrt{\left(\frac{\lambda}{ma}\right)}t + \varepsilon\right]$$

Exercises 16c

1 Find the solution of the equation $\mathrm{d}^2x/\mathrm{d}t^2 = 25x$ given that when $t = 0$, $x = 12$ and $\mathrm{d}x/\mathrm{d}t = 10$.
2 Find the solution of the equation $\mathrm{d}^2x/\mathrm{d}t^2 + 4x = 0$ given that when $t = 0$, $x = 4$ and $\mathrm{d}x/\mathrm{d}t = 6$.
3 A particle starts from rest and moves towards a fixed point O under the influence of a force which is directed towards O, and which varies as the distance of the particle from O. Initially, the particle was 10 cm from O and its acceleration was 10 cm/sec^2 towards O. Find (i) its speed when 8 cm from O, (ii) its speed at O, (iii) its distance from O after $\pi/3$ sec.
4 A particle moves so that its equation of motion is $\mathrm{d}^2x/\mathrm{d}t^2 = -16x$. Initially $v = -16$ cm/sec and $x = 3$ cm. Find (i) its speed when $x = 5$ cm, (ii) its speed when $x = 4$ cm, (iii) the value of x when its speed is 20 cm/sec.
5 Solve the equation $\mathrm{d}^2x/\mathrm{d}t^2 = 36x - 72$ given that when $t = 0$, $x = 7$ and $\mathrm{d}x/\mathrm{d}t = 32$.

Exercises 16

1 Obtain the general solution of the following differential equations:

(a) $\tan x \dfrac{\mathrm{d}y}{\mathrm{d}x} = \cot y$ (b) $\dfrac{\mathrm{d}y}{\mathrm{d}x} + y = x^2 y$

(c) $y - x\dfrac{\mathrm{d}y}{\mathrm{d}x} = xy$

2 Obtain the differential equation for which $y = Ax + A^3$ is the general solution.
3 Solve the equation

$$\left(\frac{\mathrm{d}y}{\mathrm{d}x}\right)^2 - (x + y)\frac{\mathrm{d}y}{\mathrm{d}x} + xy = 0$$

and interpret the results geometrically.
4 Find the general expression for y given that it satisfies the equation

$$y(1 + x^2)\frac{\mathrm{d}y}{\mathrm{d}x} - 2x(1 - y^2) = 0$$

5 Solve the differential equation

$$(1 + x^2)y\frac{dy}{dx} = 9 + y^2$$

given that $y = 0$ when $x = 0$. [AEB]

6 Show that the solutions of the differential equation

$$xy\left(\frac{dy}{dx}\right)^2 - (x^2 - y^2)\frac{dy}{dx} - xy = 0$$

are the two families of curves $xy = A$ and $x^2 - y^2 = B$. Show that these two sets of curves are orthogonal trajectories.

7 Given that $y = (\arcsin x)^2$, prove that $(1 - x^2)(dy/dx)^2 = 4y$. Deduce that $(1 - x^2)(d^2y/dx^2) - x\,(dy/dx) = 2$.

8 If $y = A\tan(x/2)$, where A is a constant, prove that

$$(1 + \cos x)(d^2y/dx^2) = y$$

9 Write down the general solution of the differential equation $dy/dt = -ky$, where k is a constant. A radioactive substance disintegrates at a rate proportional to its mass. If the mass remaining at time t is m, show that $m = m_0e^{-kt}$, where m_0 is the initial mass and k is a constant.

One third of the original mass of the substance disintegrates in 70 days. Calculate, correct to the nearest day, the time required for the substance to be reduced to have half its original mass. If the original mass was 100 g, calculate correct to the nearest g the mass remaining after 210 days. [JMB]

10 Find the curves such that the portion of the tangent included between the co-ordinate axes is bisected at the point of contact.

11 Prove that if $y = e^{-kt}(A\sin pt + B\cos pt)$, where k, p, A and B are constants, then

$$\frac{d^2y}{dt^2} + 2k\frac{dy}{dt} + (p^2 + k^2)y = 0$$

12 Find y in terms of x if

$$\frac{dy}{dx} = (1 + y)^2\sin^2 x\cos x$$

and $y = 2$ when $x = 0$. [AEB]

13 Given that $\dfrac{dy}{dx} = \dfrac{y + 9x}{y + x}$, use the substitution $y = zx$ to obtain an equation expressing dz/dx in terms of z and x. Solve this equation and deduce that $(3x + y)(3x - y)^2 = C$, where C is an arbitrary constant. [JMB]

14 A rectangular tank has vertical sides of depth h and a horizontal base of unit area. An inlet supplying water at a constant rate fills the tank in time T when running alone. An outlet, through which water flows at a rate proportional to the square root of the depth of water in the tank, empties the tank in time $4T$ when running alone. Show that, if x is the depth of water at time t when both outlet and inlet are running, then $dx/dt = p - q\sqrt{x}$, where $p = h/T$ and $q = \sqrt{h}/2T$. Deduce that if both inlet and outlet are running, the initially empty tank will be filled in time $4T(2\log_e 2 - 1)$. [JMB]

15 A particle is moving in a straight line and its distance at time t from a fixed point O in the line is x. Its speed is given by

$$50\frac{dx}{dt} = (40 - x)(x - 20)$$

and $x = 25$ when $t = 0$. Find an expression for x in terms of t. Find the greatest speed in the interval $20 < x < 40$ and the value of t when the greatest speed is attained. [JMB]

16 Show that the substitution $y = vx$ (where v is function of x) reduces the equation

$$x + y\frac{dy}{dx} = x\frac{dy}{dx} - y$$

to a separable equation for v and hence show that its solution is

$$\log_e(x^2 + y^2) = 2\arctan\frac{y}{x} + C$$

17 Use the substitution $y = vx$, where v is a function of x, to reduce the differential equation

$$x\frac{dy}{dx} - y = \tfrac{1}{4}x^2 - y^2$$

to a differential equation involving v and x. Hence find y as a function of x given that $y = 0$ when $x = 1$.

18 By using the substitution $y = xz$, or otherwise, solve the differential equation

$$2\frac{dy}{dx} = \frac{y}{x} + \frac{y^2}{x^2}$$ [JMB]

19 If $d^2s/dx^2 + s = 1$ and when $s = 2$, $ds/dx = 0$, prove that $(ds/dx)^2 = 2s - s^2$. If $s = 0$ when $x = 0$, prove also that $s = 1 - \cos x$.

20 Show that, by means of the substitution $y = t - x^2$, the equation

$$\frac{d^2y}{dx^2} + x^2 + y + 2 = 0$$

becomes

$$\frac{d^2t}{dx^2} + t = 0$$

Solve this latter equation and hence obtain the general solution of the original equation.

21 The rate of cooling of a body is given by the equation

$$\frac{dT}{dt} = -k(T - 10)$$

where T is the temperature in degrees Celsius, k is a constant, and t is the time in minutes.

When $t = 0$, $T = 90$ and when $t = 5$, $T = 60$. Show that $T = 41\frac{1}{4}$ when $t = 10$. [JMB]

22 The motion of a particle P, whose co-ordinates are (x, y) referred to a pair of fixed axes through a point O, satisfies the equations

$$\frac{d^2x}{dt^2} = -w^2x \qquad \frac{d^2y}{dt^2} = -w^2y$$

The initial conditions are $x = a$, $y = 0$, $dx/dt = 0$ and $dy/dt = bw$ when $t = 0$. Prove that the path of the particle is the ellipse

$$\left(\frac{x}{a}\right)^2 + \left(\frac{y}{b}\right)^2 = 1$$

(The general solutions of $d^2x/dt^2 = -w^2x$, $d^2y/dx^2 = -w^2y$ may be quoted.) [JMB]

23 A particle of mass m is projected with speed V vertically upwards from a point on horizontal ground. Its subsequent motion is subject to gravity and to a resistance kmv^2, where v is the speed and k is a constant. If its acceleration is equal to $-g - kv^2$ show that the greatest height attained is

$$\frac{1}{2k}\log_e\left(1 + k\frac{V^2}{g}\right)$$

24 Solve the differential equation $dy/dx = \cos x(\sec y - \tan y)$ given that $y = 0$ when $x = \frac{1}{2}\pi$. [AEB]

25 (a) In the differential equation

$$(x + y)\frac{dy}{dx} = x^2 + xy + x + 1$$

change the dependent variable from y to z, where $z = x + y$. Deduce the general solution of the given equation.

(b) The normal at the point $P(x, y)$ on a curve meets the x-axis at Q and N is the foot of the ordinate of P. If $NQ = \dfrac{x(1 + y^2)}{(1 + x^2)}$, find the equation of the curve, given that it passes through the point (3, 1). [JMB]

26 A particle is projected vertically upwards with a speed g/k, where g, k are constants. If the subsequent motion is subject to gravity and to a resistance to motion per unit mass of k times the speed, then the acceleration is given by

$$\frac{d^2x}{dt^2} = -g - kv$$

Find expressions for the speed v and the height x reached after time t. Also show that the greatest height H the particle can reach is given by $k^2H = g(1 - \log_e 2)$.

27 A particle moving along a straight line OX is at a distance x from O at time t and its speed is given by

$$8t^2 \frac{dx}{dt} = (1 - t^2)x^2$$

If $x = 4$ when $t = 1$ prove that $x = 8t/(1 + t^2)$. Find (i) the speed when $t = 0$; (ii) the maximum distance from 0; (iii) the maximum speed *towards* 0 for positive values of t. [JMB]

28 The equation of motion of a particle is

$$2c(1 + \sin\theta)\left(\frac{d\theta}{dt}\right)^2 = g[2(\cos\alpha - \cos\theta) - (\cos 2\alpha - \cos 2\theta)]$$

If α and θ are small, show that this equation reduces approximately to

$$2c\left(\frac{d\theta}{dt}\right)^2 = g(\alpha^2 - \theta^2)$$

If initially $t = 0$ and $\theta = \alpha$, deduce that

$$\theta = \alpha \sin\left[\sqrt{\left(\frac{g}{2c}\right)}t + \frac{\pi}{2}\right]$$

29 The displacement of a particle at time t is x, measured from a fixed point and $dx/dt = a(c^2 - x^2)$, where a and c are positive constants and $x = 0$ when $t = 0$. Prove that

$$x = c\,\frac{(e^{2act} - 1)}{(e^{2act} + 1)}$$

If $x = 3$ when $t = 1$ and $x = \frac{75}{17}$ when $t = 2$, prove that $c = 5$ and find the value of a. [JMB]

30 A mass is moving horizontally against a resistance which is proportional to its speed. If at any time t its speed is v and x is the distance moved, its equation of motion is known to be

$$P(V^2 - v^2) = 7V^2v^2 \frac{dv}{dx}$$

P, V being constant. Find an expression for x in terms of v and from this expression show that the distance covered by the mass while its speed increases from $\frac{1}{3}V$ to $\frac{3}{4}V$ is $\dfrac{7V^3}{12P}(6\log_e \frac{7}{2} - 5)$.

31 Given that

$$(1 + \sin^2 x)\frac{dy}{dx} = e^{-2y}\sin 2x$$

and $y = 1$ when $x = 0$, find the value of y when $x = \pi/2$.

32 Find the solution of the differential equation

$$\frac{dy}{dx} = \frac{x(1 - y^2)}{y(1 + x^2)}$$

for which $y = \frac{1}{2}$ when $x = 1$. [O]

33 (a) Given that $dy/dx = 9y^2 - 4$ and that $y = 1$ when $x = 2$, find an equation expressing x in terms of y.
(b) In a certain country the price p of a particular commodity increases with the time t at a rate equal to kp, where k is a positive constant. Write down a differential equation expressing this information. Show that if $p = 1$ when $t = 0$ and $p = \alpha$ when $t = 1$, then at time t, $p = \alpha^t$. [C]

34 Solve the differential equation

$$(x+2)\frac{dy}{dx} + (x+1)y = 0$$

(where $x > -2$), given that $y = 2$ when $x = 0$. Show that the maximum value of y is e, and sketch the graph of y against x for $x > -2$. Draw on your graph the tangent to the curve at its point of inflexion. Find the equation of this tangent. [LU]

35 By solving the differential equation

$$(1 - e^x)\tan y \frac{dy}{dx} = e^x$$

find an expression for y in terms of x and a constant. [JMB]

17
Introduction to co-ordinate geometry

17.1 Co-ordinates

Cartesian co-ordinates

When drawing graphs it is necessary to have two fixed reference lines. These are known as the axes, Ox and Oy, and are normally at right angles to each other. The axes enable us to locate any point P in a plane by means of its perpendicular distances from Oy and Ox.

Referring to *Figure 17.1*, the lines PL and PM, perpendicular to the axes, define the position of P. The lengths PM, PL are known as the cartesian co-ordinates of P. The length PM (x) is known as the abscissa and the length PL (y) is known as the ordinate. The pair are written in order as (x, y). We note that if P is the point (x, y) then it is also true that OL $= x$, and OM $=$ y.

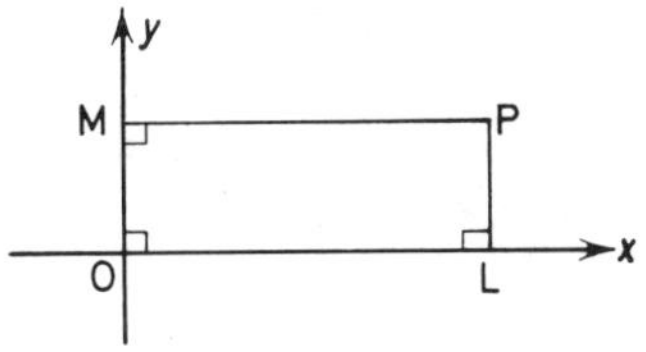

Figure 17.1

The usual sign convention is used. For points to the right of Oy the abscissa is positive, and for points to the left of Oy the abscissa is negative. For points above Ox the ordinate is positive, and for points below Ox the ordinate is negative. Thus the points A(4, 3), B(-2, 5), C(-4, -3), D(1, -3) are as shown in *Figure 17.2*.

Polar co-ordinates

The position of a point P in a plane can be described by other methods. Consider a fixed line Ox, O, the origin, being a fixed point on it. Referring to *Figure 17.3*, we see that the position of P is known if the angle POx and the distance OP are given. The angle POx (θ in *Figure 17.3*) is called the vectorial angle and is considered positive when measured in an anticlockwise direction. The distance OP (r in *Figure 17.3*) is called the radius vector and

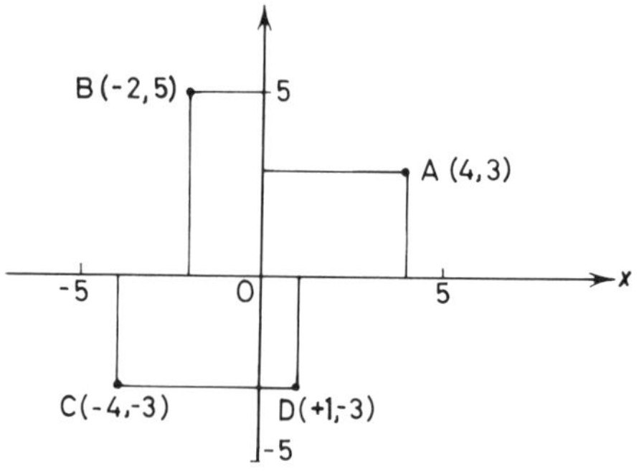

Figure 17.2

is considered positive when measured from O along the line bounding the vectorial angle and negative in the opposite direction. Then (r, θ) are known as the polar co-ordinates of P. In *Figure 17.3*, if OP is produced backwards to P′ so that OP = OP′, then P′ is the point $(r, \theta + \pi)$. Note that its polar co-ordinates could be given as $(-r, \theta)$, $(r, \theta + 3\pi)$ or in many other forms. To avoid confusion, it is usual to take r as positive and the angle θ between $-\pi$ and $+\pi$, thus P′ is the point $[r, -(\pi - \theta)]$.

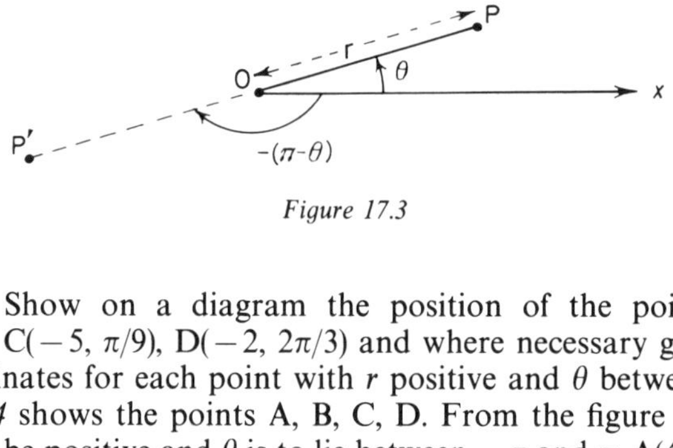

Figure 17.3

Example 1 Show on a diagram the position of the points A(4, $\pi/6$), B(3, $-5\pi/4$), C(-5, $\pi/9$), D(-2, $2\pi/3$) and where necessary give alternative polar co-ordinates for each point with r positive and θ between $-\pi$ and π.

Figure 17.4 shows the points A, B, C, D. From the figure it can be seen that, if r is to be positive and θ is to lie between $-\pi$ and π, A(4, $\pi/6$) is unaltered but B(3, $-5\pi/4$) becomes (3, $3\pi/4$), C(-5, $\pi/9$) becomes (5, $-8\pi/9$), D(-2, $2\pi/3$) becomes (2, $-\pi/3$).

The transformation from polar co-ordinates to cartesian co-ordinates or vice versa

Consider *Figure 17.5*, where the point P has cartesian co-ordinates (x, y) and polar co-ordinates (r, θ). Then

$$x = \text{ON} = \text{OP}\cos\theta = r\cos\theta \qquad (17.1)$$

$$y = \text{PN} = \text{OP}\sin\theta = r\sin\theta \qquad (17.2)$$

Example 2 Find the cartesian co-ordinates of the points P(3, $\pi/6$) and Q(5, $3\pi/4$).

P is the point $[3\cos(\pi/6), 3\sin(\pi/6)]$, i.e. (2·598, 1·5).

Q is the point $[5\cos(3\pi/4), 5\sin(3\pi/4)]$, i.e. ($-3$·536, -3·536).

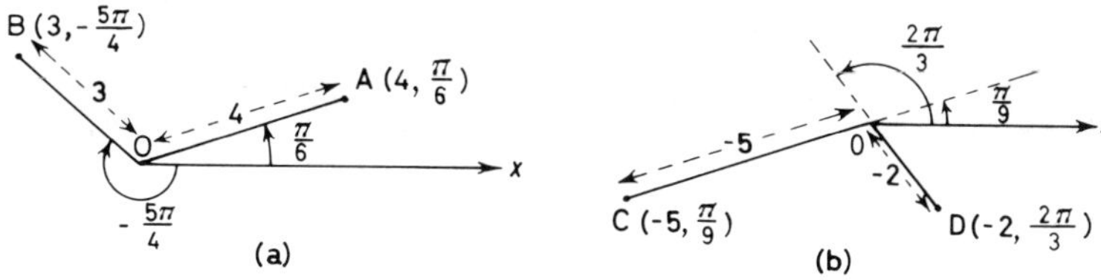

Figure 17.4

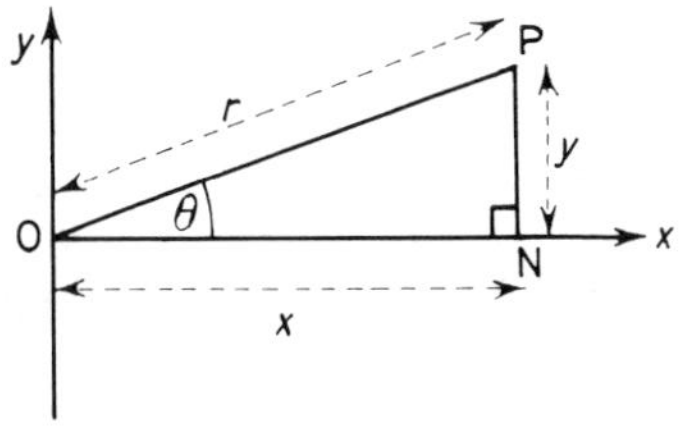

Figure 17.5

From (17.1) and (17.2), we have by squaring and adding (or direct from *Figure 17.5*)

$$r^2 = x^2 + y^2 \tag{17.3}$$

and by division

$$\tan\theta = \frac{y}{x} \tag{17.4}$$

Example 3 Find the polar co-ordinates of the points A(−4, 4), B(−3, −3).
For A,

$$r^2 = (-4)^2 + (4)^2 \quad \text{and} \quad \theta = \arctan\left(\frac{-4}{4}\right)$$

and for B,

$$r^2 = (-3)^2 + (-3)^2 \quad \text{and} \quad \theta = \arctan\left(\frac{-3}{-3}\right)$$

It is best to draw a figure to find which value of θ is required and from

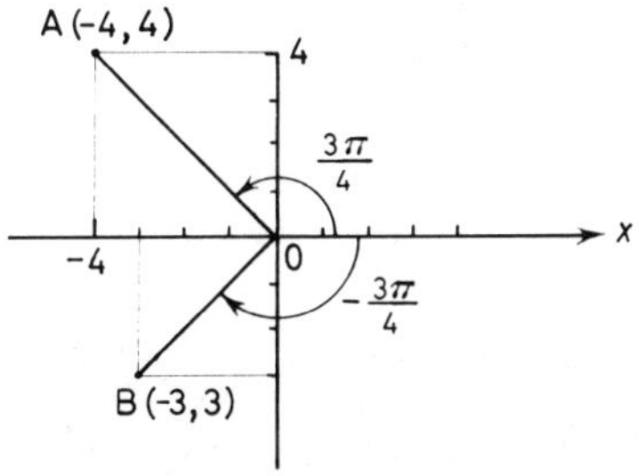

Figure 17.6

Figure 17.6 it can be seen that: A is the point $(4\sqrt{2}, 3\pi/4)$ and B is $(3\sqrt{2}, -3\pi/4)$.

Exercises 17a

1 Indicate on a diagram the positions of the points whose cartesian co-ordinates are: A(3, 4), B(−5, 2), C(0, 6), D(6, 0), E(−1, −2), F(6, −5), G(−3, 0), H(0, −3).

2 Indicate on a diagram the positions of the points whose polar co-ordinates are $A(3, \pi/4)$, $B(4, 2\pi/3)$, $C(5, -\pi/4)$, $D(3, \pi)$, $E(5, -\pi/2)$, $F(4, -5\pi/6)$, G(2, 0).

3 By means of a diagram, rewrite the polar co-ordinates of the points $A(-2, 3\pi/4)$, $B(5, 17\pi/9)$, $C(3, 4\pi)$, $D(-2, -5\pi/4)$, $E(-6, 10\pi/9)$, $F(-4, -5\pi/4)$, $G(-6, -13\pi/6)$ with the radii vectors all positive and the vectorial angles between $-\pi$ and π.

4 What are the cartesian co-ordinates of the points whose polar co-ordinates are $(3, \pi/2)$, $(4, -\pi/3)$, $(5, \pi)$, $(2, -5\pi/6)$, $(3, -\pi/2)$.

5 Find the polar co-ordinates of the points whose cartesian co-ordinates are (2, 2), (−3, −4), (0, 5), (−12, 5), (3, 0), (6, −3).

6 Find which of the following points coincide: A(3, 3), $B(-6, \pi/3)$, $C(3\sqrt{2}, 5\pi/4)$, D(3, −3), E(−3, −5·196), $F(3\sqrt{2}, -7\pi/4)$, $G(3\sqrt{2}, -5\pi/4)$, H(−3, −3), $J(-3\sqrt{2}, 3\pi/4)$.

7 The six points A, B, C, D, E, F are equally spaced on the circumference of a circle radius 2, centre the origin. If A is the point $(\sqrt{2}, \sqrt{2})$ write down the *polar* co-ordinates of the six points.

8 On which line does the point P lie if its cartesian co-ordinates are (A, B) and its polar co-ordinates are (A, B)?

9 A point P is such that its x and y co-ordinates are equal both in magnitude and sign. Plot on a diagram several possible positions of P and deduce on which line P must lie. If P is such that its x and y co-ordinates are equal in magnitude but opposite in sign on which line must P lie?

10 ABCD is a square, AC is a diagonal. If the co-ordinates of A, C are (−5, 8), (7, −4) find the co-ordinates of B and D.

17.2 The distance between two given points in terms of their cartesian co-ordinates

Let $P(x_1, y_1)$ and $Q(x_2, y_2)$ be the two given points. Draw PA, QB parallel to Oy, and QL parallel to Ox. The angle PLQ (see *Figure 17.7*) is a right angle. Then

$$QL = BA = OA - OB = x_1 - x_2$$
$$PL = PA - AL = PA - QB = y_1 - y_2$$

and since PLQ is a right angle

$$PQ = \sqrt{(PL^2 + LQ^2)}$$
$$PQ = \sqrt{[(x_1 - x_2)^2 + (y_1 - y_2)^2]} \qquad (17.5)$$

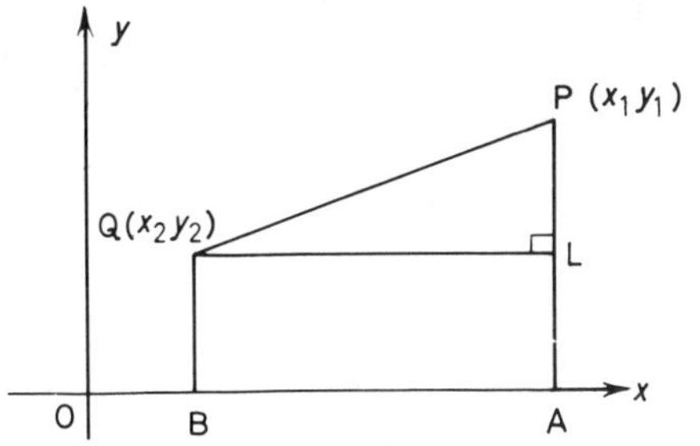

Figure 17.7

In the above case, we have assumed that all the co-ordinates are positive but, if due regard is paid to the usual sign convention, the formula is true in all cases.

Example 1 Show that the points A(5, −6), B(−3, 0), C(−1, 2) form an isosceles triangle.

Substituting in the formula (17.5), we have

$$\text{BC} = \sqrt{\{[(-3)-(-1)]^2 + [0-2]^2\}} = \sqrt{8} = 2\sqrt{2}$$
$$\text{AB} = \sqrt{\{[5-(-3)]^2 + [(-6)-0]^2\}} = \sqrt{100} = 10$$
$$\text{AC} = \sqrt{\{[5-(-1)]^2 + [(-6)-2]^2\}} = \sqrt{100} = 10$$

Since AB = AC, the triangle ABC is isosceles.

Exercises 17b

1 Show that the points A(6, 2), G(3, 6), C(−1, 3) and D(2, −1) are vertices of a square.

2 If P is the point (3, 4) and the reflections of P in Ox and Oy are A and B respectively, find the distance AB.

3 Show that the points A($\frac{3}{5}$, $\frac{8}{5}$), B($\frac{12}{5}$, $\frac{2}{5}$), C($\frac{7}{5}$, $-\frac{3}{5}$), D($-\frac{2}{5}$, $\frac{3}{5}$) are the vertices of a parallelogram.

4 Find the lengths of the sides of the triangle ABC where A, B, C are the points (0, 4), (4, 10) and (7, 8) respectively. Hence show that (a) ABC is right angled. [Hint: use Pythagoras.] (b) AB = 2BC.

5 The points A(x, 1) and B(−6, −5) are equidistant from the point C(3, −2). Find two possible values for x.

6 Find the distances between the following pairs of points:
(i) (a, b), $(-a, -b)$, (ii) $(2a, 2b)$, $(0, 0)$, (iii) $(3a, 3b)$, (a, b), (iv) $(a+b, a-b)$, $(b-a, a+b)$.

7 Show that the points A(5, 4), B(8, 1), C(6, 3) lie on a straight line. [Hint: show that AC + CB = AB.]

8 The points P(x, y), A(4, 3) and B(−1, −3) are such that PA = PB. By equating the expressions for PA^2 and PB^2, show that $10x + 12y - 15 = 0$.

9 Show, from first principles, that the distance between the two points whose polar co-ordinates are (r_1, θ_1) and (r_2, θ_2) is given by the expression $\sqrt{[r_1{}^2 + r_2{}^2 - 2r_1r_2\cos(\theta_1 - \theta_2)]}$.

10 Find the co-ordinates of the point P which is equidistant from the three points A(1, −1), B(9, 7), C(1, 7).

17.3 The co-ordinates of the points which divide the line joining two given points internally and externally in a given ratio

Internal division In *Figure 17.8*, R divides PQ *internally* in the ratio $k{:}l$.

External division In *Figure 17.9*, R divides PQ *externally* in the ratio $k{:}l$.

In *both* cases let P, Q and R be the points (x_1, y_1), (x_2, x_2) and (X, Y) respectively. Then

$$\mathrm{AC}:\mathrm{CB} = \mathrm{PM}:\mathrm{MN} = \mathrm{PR}:\mathrm{RQ} = k:l$$

hence

$$\frac{\mathrm{AC}}{\mathrm{CB}} = \frac{k}{l}$$
$$l\mathrm{AC} - k\mathrm{CB} = 0$$

Internal division

$\mathrm{AC} = X - x_1 \qquad \mathrm{CB} = x_2 - X$

hence

$$l(X - x_1) - k(x_2 - X) = 0$$

whence

$$X = \frac{lx_1 + kx_2}{l + k} \qquad (17.6)$$

similarly

$$Y = \frac{ly_1 + ky_2}{l + k} \qquad (17.7)$$

External division

$\mathrm{AC} = X - x_1 \qquad \mathrm{CB} = X - x_2$

hence

$$l(X - x_1) - k(X - x_2) = 0$$

whence

$$X = \frac{lx_1 - kx_2}{l - k} \qquad (17.8)$$

similarly

$$Y = \frac{ly_1 - ky_2}{l - k} \qquad (17.9)$$

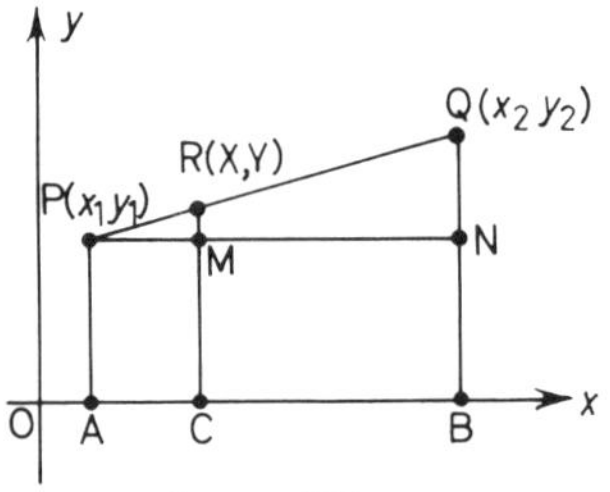

Figure 17.8

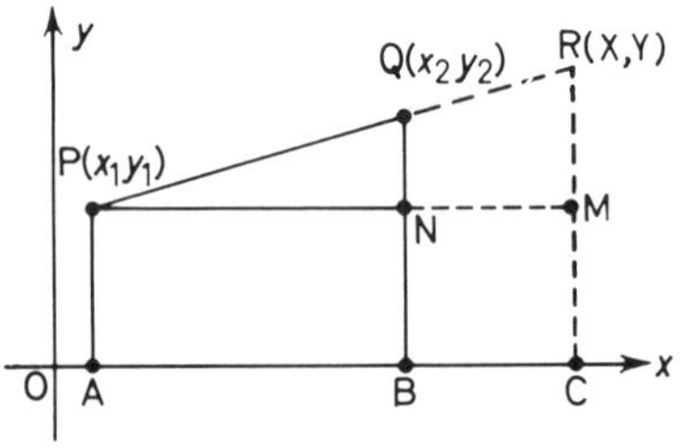

Figure 17.9

It is useful to remember that the formulae for external division are obtainable from those for internal division by changing the sign of either l or k.

Example 1 The points A and B divide the line joining P(3, 2) and Q(7, 9) internally and externally in the ratio 5:4. Find the co-ordinates of A and B.

A is the internal divisor, therefore using (17.6) and (17.7)

$$X = \frac{4 \times 3 + 5 \times 7}{4 + 5} = \frac{47}{9}$$

$$Y = \frac{4 \times 2 + 5 \times 9}{4 + 5} = \frac{53}{9} \qquad \text{A is the point } (\tfrac{47}{9}, \tfrac{53}{9})$$

B is the external divisor, therefore using (17.8) and (17.9)

$$X = \frac{4 \times 3 - 5 \times 7}{4 - 5} = 23$$

$$Y = \frac{4 \times 2 - 5 \times 9}{4 - 5} = 37 \qquad \text{B is the point } (23, 37)$$

Example 2 A, B, C are respectively the three points (−5, 2), (3, 4), (7, 5). Find (i) the ratio in which B divides AC and (ii) the ratio in which C divides AB.

(i) Let $k{:}l$ be the required ratio. Then using (17.6) and (17.7) and noting that (x_1, y_1) and (x_2, y_2) are the points (−5, 2) and (7, 5) respectively, we have

$$\frac{7k + (-5)l}{k + l} = 3 \quad \text{and} \quad \frac{5k + 2l}{k + l} = 4$$

From the first of these two equations

$$7k - 5l = 3k + 3l$$
$$4k = 8l$$

hence

$$\frac{k}{l} = \frac{2}{1}$$

As a check, substitute $k = 2$, $l = 1$ in the second equation

$$\text{LHS} = \frac{5 \times 2 + 2 \times 1}{2 + 1} = \frac{12}{3} = 4$$

(ii) Again let $k{:}l$ be the required ratio and again use (17.6) and (17.7) but this time (x_1, y_1) and (x_2, y_2) are the points (−5, 2) and (3, 4) respectively, hence

$$\frac{3k + (-5)l}{k + l} = 7 \qquad \frac{4k + 2l}{k + l} = 5$$

From the first of these two equations

$$3k - 5l = 7k + 7l$$
$$-4k = 12l$$
$$\frac{k}{l} = -\frac{3}{1}$$

As a check, substitute $k = -3$, $l = 1$ in the second equation

$$\text{LHS} = \frac{4(-3) + 2 \times 1}{-3 + 1} = \frac{-10}{-2} = 5$$

Note that in part (ii) the formulae for the internal divisors were used but gave a negative value for the ratio, indicating that C is an external divisor of AB.

Exercises 17c

1 A and B are the points (3, 5) and (−5, −7) respectively. Find the co-ordinates of the points which divide AB internally and externally in the ratio 3:1.

2 Find the co-ordinates of the mid-point of the line joining the points A(5, 6), B(11, 2).

3 P and Q are the points dividing the line joining A(−3, −4), B(5, 12) internally and externally in the ratio 5:3. Find the co-ordinates of P and Q.

4 A, B, C are the points (5, −3), (−4, 9), (14, −15) respectively. Given that ABC is a straight line, find the ratios in which (a) B divides AC, (b) A divides BC and (c) C divides AB.

5 The line joining the points A(3, 4) and B(7, 6) meets the line joining C(1, 3) and D(11, 8) at the point P. Given P is the mid-point of AB, find its co-ordinates and hence find the ratio CP:PD.

6 The three points A(5, 6), B(−3, 2), C(−8, −5) form a triangle. Find the co-ordinates of the A′, the mid-point of BC. If G is a point on AA′ such that AG:GA′ = 2:1, find the co-ordinates of G.

7 The line joining A(a, b) and B(p, q) is divided into six equal parts by the points P_1, P_2, P_3, P_4, P_5. Find the co-ordinates of P_2 and P_5.

8 The two points A(4, 3) and B(8, −6) together with the origin O form a triangle OAB. Find the co-ordinates of the point P in which the external bisector of AOB meets AB. [Hint: find the lengths OA and OB; then the external bisector of an angle of a triangle divides the opposite side externally in the ratio of the sides containing the angle.] Deduce that the internal bisector of the angle AOB is the x-axis.

9 If A(x_1, y_1), B(x_2, y_2), C(x_3, y_3) are the vertices of the triangle ABC write down the co-ordinates of A′ the mid-point of BC. Hence find the co-ordinates ($\bar{x}$, $\bar{y}$) of G the point which divides AA′ internally in the ratio 2:1 (G is known as the centroid).

10 Find in what ratio the point ($4b - 2a$, $9c - a$) divides the line joining the points ($a + b$, $3c + 5a$) and ($5b - 3a$, $11c - 3a$).

17.4 The area of a triangle in terms of the co-ordinates of its vertices

Let the triangle be ABC, where A, B and C are the points (x_1, y_1), (x_2, y_2), (x_3, y_3) respectively

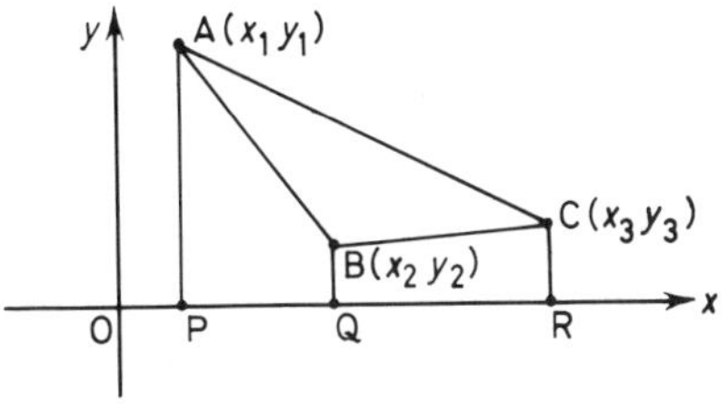

Figure 17.10

Draw AP, BQ, CR, perpendicular to Ox (see *Figure 17.10*). Then

$$\begin{aligned}\text{area } \triangle\text{ABC} &= \text{area of trapezium APRC} - \text{area of trapezium APQB}\\ &\quad - \text{area of trapezium BQRC}\\ &= \tfrac{1}{2}(\text{AP} + \text{CR}).\text{PR} - \tfrac{1}{2}(\text{AP} + \text{BQ}).\text{PQ}\\ &\quad - \tfrac{1}{2}(\text{BQ} + \text{CR}).\text{QR}\\ &= \tfrac{1}{2}(y_1 + y_3)(x_3 - x_1) - \tfrac{1}{2}(y_1 + y_2)(x_2 - x_1)\\ &\quad - \tfrac{1}{2}(y_2 + y_3)(x_3 - x_2)\end{aligned}$$

which when simplified gives

$$\text{area } \triangle\text{ABC} = \tfrac{1}{2}(x_1y_2 - x_2y_1 + x_2y_3 - x_3y_2 + x_3y_1 - x_1y_3) \qquad (17.10)$$

The numerical value for the area is independent of the order in which the vertices are taken. However, if the order is such that, on going round the triangle, the area is always on the left-hand side the area will be positive.

Condition for three points to be collinear

If the three points A(x_1, y_1), B(x_2, y_2) and C(x_3, y_3) are collinear, then the area of the triangle ABC is zero. Hence from (17.10), the condition for the three points to be collinear is

$$x_1y_2 - x_2y_1 + x_2y_3 - x_3y_2 + x_3y_1 - x_1y_3 = 0 \qquad (17.11)$$

Example 1 Given the four points A(3, 4), B(9, 7), C(7, 6) and D(5, −3), show that A, B and C are collinear and find the area of the triangle ABD.

Consider the area of ABC by substituting the co-ordinates of A, B, C in (17.10):

$$\begin{aligned}\text{area } \triangle\text{ABC} &= \tfrac{1}{2}(3 \times 7 - 4 \times 9 + 9 \times 6 - 7 \times 7 + 7 \times 4 - 6 \times 3)\\ &= \tfrac{1}{2}(21 - 36 + 54 - 49 + 28 - 18) = 0\end{aligned}$$

Hence A, B C are collinear.

The area of △ABD, using the co-ordinates of A, B, D in (17.10), is given by

$$\begin{aligned}\text{area } \triangle\text{ABD} &= \tfrac{1}{2}[3 \times 7 - 4 \times 9 + 9 \times (-3) - 7 \times 5 + 5 \times 4 - (-3) \times 3]\\ &= \tfrac{1}{2}(21 - 36 - 27 - 35 + 20 + 9)\\ &= 24 \text{ square units} \quad \text{(ignoring the negative sign)}\end{aligned}$$

Exercises 17d

1 Find the area of the triangle ABC where A, B and C are the points (5, 6), (3, 2), (8, −1).

2 Show that the points A(1, 5), B(−3, 9) and C(−2, 8) are collinear.

3 Find the areas of the triangles whose vertices are

(i) (3, 4), (5, 6), (−2, 0)

(ii) (0, 0), (1, −2), (3, −1)

(iii) (2, −7), (7, 3), (9, 7)

(iv) (4, 7), (0, 2), (−3, 0)

4 A, B, C are the points (0, 4), (4, 10), (7, 8) respectively. Using Pythagoras' theorem, prove that angle ABC is a right angle. Find the area of the triangle ABC by means of the formula (17.10) and verify your result by using the formula for the area $\frac{1}{2}$AB . BC.

5 Show that the four points A(−7, 5), B(1, 1), C(5, −1), D(13, −5) all lie on a straight line.

6 If the points A(5, 6), P(x, y) and B(2, 3) are collinear, show that $x - y + 1 = 0$.

7 The points A(3, 4), B(5, 3), C(−1, −1) and D(−3, 0) form a quadrilateral. Show that the mid-point P of AC lies on the line BD. Show also that the area of the triangle PAB is equal to the area of the triangle PCD.

8 Find the area of the quadrilateral ABCD, where A, B, C and D are the points (1, 1), (5, 4), (4, −1) and (−3, −12) respectively.

9 The four points A(0, 0), B(5, 1), C(−4, 4) and D(−1, −5) form a quadrilateral. Find the areas of the triangles ABC and ACD and hence find the area of the quadrilateral by adding the two results. Draw a figure and explain why the sum of the areas of triangles ABD and CBD do not equal the area of the quadrilateral.

10 Show that the condition $\dfrac{y_1 - y_2}{x_1 - x_2} = \dfrac{y_2 - y_3}{x_2 - x_3}$ is equivalent to the condition (17.11) for the three points (x_1, y_1), (x_2, y_2), (x_3, y_3) to be collinear.

17.5 Loci

If a curve can be defined by a geometrical property common to all points on it, then there will be an algebraic relation which is satisfied by the co-ordinates of all points on the curve. Such an algebraic relation is called the equation of the curve. Conversely, all points whose co-ordinates satisfy a given algebraic relation are on a curve known as the locus of the given equation.

Example 1 A given circle has its centre C at the point (3, 4) and its radius is 5 units. Find its equation.

In order to find the equation of the circle, we have to find the algebraic relation satisfied by all points on the curve.

Let P(x, y) be any point on the curve, then PC is a radius of the circle. Hence

$$PC = 5$$

thus

$$\sqrt{[(x-3)^2+(y-4)^2]} = 5$$

or

$$(x-3)^2+(y-4)^2 = 25$$

which is the required equation.

Example 2 A point C moves so that its distances from two fixed points A(5, 3) and B(7, 4) are always equal. Find the equation of the locus of C.

Let C be the point (x, y), then CA = CB, therefore

$$CA^2 = CB^2$$

that is

$$(x-5)^2+(y-3)^2 = (x-7)^2+(y-4)^2 \quad \text{(see (17.5))}$$

$$x^2-10x+25+y^2-6y+9 = x^2-14x+49+y^2-8y+16$$

thus $4x+2y = 31$, which is the required equation.

Example 3 A circle of radius 6 units passes through O, the origin of co-ordinates, and has the y-axis as a diameter. Find its polar equation.

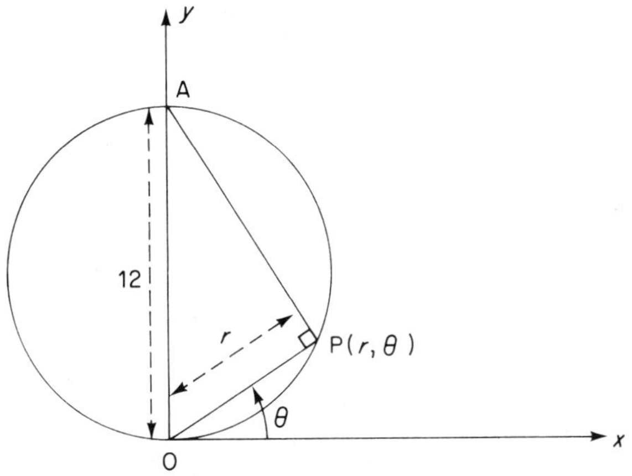

Figure 17.11

Since the polar equation is required, let any point P on the circle have polar co-ordinates (r, θ) (see *Figure 17.11*). Since OA is a diameter, angle OPA is a right angle, hence

$$OP/OA = \cos POA$$

$$OP = OA \cos POA$$

$$= 12 \cos(90° - \theta)$$

$$r = 12 \sin \theta$$

which is the required equation.

The equation of the curve need not be a direct relation between the co-

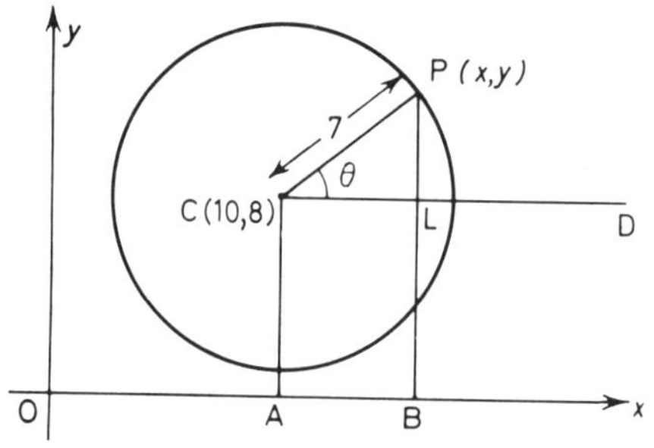

Figure 17.12

ordinates (r, θ) or (x, y) of any point on the curve. The co-ordinates can be obtained in terms of a third variable known as a parameter.

Example 4 A circle has its centre C at the point (10, 8) and its radius is 7 units. Find its equation.

Referring to *Figure 17.12*, CLD is parallel to Ox. P(x, y) is any point on the circle and PCD $= \theta$. As θ varies, P describes the circle.

$$\begin{aligned} y = \text{PB} &= \text{PL} + \text{LB} = \text{PL} + \text{CA} \\ &= \text{PC} \sin \theta + 8 \\ &= 7 \sin \theta + 8 \\ x = \text{OB} &= \text{OA} + \text{AB} = \text{OA} + \text{CL} \\ &= 10 + \text{PC} \cos \theta \\ &= 10 + 7 \cos \theta \end{aligned}$$

Hence

$$x = 7 \cos \theta + 10 \qquad y = 7 \sin \theta + 8 \qquad \text{(i)}$$

are the parametric equations of the circle and for any value of θ, the equations (i) give the x and y co-ordinates of a point on the circle.

The x, y equation can be obtained by eliminating the parameter θ. From (i)

$$x - 10 = 7 \cos \theta \qquad y - 8 = 7 \sin \theta$$

Squaring and adding gives

$$(x - 10)^2 + (y - 8)^2 = 49 \quad (\cos^2 \theta + \sin^2 \theta = 1)$$

which is of the same form as the circle in Example 1.

Exercises 17e

1 Find the equation of the circle centre (3, −4) radius 7.
2 A point P moves so that PA = 2PB, where A, B are the fixed points (−2, 1), (5, 6) respectively. Find the locus of P.
3 A(3, 2) and B(6, 4) are two fixed points and the point C moves so that the angle ACB is always a right angle. Using Pythagoras' theorem find the locus of C.

4 A circle of radius 4 units passes through O the origin and has the y-axis as a diameter. Find its polar equation.

5 Find the equation of the circle on AB as diameter, where A and B are the points $(-3, -4)$ and $(7, 20)$.

6 Find the equations of the curves whose parametric equations are (i) $x = ct$, $y = c/t$, (ii) $x = a\cos\theta$, $y = b\sin\theta$.

7 A point P moves so that its perpendicular distance from the y-axis is always equal to its distance from the point (2, 3). Find the equation of the locus of P.

8 A point P moves so that its distance from the axis of x is half its distance from the origin. Find the locus of P.

9 A point P moves along a line parallel to the axis of x at a distance 6 units from it. Find the polar equation of the locus of P.

10 A(0, 2) and B(0, -2) are two fixed points. The point P moves so that $PA + PB = 8$. Find the equation of the locus of P. [Hint: use the result (17.5) but before simplifying, rewrite the given condition in the form $PA = 8 - PB$.]

17.6 The points of intersection of two loci

In order to find the points of intersection of two loci, we note that, where they intersect, there is a point common to both curves. The co-ordinates of this point will satisfy the equations of both curves simultaneously. Thus if the equations are solved simultaneously, the solutions will be the co-ordinates of the common points.

Example 1 Find the point of intersection of the two loci $3x - y - 5 = 0$ and $12x + y - 25 = 0$.

$$3x - y - 5 = 0$$
$$12x + y - 25 = 0$$

Adding these equations eliminates y and gives

$$15x - 30 = 0$$

Thus $x = 2$, and by substituting in the first equation $y = 1$. The required point is (2, 1).

Example 2 Find the points of intersection of the circle

$$(x - 3)^2 + (y - 4)^2 = 25$$

(see Section 17.5, Example 1) and the locus $y + x - 12 = 0$.

The equation of the circle simplifies to

$$x^2 + y^2 - 6x - 8y = 0 \qquad \text{(i)}$$

and from the second equation

$$y = 12 - x \qquad \text{(ii)}$$

Substituting (ii) in (i)

$$x^2 + (12 - x)^2 - 6x - 8(12 - x) = 0$$
$$2x^2 - 22x + 48 = 0$$
$$2(x - 3)(x - 8) = 0$$
$$x = 3 \quad \text{or} \quad 8$$

and substituting in (ii) the two points are (3, 9) and (8, 4).

Example 3 Find the points of intersection of the two circles $r = 12\cos\theta$ and $r = 6$.

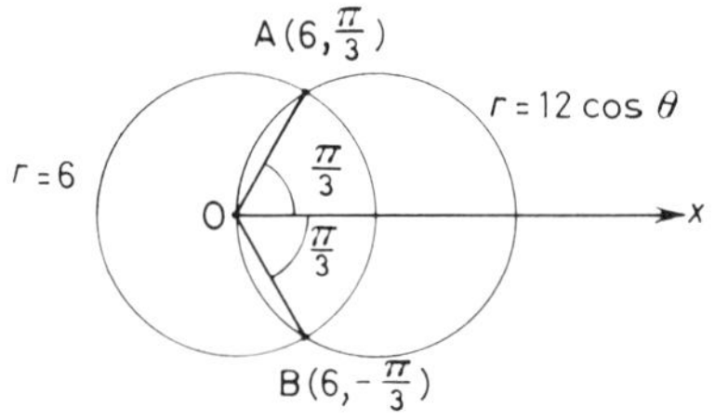

Figure 17.13

Since $r = 12\cos\theta$ and $r = 6$, at the points of intersection $6 = 12\cos\theta$, thus $\cos\theta = \frac{1}{2}$. Therefore

$$\theta = \pi/3 \quad \text{or} \quad -\pi/3$$

Hence the required points are A(6, $\pi/3$) and B(6, $-\pi/3$) (see *Figure 17.13*).

Exercises 17f

1 Find the points of intersection of the two loci $3x + 2y - 1 = 0$, $2x - 3y + 21 = 0$.
2 Find the points of intersection of the circle $(x - 5)^2 + (y - 6)^2 = 49$ and the locus $y + x - 18 = 0$.
3 Find the points of intersection of the two circles $x^2 + (y + 1)^2 = 2$ and $(x + 1)^2 + (y + 2)^2 = 4$.
4 Find the points of intersection of the circle $r = 10$ and the locus $r\cos\theta = 5\sqrt{3}$.
5 Show that the three loci $x + y + 1 = 0$, $3x + 2y + 6 = 0$, $2x + 5y - 7 = 0$ have a common point.
6 Find the point or points common to the two loci whose equations are $x^2 + y^2 + 6x + 8y = 0$ and $4x + 3y - 1 = 0$.
7 Show that the two circles $(x - 3)^2 + (y - 4)^2 = 25$ and $(x - 1)^2 + (y - \frac{5}{2})^2 = 56\frac{1}{4}$ touch each other.
8 Find the points common to the two loci whose equations are $r = 6\cos\theta$, $r = 6\sqrt{3}\sin\theta$.
9 Find the points common to the two loci whose equations are $y^2 - 6y - 4x + 1 = 0$, $2y - x - 11 = 0$.
10 Find the point of intersection of the loci $x - y = 3$, $x + 3y = 7$. Show that the locus whose equation is $x^2 - xy - 3x = 0$ passes through this point.

17.7 Change of origin

Let Ox, Oy be the original axes, O′ the new origin and (h, k) the co-ordinates of O′ referred to the original axes.

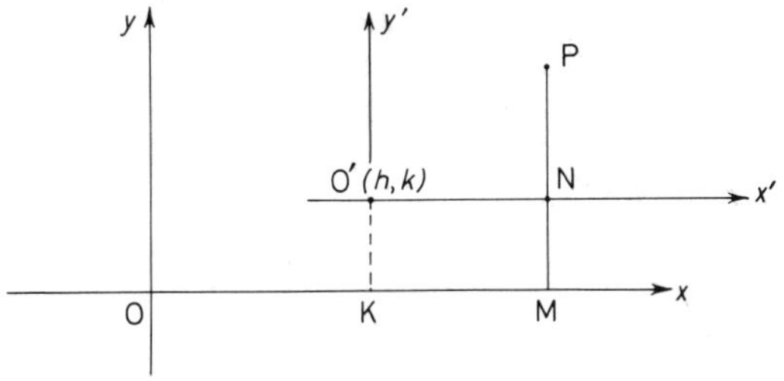

Figure 17.14

Through O′ draw O′x' and O′y' parallel to and in the same sense as Ox and Oy respectively (see *Figure 17.14*).

Suppose P is any point whose co-ordinates referred to the original axes are (x, y). We require to find its co-ordinates, say (X, Y), referred to the new axes.

Draw PM parallel to Oy, cutting Ox in M and O′x' in N. Then

$$\begin{aligned} x = \text{OM} &= \text{OK} + \text{KM} \quad \text{(see } \textit{Figure 17.14}) \\ &= \text{OK} + \text{O'N} \\ &= h + X \end{aligned}$$

and

$$\begin{aligned} y = \text{MP} &= \text{MN} + \text{NP} \\ &= \text{KO'} + \text{NP} \\ &= k + Y \end{aligned}$$

The old co-ordinates are given in terms of the new co-ordinates by the two equations.

$$x = h + X \qquad y = k + Y \tag{17.12}$$

If we use the equations (17.12) to substitute for x and y in a given equation, the equation of the curve referred to the new axes will be obtained.

Example 1 Find the equation of the straight line $3x + 2y + 1 = 0$ referred to axes through the point $(1, -2)$.

In this case, $h = 1$, $k = -2$. Therefore from (17.12)

$$x = 1 + X \qquad y = -2 + Y$$

Substituting in the given equation, we have

$$3x + 2y + 1 = 0$$

which becomes

$$3(1 + X) + 2(-2 + Y) + 1 = 0$$
$$3X + 2Y = 0$$

which is a straight line through the new origin.

Exercises 17g

Change the origin of co-ordinates in each of the following cases:

	Original equation	*New origin*
1	$3x - 2y + 4 = 0$	$(3, 2)$
2	$5x + y - 7 = 0$	$(-2, -5)$
3	$2y - 5x - 3 = 0$	$(1, -1)$
4	$x + 5y - 2 = 0$	$(-3, 1)$
5	$2x - 3y = 0$	$(5, 2)$
6	$y^2 = 4a(x - 1)$	$(1, 0)$
7	$\dfrac{(x-1)^2}{4} + \dfrac{(y-3)^2}{16} = 1$	$(1, 3)$
8	$(x - 3)^2 + (y - 3)^2 = 25$	$(3, 3)$
9	$\dfrac{(x+1)^2}{4} - \dfrac{(y-2)^2}{16} = 1$	$(-1, 2)$
10	$x^2 + y^2 - 8x + 6y = 0$	$(4, -3)$

17.8 Loci:polar co-ordinates

We have seen in Section 17.1 that we can describe the position of a point P by its cartesian co-ordinates (x, y) or its polar co-ordinates (r, θ). The two sets are related:

$$x = r\cos\theta \qquad y = r\sin\theta$$
$$r = \sqrt{(x^2 + y^2)} \qquad \tan\theta = y/x.$$

The cartesian equation of a curve expresses the relationship between the cartesian co-ordinates of the points which lie on the curve. Its polar equation expresses the relationship which must exist between the polar co-ordinates of the points on the curve.

If the equation of a locus is known in cartesian co-ordinates, it may be found in polar form and vice versa.

Example 1 Find the polar equation of the line $y = mx$.

$y = mx$ is a line through the origin with gradient m (see Section 18.2). If $y = mx$, then $y/x = m$, that is $\tan\theta = m$. Therefore

$$\theta = \arctan m = \text{a constant}$$

The line with polar equation $\theta = \alpha$ is a line through the origin with gradient $\tan\alpha$.

Example 2 Find the cartesian equation of the line $r = p\sec(\theta - \alpha)$ and give a geometrical interpretation of p and α.

$$r = p\sec(\theta - \alpha)$$

therefore

$$\begin{aligned} p &= r\cos(\theta - \alpha) \\ &= r\cos\theta\cos\alpha + r\sin\theta\sin\alpha \end{aligned}$$

Therefore, in cartesian form, the line has the equation

$$x\cos\alpha + y\sin\alpha = p$$

Thus p is the length of the perpendicular from the origin to the line, and α is the angle that this perpendicular makes with the x-axis. See *Figure 18.5* and equation (18.8).

Example 3 Find the polar equation of the circle $x^2 + y^2 = a^2$.

$x^2 + y^2 = a^2$ is the equation of a circle with centre the origin and radius a (see Section 19.1).

Its polar equation is $r^2 = a^2$, i.e. $r = a$ represents a circle of radius a and centre the origin.

Example 4 Find the polar equation of the circle centre $(a, 0)$ and radius a.

Its cartesian equation is $(x - a)^2 + y^2 = a^2$ (see Section 19.1), that is

$$x^2 + y^2 - 2ax = 0$$

Its polar equation is

$$\begin{aligned} r^2\cos^2\theta + r^2\sin^2\theta - 2ar\cos\theta &= 0 \\ r^2 - 2ar\cos\theta &= 0 \\ r &= 2a\cos\theta \end{aligned}$$

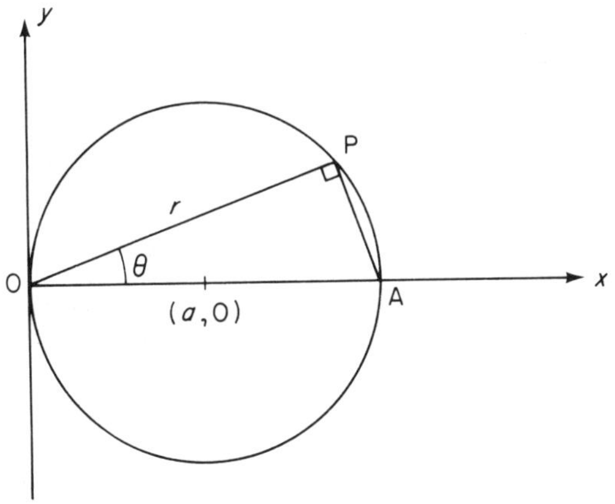

Figure 17.15

We could obtain this from first principles. In *Figure 17.15*, OA is a diameter of the circle and $P(r, \theta)$ is any point on it.

Thus, since $\angle OPA = 90°$, $OP/OA = \cos\theta$. Therefore

$$\frac{r}{2a} = \cos\theta$$

$$r = 2a\cos\theta \quad \text{as before}$$

Example 5 Sketch the curve $r = a\theta$ $(a > 0)$.

As θ increases, r increases. The curve winds round and round the origin. It is called an Archimedean spiral. It is sketched in *Figure 17.16*.

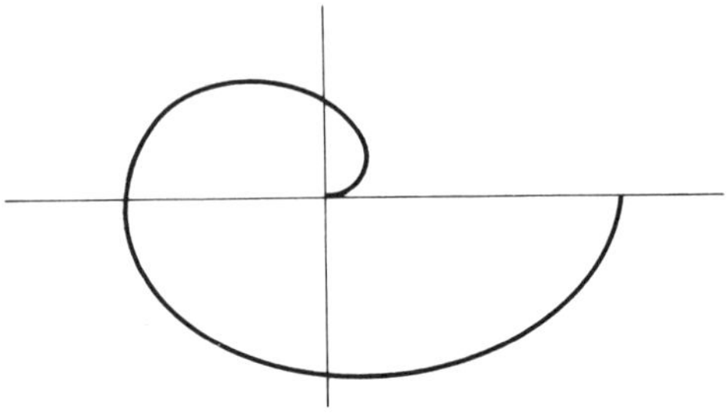

Figure 17.16

Example 6 Sketch the curve $r = a(1 + \cos\theta)$.

Since $\cos(-\theta) = \cos\theta$, the curve is symmetrical about the line $\theta = 0$. r is defined for all values of θ. It cannot exceed $2a$. We can easily obtain the following table of values

θ	0	$\pi/6$	$\pi/3$	$\pi/2$	$2\pi/3$	$5\pi/6$	π
$\cos\theta$	1	0·87	0·5	0	−0·5	−0·87	−1
r	$2a$	$1{\cdot}87a$	$1{\cdot}5a$	a	$0{\cdot}5a$	$0{\cdot}13a$	0

The curve is sketched in *Figure 17.17*. It is called a cardioid.

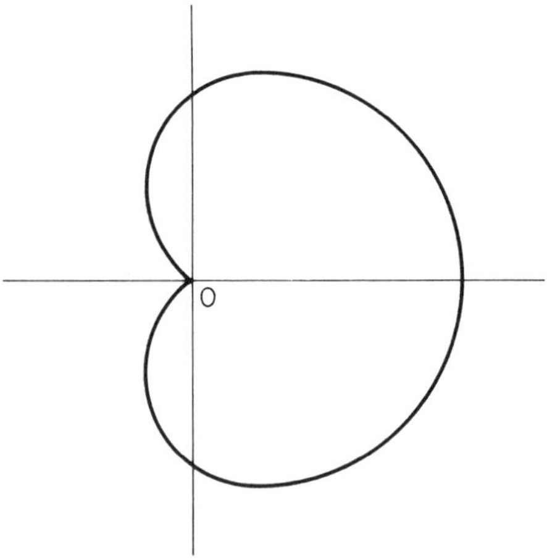

Figure 17.17

Exercises 17h

Find the polar equations of the following curves:

1 $x^2 + y^2 = 9$ **2** $x^2 + (y - 3)^2 = 9$ **3** $x^2 + xy = 1$

4 $xy = c^2$ **5** $(x^2 - y^2)^2 = c^2(x^2 + y^2)$

Find the cartesian equations of the following curves:

6 $\theta = \pi/4$ **7** $\tan\theta = \sqrt{3}$ **8** $r = 3$

9 $r\sin\theta = 5$ **10** $r(1 + \cos\theta) = 3$ **11** $r^2\cos 2\theta = a^2$

Sketch the curves

12 $r = 2$ **13** $\theta = \pi/6$ **14** $r = 4\sec\theta$

15 $r = 2 + \cos\theta$ **16** $r\sin\theta = r\cos\theta$ **17** $r = e^{a\theta}(\theta > 0)$

Find the points of intersection of the following curves:

18 $r\cos\theta = 1,\ \theta = \pi/4$ **19** $r = 4,\ \theta = \pi/3$

20 $r = 2a(1 + \cos\theta),\ 2r\cos\theta + a = 0$

Exercises 17

1 Find the cartesian co-ordinates of the points whose polar co-ordinates are $(6, -\pi/2)$, $(\sqrt{2}, 3\pi/4)$, $(2, \pi/6)$, $(8, \pi/2)$, $(\sqrt{2}, -3\pi/4)$, $(5, \pi)$.

2 Find the polar co-ordinates of the points whose cartesian co-ordinates are $(5, -5)$, $(-\sqrt{3}, -1)$, $(-3, 3)$, $(1, \sqrt{3})$, $(0, 2)$, $(3, 0)$. Draw a diagram of the points.

3 Given the cartesian co-ordinates of A, B, C, D are $(4, 4)$, $(-4, -4)$, $(-5, 5)$, $(3, -1)$ and the polar co-ordinates of E, F, G, H, J are $(2, \pi/6)$, $(-4\sqrt{2}, 5\pi/4)$, $(4\sqrt{2}, -5\pi/4)$, $(-5\sqrt{2}, -\pi/4)$, $(4\sqrt{2}, 11\pi/4)$, find which points coincide.

4 Find the lengths of the sides of the quadrilateral whose vertices are the points A(5, 3), B(6, 7), C(8, 1), D(5, −3). Also find its area.

5 Show that the four points A(6, 7), B(7, 10), C(0, −3), D(−1, −6) form a parallelogram.

6 Show that the three points A(7, 4), B(10, 2), C(6, −4) form a right-angled triangle. Find the co-ordinates of a point D such that ABCD is a rectangle.

7 Rewrite the following polar co-ordinates with the radii vectors all positive and the vectorial angles between $-\pi$ and π: $A(-5, \pi)$, $B(-3, -3\pi/2)$, $C(-5, \frac{7}{2}\pi)$, $D(3, 6\pi)$, $E(-5, 3\pi)$, $F(-1, \frac{7}{4}\pi)$, $G(3, \frac{9}{4}\pi)$.

8 Find the points P and Q which divide the line joining A(3, 2) and B(10, 16) internally and externally in the ratio 3:4.

9 Show that the four points A(3, 4), B(9, 13), C(11, 16) and D(15, 22) all lie on a line. Find the ratios in which B and D divide the line joining A and C.

10 Find the equation of the locus of a point which moves so that its distance from the x-axis is twice its distance from the point (2, −3).

11 The points A(3, 4), B($2a$, 5), C(6, a) form a triangle whose area is $19\frac{1}{2}$ square units. Find the two possible values of a.

12 Find the co-ordinates of the point which is equidistant from the three points A(3, 4), B(13, 6), C(3, 4).

13 The polar co-ordinates of the vertices of a triangle are given by the following table.

θ	$\pi/6$	$-\pi/3$	$2\pi/3$
r	12	16	9

Find the lengths of the sides of the triangle and its area.

14 Show that the following points A(5, 6), $B(-1, \frac{6}{5})$, C(−5, −2) are collinear and find the ratios (i) AB:BC, (ii) AC:CB, (iii) BA:AC.

15 Find the co-ordinates of the centroid and of the circumcentre of the triangle ABC, where A, B, C are the points $(-2, -3)$, $(8, 11)$ and $(-4, 9)$ respectively.

16 Find the area of the quadrilateral whose vertices are the points A(5, 4), B(8, 5), C(6, −2), D(−3, −1) respectively.

17 Find the equation of the locus of a point which moves so that the sum of the squares of its distances from the points (3, 0) and (−3, 0) is equal to 72 units.

18 Find the distance from the origin to the point on the x-axis which is equidistant from the points P(−2, 2) and Q(4, 4). [LU]

19 The straight lines $3x + by + 1 = 0$ and $ax + 6y + 1 = 0$ intersect at the point (5, 4). Find the values of a and b. If the first line meets the x-axis at A and the second meets the y-axis at B, find the length AB.

20 O is the origin of co-ordinates and B is the point (0, 6). Find the polar equation of the circle on OB as diameter.

21 The line $3y = ax + 9$ touches the curve $y^2 = 4x$. Find the value of a.

22 Show that the co-ordinates of the point common to the curve $y^2 = 4ax$ and the line $ty - x = at^2$ are $(at^2, 2at)$.

23 s and s' are two circles of radii 1 and 3 respectively and centres A(0, 0) and B(−1, 3) respectively. If s and s' meet at the points P and Q, show that $\angle$APB = AQB = 90°.

24 A point moves so that its distance from the axis of x is equal to its distance from the point (1, 1). Find the equation of its locus.

25 Given that P is the point (4, 7), write down the co-ordinates of the points which are (i) the reflection of P in the x-axis, (ii) the reflection of P in the line $y = x$, (iii) the reflection of P in the line $y = -x$.

26 Find the equation of the loci whose parametric equations are (i) $x = 4t^2$, $y = 16t$; (ii) $x = 4 + t$, $y = 6 - 1/t$.

27 Show that the point P with co-ordinates $[(1-k)x_1 + kx_2, (1-k)y_1 + ky_2]$ lies on the line joining $A(x_1, y_2)$ and $B(x_2, y_2)$ and that AP = kAB.

28 The points A, B, C have the co-ordinates (2, 3), (−11, 8) and (−4, −5) respectively. The point D is such that ABCD is a parallelogram having AC as diagonal. Find the co-ordinates of the mid-point of AC and deduce the co-ordinates of D.

29 A variable line meets the axes at A, B. O is the origin. If AB moves so that the area of $\triangle$AOB is constant, find the locus of the mid-point of AB.

30 A point P moves along the straight line which passes through the point A(5, 0) and makes an angle of 45° with the x-axis. Find the equation of the locus of P. [Hint: use the sine rule on the triangle OAP.]

18

The straight line

18.1 The equation of a straight line parallel to one of the co-ordinate axes

In *Figure 18.1*, let $P(x, y)$ be any point on the line. Since the ordinate PN is equal to OA for all positions of P then

$$y = b \tag{18.1}$$

is true for all points on the line and is the required equation.

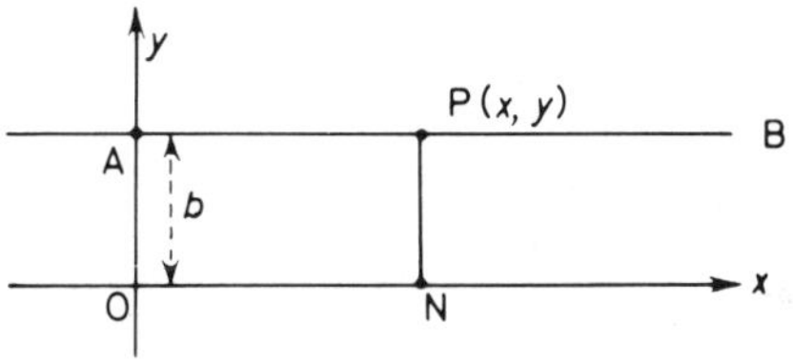

Figure 18.1

Similarly, $x = a$ is the equation of a line parallel to Oy and distance a from it. In particular, the axes Ox and Oy have the equations $y = 0$ and $x = 0$ respectively.

18.2 The equation of any straight line in terms of its slope and its intercept on the *y*-axis

Consider *Figure 18.2*. ABP is the straight line, $OB = c$, and angle $BAO = \theta$. $P(x, y)$ is any point on the line, PN is its ordinate and BM is parallel to Ox. Then

$$\angle PBM = \angle BAO = \theta$$

thus

$$PM = BM \tan\theta$$
$$y - c = x \tan\theta$$
$$y = x \tan\theta + c$$

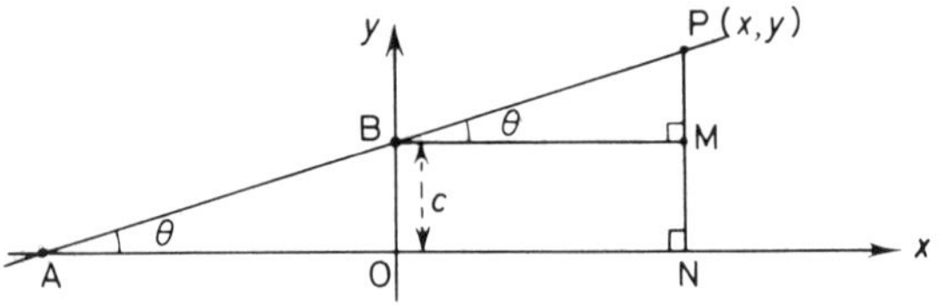

Figure 18.2

$\tan\theta$ is the slope of the line and is generally denoted by m. Hence the required equation is

$$y = mx + c \tag{18.2}$$

Example 1 Write down the equation of the line which makes an angle of 45° with Ox and cuts Oy at a distance of 3 units above the origin.

From (18.2) the required equation is

$$y = x \tan 45^\circ + 3$$

that is

$$y = x + 3 \quad \text{(see } \textit{Figure 18.3}\text{)}$$

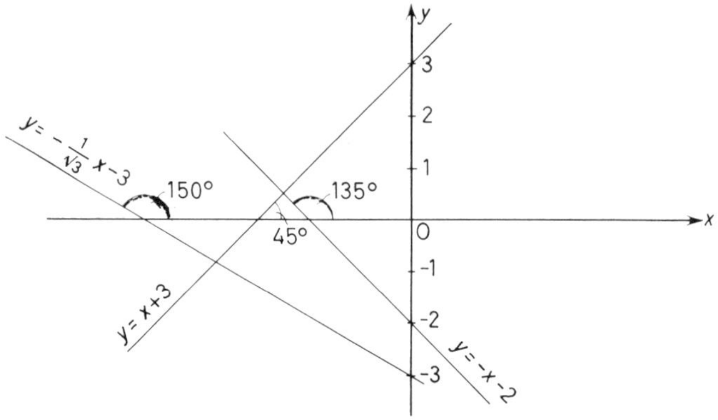

Figure 18.3

Example 2 Write down the equation of the line which makes an angle of 150° with Ox and an intercept of -3 units on Oy.

In this case $c = -3$ and $m = \tan 150^\circ = -1/\sqrt{3}$. Hence the required equation is

$$y = -\frac{1}{\sqrt{3}}x - 3 \quad \text{(see } \textit{Figure 18.3}\text{)}$$

Example 3 Sketch the line whose equation is $y = -x - 2$.

Comparing $y = -x - 2$ with $y = mx + c$, we have $\tan\theta = m = -1$ and $c = -2$; hence $\theta = 135^\circ$ and $c = -2$.

Thus the line makes an angle of 135° with the positive direction of Ox and cuts Oy at a distance of 2 units *below* the origin (see *Figure 18.3*).

It will be noticed that all the equations of a straight line have so far been of the first degree in x and y. We shall now prove that any equation of the first degree in x and y represents a straight line.

18.3 Any equation of the first degree in x and y represents a straight line

The most general form of an equation of the first degree is

$$Ax + By + C = 0 \tag{18.3}$$

In order to prove that this equation represents a straight line, it is sufficient to show that the area of the triangle formed by joining any three points on the locus is zero.

Let (x_1, y_1), (x_2, y_2), (x_3, y_3) be any three points on the locus. Since the points are on the locus of the equation (18.3), their co-ordinates must satisfy the equation; thus

$$Ax_1 + By_1 + C = 0$$
$$Ax_2 + By_2 + C = 0$$
$$Ax_3 + By_3 + C = 0$$

Subtracting the first of these equations from the second and third in turn, we obtain

$$A(x_2 - x_1) + B(y_2 - y_1) = 0$$
$$A(x_3 - x_1) + B(y_3 - y_1) = 0$$

By considering the value of the ratio A/B obtained from each of these equations, we have

$$-\frac{(y_2 - y_1)}{(x_2 - x_1)} = \frac{A}{B} = -\frac{(y_3 - y_1)}{(x_3 - x_1)}$$

thus

$$(y_2 - y_1)(x_3 - x_1) = (y_3 - y_1)(x_2 - x_1)$$

whence

$$x_1 y_2 - x_2 y_1 + x_2 y_3 - x_3 y_2 + x_3 y_1 - x_1 y_3 = 0$$

which (see (17.11)) proves that the area of the triangle formed by the three points is zero. Hence the locus is a straight line.

The equation $Ax + By + C = 0$ *appears* to involve three separate constants, but we can divide throughout the equation by any of A, B or C (which is not zero). Dividing by B we have

$$\frac{A}{B}x + y + \frac{C}{B} = 0$$

that is

$$y = -\frac{A}{B}x - \frac{C}{B}$$

from which it is clear that only the values of the *two* ratios A : B : C are required. Comparing this with

$$y = mx + c \quad \text{(see (18.2))}$$

we have

$$m = -\frac{A}{B} \qquad c = -\frac{C}{B} \tag{18.4}$$

Example 1 Find the slopes of the lines $3x + 12y - 3 = 0$, $5x - 2y - 4 = 0$ and the lengths of their intercepts on Oy.

Rewriting the two equations, we have

$$12y = -3x + 3 \quad \text{and} \quad 2y = 5x - 4$$

thus

$$y = -\tfrac{1}{4}x + \tfrac{1}{4} \quad \text{and} \quad y = \tfrac{5}{2}x - 2$$

On comparing these equations with $y = mx + c$, we see that their slopes are

$$\tan\theta_1 = -\tfrac{1}{4} \quad (\theta_1 = 165^\circ 58')$$
$$\tan\theta_2 = \tfrac{5}{2} \quad (\theta_2 = 68^\circ 12')$$

and their intercepts on Oy are

$$c_1 = \tfrac{1}{4} \qquad c_2 = -2$$

Exercises 18a

1 Write down the equations of the following lines:
(i) the line making an angle of 20° with Ox and an intercept of +5 units on Oy
(ii) the line making an angle of 150° with Oy and an intercept of −4 units on Oy
(iii) the line through the origin making an angle of 50° with Ox
(iv) the line through the origin making an angle of 20° with Oy.

2 Find the equations of
(i) the line through the origin parallel to the line $y = 5x + 2$
(ii) the line through the origin parallel to the line $3x + 2y + 4 = 0$
(iii) the line which makes an intercept of +4 units on Ox and whose slope is $\frac{3}{4}$
(iv) the line parallel to Ox and passing through the point of intersection of $2y - 3x + 4 = 0$ and the y-axis.

3 Find the slopes, and the lengths of the intercepts the following lines make on Oy: (i) $3y - 4x + 6 = 0$, (ii) $2y + 2x - 3 = 0$, (iii) $5x - 3y = 0$, (iv) $y - 6 = 0$.

4 Plot the two points A(0, 5) and B(3, 9) on a diagram. Hence show that the slope of the line AB is $\frac{4}{3}$. Write down the equation of AB. What is the slope of the line if A and B are the points (x_1, y_1) and (x_2, y_2) respectively?

5 Plot the two points A(0, 3) and B(−2, 7) on a diagram. Hence show that the slope of the line AB is −2. Write down the equation of AB. If A and B are the points (x_1, y_1) and (x_2, y_2), what is the slope of AB? Does this agree with the result in question 4?

18.4 Useful forms of the equation of a straight line

The equation of the straight line making intercepts a and b on Ox and Oy respectively

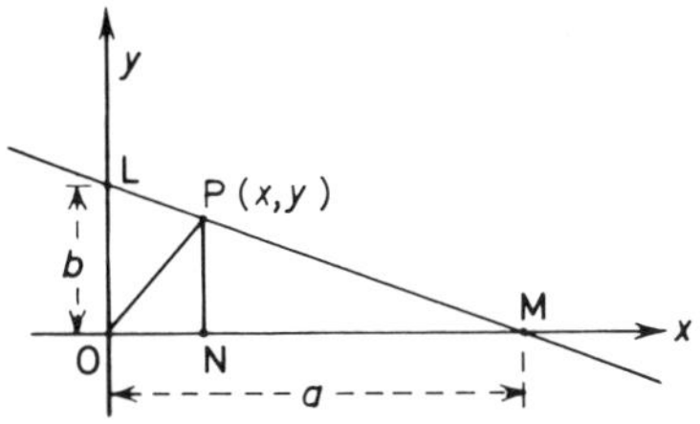

Figure 18.4

P(x, y) is any point on the line, PN is the ordinate (see *Figure 18.4*). Now

$$\triangle MPO + \triangle LPO = \triangle LOM$$

Hence

$$\tfrac{1}{2}ay + \tfrac{1}{2}bx = \tfrac{1}{2}ab$$

Divide throughout by $\tfrac{1}{2}ab$, then

$$\frac{x}{a} + \frac{y}{b} = 1 \qquad (18.5)$$

This is called the intercept form of the equation of the straight line.

Example 1 Write down the equations of the lines making intercepts a and b on Ox and Oy respectively, and rewrite the equations in the form $y = mx + c$, where (i) $a = 2$, $b = 3$; (ii) $a = -1$, $b = 4$; (iii) $a = 5$, $b = -2$; (iv) $a = -3$, $b = -4$.

(i) $x/2 + y/3 = 1$, thus $3x + 2y = 6$, that is

$$y = -\tfrac{3}{2}x + 3$$

(ii) $x/-1 + y/4 = 1$, thus $4x - y = -4$, that is

$$y = 4x + 4$$

(iii) $x/5 + y/-2 = 1$, thus $-2x + 5y = -10$, that is

$$y = \tfrac{2}{5}x - 2$$

(iv) $x/-3 + y/-4 = 1$, thus $-4x - 3y = 12$, that is

$$y = -\tfrac{4}{3}x - 4$$

The equation of a straight line with given slope m and passing through a given point P(x_1, y_1)

The equation of a line with slope m is by (18.2) $y = mx + c$. In this case, c is unknown, but since P(x_1, y_1) lies on the line, $y_1 = mx_1 + c$ and so, by subtraction,

$$y - y_1 = m(x - x_1) \qquad (18.6)$$

which is the required equation.

Example 2 Write down the equation of the line with slope $-\frac{2}{3}$ and passing through the point $(-3, 4)$, and simplify the equation.

By (18.6) the equation is $(y-4) = -\frac{2}{3}(x+3)$, that is

$$3y - 12 = -2x - 6$$
$$3y + 2x - 6 = 0$$

The equation of a line passing through two given points (x_1, y_1) and (x_2, y_2)

The equation of a line passing through a given point (x_1, y_1) is by (18.6) $y - y_1 = m(x - x_1)$. In this case, m is unknown. However, (x_2, y_2) also lies on the line. Hence on substituting in the equation, we have $y_2 - y_1 = m(x_2 - x_1)$, whence by division

$$\frac{y - y_1}{y_2 - y_1} = \frac{x - x_1}{x_2 - x_1} \tag{18.7}$$

which is the required equation.

Example 3 What is the simplified form of the equation of the lines through the following pairs of points (i) $(3, 4)$, $(6, 8)$; (ii) $(5, 3)$, $(7, -2)$?

(i) On substituting in (18.7), we have

$$\frac{y-4}{8-4} = \frac{x-3}{6-3}$$
$$\frac{y-4}{4} = \frac{x-3}{3}$$
$$3y - 12 = 4x - 12$$
$$3y = 4x$$

(ii) On substituting in (18.7), we have

$$\frac{y-3}{-2-3} = \frac{x-5}{7-5}$$
$$\frac{y-3}{-5} = \frac{x-5}{2}$$
$$2y - 6 = -5x + 25$$
$$2y + 5x - 31 = 0$$

The equation of a line such that the length of the perpendicular from the origin to the line is p, and the angle that perpendicular makes with Ox is α

Referring to *Figure 18.5*, PO is the perpendicular from the origin to the line. Since $\angle \text{OPM} = 90°$

$$\frac{\text{OM}}{\text{OP}} = \sec\alpha$$
$$\text{OM} = \text{OP}\sec\alpha = p\sec\alpha \tag{i}$$

Also

$$\angle \text{LOP} = 90° - \alpha$$

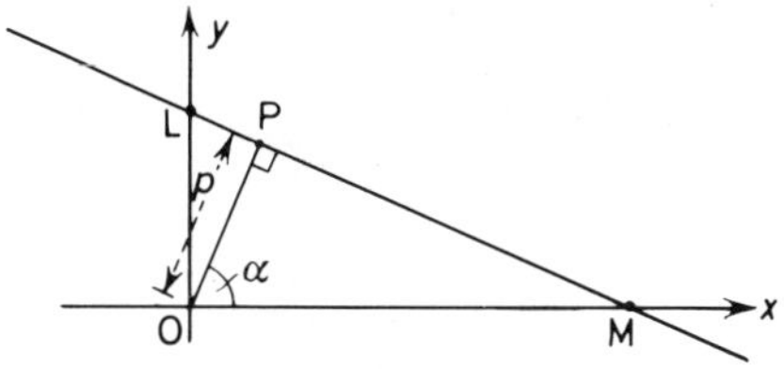

Figure 18.5

so that

$$\frac{\text{OL}}{\text{OP}} = \sec(90° - \alpha)$$

$$\text{OL} = \text{OP}\sec(90° - \alpha) = p\sec(90 - \alpha) \qquad \text{(ii)}$$

But OM, OL are the intercepts the line makes on the axes Ox and Oy. Hence by (18.5) its equation is

$$\frac{x}{p\sec\alpha} + \frac{y}{p\sec(90° - \alpha)} = 1$$

$$x\cos\alpha + y\cos(90° - \alpha) = p$$

$$x\cos\alpha + \text{y}\sin\alpha = p \qquad (18.8)$$

This is called the perpendicular form of the equation of a straight line.

Example 4 If in *Figure 18.5* OP = 5 units and ∠POL = 20°, find the equation of the line LM. In (18.8)

$$\alpha = \angle\text{POM} = 90° - \angle\text{POL}$$
$$= 90° - 20°$$
$$= 70°$$

and

$$p = \text{OP} = 5 \text{ units}$$

Hence LM is the line $x\cos 70° + y\sin 70° = 5$.

Example 5 If in *Figure 18.5* the equation of PLM is $3x + 4y - 12 = 0$, put this equation into both (i) the intercept and (ii) the perpendicular form.

(i) $$3x + 4y - 12 = 0$$

If we divide by 12,

$$\frac{x}{4} + \frac{y}{3} = 1 \quad \text{as required}$$

(ii) If instead we divide by $\sqrt{(3^2 + 4^2)} = 5$ then

$$\tfrac{3}{5}x + \tfrac{4}{5}y = \tfrac{1}{5}$$

which is the perpendicular form because $\frac{3}{5}$, $\frac{4}{5}$ are the cosine and sine of the same angle since the sum of their squares is unity.

Exercises 18b

1 Find the equations of the following lines:
(i) passing through the points (5, 3), (−2, 1)
(ii) passing through the points (6, −2), (3, 7)
(iii) making an angle of 135° with Ox and passing through the point (−2, 5)
(iv) parallel to $2y + 3x - 4 = 0$ and passing through the point (5, −2)
(v) passing through the points (−5, 0), (0, −2)
(vi) such that the length of the perpendicular to the line from the origin is 6 units and the perpendicular makes an angle of 45° with Ox.

2 Write down the equation of the line making intercepts of −5, +3 on the x- and y-axes respectively. Put the equation into the perpendicular form.

3 Find the slope of the line through the points (5, 3), (7, −2). Also find (i) the perpendicular form, (ii) the intercept form, of its equation.

4 O is the origin and a line OA of length $2a$ makes an angle α with the x-axis. Find the equation of the perpendicular bisector of OA.

5 A line makes an obtuse angle θ with the positive direction of Ox. If α is the angle the perpendicular to the line from the origin makes with the positive direction of Ox, find the relation between α and θ. Does the same relation hold when θ is acute?

18.5 The co-ordinates of the point of intersection of two straight lines

Let the equations of the lines be

$$a_1x + b_1y + c_1 = 0 \qquad a_2x + b_2y + c_2 = 0$$

As we have noted in Section 17.6, to obtain the point of intersection we solve these equations simultaneously. Eliminating x and then y, we obtain

$$\frac{x}{b_1c_2 - b_2c_1} = \frac{-y}{a_1c_2 - a_2c_1} = \frac{1}{a_1b_2 - a_2b_1} \tag{18.9}$$

The values of x, y for the point of intersection are

$$x = \frac{b_1c_2 - b_2c_1}{a_1b_2 - a_2b_1} \qquad y = -\frac{a_1c_2 - a_2c_1}{a_1b_2 - a_2b_1} \tag{18.10}$$

We note that if $a_1b_2 - a_2b_1 = 0$, there is no solution. But this may be written

$$a_1b_2 = a_2b_1$$

that is

$$\frac{a_1}{b_1} = \frac{a_2}{b_2}$$

and thus the slopes (see (18.4)) of the two lines are equal. That is, the lines are parallel and have no point of intersection.

If in addition either $b_1c_2 - b_2c_1 = 0$ or $a_1c_2 - a_2c_1 = 0$, then

$$\frac{a_1}{a_2} = \frac{b_1}{b_2} = \frac{c_1}{c_2} \quad (= k \text{ say}) \tag{18.11}$$

and from this we see that all three of the quantities in (18.9) are zero and x and y are indeterminate. The geometrical explanation is that, since from (18.11) $a_1 = ka_2$, $b_1 = kb_2$, $c_1 = kc_2$, one equation is a multiple of the other and the lines coincide.

Example 1 Find a general solution for the point of intersection of the lines $ax + 5y + b = 0$, $2x + y + 3 = 0$ and discuss the three cases (i) $a \neq 10$; (ii) $a = 10$, $b \neq 15$; (iii) $a = 10$, $b = 15$.

From (18.9) the solution is given by

$$\frac{x}{15 - b} = \frac{-y}{3a - 2b} = \frac{1}{a - 10}$$

$$x = \frac{15 - b}{a - 10} \qquad y = \frac{-3a + 2b}{a - 10}$$

Case (i): since $a \neq 10$,

$$a - 10 \neq 0$$

and hence a real point of intersection exists.

Case (ii): since $a = 10$, $b \neq 15$,

$$a - 10 = 0 \quad \text{but} \quad 15 - b \neq 0$$

hence the lines are separate but parallel.

Case (iii): if $a = 10$, $b = 15$, then the equations are

$$10x + 5y + 15 = 0$$
$$2x + y + 3 = 0$$

and the first equation is a multiple of the second and the lines are not distinct.

18.6 The positive and negative sides of a line

Consider any straight line and let its equation be $ax + by + c = 0$. Let $P(x_1, y_1)$ be any point and let the line through P parallel to the y-axis cut the line in the point Q whose co-ordinates are (x_1, y_2). Then it is clear from *Figure 18.6* that, so long as P remains on the same side of the straight line, PQ is always drawn in the same direction. If P is a point on the other side of the line, then PQ is drawn in the opposite direction. That is, PQ is *positive*

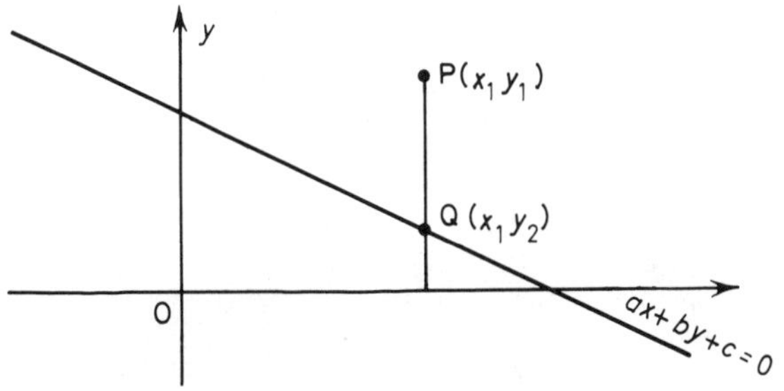

Figure 18.6

for all points on one side of the line and *negative* for all points on the other side.

Now $$PQ = y_1 - y_2 \tag{i}$$

and since the point $Q(x_1, y_2)$ lies on the line

$$ax_1 + by_2 + c = 0$$

$$y_2 = -\frac{ax_1 + c}{b} \tag{ii}$$

From (i) and (ii)

$$PQ = y_1 - \left(-\frac{ax_1 + c}{b}\right)$$

$$= \frac{ax_1 + by_1 + c}{b} \tag{iii}$$

Since the sign of PQ changes as P crosses the line, it follows from (iii) that the sign of $ax_1 + by_1 + c$ (b is fixed) must alter as P crosses the line. Thus the line divides the co-ordinate plane into two parts such that $ax + by + c$ is greater than or less than zero. If $c > 0$, the origin is on the positive side of the line. If $c < 0$, the origin is on the negative side of the line.

Example 1 The co-ordinates (x, y) of a point P satisfy all three of the inequalities

$$2x + 3y - 6 > 0$$
$$x - y + 6 > 0$$
$$y + 5x - 20 < 0$$

Draw a diagram to show the area within which P must lie.

Figure 18.7 shows the required area.

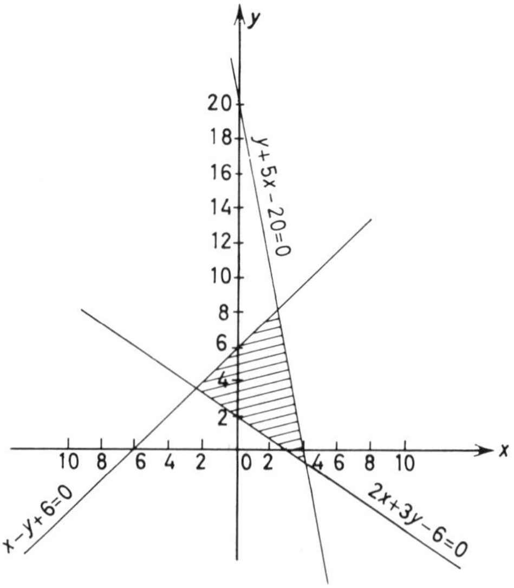

Figure 18.7

Exercises 18c

1 Show that the three lines $3x-2y+1=0$, $x+2y+3=0$, $7x-2y+5=0$ pass through the same point.

2 Find the points of intersection of the following pairs of lines:
(a) $2x+3y-13=0$, $3x-y-3=0$
(b) $2x+y-2=0$, $4x+5y+5=0$
(c) $x/a+y/b=1$, $x/b+y/a=1$
(d) $y=4x$, $y=3x+2$.

3 Show that the following three lines do not form a triangle: (i) $18x-12y+9=0$, (ii) $12x+8y-6=0$, (iii) $8y=12x+6$.

4 What must be the value of k in order that the lines (i) $2x+y-3=0$, (ii) $kx+3y+1=0$, (iii) $x+y+7=0$ may meet in a point. Discuss the cases when $k=3$ and $k=6$.

5 Find the equation of the line joining the origin to the point of intersection of the lines $x+2y+5=0$, $3x-2y-13=0$.

6 A point P has co-ordinates (x, y). Draw separate sketches showing the area in which P must lie in the following cases: (i) $3x+2y-1>0$, (ii) $2x-y-4<0$, (iii) $x+y+5>0$, (iv) $2x-y+7<0$.

7 A point $P(x, y)$ is such that its co-ordinates satisfy all the inequalities $x+y-6\leqslant 0$, $y-2x+6\geqslant 0$, $4x+y\geqslant 0$, $y-2x-6\leqslant 0$. Show in a sketch the area within which P must lie.

8 If in question 7 all the inequality signs are reversed, in which area must $P(x, y)$ lie?

9 Show that for one particular value of k all the lines $kx-3y-3=0$, $-x+2y-k=0$, $9x-4y-11=0$, $13x-ky-19=0$ pass through one point and state the co-ordinates of the point.

10 A point $P(x, y)$ lies within the triangle formed by the three lines $5x+7y-35=0$, $4x-11y-28=0$, $14x+3y+68=0$. Write down with the aid of a sketch the three inequalities its co-ordinates must satisfy.

18.7 The angle between two straight lines

If the equations of the lines are given in the perpendicular form (see (18.8))

$$x\cos\alpha+y\sin\alpha-p_1=0 \qquad x\cos\beta+y\sin\beta-p_2=0$$

then, referring to *Figure 18.8*, the required angle is either $\alpha-\beta$ or $\pi-(\alpha-\beta)$. Because α and β are the angles the perpendiculars from the origin to the two lines make with Ox, then the angle between the perpendiculars is $\alpha-\beta$. Also the angle between any two lines is equal or supplementary to the angle between two lines perpendicular to them. Hence the required angle is

$$(\alpha-\beta) \quad \text{or} \quad \pi-(\alpha-\beta) \tag{18.12}$$

If the equations of the lines are given in the form $y=m_1x+c_1$, $y=m_2x+c_2$ and θ_1, θ_2 are the angles the lines make with Ox, then

$$m_1=\tan\theta_1 \qquad m_2=\tan\theta_2 \tag{i}$$

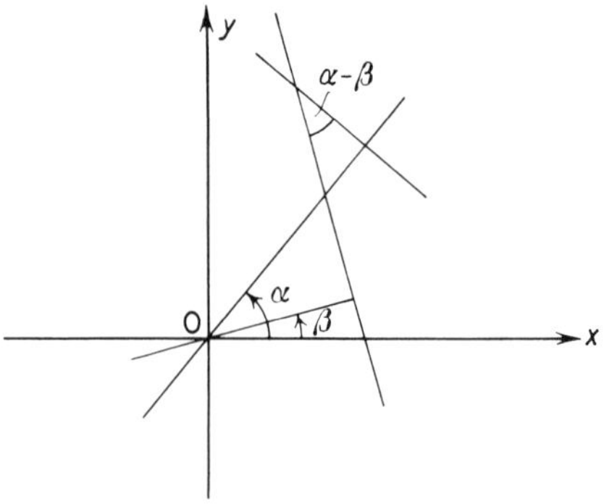

Figure 18.8

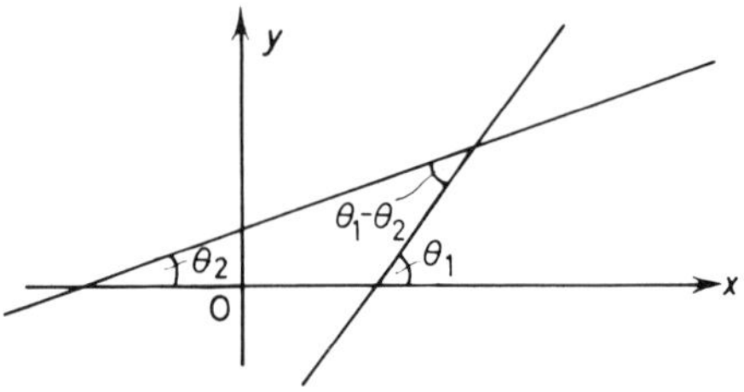

Figure 18.9

Referring to *Figure 18.9*, it can be seen that the required angle is $\theta_1 - \theta_2$. Now

$$\tan(\theta_1 - \theta_2) = \frac{\tan\theta_1 - \tan\theta_2}{1 + \tan\theta_1 \tan\theta_2}$$

$$= \frac{m_1 - m_2}{1 + m_1 m_2} \quad \text{from (i)}$$

Thus the required angle is

$$\arctan\left(\frac{m_1 - m_2}{1 + m_1 m_2}\right) \tag{18.13}$$

If the equations of the lines are given in the form $a_1x + b_1y + c_1 = 0$, $a_2x + b_2y + c_2 = 0$, then from (18.4) $m_1 = -a_1/b_1$, $m_2 = -a_2/b_2$. Hence the required angle is

$$\arctan\left(\frac{-\dfrac{a_1}{b_1} + \dfrac{a_2}{b_2}}{1 + \dfrac{a_1 a_2}{b_1 b_2}}\right)$$

that is

$$\arctan\left(\frac{a_2 b_1 - a_1 b_2}{a_1 a_2 + b_1 b_2}\right) \tag{18.14}$$

Referring to (18.13) it should be noted that if

$$m_1 m_2 = -1 \tag{18.15}$$

that is

$$m_1 = -\frac{1}{m_2}$$

then the angle between the lines is $\pi/2$, and the lines are perpendicular.

Similarly from (18.14), the lines are perpendicular if

$$a_1 a_2 + b_1 b_2 = 0 \tag{18.16}$$

Example 1 Find the angle between the lines $y = \frac{1}{3}x + \frac{4}{3}$, $y = \frac{1}{2}x + \frac{5}{8}$.

The slopes of the lines are $m_1 = \frac{1}{3}$ and $m_2 = \frac{1}{2}$. Hence from (18.13), the required angle is

$$\arctan\left(\frac{\frac{1}{3} - \frac{1}{2}}{1 + \frac{1}{3} \times \frac{1}{2}}\right) = \arctan(-\tfrac{1}{7})$$

The negative sign indicates that the *obtuse* angle between the lines is being found; hence the required *acute* angle is $\arctan(\frac{1}{7}) = 8°8'$.

Example 2 Find the equation of the line which is perpendicular to the line $2x + 3y - 1 = 0$ and passes through the point (4, 3).

From (18.4) the slope of the given line is

$$m_1 = -\tfrac{2}{3}$$

Hence the slope of a perpendicular line is

$$\begin{aligned} m_2 &= -\frac{1}{-\frac{2}{3}} \\ &= \tfrac{3}{2} \end{aligned}$$

Since the required line passes through the point (4, 3), its equation is (see (18.6))

$$(y - 3) = \tfrac{3}{2}(x - 4)$$

that is

$$2y - 3x + 6 = 0$$

Example 3 OA and OB are the equal sides of an isosceles triangle lying in the first quadrant. OA and OB make angles θ_1 and θ_2 respectively with Ox. Show that the gradient of the bisector of the acute angle AOB is $\operatorname{cosec}\theta - \cot\theta$, where $\theta = \theta_1 + \theta_2$.

Referring to *Figure 18.10*, let $\angle \text{AOP} = \alpha = \angle \text{POB}$. Hence

$$\begin{aligned} \angle \text{PO}x &= \theta_1 + \alpha \quad \text{or} \quad \theta_2 - \alpha \\ 2\angle \text{PO}x &= \theta_1 + \theta_2 \\ &= \theta \quad \text{(given)} \end{aligned}$$

Therefore

$$\angle \text{PO}x = \tfrac{1}{2}\theta$$

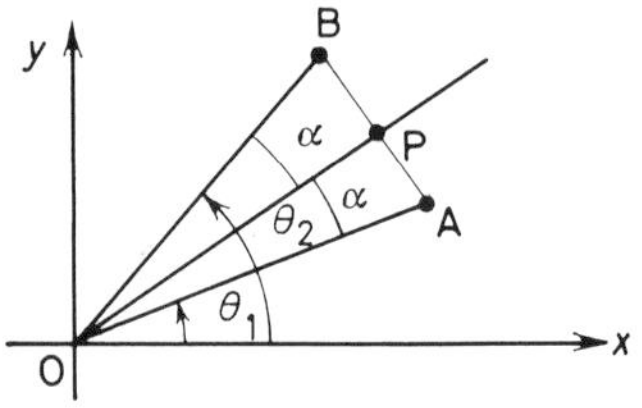

Figure 18.10

Now

$$\begin{aligned}\text{gradient of PO} &= \tan \angle \text{PO}x \\ &= \tan \tfrac{1}{2}\theta \\ &= \frac{\sin \frac{1}{2}\theta}{\cos \frac{1}{2}\theta} \\ &= \frac{2\sin^2 \frac{1}{2}\theta}{2\sin \frac{1}{2}\theta \cos \frac{1}{2}\theta} \\ &= \frac{1-\cos\theta}{\sin\theta} \\ &= \operatorname{cosec}\theta - \cot\theta\end{aligned}$$

Exercises 18d

1 Find the acute angles between the following pairs of lines:
(a) $y = 3x + 4$, $y = 2x - 1$
(b) $3x + 4y + 7 = 0$, $4x - 5y + 2 = 0$
(c) $x\cos(\alpha - \pi/4) + y\sin(\alpha - \pi/4) = 3$, $x\cos(\alpha + \pi/4) + y\sin(\alpha + \pi/4) = 7$
(d) $5x - 6y + 7 = 0$, $6x + 5y - 3 = 0$.

2 Find the equation of the line perpendicular to the line $3x + 2y + 4 = 0$ and passing through the point (5, 6).

3 Find the equation of the line passing through the points (1, 4) and (−2, 7) and show that it is perpendicular to the line $x - y + 3 = 0$.

4 Given the two lines $x + y + 7 = 0$ and $\sqrt{3}x - y + 5 = 0$, put their equations into the perpendicular form and hence find the acute angle between them. Verify your result by using (18.14).

5 Find the equations of the sides of the triangle ABC where A, B, C are the points (5, 7), (3, 3), (7, 1) respectively. Hence show that the triangle ABC has angles of 90°, 45° and 45°. Verify this result by finding the lengths of the sides of the triangle.

18.8 The perpendicular distance of a point from a straight line

Let $P(x_1, y_1)$ be the point and the equation of the line be

$$x\cos\alpha + y\sin\alpha = p \tag{18.17}$$

Referring to *Figure 18.11*, any line parallel to the given line is

$$x\cos\alpha + y\sin\alpha = p'$$

and this passes through the given point $P(x_1, y_1)$ if

$$x_1 \cos\alpha + y_1 \sin\alpha = p' \tag{18.18}$$

Now the required distance is

$$\begin{aligned} PC &= AB \\ &= OB - OA \\ &= p' - p \\ &= x_1 \cos\alpha + y_1 \sin\alpha - p \quad \text{(from (18.8))} \end{aligned}$$

Hence the required result is

$$x_1 \cos\alpha + y_1 \sin\alpha - p \tag{18.19}$$

If the equation of the line is given as $ax + by + c = 0$, it may be rewritten

$$\frac{a}{\sqrt{(a^2 + b^2)}}x + \frac{b}{\sqrt{(a^2 + b^2)}}y + \frac{c}{\sqrt{(a^2 + b^2)}} = 0$$

which is in the perpendicular form because

$$\frac{a}{\sqrt{(a^2 + b^2)}} \quad \text{and} \quad \frac{b}{\sqrt{(a^2 + b^2)}}$$

are the sine and cosine of an angle since the sum of their squares is unity. Hence from (18.19) the length of the perpendicular is

$$\frac{ax_1 + by_1 + c}{\sqrt{(a^2 + b^2)}} \tag{18.20}$$

If the denominator $\sqrt{(a^2 + b^2)}$ is always taken as positive, then the length of the perpendicular from any point on the positive side of the line will be positive and from any point on the negative side of the line it will be negative (see Section 18.6).

Example 1 Find the length of the perpendicular from the point P(2, −4) to the line $3x + 2y - 5 = 0$ and state which side of the line P is on.

From (18.20) the length of the perpendicular is

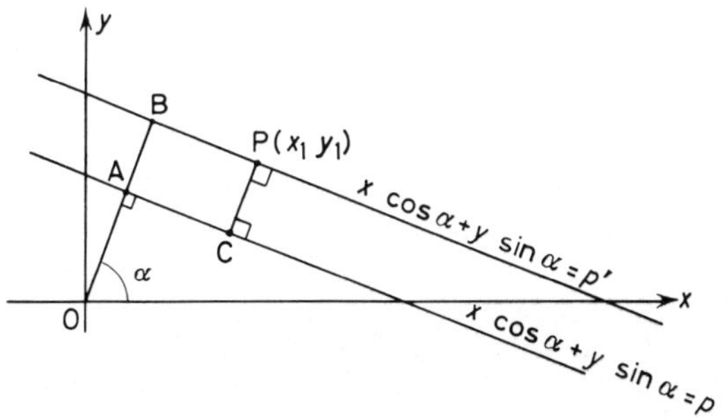

Figure 18.11

$$\frac{3\times(2)+2\times(-4)-5}{\sqrt{(3^2+2^2)}}=-\frac{7}{\sqrt{13}}$$

Thus the length of the perpendicular is $7/\sqrt{13}$ and it is on the negative side of the line. Substituting the co-ordinates (0, 0) of the origin in the equation of the line $3x+2y-5$, we obtain -5. Thus the origin is also on the negative side of the line. Hence P is on the same side of the line as the origin.

Exercises 18e

1 Find the lengths of the perpendiculars from the point P to the line L in the following cases:
(a) P(3, 4), $L \equiv 3x+4y+6=0$
(b) P(−2, −1), $L \equiv 3x+4y+6=0$
(c) P(5, 6), $L \equiv 3x-6=0$
(d) P(−3, −2), $L \equiv 5x-12y+1=0$.

2 Find the lengths of the perpendiculars from P(3, 4) to the two lines $7x+24y-1=0$ and $3x+4y-36=0$ and state on which sides of the lines the point P is situated.

3 Show that the point P(1, 1) is equidistant from the three lines $5x+12y+9=0$, $3x+4y-17=0$, $3x-4y-9=0$. Is P the incentre of the triangle formed by the three lines?

4 The point P(a, 2) is equidistant from two lines $4x-3y+7=0$ and $7x+24y-30=0$; find the value of a.

5 Find the incentre of the triangle formed by the three lines $12x-5y+9=0$, $3x+4y-27=0$, $5x+12y-45=0$.

18.9 The equation of a straight line through the point of intersection of two given straight lines

Let the equations of the straight lines be

$$a_1x+b_1y+c_1=0 \qquad a_2x+b_2y+c_2=0 \qquad (18.21)$$

Consider the equation

$$(a_1x+b_1y+c_1)+\lambda(a_2x+b_2y+c_2)=0 \qquad (18.22)$$

where λ is any constant.

This is the equation of a straight line since it is of the first degree in x and y. Further, it is satisfied by the co-ordinates of the common point of the two given lines, since these co-ordinates satisfy simultaneously the equations (18.21) and hence must satisfy the equation (18.22). Thus it is the required line.

This is an example of a more general device used in co-ordinate geometry; namely if $S_1=0$ and $S_2=0$ are the equations of any two loci, then $S_1+\lambda S_2=0$ is the equation of a locus through the common points of $S_1=0$ and $S_2=0$.

In the case of the straight line (18.22), a second condition is required to find λ.

Example 1 Find the equation of the line drawn through the point of intersection of the lines $3x + 2y - 3 = 0$, $5x - y + 8 = 0$ which also passes through the point (2, 2).

Any line through the point of intersection is given by $(3x + 2y - 3) + \lambda(5x - y + 8) = 0$.

If this passes through the point (2, 2), its co-ordinates satisfy the equation and hence

$$(6 + 4 - 3) + \lambda(10 - 2 + 8) = 0$$
$$7 + 16\lambda = 0$$
$$\lambda = -\tfrac{7}{16}$$

Hence the required equation is

$$(3x + 2y - 3) - \tfrac{7}{16}(5x - y + 8) = 0$$
$$16(3x + 2y - 3) - 7(5x - y + 8) = 0$$
$$13x + 39y - 104 = 0$$

that is

$$x + 3y - 8 = 0$$

Example 2 Find the equation of the line drawn through the point of intersection of $3x - y - 13 = 0$ and $x - 4y + 3 = 0$ and which is perpendicular to $5y + 2x = 0$.

Any line through the point of intersection is given by $(3x - y - 13) + \lambda(x - 4y + 3) = 0$, that is $(3 + \lambda)x - (1 + 4\lambda)y - 13 + 3\lambda = 0$. Its slope is $\dfrac{3 + \lambda}{1 + 4\lambda}$. Hence it is perpendicular to $5y + 2x = 0$ whose slope is $-\tfrac{2}{5}$ if

$$\frac{3 + \lambda}{1 + 4\lambda} \times -\frac{2}{5} = -1 \quad \text{(see (18.15))}$$

whence

$$6 + 2\lambda = 5 + 20\lambda$$
$$\lambda = \tfrac{1}{18}$$

Therefore the required equation is

$$(3x - y - 13) + \tfrac{1}{18}(x - 4y + 3) = 0$$
$$55x - 22y - 231 = 0$$

that is

$$5x - 2y - 21 = 0$$

In the particular cases when λ has either of the values

$$\pm\frac{\sqrt{(a_1^2 + b_1^2)}}{\sqrt{(a_2^2 + b_2^2)}}$$

equation (18.22) may be written

$$(a_1x + b_1y + c_1) = \pm\frac{\sqrt{(a_1^2 + b_1^2)}}{\sqrt{(a_2^2 + b_2^2)}}(a_2x + b_2y + c_2)$$

that is

$$\frac{a_1x + b_1y + c_1}{\sqrt{(a_1{}^2 + b_1{}^2)}} = \pm\frac{a_2x + b_2y + c_2}{\sqrt{(a_2{}^2 + b_2{}^2)}} \tag{18.23}$$

which shows that the perpendiculars from the point (x, y) to either of the lines $a_1x + b_1y + c_1 = 0$ or $a_2x + b_2y + c_2 = 0$ are equal in magnitude. Hence (18.23) gives the equations of the bisectors of the angles between the lines.

To distinguish between the two bisectors, write the equations of the lines with their constants both positive and take both the denominators positive. Then taking the positive sign in (18.23) gives the bisector of the angle in which the origin lies.

Example 3 Write down the equations of the bisectors between the lines $3x - y - 2 = 0$ and $2x - 2y + 7 = 0$.

Rewriting the first equation $-3x + y + 2 = 0$, the equations of the bisectors are given by

$$\frac{-3x + y + 2}{\sqrt{10}} = \pm\frac{2x - 2y + 7}{\sqrt{8}}$$

Thus $2\sqrt{2}(-3x + y + 2) = +\sqrt{10}(2x - 2y + 7)$ is the equation of the bisector of that angle in which the origin lies; that is

$$(2\sqrt{5} + 6)x - 2(1 + \sqrt{5})y + 7\sqrt{5} - 4 = 0$$

and the other bisector is

$$(2\sqrt{5} - 6)x + 2(1 - \sqrt{5})y + 7\sqrt{5} + 4 = 0$$

Exercises 18f

1 A line passes through the point of intersection of the lines $3x + 2y - 1 = 0$ and $5x + 6y + 1 = 0$. Find its equation in the following cases: (a) if it also passes through the origin, (b) if it is perpendicular to $4x - y = 0$, (c) if it is parallel to $2x + 3y - 1 = 0$.

2 Find the equations of the bisectors of the angles between the lines $2x + 4y - 3 = 0$ and $2x - y + 7 = 0$ and verify that the bisectors are at right angles to one another.

3 Show that one of each pair of bisectors of the angles between the lines $3x - 4y - 4 = 0$, $12x - 5y + 6 = 0$, $7x + 24y - 56 = 0$ taken in pairs, pass through the point (1, 1).

4 Two lines through the origin have a combined equation $2y^2 - xy - 6x^2 = 0$. Factorise this in order to find the separate equations of the two lines and hence show that the combined equation of the internal bisectors of the angles between the two lines is $x^2 - 16xy - y^2 = 0$.

5 Find the equation of the two lines through the point of intersection of the lines $3x + 2y - 1 = 0$ and $2x - y + 7 = 0$ which are also (a) perpendicular to $3x + 2y - 1 = 0$, (b) perpendicular to $2x - y + 7 = 0$.

Exercises 18

1 P, Q, R are the three points with co-ordinates (1, 0), (2, −4), (−5, −2) respectively. Find
(a) the equations of PQ, QR, PR
(b) the equation of the line through P perpendicular to QR
(c) the equation of the line through Q perpendicular to PR
(d) the point of intersection of the lines (b) and (c)
(e) the area of the triangle PQR.

2 Find the equation of the perpendicular bisector of AB where A, B are the points (3, 2), (5, 1) respectively.

3 Sketch on the same diagram the lines whose equations are
(a) $y = 3x$
(b) $y = -3x$
(c) $2x + 3y - 12 = 0$
(d) $3x - 5y + 75 = 0$
(e) $x - 7 = 0$
(f) $y + 8 = 0$.

4 Find the equation of the lines through the point (6, 5) which are (a) perpendicular to $3x - 4y = 0$, (b) parallel to $3x - 4y = 0$.

5 Find the equation of the line joining the points (3, 6), (5, 7) and show that it is perpendicular to the line joining the points (−3, 4), (−2, 2).

6 Rewrite the equation of the line $5x - 4y - 20 = 0$ in (a) intercept form, (b) perpendicular form.

7 The co-ordinates of the three points L, M, N are (a, a), $(-a, -a)$ and $(0, -a)$ respectively. A point X is taken on MN such that the ratio of MX to XN is t:1 and a point Y is taken on LX such that the ratio of LY to YX is also t:1. Prove that the co-ordinates of X and Y are respectively

$$\left[-\frac{a}{1+t}, -a\right] \text{ and } \left[\frac{a}{(1+t)^2}, \frac{a(1-t)}{1+t}\right]$$ [JMB, part]

8 Show that the three equations $x + 2y - k = 0$, $x + ky - 2 = 0$ and $kx + 4y - 4 = 0$ are consistent when $k = -4$ or 2. Give a geometrical interpretation in either case. Discuss the case when $k = -2$.

9 A straight line parallel to the line $2x + y = 0$ intersects the x-axis at A and the y-axis at B. The perpendicular bisector of AB cuts the y-axis at C. Prove that the gradient of the line AC is $-\frac{3}{4}$. Find also the tangent of the acute angle between the line AC and the bisector of the angle AOB, where O is the origin. [JMB]

10 The lines L_1, L_2, L_3 have the equations $x + y + 1 = 0$, $y + 2x + 2 = 0$, $3y - 9x + 11 = 0$ and meet the y-axis at the points A, B, C respectively. If D is the point (−2, −3), prove that DA, DB and DC are perpendicular to L_1, L_2, L_3 respectively.

11 Find the co-ordinates of the foot of the perpendicular from the point (x_1, y_1) to the line $ax + by + c = 0$ and deduce that the co-ordinates of the image of the point in the line are

$$\left[x_1 - \frac{2a}{a^2+b^2}(ax_1 + by_1 + c),\quad y_1 - \frac{2b}{a^2+b^2}(ax_1 + by_1 + c)\right]$$

[JMB, part]

12 The vertices B, C of a triangle ABC lie on the lines $3y = 4x$, $y = 0$ respectively, and the side BC passes through the point $(\frac{2}{3}, \frac{2}{3})$. If ABOC is a rhombus where O is the origin of co-ordinates, find the equation of the line BC and prove that the co-ordinates of A are $(\frac{8}{5}, \frac{4}{5})$. [LU]

13 A point P lies in the plane of the triangle ABC and G is the centroid of this triangle. Prove that

$$PA^2 + PB^2 + PC^2 = 3PG^2 + \tfrac{1}{3}(BC^2 + CA^2 + AB^2)$$

What is the least value of $PA^2 + PB^2 + PC^2$ as P varies in the plane?

14 Given the four points $(a/m_r, am_r)$, $r = 1, 2, 3, 4$, find the condition that the line joining any two of the four points is perpendicular to the line joining the other two.

15 A, B, C are the points $(-4, -2)$, $(3, 1)$, $(-2, 6)$. D is a point on the opposite side of AB to C which moves so that the area of the triangle ADB is always 50 square units. Find the equation of the locus of D. If CD meets AB at the point Q verify that the ratio CQ:QD is constant for all positions of D.

16 A, B, C, D are the points (x_1, y_1), (x_2, y_2), (x_3, y_3), (x_4, y_4). Show that ABCD is a parallelogram provided that $x_1 + x_3 = x_2 + x_4$ and $y_1 + y_3 = y_2 + y_4$. Show also that the parallelogram is a rectangle if

$$x_1x_3 + y_1y_3 = x_2x_4 + y_2y_4$$

17 Without drawing a figure, determine whether the point (4, 3) is inside or outside the triangle formed by the lines $y = x + 6$, $3y + 4x - 24 = 0$, $y + 8 = 0$.

18 The altitudes AD, BE, CF of a triangle ABC are $x + y = 0$, $x - 4y = 0$ and $2x - y = 0$ respectively. If the co-ordinates of A are $(k, -k)$, find the co-ordinates of B and C. Find also the locus of the centroid of the triangle ABC as k varies.

19 Find the equations of the lines through the point (2, 3) which make angles of 45° with the line $x - 2y = 1$. [LU]

20 Obtain the equation of the straight line through the point $P(h, k)$ perpendicular to the line $ax + by + c = 0$. PA, PB are the perpendiculars from P to the lines $y = x$, $y = \frac{1}{2}x$. Find the co-ordinates of M, the middle point of AB, and show that if P moves on the line $5x + 4y + 10 = 0$ then M will move on the line $x - 7y = 5$. [WJC]

21 Show that the area of the parallelogram formed by the lines $3x + 4y = 7p$, $3x + 4y = 7q$, $4x + 3y = 7r$, $4x + 3y = 7s$ is $7(p - q)(r - s)$.

22 The length of the perpendicular to a line from the origin is 5 units. The line passes through the point (3, 5). Find its equation.

23 Show that *any* point on the line $4x + 7y - 26 = 0$ is equidistant from the two lines $3x + 4y - 12 = 0$ and $5x + 12y - 52 = 0$.

24 Find the values of k for which the lines $2x + ky + 4 = 0$, $4x - y - 2k = 0$, $3x + y - 1 = 0$ are concurrent. [JMB, part]

25 Show that the line $y(m + 1) = x(m - 1) + 4$ always passes through a fixed point and find the co-ordinates of that point.

26 OA and OB are the equal sides of an isosceles triangle lying in the first quadrant. The slopes of OA and OB are $\frac{7}{17}$ and 1 and the length of the perpendicular from O to AB is $\sqrt{13}$. Find the equation of AB. (Use the result of Example 3 in Section 18.7.)

27 The points P, Q are such that the line $x \cos \alpha + y \sin \alpha = p$ is the perpendicular bisector of PQ. If the co-ordinates of P are (x, y) find the co-ordinates of Q. If the line is fixed find the locus of Q as P moves along the y-axis.

28 The vertices O, A, B of a square OABC are the points (0, 0), (1, 0) and (1, 1) respectively. P is a variable point on the side BC. OP produced meets AB produced at Q and a line through B parallel to CQ meets OP at R. Prove that R lies on the diagonal AC when $\text{CP} = (\sqrt{5} - 1)/2$. Find the equation of the locus of R as P varies and give a rough sketch of the locus. [LU]

29 (a) Find the gradients of the bisectors of the angle between the lines $y - 7x = 0$, $x + y = 0$.

(b) If A, B, C and D are the points on the x-axis with abscissae 2, 4, 6 and 8 respectively, find the co-ordinates of the two points P and Q in the first quadrant which are such that

$$\tan \text{APB} = \tan \text{CPD} = \tan \text{AQB} = \tan \text{CQD} = \tfrac{1}{2}$$ [JMB]

30 Find the condition that the lines

$$a_1x + b_1y + c_1 = 0$$
$$a_2x + b_2y + c_2 = 0$$
$$a_3x + b_3y + c_3 = 0$$

are concurrent.

19
The circle

19.1 The equation of a circle

Let C(a, b) be the centre and r the radius of the circle. Let P(x, y) be any point on the circumference of the circle, then (see *Figure 19.1*)

$$CP = r$$
$$CP^2 = r^2$$

Now referring to (17.5), which gives an expression for the distance between two points, we have

$$(x-a)^2 + (y-b)^2 = r^2 \tag{19.1}$$

which is the required equation.

If we let $a = b = 0$, the centre of the circle will be the origin and the equation reduces to

$$x^2 + y^2 = r^2 \tag{19.2}$$

Equation (19.1) may be written

$$x^2 + y^2 - 2ax - 2by + a^2 + b^2 = r^2$$

The equation of a circle is thus of the form

$$x^2 + y^2 + 2gx + 2fy + c = 0 \tag{19.3}$$

where g, f, c are constants. Conversely, (19.3) can be rewritten

$$x^2 + 2gx + g^2 + y^2 + 2fy + f^2 = g^2 + f^2 - c$$

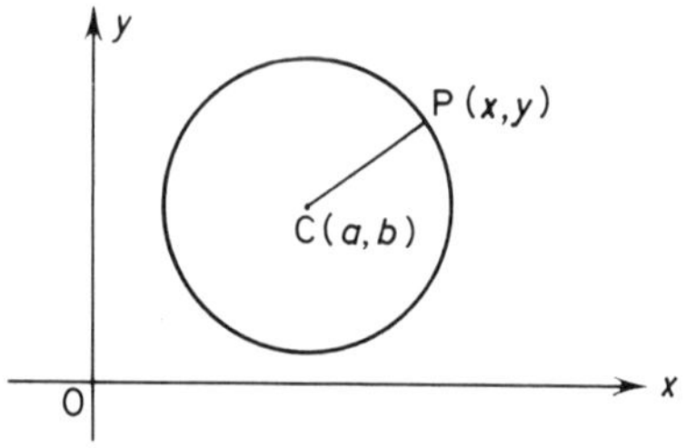

Figure 19.1

that is

$$(x+g)^2+(y+f)^2=g^2+f^2-c$$

Comparing this with (19.1), we see that (19.3) represents

$$\text{a circle centre } (-g, -f), \text{ radius } \sqrt{(g^2+f^2-c)} \qquad (19.4)$$

In general, the equation of a circle is such that

(i) the coefficients of x^2 and y^2 are equal;
(ii) there is no term in xy.

Example 1 Find the equation of the circle centre $(-3, 4)$, radius 7.
The equation is

$$(x+3)^2+(y-4)^2=7^2$$

$$x^2+y^2+6x-8y-24=0$$

Example 2 Find the centre and radius of the circle $4x^2+4y^2-12x+5=0$.
In order to put the given equation into the standard form (19.1), it is first necessary to divide throughout by 4, thus

$$x^2+y^2-3x+\tfrac{5}{4}=0$$

that is

$$x^2-3x+(-\tfrac{3}{2})^2+y^2=(-\tfrac{3}{2})^2-\tfrac{5}{4}$$

whence

$$(x-\tfrac{3}{2})^2+y^2=1$$

Thus the circle has centre $(\tfrac{3}{2}, 0)$, radius 1.

Example 3 Find the equation of the circle centre $(+4, -7)$ which touches the line $3x+4y-9=0$.
Since the line is a tangent, then the radius of the circle is equal to the perpendicular distance from the centre to the line. Thus

$$\begin{aligned}\text{radius} &= \frac{3(4)+4(-7)-9}{\sqrt{(3^2+4^2)}}\\ &= \frac{-25}{5}\\ &= -5\end{aligned}$$

Thus the equation of the circle is

$$(x-4)^2+(y+7)^2=25$$

that is

$$x^2+y^2-8x+14y+40=0$$

Example 4 Find the equation of the circle with AB as diameter. A, B are the points (x_1, y_1), (x_2, y_2).
Let $P(x, y)$ be any other point on the circumference of the circle (see *Figure 19.2*).

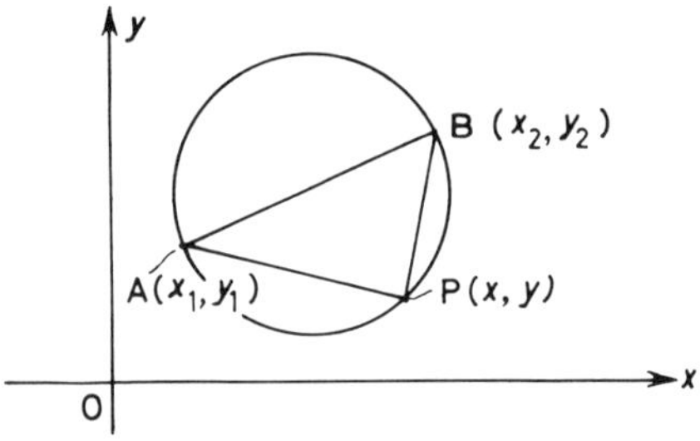

Figure 19.2

The slopes of AP and BP are respectively

$$\frac{y-y_1}{x-x_1} \quad \text{and} \quad \frac{y-y_2}{x-x_2}$$

Since AB is a diameter, $\angle$APB = 90°; thus AP and PB are perpendicular; hence by (18.15) the product of their slopes is -1. Thus

$$\left(\frac{y-y_1}{x-x_1}\right)\left(\frac{y-y_2}{x-x_2}\right) = -1$$

or

$$(x-x_1)(x-x_2)+(y-y_1)(y-y_2)=0 \qquad (19.5)$$

which is the condition satisfied by the co-ordinates of *any* point on the circle and is therefore the required equation.

Exercises 19a

1 Write down the equations of the following circles: (i) centre (3, 7) radius 5, (ii) centre $(-3, -7)$ radius 6, (iii) centre (5, 0) radius 5, (iv) centre $(0, -3)$ radius 4.

2 Find the centre and radius of the following circles:
(i) $x^2+y^2+2x+6y+6=0$
(ii) $9x^2+9y^2+27x+12y+19=0$
(iii) $x^2+y^2-5x=0$
(iv) $4x^2+4y^2-28y+33=0$
(v) $x^2+y^2-2ax+2by+2b^2=0$ (a, b constant)
(iv) $x^2+y^2+2ax-2ay=0$ (a constant).

3 Find the equation of the circle centre $(7, -6)$ which touches the line $3x-4y+5=0$.

4 Find the equation to the circle which has the points (3, 2) $(0, -1)$ as ends of a diameter.

5 Find the equation of the circle centre $(3, -2)$ touching the line $x+y-3=0$.

6 Show that the circle $x^2+y^2-2x-2y+1=0$ touches both Ox and Oy.

7 Show that the circle $x^2+y^2-2ax-2ay+a^2=0$ (a is a positive constant) lies wholly in the first quadrant and touches both Ox and Oy.

8 Use the result of question 7 to find the equation of the circle lying in the first quadrant which touches both axes and also the line $5x+12y-52=0$.

9 Find the equation to the diameter of the circle $x^2 + y^2 - 8x + 6y + 21 = 0$ which when produced passes through the point (2, 5).

10 Find the equation of the circle which passes through the point (1, 1) has a radius of $\frac{1}{2}\sqrt{10}$, and whose centre lies on the line $y = 3x - 7$.

19.2 The equation of a circle through three non-collinear points

Let the equation of the circle be $x^2 + y^2 + 2gx + 2fy + c = 0$ and the three points be (x_1, y_1), (x_2, y_2), (x_3, y_3). Since the circle passes through all three points, the co-ordinates of each point must satisfy the equation of the circle. Hence

$$\begin{aligned} x_1^2 + y_1^2 + 2gx_1 + 2fy_1 + c &= 0 \\ x_2^2 + y_2^2 + 2gx_2 + 2fy_2 + c &= 0 \\ x_3^2 + y_3^2 + 2gx_3 + 2fy_3 + c &= 0 \end{aligned}$$

are three simultaneous equations which can be solved for g, f and c.

Example 1 Find the equation of the circle through the points (6, 1), (3, 2), (2, 3).

Let the equation of the circle be $x^2 + y^2 + 2gx + 2fy + c = 0$. Then since (6, 1) lies on the circle

$$36 + 1 + 12g + 2f + c = 0$$

Similarly

$$9 + 4 + 6g + 4f + c = 0$$

and

$$4 + 9 + 4g + 6f + c = 0$$

Solving these simultaneous equations, we have

$$f = -6 \qquad g = -6 \qquad c = 47$$

Hence the required equation is

$$x^2 + y^2 - 12x - 12y + 47 = 0$$

Exercises 19b

Find the equation of the circles passing through the following points and state the length of the respective radii:

1 (0, 0), (3, 1) and (5, 5)
2 (5, 0), (6, 0) and (8, 1)
3 (3, 2), (1, 1) and (1, 0)
4 (2, 1), (−2, 5) and (−3, 2)
5 Find the equation and radius of the circumcircle of the triangle formed by the three lines $2y - 9x + 26 = 0$, $9y + 2x + 32 = 0$, $11y - 7x - 27 = 0$.

19.3 The equation of the tangent at the point (x_1, y_1) on the circle $x^2 + y^2 + 2gx + 2fy + c = 0$

Differentiating the equation with respect to x, we have

$$2x + 2y\frac{dy}{dx} + 2g + 2f\frac{dy}{dx} = 0$$

therefore

$$\frac{dy}{dx} = -\frac{(x+g)}{(y+f)}$$

Hence the gradient of the tangent at the point (x_1, y_1) is

$$-\frac{(x_1+g)}{(y_1+f)}$$

Thus by (18.6) the equation of the tangent is

$$(y - y_1) = -\frac{(x_1+g)}{(y_1+f)}(x - x_1)$$

$$yy_1 + yf - {y_1}^2 - y_1 f = -xx_1 + {x_1}^2 - gx + gx_1$$
$$xx_1 + yy_1 + gx + fy = {x_1}^2 + {y_1}^2 + gx_1 + fy_1$$

Now add $gx_1 + fy_1 + c$ to both sides to obtain

$$xx_1 + yy_1 + g(x + x_1) + f(y + y_1) + c = {x_1}^2 + {y_1}^2 + 2gx_1 + 2fy_1 + c$$
$$= 0$$

because (x_1, y_1) lies on the circle. Hence the required equation is

$$xx_1 + yy_1 + g(x + x_1) + f(y + y_1) + c = 0 \qquad (19.6)^*$$

Example 1 Find the equation of the tangent at the point $(-\frac{3}{2}, 1)$ on the circle $4x^2 + 4y^2 - 12x + 24y - 55 = 0$.

When using (19.6) to find the equation of the tangent, it is not necessary to reduce the coefficient of x^2 and y^2 to unity, as it was when the centre and radius had to be found (see Section 19.1, Example 2).

Hence the equation of the tangent is

$$4x(-\tfrac{3}{2}) + 4y \,.\, 1 - 6(x - \tfrac{3}{2}) + 12(y + 1) - 55 = 0$$

that is

$$8y - 6x - 17 = 0$$

Example 2 Find the equation of the tangent at the point (1, 0) on the circle $x^2 + y^2 - 5x - y + 4 = 0$.

The equation of the tangent is

$$x \,.\, 1 + y \,.\, 0 - \tfrac{5}{2}(x + 1) - \tfrac{1}{2}(y + 0) + 4 = 0$$

* Note that this equation can be obtained from the general equation of a circle by replacing x^2 by xx_1, y^2 by yy_1, $2x$ by $(x + x_1)$, and $2y$ by $(y + y_1)$.

that is

$$3x + y - 3 = 0$$

Exercises 19c

Find the equations of the tangents to the following circles at the given points:

1 $x^2 + y^2 - 10y = 0$; (3, 9)
2 $2x^2 + 2y^2 + x - 11y - 1 = 0$; (−2, 5)
3 $x^2 + y^2 + 3x - 3y - 38 = 0$; (−7, −2)
4 $9x^2 + 9y^2 - 12x + 42y - 236 = 0$; $(-2, \frac{8}{3})$
5 Verify that the point (8, 6) is common to both the circles $x^2 + y^2 - 11x - 7y + 30 = 0$ and $x^2 + y^2 - x + 3y - 110 = 0$. Find the equations of the tangents to each of the circles at the point (8, 6) and hence deduce that the circles touch each other.

19.4 The length of the tangent from a point P(*X*, *Y*) outside the circle $x^2 + y^2 + 2gx + 2fy + c = 0$

Referring to *Figure 19.3*, C is the centre of the circle and T the point of contact of the tangent. PT is perpendicular to the radius TC.

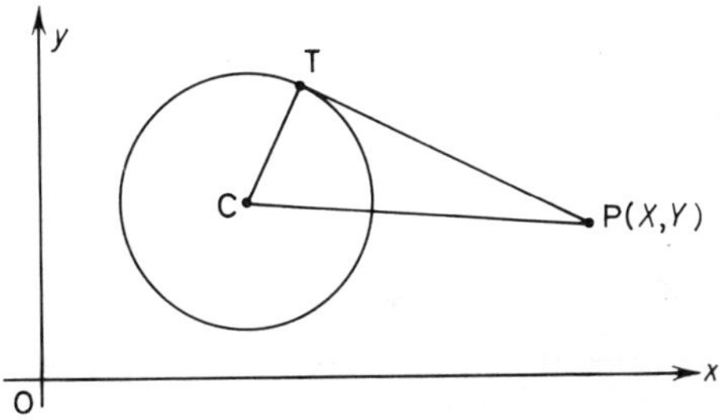

Figure 19.3

Hence

$$PT^2 = PC^2 - CT^2 \qquad \text{(i)}$$

From (19.4) C is the point $(-g, -f)$ and P is the point (X, Y), therefore

$$PC^2 = (X + g)^2 + (Y + f)^2 \qquad \text{(ii)}$$

Also from (19.4)

$$TC = \text{radius} = \sqrt{(g^2 + f^2 - c)} \qquad \text{(iii)}$$

Hence substituting (iii) and (ii) in (i)

$$PT^2 = (X + g)^2 + (Y + f^2) - (g^2 + f^2 - c)$$
$$= X^2 + Y^2 + 2gX + 2fY + c \qquad (19.7)$$

Thus the square of the length of the tangent is obtained by substituting the co-ordinates of the point in the left-hand side of the equation of the circle. Note that if PT^2 has a negative value, it indicates that P is inside the circle.

Example 1 Find the length of the tangent from the point (5, 6) to the circle $x^2 + y^2 + 2x + 4y - 21 = 0$.

From (19.7)

$$PT^2 = 5^2 + 6^2 + 2 \times 5 + 4 \times 6 - 21$$
$$= 74$$

hence $PT = \sqrt{74}$.

19.5 The points of intersection of the straight line $y = mx + c$ and the circle $x^2 + y^2 = r^2$

The co-ordinates of the points of intersection will satisfy the equations of the line and the circle simultaneously:

$$x^2 + y^2 = r^2 \qquad \text{(i)}$$
$$y = mx + c \qquad \text{(ii)}$$

Substituting in (i) from (ii), we have

$$x^2 + (mx + c)^2 = r^2$$

or

$$(1 + m^2)x^2 + 2mcx + c^2 - r^2 = 0 \qquad (19.8)$$

This equation has real, coincident or complex roots according to whether the discriminant of this quadratic, namely

$$(2mc)^2 - 4(1 + m^2)(c^2 - r^2) = 4[r^2(1 + m^2) - c^2]$$

is positive, zero or negative, i.e. according as c^2 is less than, equal to, or greater than $r^2(1 + m^2)$.

Example 1 Find for what values of c the line $y = 2x + c$ meets the circle $x^2 + y^2 = 9$ in two real, coincident and imaginary points. Illustrate with a diagram.

Substituting $y = 2x + c$ in the equation of the circle $x^2 + y^2 = 9$, we obtain

$$x^2 + (2x + c)^2 = 9$$
$$5x^2 + 4cx + c^2 - 9 = 0$$

The discriminant of this quadratic is $16c^2 - 20(c^2 - 9) = 180 - 4c^2$. Hence

if $180 - 4c^2 > 0$ i.e. $c^2 < 45$ we have real roots
if $180 - 4c^2 = 0$ i.e. $c^2 = 45$ we have coincident roots
if $180 - 4c^2 < 0$ i.e. $c^2 > 45$ we have complex roots

These three cases are illustrated in *Figure 19.4*.

The geometrical interpretation of these results is quite generally: if $c^2 < r^2(1 + m^2)$, the line cuts the circle in two distinct points; if $c^2 = r^2(1 + m^2)$, the line is a tangent to the circle; if $c^2 > r^2(1 + m^2)$, the line and circle do not meet.

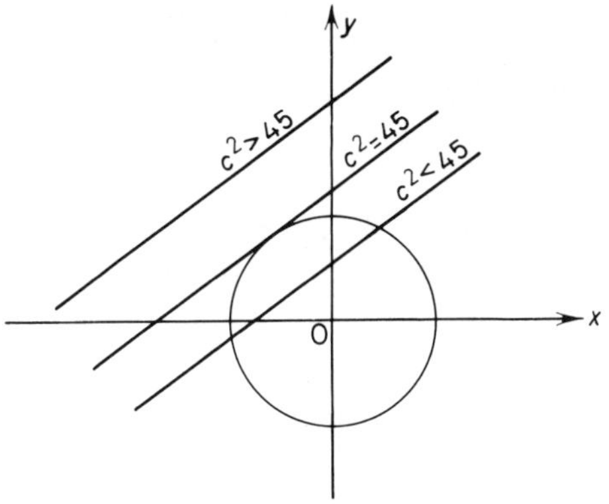

Figure 19.4

Referring back to (19.8), the condition for $y = mx + c$ to be a tangent to the circle $x^2 + y^2 = r^2$ is $c^2 = r^2(1 + m^2)$. That is

$$c = \pm r\sqrt{(1 + m^2)}$$

Thus the two lines

$$y = mx \pm r\sqrt{(1 + m^2)} \tag{19.9}$$

are always tangents to the circle $x^2 + y^2 = r^2$.

Example 2 Find the equations of the tangents to the circle $x^2 + y^2 = 25$ which pass through the point (15, −5).

The radius of the given circle is 5 units. Hence from (19.9), $y = mx \pm 5\sqrt{(1 + m^2)}$ are always tangents to the circle. These lines pass through the point (15, −5) if these co-ordinates satisfy the equation:

$$-5 = 15m \pm 5\sqrt{(1 + m^2)}$$

$$(-5 - 15m) = \pm 5\sqrt{(1 + m^2)}$$

On removing the common factor 5 and squaring, we have

$$(1 + 3m)^2 = 1 + m^2$$

$$8m^2 + 6m = 0$$

which has two roots $m = -\frac{3}{4}$ or 0. Hence the two tangents are

$$y = -\tfrac{3}{4}x + 5\sqrt{(1 + \tfrac{9}{16})}$$

and

$$y = 0 - 5\sqrt{(1 + 0)}$$

that is $4y + 3x = 25$ and $y = -5$.

Exercises 19d

1 Find the lengths of the tangents from the point (5, −2) to (i) the circle $x^2 + y^2 + 2x - 3y = 0$ and (ii) the circle $x^2 + y^2 - x - 5y + 9 = 0$.

2 Find if the points A(2, −1), B(−2, −1), C(3, −2) are inside, outside, or on the circle $x^2 + y^2 - 2x + y - 5 = 0$.

3 The length of the tangent from the point (3, 2) to the circle $x^2 + y^2 - 2x - 3y + k = 0$ is 9 units. Find the value of k.
4 Find the points of intersection of the line $x + y - 3 = 0$ and the circle $x^2 + y^2 + x - 5y + 4 = 0$.
5 Find the equations of the tangents to the circle $x^2 + y^2 = 289$ which are parallel to the line $8x - 15y = 0$.
6 Write down the equation of the tangent to the circle $x^2 + y^2 - 3x + 5y = 0$ at the point (0, 0).
7 Show that the line $3x - 4y - 10 = 0$ is a common tangent of the two circles $x^2 + y^2 = 4$ and $x^2 + y^2 - 22x - 24y + 240 = 0$.
8 Given the three circles

$$x^2 + y^2 - 16x + 60 = 0$$
$$x^2 + y^2 - 12x + 20 = 0$$
$$x^2 + y^2 - 16x - 12y + 84 = 0$$

find (i) the co-ordinates of a point such that the lengths of the tangents from it to each of the three circles are equal, (ii) the length of each tangent. [LU]
9 (i) Find the radius and co-ordinates of the centre of the circle $x^2 + y^2 - 2x - 6y + 6 = 0$. (ii) If the line $x = 2y$ meets the circle $x^2 + y^2 - 8x + 6y - 15 = 0$ at the points P, Q, find the co-ordinates of P and Q and the equation to the circle passing through P, Q and the point (1, 1). [LU]
10 A circle touches the y-axis at (0, 3) and passes through (9, 0). Find its equation. Find also the equation of the other tangent from the origin. [WJC]

Exercises 19

1 A circle, the co-ordinates of whose centre are both positive, touches both axes of co-ordinates. If it also touches the line $3x + 4y - 60 = 0$ find its equation and the co-ordinates of its point of contact with this line.
2 Show that the distance between the centres of the following circles is equal to the sum of their radii: $x^2 + y^2 - 2x - 4y - 20 = 0$, $x^2 + y^2 - 26x - 22y + 190 = 0$.
3 Prove that the line $3x + 4y = 13$ is a tangent to the circle $x^2 + y^2 - 2x - 3 = 0$ and find the equations of the two tangents perpendicular to this one. [LU]
4 Find the radii and the co-ordinates of the centres of the two circles which touch the x-axis and which pass through the points $(3, -2)$ and $(2, -1)$.
5 Determine the two values of c for which the line $3x + 4y + c = 0$ is a tangent to the circle $x^2 + y^2 - 6x - 2y - 15 = 0$.
6 Show that for all values of θ the line $x\cos\theta + y\sin\theta = a$ is a tangent to the circle $x^2 + y^2 = a^2$ and find the point of contact in terms of θ.
7 State the equation of the straight line which has gradient m and which passes through the point P(0, 18). Show that this line is a tangent to the circle, centre C(4, 6) and radius 10, provided that m satisfies the equation $21m^2 - 24m - 11 = 0$. Find the product of the gradients of the tangents from P to this circle. [AEB]

8 A, B are the points of contact of the tangents from the point P(1, 1) to the circle $x^2 + y^2 - 4x - 6y + 12 = 0$. Find the centre and radius of the circle and the length PA. Hence if the chord AB subtends an angle 2θ at the centre of the circle find the values of $\tan\theta$.

9 Find the centres and radii of the circles $x^2 + y^2 + 8x + 10y - 4 = 0$ and $x^2 + y^2 - 2x - 4 = 0$. Find also the distance between their centres and hence (i) show that the circles intersect at right angles, (ii) find the length of their common chord. [AEB]

10 Show that the pair of tangents from the point (23, 7) to the circle $x^2 + y^2 = 289$ are mutually perpendicular.

11 Show that the line $x - 5 = 0$ always cuts the circle $x^2 + y^2 - (6 + \lambda)x - 6y + (5\lambda - 11) = 0$ in the same two points, whatever the value of λ. Find the co-ordinates of these points.

12 Find the equation of the circle that passes through the points (0, 1), (0, 4), (2, 5). Show that the axis of x is a tangent to this circle and determine the equation of the other tangent which passes through the origin. [JMB]

13 Mark the three points A(0, 2), B(0, −2), C(−4, 2) in a sketch and write down the co-ordinates of the centre, the length of the radius and the equation of the circle through the three points. Show that the line $x + y + 6 = 0$ is a tangent to the circle. Also obtain the equation of a second circle that passes through the two points A, B and touches the line $x + y + 6 = 0$. [JMB]

14 A circle touches both the x-axis and the line $4x - 3y + 4 = 0$. Its centre is in the first quadrant and lies on the line $x - y - 1 = 0$. Prove that its equation is $x^2 + y^2 - 6x - 4y + 9 = 0$. [JMB, part]

15 If $P(x_1, y_1)$ is a point outside the circle $x^2 + y^2 + 2gx + 2fy + c = 0$ show that the length of the tangent PT from P to the circle is given by

$$PT^2 = x_1{}^2 + y_1{}^2 + 2gx_1 + 2fy_1 + c$$

Two circles have centres A(1, 3) and B(6, 8) and intersect at C(2, 6) and D. Find the equation of each of the circles and that of the line CD. The tangents to the circles from a point P are of equal length. Verify that P lies on CD. [JMB]

16 Find the equation of the circle which has as the ends of a diameter the points where the line $x - y = 1$ meets the locus $x^2 + 2y^2 - 4x - 4y + 4 = 0$. [Hint: show that the equation (19.5) can be written $x^2 + y^2 - x(x_1 + x_2) - y(y_1 + y_2) + x_1x_2 + y_1y_2 = 0$ and remember that the sum of the roots of a quadratic equation equal '$-b/a$' and the product 'c/a'.]

17 The circles whose equations are $x^2 + y^2 - x + 6y + 7 = 0$ and $x^2 + y^2 + 2x + 2y - 2 = 0$ intersect at the points A and B. Find (i) the equation of the line AB, (ii) the co-ordinates of A and B. Show that the two given circles intersect at right angles and obtain the equation of the circle which passes through A and B and which also passes through the centres of the two circles. [AEB]

18 Prove that the circles $x^2 + y^2 - 20x - 14y + 113 = 0$ and $4x^2 + 4y^2 + 16x - 16y - 49 = 0$ lie entirely outside each other and find the length of the shortest distance from a point on one circle to a point on the other.

19 Prove that the circles $x^2 + y^2 - 10x - 8y - 59 = 0$ and $x^2 + y^2 - 16x - 16y + 119 = 0$ lie one entirely inside the other and find the length of the shortest distance from a point on one circle to a point on the other.

20 Prove that the circle which has as a diameter the common chord of the two circles $x^2 + y^2 - 14x - 6y + 33 = 0$, $x^2 + y^2 + 2x - 6y - 15 = 0$ touches the axes of co-ordinates.

21 The line joining the points (x_1, y_1) and (x_2, y_2) is a diameter of a circle. Show that the equation of the circle is

$$(x - x_1)(x - x_2) + (y - y_1)(y - y_2) = 0$$

The points A, B and C have co-ordinates $(0, -14)$, $(-5, 1)$ and $(7, -5)$ respectively. The perpendicular from A to BC meets BC at L. Find (i) the equation of the circle through A, L and C, (ii) the ratio $\triangle ALC : \triangle ALB$. Find also, without using tables, the tangent of the angle between the line BC and the tangent at L to the circle through A, L and C. [AEB]

22 The two circles $x^2 + y^2 + 2\lambda x + 3 = 0$ and $x^2 + y^2 + 2\lambda y - 3 = 0$ have centres C_1 and C_2 respectively. If P is one of their points of intersection show that $C_1C_2{}^2 = C_1P^2 + C_2P^2$ for all values of λ.

23 The equations of the sides of a triangle are $x + y - 4 = 0$, $x - y - 4 = 0$, $2x + y - 5 = 0$. Prove that for all numerical values of p and q the equation $p(x + y - 4)(2x + y - 5) + q(x - y - 4)(2x + y - 5) = (x - y - 4) \times (x + y - 4)$ represents a curve passing through the vertices of this triangle.

Find the values of p and q which make this curve a circle and so determine the centre and radius of the circumcircle of the triangle. [JMB]

24 A circle with centre at the point $P(h, k)$ touches the y-axis, and passes through the point $S(2, 0)$. Show that P lies on the curve $y^2 = 4x - 4$, and sketch this curve. Show that the straight line joining $P(h, k)$ to the point $Q(2 + h, 0)$ cuts the curve $y^2 = 4x - 4$ at right angles at P. [LU]

25 Find the equation of the circle circumscribing the triangle whose sides are $x = 0$, $y = 0$, $lx + my = 1$. If l and m can vary so that $l^2 + m^2 = 4l^2m^2$ find the locus of the centre of the circle. [Hint: if $lx + my = 1$ meets the axes at P, Q then PQ is a diameter of the required circle.]

26 Prove that the circles $x^2 + y^2 + 2x - 8y + 8 = 0$ and $x^2 + y^2 + 10x - 2y + 22 = 0$ touch one another. Find (i) the point of contact, (ii) the equation to the common tangent at this point and (iii) the area of the triangle enclosed by this common tangent, the line of centres and the y-axis. [LU]

27 Find the co-ordinates of the centre and the radius of the circle $x^2 + y^2 - 4x - 2y + 4 = 0$. Find the equations of the tangents to this circle from the origin. Show that the line $5x + 12y = 35$ is a tangent to the circle and find the co-ordinates of the centre of the circle which is the reflection of the given circle in this line. [JMB]

28 Two circles, C_1 and C_2, have equations $x^2 + y^2 - 4x - 8y - 5 = 0$ and $x^2 + y^2 - 6x - 10y + 9 = 0$, respectively. Find the x co-ordinates of the points P and Q at which the line $y = 0$ cuts C_1, and show that this line touches C_2. Find the tangent of the acute angle made by the line $y = 0$ with the tangents to C_1 at P and Q. Show that, for all values of the constant λ, the circle C_3 whose equation is

$$\lambda(x^2 + y^2 - 4x - 8y - 5) + x^2 + y^2 - 6x - 10y + 9 = 0$$

passes through the points of intersection of C_1 and C_2. Find the two possible values of λ for which the line $y = 0$ is a tangent to C_3. [JMB]

29 Find the condition that the two circles $x^2 + y^2 + 2g_1x + 2f_1y + c_1 = 0$, $x^2 + y^2 + 2g_2x + 2f_2y + c_2 = 0$ may touch, and prove that, if they touch, the point of contact lies on each of the lines

$$2(g_1 - g_2)x + 2(f_1 - f_2)y + c_1 - c_2 = 0$$
$$(f_1 - f_2)x - (g_1 - g_2)y + f_1g_2 - f_2g_1 = 0$$

30 Find the equation of the circle with centre at the point (2, 3) and radius 5. Find the equation of the tangent at the point (5, 7) and verify that it is parallel to the diameter through the point (−2, 6). Write down the co-ordinates of the point of contact of the other tangent parallel to this diameter. [JMB]

20
The parabola, ellipse and hyperbola

20.1 Introduction

The locus of a point P(x, y) which moves so that the ratio of its distance from a fixed point S (the focus) to its distance from a fixed straight line ZQ (the directrix) is a constant (e, called the eccentricity), has different forms according to the value of e. The locus is known as a parabola when $e = 1$, an ellipse when $e < 1$ and a hyperbola when $e > 1$. We shall see in the sections that follow that each of these is given by a second-degree equation in x and y.

20.2 The parabola ($e = 1$)

Let SZ be the line through the focus perpendicular to the directrix ZQ (see *Figure 20.1*).

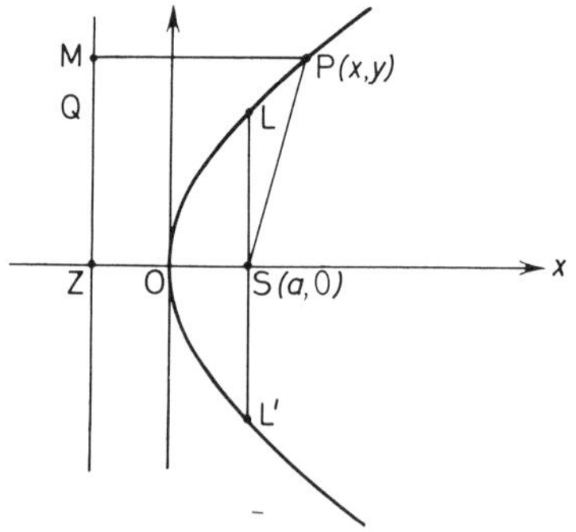

Figure 20.1

By definition, the locus passes through the point midway between S and Z. The form of the equation of the locus depends on the choice of axes. The simplest form of the equation is obtained by taking the origin O as the point midway between S and Z and axes perpendicular and parallel to ZQ.

Let SO = OZ = a referred to these axes. The focus S is the point (a, 0) and the directrix ZQ is the line $x = -a$. If P(x, y) is any point on the locus, PS = PM; therefore

$$\sqrt{[(x-a)^2 + y^2]} = x + a$$
$$(x-a)^2 + y^2 = (x+a)^2$$

which gives

$$y^2 = 4ax \tag{20.1}$$

This is the simplest form of the equation of a parabola and is obtained because of our choice of axes.

To trace the parabola (assuming $a > 0$), we first observe that y is not defined if x is negative, so that the curve lies wholly to the right of the origin. Since we can rewrite the equation $y = \pm 2\sqrt{ax}$, the curve is symmetrical about Ox, which is often referred to as the axis of the parabola. If x is zero, $y^2 = 0$, showing that the y-axis meets the curve in two coincident points at (0, 0), known as the vertex. Hence the y-axis is the tangent at the vertex. The general shape is shown in *Figure 20.1*.

The double ordinate LSL′ through the focus is known as the latus rectum. Since the abscissa of the point L is $x = a$, we find, by substitution in (20.1), that the ordinate LS has length $2a$. Hence

$$\text{LSL}' = 2\text{LS} = 4a \tag{20.2}$$

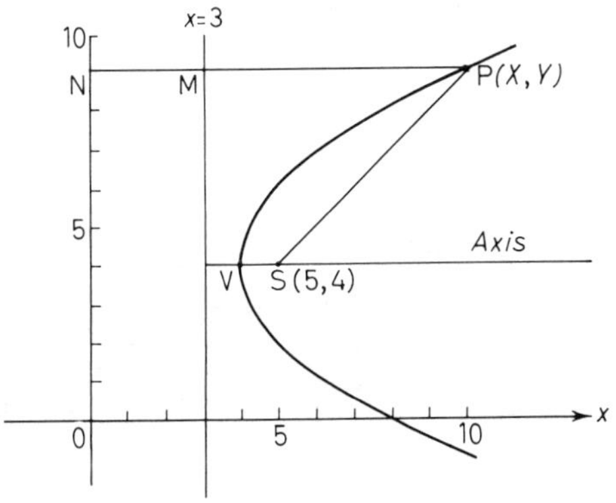

Figure 20.2

Example 1 Find the equation of the parabola with focus (5, 4) and directrix $x = 3$.

Refer to *Figure 20.2*. Let P(X, Y) be any point on the parabola, then P is equidistant from the focus and the directrix. Hence

$$\text{SP} = \text{PM} = \text{PN} - \text{MN}$$
$$\sqrt{[(X-5)^2 + (Y-4)^2]} = X - 3$$
$$(X-5)^2 + (Y-4)^2 = (X-3)^2$$
$$X^2 - 10X + 25 + Y^2 - 8Y + 16 = X^2 - 6X + 9$$
$$Y^2 - 8Y - 4X + 32 = 0$$

The required equation is thus

$$y^2 - 8y - 4x + 32 = 0$$

This may be rewritten

$$(y-4)^2 = 4(x-4)$$

Referring to *Figure 20.2*, the vertex V is the point (4, 4). If the origin of co-ordinates is moved to this point, the equation becomes $y^2 = 4x$, which is the same as (20.1) with $a = 1$.

Example 2 Find the equation of the parabola with focus $(-3, 2)$ and directrix $x - y + 1 = 0$.

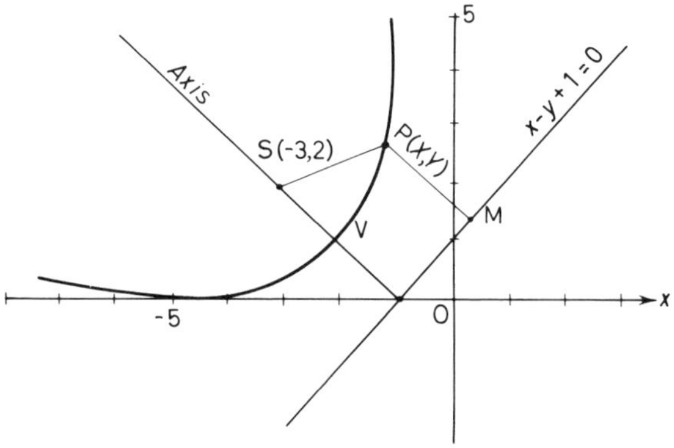

Figure 20.3

Let P(X, Y) be any point on the parabola (see *Figure 20.3*). Then P is equidistant from the focus and the directrix. Hence

$$\text{SP} = \text{PM}$$

$$\sqrt{[(X+3)^2 + (Y-2)^2]} = \frac{X - Y + 1}{\sqrt{2}} \quad \text{(see (17.5) and (18.20))}$$

$$2(X^2 + 6X + 9 + Y^2 - 4Y + 4) = X^2 - 2XY + Y^2 + 2X - 2Y + 1$$

that is

$$X^2 + 2XY + Y^2 + 10X - 6Y + 25 = 0$$

Hence the required equation is

$$x^2 + 2xy + y^2 + 10x - 6y + 25 = 0$$

To reduce this equation to its simplest form (see (20.1)), we should need to change the origin and rotate the axes. The latter is beyond the scope of this book.

Example 3 A telephone wire hangs from two points P, Q 60 m apart. P, Q are on the same level. The mid-point of the telephone wire is 3 m below the level of PQ. Assuming that it hangs in the form of a parabola*, find its equation.

* The true shape of the hanging wire is a catenary but its approximation to a parabola is often of practical use.

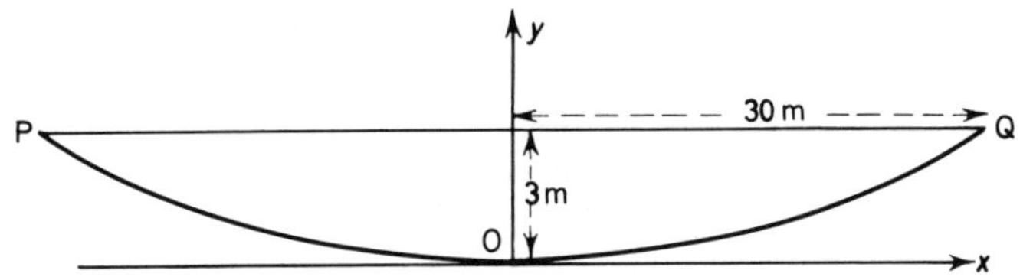

Figure 20.4

With axes as shown in *Figure 20.4*, the required equation is of the form $x^2 = 4ay$.

The point Q has co-ordinates (30, 3) and lies on the curve, so that

$$30^2 = 4 \times a \times 3$$

therefore

$$75 = a$$

The required equation is thus $x^2 = 300y$.

20.3 The equations of the tangent and normal at the point (x_1, y_1) on the parabola $y^2 = 4ax$

Differentiating the equation of the parabola with respect to x, we have

$$2y\frac{dy}{dx} = 4a$$

Hence the gradient of the tangent at the point (x_1, y_1) is $2a/y_1$ and the equation of the tangent is

$$(y - y_1) = \frac{2a}{y_1}(x - x_1)$$

or

$$yy_1 - y_1{}^2 = 2ax - 2ax_1$$

However, since (x_1, y_1) lies on the curve, $y_1{}^2 = 4ax_1$. Hence

$$yy_1 - 4ax_1 = 2ax - 2ax_1$$

and the required equation is

$$yy_1 = 2a(x + x_1) \tag{20.3}$$

It should be noted that the equation of the tangent can be obtained from the original equation of the parabola by replacing y^2 by yy_1 and $4ax$ by $2a(x + x_1)$. This is a similar rule to the one used for a tangent to a circle (see Section 19.3).

The normal to a curve at a point is the line passing through the point and perpendicular to the tangent at the point. Hence, since the slope of the tangent is $2a/y_1$ (see (20.3)), the slope of the normal is $-y_1/2a$. Hence the equation of the normal is

$$y - y_1 = \frac{-y_1}{2a}(x - x_1) \tag{20.4}$$

Example 1 Find the equations of the tangents to the parabola $y^2 = 48x$ at the points (3, 12), (48, −48). Show that these tangents are at right angles and find their point of intersection.

Here $4a = 48$, therefore $a = 12$. For the tangent at the point (3, 12), $x_1 = 3$, $y_1 = 12$. Hence substituting in (20.3), we have

$$y \times 12 = 24(x + 3)$$
$$y = 2x + 6 \tag{i}$$

Similarly for the tangent at the point (48, −48)

$$y(-48) = 24(x + 48)$$
$$y = -\tfrac{1}{2}x - 24 \tag{ii}$$

From (i) and (ii), the slopes of the tangents are 2 and $-\frac{1}{2}$ and the product of these is −1. Hence by (18.15), the tangents are at right angles.

From (i) and (ii), at the point of intersection

$$2x + 6 = y = -\tfrac{1}{2}x - 24$$
$$4x + 12 = -x - 48$$
$$x = -12 \quad \text{and} \quad y = -18$$

Note that, since $a = 12$, $x = -12$ is the equation of the directrix and this point lies on the directrix.

Exercises 20a

1 Sketch the following parabolas showing foci and directrices: (i) $y^2 = 8x$, (ii) $y^2 = -24x$, (iii) $x^2 = -y$, (iv) $x^2 = 12y$, (v) $3y^2 + 8x = 0$.
2 The parabola $y^2 = 4ax$ passes through the point (2, −4). Find the co-ordinates of the focus.
3 A rod rests on two horizontal supports 12 m apart and the maximum sag is 0.1 m. If the supports are at the same level and the rod is in the shape of a parabola find its equation in its simplest form.
4 Find the equations of the tangent and normal (i) to the parabola $y^2 = 4x$ at the point (1, 2), (ii) to the parabola $x^2 = -12y$ at the point (−6, −3).
5 The normal to the parabola $y^2 = 12x$ at the point (3, 6) is produced to meet the curve again at the point Q. Find the co-ordinates of Q.
6 Find the equations of the parabolas with the following foci and directrices: (i) focus (2, 1), directrix $x = -3$; (ii) focus (0, 0), directrix $x + y = 4$; (iii) focus (−2, −3), directrix $3x + 4y - 3 = 0$.
7 The normal at a point P(2, 4) on the parabola $y^2 = 8x$ meets the axis of x at G. N is the foot of the perpendicular from P to the axis. Prove that NG = 4 units.
8 PSQ is a chord of the parabola $y^2 = 24x$. S is the focus and P is the point ($\frac{3}{2}$, 6). Find the co-ordinates of the point Q, and show that the tangents at P and Q are at right angles.
9 A circle with centre (3, 0) and radius 6 units meets the parabola $y^2 = 12x$

at the points P, Q. Prove that the tangents to the parabola at P and Q meet on the circle.

10 The tangent to the parabola $x^2 = 8y$ at the point P(12, 18) meets the tangent at the vertex at the point V. If S is the focus, prove that SV and VP are perpendicular.

20.4 The points of intersection of the line $y = mx + c$ and the parabola $y^2 = 4ax$

In order to find the points of intersection, we solve the two equations simultaneously.

From $y = mx + c$ and $y^2 = 4ax$, we have

$$(mx + c)^2 = 4ax$$

or

$$m^2x^2 + 2(mc - 2a)x + c^2 = 0 \tag{20.5}$$

The discriminant of this quadratic equation is

$$[2(mc - 2a)]^2 - 4m^2c^2$$

or

$$8a^2 - 8amc = 8a(a - mc)$$

Thus the quadratic equation (20.5) has real, equal or complex roots according as $8a^2 - 8amc$ is greater than, equal to or less than zero.

Thus if $c < a/m$, the line meets the parabola in real points; if $c > a/m$, the line does not meet the parabola; if $c = a/m$, the line touches the parabola. Thus

$$y = mx + \frac{a}{m} \tag{20.6}$$

touches the parabola $y^2 = 4ax$ for all values of m.

Example 1 Find the equation of the tangent to the parabola $y^2 = -12x$ which is parallel to the line $y + x = 5$.

Since the tangent is parallel to the line $y + x = 5$, it has the same slope as this line. Hence $m = -1$. And since $y^2 = -12x$ is the equation of the parabola, $4a = -12$, therefore $a = -3$. Substituting for a and m in (20.6), the required equation is

$$y = (-1)x + \frac{-3}{-1}$$

that is

$$y + x = 3$$

Example 2 Show that the point of intersection of two perpendicular tangents to a parabola always lies on the directrix.

Let the equation of the parabola be $y^2 = 4ax$. Then the line $y = mx + a/m$ is always a tangent. If in place of m we write $-1/m$, then the line

$y = -x/m - am$ is also a tangent and is perpendicular to $y = mx + a/m$. By subtraction, the abscissa of the point of intersection of these two tangents is given by

$$\left(mx + \frac{a}{m}\right) - \left(-\frac{1}{m}x - am\right) = 0$$

$$\left(m + \frac{1}{m}\right)x + a\left(\frac{1}{m} + m\right) = 0$$

that is by $x + a = 0$ and this is the equation of the directrix.

Example 3 Find the equations of the tangents from the point (2, 4) to the parabola $y^2 = 6x$.

The equation of the parabola is $y^2 = 6x$, hence $4a = 6$, i.e. $a = \frac{3}{2}$. Hence any tangent to the parabola is of the form $y = mx + 3/2m$. This tangent passes through the point (2, 4) if $4 = 2m + 3/2m$, that is

$$4m^2 - 8m + 3 = 0$$

$$(2m - 1)(2m - 3) = 0$$

$$m = \tfrac{1}{2} \quad \text{or} \quad \tfrac{3}{2}$$

Therefore, the tangents from the point (2, 4) are

$$y = \tfrac{1}{2}x + \frac{\frac{3}{2}}{\frac{1}{2}} \quad \text{i.e.} \quad 2y = x + 6$$

$$y = \tfrac{3}{2}x + \frac{\frac{3}{2}}{\frac{3}{2}} \quad \text{i.e.} \quad 2y = 3x + 2$$

Exercises 20b

1 The tangent to a parabola at any point P meets the directrix at R. If S is the focus, prove that ∠RSP is a right angle.

2 The tangent to a parabola at any point P meets the axis of the parabola at T. PN is drawn perpendicular to the axis to meet it at N and V is the vertex. Prove that TV = VN.

3 P is any point on a parabola whose focus is S. PM is drawn parallel to the axis of the parabola. Prove that the tangent at P bisects ∠SPM.

4 Show that the equations of the tangents from the point (4, 6) to the parabola $y^2 = 5x$ are $4y = 5x + 4$ and $4y = x + 20$.

5 A point source of light is placed at the focus of a parabolic mirror. Show that all the rays will be reflected parallel to the axis of the parabola.

20.5 Parametric equations of the parabola

For all values of t, the equation $y^2 = 4ax$ is always satisfied by

$$x = at^2 \qquad y = 2at \tag{20.7}$$

These are known as the parametric equations of the parabola. $(at^2, 2at)$ can be used as a general point on the parabola $y^2 = 4ax$; t has any value.

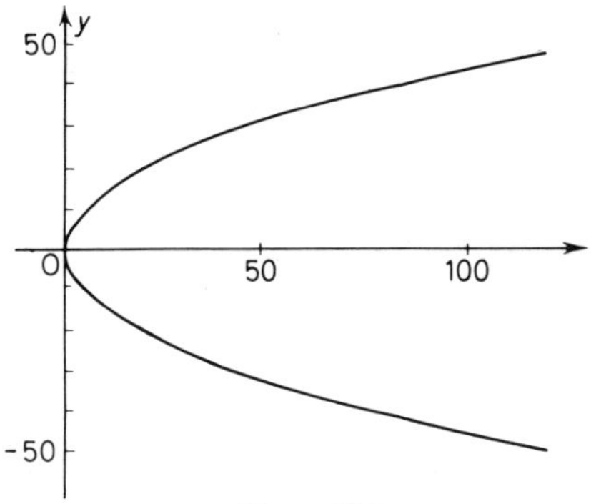

Figure 20.5

Substituting the co-ordinates of the general point $(at^2, 2at)$ in (20.3), we have

$$y.2at = 2a(x + at^2)$$

$$ty = x + at^2 \tag{20.8}$$

which is the equation of the tangent at $(at^2, 2at)$. Also

$$y - 2at = \frac{-2at}{2a}(x - at^2)$$

that is

$$y + tx = 2at + at^3 \tag{20.9}$$

is the equation of the normal at $(at^2, 2at)$.

Example 1 Sketch the parabola whose parametric equations are $x = 5t^2$, $y = 10t$.

This example can be done immediately by eliminating t and obtaining the cartesian equation of the curve, $y^2 = 20x$. However, to illustrate the method of sketching from parametric equations we proceed as follows.

We let t have the values $-5, -4, \ldots, +5$ and find the corresponding values of x and y from the given equations. Thus we can construct the following table.

t	-5	-4	-3	-2	-1	0	1	2	3	4	5
x	125	80	45	20	5	0	5	20	45	80	125
y	-50	-40	-30	-20	-10	0	10	20	30	40	50

The last two lines of the table enable us to plot the points and hence sketch the curve (*Figure* 20.5).

Example 2 The tangent to the parabola $y^2 = 4ax$ at the point P meets the directrix at Q. M is the mid-point of PQ. Find the co-ordinates of M in terms of the parameter of the point P and the locus of M as P moves on the parabola.

Referring to *Figure 20.6*, let P be the point $(ap^2, 2ap)$. Then by (20.8) the equation of the tangent at P is

$$y = \frac{x}{p} + ap$$

Therefore, Q has co-ordinates

$$\left(-a, \frac{-a}{p} + ap\right)$$

and so M has co-ordinates

$$\left[\frac{a}{2}(p^2 - 1), \frac{a}{2}\left(3p - \frac{1}{p}\right)\right]$$

If M is the point (X, Y), we have

$$X = \frac{a}{2}(p^2 - 1) \quad \text{and} \quad Y = \frac{a}{2}\left(3p - \frac{1}{p}\right)$$

We obtain the locus of M by eliminating p from these equations.

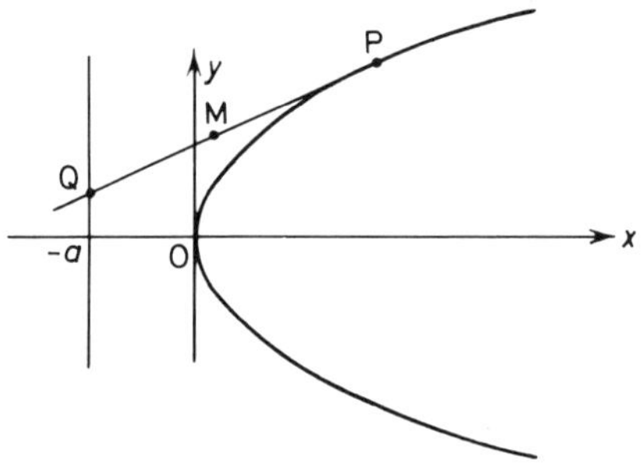

Figure 20.6

Thus we have

$$p^2 = (2X + a)/a$$

therefore

$$Y = \frac{a}{2}\left(\frac{3p^2 - 1}{p}\right)$$

$$= \frac{a}{2}\frac{\left(\dfrac{6X + 3a}{a} - 1\right)}{\sqrt{\left(\dfrac{2X + a}{a}\right)}} = \frac{(6X + 2a)}{2\sqrt{(2X + a)}}\sqrt{a}$$

The locus of M thus has the equation

$$2y\sqrt{(2x + a)} = (6x + 2a)\sqrt{a}$$

that is

$$y^2(2x + a) = a(3x + a)^2$$

Example 3 PSP′ is a focal chord of a parabola, S the focus. If P is the point $(at^2, 2at)$, find the point P′ and hence show that the tangents at P and P′ are at right angles.

Since P is the point $(at^2, 2at)$, the cartesian equation of the curve is $y^2 = 4ax$. Hence S is the point $(a, 0)$.

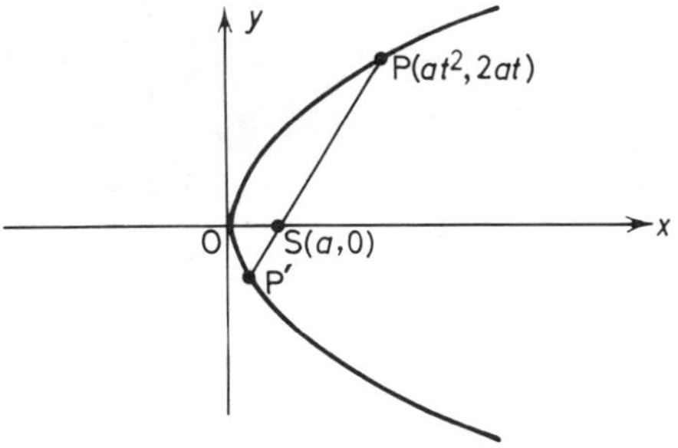

Figure 20.7

Let P′ be the point $(at'^2, 2at')$. Since PSP′ is a straight line (see *Figure 20.7*), the slopes of PS and SP′ are the same. Therefore

$$\frac{2at}{at^2 - a} = \frac{-2at'}{a - at'^2}$$
$$2a^2t - 2a^2tt'^2 = -2a^2t't^2 + 2a^2t'$$

Simplifying, we have

$$t - t' = tt'^2 - t't^2$$
$$(t - t') = tt'(t' - t)$$

Since $t \neq t'$

$$-1 = tt'$$
$$t' = \frac{-1}{t}$$

Therefore, P′ is the point $(a/t^2, -2a/t)$.

From (20.8) the slope of the tangent at $P(at^2, 2at)$ is $1/t$. Hence the slope of the tangent at P′ is $1/(-1/t)$, i.e. $-t$, and the product of these two slopes is -1. Hence the tangents at P, P′ are at right angles.

Exercises 20c

1 Sketch the following parabolas: (i) $x = t^2$, $y = 2t$, (ii) $x = 10t$, $y = 5t^2 - 3$, (iii) $x = 3t^2 + 4$, $y = -6t$, (iv) $x = -8t - 2$, $y = 4t^2 + 1$.

2 Find the length of the latus rectum of the parabola $4x = t^2$, $2y = t$.

3 Find the equations of the tangent and normal at any point on the parabola $x = 6t$, $y = 3t^2$.

4 Show that the point of intersection P of the tangents at the points $A(at_1^2, 2at_1)$, $B(at_2^2, 2at_2)$ on the parabola $y^2 = 4ax$ has co-ordinates $X = at_1t_2$, $Y = a(t_1 + t_2)$. If $t_1 - t_2 = 3$, find the locus of P as A and B vary.

5 The normal at any point P on the parabola $y^2 = 12x$ meets the axis of the parabola at G. Show that the co-ordinates of M, the mid-point of PG, are $(3 + 3t^2, 3t)$. Hence show that the locus of M as P moves round the parabola is $y^2 = 3x - 9$.

Exercises 20d

1 A point moves in such a way that its distance from the point S(5, 12) is always equal to its perpendicular distance from the line $y = 13$. Show that the equation of its locus takes the form $y = ax - bx^2$ and find the constants a, b.

Show that the curve passes through the origin O; find the equation of the tangent at that point; and show that the tangent bisects the angle between OS and the positive direction of the y-axis. [WJC]

2 Obtain the equation of the normal to the parabola $y^2 = 4ax$ at the point $(at^2, 2at)$. The normal at a point P makes an angle of 60° with the x-axis and meets the parabola again at the point Q. Show that PQ = $32a/3$.

3 Prove that the line $ax + by + c + \lambda(a'x + b'y + c') = 0$ is a tangent to the parabola $y^2 = 4x$ if $\lambda^2(a'c' - b'^2) + \lambda(ac' + a'c - 2bb') + ac - b^2 = 0$. Hence, or otherwise, find the equations of the two tangents to the parabola $y^2 = 4x$ which pass through the intersection of the lines $x - y + 1 = 0$, $2x + 3y - 5 = 0$.

4 The points P, Q on the parabola $y^2 = 4ax$ have co-ordinates $(ap^2, 2ap)$, $(aq^2, 2aq)$ respectively. Show that if PQ passes through the focus $(a, 0)$ of the parabola, then $pq = -1$. Express the co-ordinates of the mid-point M of the chord PQ as functions of pq and $p + q$, and find the equation of the locus of the mid-points M of all focal chords. Show that the locus is another parabola and state the co-ordinates of its vertex and focus. Give on one diagram a rough sketch of this locus and of the given parabola. [JMB]

5 Prove that the equation of the chord joining the points $(at_1{}^2, 2at_1)$, $(at_2{}^2, 2at_2)$ on the parabola $y^2 = 4ax$ is $y(t_1 + t_2) - 2x = 2at_1t_2$. A variable chord of this parabola always passes through the point $(4a, 0)$. Snow that the locus of the middle point of the chord is the parabola $y^2 = 2a(x - 4a)$.

6 The tangent and normal at $P(at^2, 2at)$, a point on the parabola $y^2 = 4ax$, meet the x-axis at T and G respectively. Prove that P, T and G are equidistant from the point $(a, 0)$. Hence prove that the tangent at P to the parabola is inclined to the tangent at P to the circle through P, T and G at an angle $\tan^{-1} t$. [LU]

7 The chord joining two variable points A, B on a parabola always passes through a fixed point on the axis. Show that the locus of the point of intersection of the normals at A and B is another parabola.

8 The parabolas $y^2 = 4x$ and $(y + 4\lambda)^2 = -4(x - 4\lambda^2 - 2)$ meet at the points A and B. Show that the line AB passes through the focus of the first parabola for all values of λ.

9 The line $y = k(x - 2)$ meets the parabola $y^2 = -8x$ in the two points P and Q. Find the co-ordinates of the mid-point M of PQ in terms of k. Hence show that as k varies M lies on the parabola $y^2 = 4(2 - x)$.

10 Prove that, in a parabola, the portion of any tangent between the point of contact and the axis of the curve is bisected by the tangent at the vertex.

20.6 The ellipse ($e < 1$)

We recall that S is the focus and ZQ the directrix. If P(x, y) is any point on the curve and PM is perpendicular to ZQ, then

$$SP = ePM$$

Take Z′SZ perpendicular to the directrix ZQ (see *Figure* 20.8). Let the points A, A′ divide SZ internally and externally in the ratio e:1. Thus A, A′ are points on the ellipse.

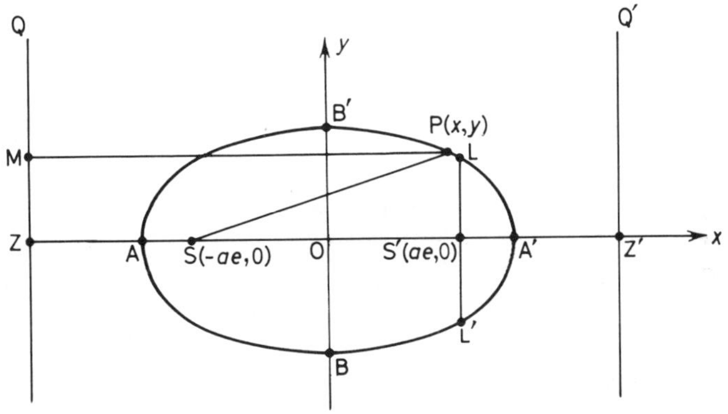

Figure 20.8

The form of the equation, like the parabola, depends on the choice of axes. The simplest form is obtained by taking the origin O as the mid-point of AA′ and axes perpendicular and parallel to AA′.

Let AA′ = $2a$, then OA = OA′ = a. Since A, A′ are points on the locus, by definition, SA = eAZ, SA′ = eA′Z. Hence

$$SA' - SA = e(A'Z - AZ) = eAA'$$

that is

$$(OS + OA') - (OA - OS) = 2ae$$
$$2OS = 2ae \quad (OA = OA')$$
$$OS = ae \tag{20.10}$$

Thus the focus S is the point ($-ae$, 0). Also

$$SA' + SA = e(A'Z + AZ)$$

that is

$$AA' = e[(OA' + OZ) + (OZ - OA)]$$
$$2a = 2eOZ \quad (OA' = OA)$$
$$OZ = a/e \tag{20.11}$$

Thus the directrix ZQ is the line $x = -a/e$. Now PS = ePM, hence

$$(x + ae)^2 + y^2 = e^2\left(x + \frac{a}{e}\right)^2$$

$$x^2 + 2aex + a^2e^2 + y^2 = e^2x^2 + 2aex + a^2$$

$$x^2(1 - e^2) + y^2 = a^2(1 - e^2)$$

that is

$$\frac{x^2}{a^2} + \frac{y^2}{a^2(1 - e^2)} = 1$$

and writing

$$b^2 = a^2(1 - e^2) \tag{20.12}$$

the equation becomes

$$\frac{x^2}{a^2} + \frac{y^2}{b^2} = 1 \tag{20.13}$$

To trace the ellipse we note that, since only even powers of both x and y occur in the equation, the curve is symmetrical about both axes.

Also, since the equation can be rewritten

$$x^2 = a^2\left(1 - \frac{y^2}{b^2}\right) \qquad \text{then} \quad -b \leqslant y \leqslant b$$

or

$$y^2 = b^2\left(1 - \frac{x^2}{a^2}\right) \qquad \text{then} \quad -a \leqslant x \leqslant a$$

The symmetry of the curve enables us to deduce the existence of a second focus S′(ae, 0) and a second directrix Z′Q′($x = a/e$).

To summarise, the curve

$$\frac{x^2}{a^2} + \frac{y^2}{b^2} = 1$$

is an ellipse of eccentricity $e < 1$ given by the equation $b^2 = a^2(1 - e^2)$. The foci are the points ($\pm ae$, 0), the directrices the lines $x = \pm a/e$. AA′ = $2a$ is the major axis, BB′ = $2b$ is the minor axis, and O is the centre of the ellipse. The chord LS′L′ through S′ perpendicular to the major axis is known as the latus rectum.

Example 1 Find (i) the eccentricity, (ii) the co-ordinates of the foci, (iii) the equations of the directrices of the ellipse $x^2/25 + y^2/16 = 1$.

(i) Comparing the given equation with (20.13), we have $a = 5, b = 4$. Substituting in (20.12), we obtain

$$16 = 25(1 - e^2)$$

$$25e^2 = 9$$

$$e = \tfrac{3}{5}$$

(ii) The co-ordinates of the foci are $(\pm ae, 0)$, that is $(\pm 3, 0)$.
(iii) The equations of the directrices are $x = \pm a/e$, that is $x = \pm \frac{25}{3}$.

Example 2 Show that the length of the latus rectum of the ellipse $x^2/a^2 + y^2/b^2 = 1$ is $2b^2/a$.

Referring to *Figure 20.8*, the latus rectum is the line through the focus S′(ae, 0) perpendicular to the major axis AA′. Hence its equation is $x = ae$. The ordinate of the point L(LS′) is therefore obtained by solving the two equations

$$x = ae \qquad \frac{x^2}{a^2} + \frac{y^2}{b^2} = 1$$

Thus

$$\frac{a^2e^2}{a^2} + \frac{y^2}{b^2} = 1$$

$$y^2 = b^2(1 - e^2)$$

Now $\text{LS}' = y = b\sqrt{(1 - e^2)}$ and from (20.12) $\sqrt{(1 - e^2)} = b/a$, therefore $\text{LS}' = b^2/a$ and so the latus rectum $2\text{LS}' = 2b^2/a$.

Exercises 20e

Find (i) the eccentricities, (ii) the co-ordinates of the foci, (iii) the equations of the directrices, (iv) the areas* and (v) sketch the ellipses:

1 $\dfrac{x^2}{100} + \dfrac{y^2}{64} = 1$ **2** $\dfrac{x^2}{64} + \dfrac{y^2}{100} = 1$

3 $\dfrac{4x^2}{25} + \dfrac{4y^2}{9} = 1$ **4** $\dfrac{x^2}{6} + \dfrac{y^2}{4} = 1$

5 $2x^2 + y^2 = 2$ **6** $\dfrac{(x-1)^2}{25} + \dfrac{(y+2)^2}{16} = 1$

7 Find the length of the latus rectum of the ellipse $x^2/169 + y^2/144 = 1$. Hence find the co-ordinates of the four points in which the latera recta meet the ellipse. Verify that these co-ordinates satisfy the equation of the ellipse.

8 Find the equation of the ellipse which has the co-ordinate axes as its principal axes and passes through the points $(-1, 3)$, $(2, -1)$. Find also its eccentricity.

9 An ellipse has eccentricity $e = \frac{4}{5}$. Its foci are the points $(0, \pm 4)$. Find the lengths of its semi-major and semi-minor axes and hence write down its equation.

10 An ellipse of eccentricity $\frac{2}{3}$ has the points (3, 2), (7, 2) as foci. Find the lengths of the major and minor axes, the equations of the directrices, the co-ordinates of its centre, and the equation of the curve.

* Area of ellipse $= 4\displaystyle\int_0^a y\,dx = \frac{4b}{a}\int_0^a \sqrt{(a^2 - x^2)}\,dx = \pi ab$. Verify this result.

20.7 The equations of the tangent and normal at the point (x_1, y_1) on the ellipse $x^2/a^2 + y^2/b^2 = 1$

Differentiating the equation of the ellipse with respect to x

$$\frac{2x}{a^2} + \frac{2y}{b^2}\frac{dy}{dx} = 0$$

Hence the gradient of the tangent at the point (x_1, y_1) is $-b^2x_1/a^2y_1$, and the equation of the tangent is

$$y - y_1 = \frac{-b^2x_1}{a^2y_1}(x - x_1)$$

$$\frac{yy_1}{b^2} - \frac{{y_1}^2}{b^2} = \frac{-xx_1}{a^2} + \frac{{x_1}^2}{a^2}$$

that is

$$\frac{yy_1}{b^2} + \frac{xx_1}{a^2} = \frac{{x_1}^2}{a^2} + \frac{{y_1}^2}{b^2}$$

or, since (x_1, y_1) lies on the ellipse,

$$\frac{yy_1}{b^2} + \frac{xx_1}{a^2} = 1 \tag{20.14}$$

We again note that the equation of the tangent is obtained from the equation of the curve by replacing x^2 by xx_1 and y^2 by yy_1.

Since the normal is perpendicular to the tangent and passes through (x_1, y_1), its equation is

$$(y - y_1) = \frac{a^2y_1}{b^2x_1}(x - x_1)$$

or

$$\frac{y - y_1}{y_1/b^2} = \frac{x - x_1}{x_1/a^2} \tag{20.15}$$

Example 1 Find the equation of the tangent and normal to the ellipse $3x^2 + 14y^2 = 138$ at the point $(-2, 3)$.

The equation of the tangent is by (20.14)

$$3x(-2) + 14y(3) = 138$$

that is

$$7y - x = 23$$

To find the equation of the normal, instead of finding 'a' and 'b' and using (20.15) we can proceed as follows.

The slope of the tangent is $+\frac{1}{7}$, hence the slope of the normal is -7. Since the normal also passes through the point $(-2, 3)$, its equation is

$$(y - 3) = -7(x + 2)$$

that is

$$y + 7x + 11 = 0$$

20.8 The points of intersection of the line $y = mx + c$ and the ellipse $x^2/a^2 + y^2/b^2 = 1$

In order to find the points of intersection, we solve the two equations simultaneously. Substituting $y = mx + c$ in $x^2/a^2 + y^2/b^2 = 1$, we have

$$\frac{x^2}{a^2} + \frac{(mx + c)^2}{b^2} = 1$$

or

$$x^2(a^2m^2 + b^2) + 2cma^2x + a^2(c^2 - b^2) = 0 \qquad (20.16)$$

The discriminant of this quadratic is

$$(2cma^2)^2 - 4[a^2(c^2 - b^2)(a^2m^2 + b^2)] \quad \text{or} \quad 4a^2b^2(b^2 + a^2m^2 - c^2)$$

Thus the quadratic (20.16) has real, equal or complex roots according as c^2 is less than, equal to, or greater than $b^2 + a^2m^2$.

If $c^2 = a^2m^2 + b^2$, the line is a tangent to the ellipse; thus the lines

$$y = mx \pm \sqrt{(a^2m^2 + b^2)} \qquad (20.17)$$

always touch the ellipse.

Example 1 Find the equations of the tangents to the ellipse $x^2 + 2y^2 = 19$ which are parallel to the line $x + 6y = 5$.

Since the tangents are parallel to the line they have the same slope as the line, that is $m = -\frac{1}{6}$; hence by (20.17) the required equations are

$$y = -\tfrac{1}{6}x \pm \sqrt{(a^2\tfrac{1}{36} + b^2)}$$

Rewriting the given equation of the ellipse in the form $x^2/19 + y^2/\frac{19}{2} = 1$, we have $a^2 = 19$, $b^2 = \frac{19}{2}$. Hence the equations of the tangents are

$$y = \tfrac{1}{6}x \pm \sqrt{(19.\tfrac{1}{36} + \tfrac{19}{2})}$$
$$= -\tfrac{1}{6}x \pm \tfrac{19}{6}$$

that is

$$6y + x = \pm 19$$

Example 2 The pair of tangents from the point P to the ellipse $x^2/a^2 + y^2/b^2 = 1$ are always at right angles. Show that the locus of P is the circle $x^2 + y^2 = a^2 + b^2$.

The line $y = mx \pm \sqrt{(a^2m^2 + b^2)}$ is always a tangent to the given ellipse. This line passes through a point $P(X, Y)$ if

$$Y = mX \pm \sqrt{(a^2m^2 + b^2)}$$

Since X, Y, a, b are given, this is a quadratic equation in m giving the slopes of the two tangents from P to the ellipse. This quadratic equation can be written in the form

$$(Y - mX)^2 = a^2m^2 + b^2$$

that is

$$m^2(X^2 - a^2) - 2mXY + (Y^2 - b^2) = 0$$

The two tangents are at right angles if their slopes are m and $-1/m$, that is if the product of their slopes is -1. Thus for perpendicular tangents, the product of the roots of this equation must be -1. So the required condition is

$$\frac{Y^2 - b^2}{X^2 - a^2} = -1$$

that is

$$X^2 + Y^2 = a^2 + b^2$$

which is the condition for the point P(X, Y) to lie on the circle $x^2 + y^2 = a^2 + b^2$, which is known as the director circle of the ellipse.

Exercises 20f

Find the equations of the tangents and normals to the following ellipses at the points stated:

1 $3x^2 + 2y^2 = 30$, (2, 3)
2 $4x^2 + 5y^2 = 24$, (1, 2)
3 $a^2x^2 + b^2y^2 = 2a^2b^2$, $(-b, a)$

Write down the equations of the tangents to the following ellipses, with the given gradients:

4 $x^2/3 + y^2/2 = 1$, gradient 2
5 $x^2 + 2y^2 = 8$, gradient 2
6 $4x^2 + 5y^2 = 20$, gradient 3
7 Show that the pair of tangents from the point (3, 4) to the ellipse $x^2/16 + y^2/9 = 1$ are at right angles.
8 The normals to the ellipse $x^2 + 4y^2 = 100$ at the points A(6, 4) and B(8, 3) meet at N. If P is the mid-point of AB and O is the origin, show that OP is perpendicular to ON.
9 Show that the slopes of the tangents from the point (h, k) to the ellipse $x^2/a^2 \pm y^2/b^2 = 1$ are given by the quadratic equation

$$m^2(h^2 - a^2) - 2mhk + (k^2 - b^2) = 0 \quad \text{[see Example 2]}$$

By considering the condition for these roots to be complex, show that (h, k) lies inside the ellipse if $h^2/a^2 + k^2/b^2 - 1$ is less than zero.
10 Find the locus of a point P which moves so that the sum of its distances from two fixed points A and B 8 units apart is always 14 units. Take AB and its perpendicular bisector as the axes of x and y respectively.

20.9 Parametric equations of an ellipse

For all values of θ, the equation $x^2/a^2 + y^2/b^2 = 1$ is always satisfied by

$$x = a\cos\theta \qquad y = b\sin\theta \tag{20.18}$$

These are parametric equations of the ellipse. $(a\cos\theta, b\sin\theta)$ can be used as a general point on the ellipse $x^2/a^2 + y^2/b^2 = 1$. θ is often referred to as the eccentric angle.

Example 1 Find the equations of the tangent and normal at any point on the ellipse $x^2/a^2 + y^2/b^2 = 1$.

Any point on the ellipse is $(a\cos\theta, b\sin\theta)$. Hence by (20.14) the equation of the tangent is

$$\frac{xa\cos\theta}{a^2} + \frac{yb\sin\theta}{b^2} = 1$$

that is

$$\frac{x\cos\theta}{a} + \frac{y\sin\theta}{b} = 1$$

The slope of the tangent is $(-b/a)\cot\theta$, hence the slope of the normal is $(a/b)\tan\theta$ and the equation of the normal is

$$by\cos\theta - b^2\sin\theta\cos\theta = ax\sin\theta - a^2\sin\theta\cos\theta$$

$$(y - b\sin\theta) = (a/b)\tan\theta(x - a\cos\theta)$$

that is

$$by\cos\theta - ax\sin\theta = (b^2 - a^2)\sin\theta\cos\theta$$

Example 2 If S and S′ are the foci of an ellipse and P any point on its circumference, show that SP + PS′ = $2a$, where $2a$ is the length of the major axis.

Let the equation of the ellipse be $x^2/a^2 + y^2/b^2 = 1$.

Any point P on it has co-ordinates $(a\cos\theta, b\sin\theta)$ and the foci are $S(-ae, 0)$, $S'(ae, 0)$. Hence

$$\begin{aligned}
SP^2 &= (a\cos\theta + ae)^2 + b^2\sin^2\theta \\
&= a^2\cos^2\theta + 2a^2e\cos\theta + a^2e^2 + a^2(1 - e^2)\sin^2\theta \\
&= a^2(\cos^2\theta + \sin^2\theta) + 2a^2e\cos\theta + a^2e^2(1 - \sin^2\theta) \\
&= a^2 + 2a^2e\cos\theta + a^2e^2\cos^2\theta \\
&= a^2(1 + e\cos\theta)^2
\end{aligned}$$

Hence

$$SP = a(1 + e\cos\theta) \qquad (20.19)$$

Similarly,

$$S'P = a(1 - e\cos\theta) \qquad (20.20)$$

whence

$$SP + S'P = 2a \qquad (20.21)$$

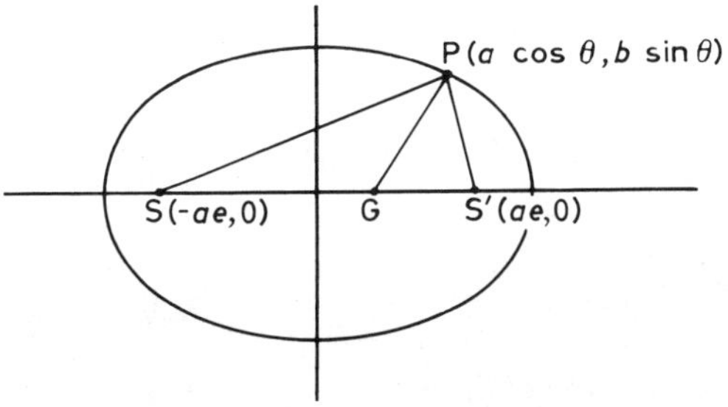

Figure 20.9

Example 3 If PG is the normal at P, show that PG bisects the angle SPS′, where S, S′ are the foci (see *Figure 20.9*).

Let P be the point $(a\cos\theta, b\sin\theta)$. By equation (20.15), the equation of PG is

$$\frac{y - b\sin\theta}{b\sin\theta/b^2} = \frac{x - a\cos\theta}{a\cos\theta/a^2}$$

that is

$$\frac{\cos\theta}{a}(y - b\sin\theta) = \frac{\sin\theta}{b}(x - a\cos\theta)$$

This meets the x-axis where $y = 0$, so that

$$-\frac{b\cos\theta\sin\theta}{a} = \frac{\sin\theta}{b}(x - a\cos\theta)$$

$$-\frac{b^2\cos\theta}{a} = x - a\cos\theta$$

$$-\frac{b^2 - a^2}{a}\cos\theta = x$$

Therefore

$$ae^2\cos\theta = x \quad (\text{since } b^2 = a^2(1 - e))$$

Referring to *Figure 20.9*,

$$SG = ae^2\cos\theta + ae$$

$$S'G = ae - ae^2\cos\theta$$

Hence

$$\frac{SG}{S'G} = \frac{1 + e\cos\theta}{1 - e\cos\theta} = \frac{SP}{S'P} \quad (\text{from (20.19), (20.20)})$$

Therefore, PG is the internal bisector of the angle SPS′.

Exercises 20g

1 Find the equation of the tangent at any point $(a\cos\theta, b\sin\theta)$ on an ellipse $x^2/a^2 + y^2/b^2 = 1$. Hence show that the equation of the normal can be written in the form

$$\frac{ax}{\cos\theta} - \frac{by}{\sin\theta} = a^2 - b^2$$

2 Show that the equation of the line joining two points whose eccentric angles are θ and ϕ is given by $(x/a)\cos\frac{1}{2}(\theta + \phi) + (y/b)\sin\frac{1}{2}(\theta + \phi) = \cos\frac{1}{2}(\theta - \phi)$. Deduce the equation of the tangent at the point θ.

3 PG, PN and PT are respectively the normal, the ordinate and the tangent at P any point on an ellipse. Also, if G, N, T are the points where they cut the major axis prove that (i) $ON.OT = OA^2$ and (ii) $OG = e^2ON$ (O is the origin).

4 $Q(-a\sin\theta, b\cos\theta)$ and $Q'(a\sin\theta, -b\cos\theta)$ are any two points on an ellipse $x^2/a^2 + y^2/b^2 = 1$. Show that QQ′ passes through the origin.

5 Show that the tangents to the ellipse $x^2/a^2 + y^2/b^2 = 1$ at points whose eccentric angles differ by 90° meet on the ellipse $x^2/a^2 + y^2/b^2 = 2$.

Exercises 20h

1 Find the equation of the tangent to the ellipse $4x^2 + 9y^2 = 72$ at the point (3, 2). Also find the equations of the tangent perpendicular to this one.

2 The tangent at the point θ to the curve $x = a\cos\theta$, $y = b\sin\theta$ meets the x-axis at A and the y-axis at B. If O is the origin, find the minimum area of triangle AOB.

3 In the preceding question, find the locus of the mid-point of AB.

4 Plot the points on the curve given by the equations $x = \cos t$, $y = \cos 2t$ for the values 0°, 30°, 60°, ..., 180° of t and sketch the curve. Prove that the distance of any point of the curve from the point $(0, -\frac{7}{8})$ is the same as its distance from the line $y = -\frac{9}{8}$. [LU]

5 Show that, for every value of ϕ, the point $P(a\cos\phi, b\sin\phi)$ lies on the ellipse $x^2/a^2 + y^2/b^2 = 1$. Obtain the equation of the tangent at P in the form $(x/a)\cos\phi + (y/b)\sin\phi = 1$. If the tangent at P meets the axes in TT′ and the diameter through P meets the ellipse again at P′, show that

$$\tan \text{TP'T'} = 2\text{OT}.\text{OT'}/(a^2 + b^2 + \text{OP}^2) \quad \text{(O being the origin)}$$

[WJC]

6 (a) Find the equations of the tangents of gradient $\frac{1}{2}$ to the ellipse $x^2 + 6y^2 = 15$.

(b) If the normal at a variable point P on the ellipse $x^2/a^2 + y^2/b^2 = 1$ meets the x-axis in Q, show that the locus of the mid-point of PQ is an ellipse concentric with the given ellipse. Find the eccentricity of this ellipse if that of the given ellipse is $\frac{1}{4}$. [JMB]

7 Show that the equation of the tangent to the ellipse $x^2/a^2 + y^2/b^2 = 1$ at the point $P(a\cos\theta, b\sin\theta)$ is $(x/a)\cos\theta + (y/b)\sin\theta = 1$. If R, R′ are the feet of the perpendiculars from the foci S, S′ on to the tangent at P, prove that SR.S′R′ $= b^2$. Show also that

$$\frac{\text{RR'}}{\text{SS'}} = \frac{a}{\sqrt{(a^2 + b^2\cot^2\theta)}}$$

[JMB]

8 The tangent and normal at the point $P(a\cos\theta, b\sin\theta)$ on the ellipse $x^2/a^2 + y^2/b^2 = 1$ meet the axis of x at $(x_1, 0)$ and $(x_2, 0)$ respectively. If θ is small show that $x_1 = a + \frac{1}{2}a\theta^2$ approximately. Find a similar approximation for x_2. [JMB, part]

9 A perpendicular is drawn, from the point $(0, -b)$ on the ellipse $x^2/a^2 + y^2/b^2 = 1$, to the tangent at any point $P(a\cos\theta, b\sin\theta)$ on the same ellipse. Write down an expression for the length of this perpendicular, and prove that the length has a stationary value when P is at either end of the minor axis, but has no other stationary value unless $2b^2 < a^2$. [LU]

10 Find the equation of the normal to the ellipse $x^2/a^2 + y^2/b^2 = 1$ at the point P whose eccentric angle is θ. The tangent and normal at P cuts the y-axis at T and G respectively. Prove that the circle on TG as diameter passes through the foci. Find the centre and radius of this circle. [SUJB]

20.10 The hyperbola ($e > 1$)

We obtain the simplest equation of the hyperbola in a similar manner to that used in Section 20.6 to obtain the equation of the ellipse. Referring to *Figure 20.10*, S is the focus, ZQ the directrix, P(x, y) any point on the curve and PM is perpendicular to ZQ.

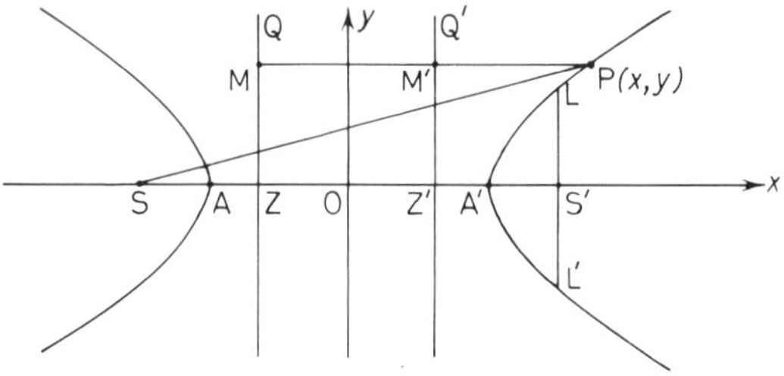

Figure 20.10

SZS′ is perpendicular to the directrix ZQ. A and A′ are the points dividing SZ internally and externally in the ratio $e:1$. Thus A, A′ are points on the hyperbola. O is the mid-point of AA′ and the axes are as shown in *Figure 20.10*. Let AA′ $= 2a$.

Following the method used for the ellipse, OS $= ae$, OZ $= a/e$; thus S is the point $(-ae, 0)$ and ZQ is the line $x = -a/e$.

$$\mathrm{PS} = e\mathrm{PM}$$

$$(x + ae)^2 + y^2 = e^2(x + a/e)^2$$

$$x^2 + 2aex + a^2e^2 + y^2 = e^2x^2 + 2aex + a^2$$

$$a^2(e^2 - 1) = x^2(e^2 - 1) - y^2$$

thus

$$\frac{x^2}{a^2} - \frac{y^2}{a^2(e^2 - 1)} = 1$$

and writing $b^2 = a^2(e^2 - 1)$ the equation becomes

$$\frac{x^2}{a^2} - \frac{y^2}{b^2} = 1 \qquad (20.22)$$

To trace the hyperbola, we note that only even powers of x and y occur in the equation. Hence the curve is symmetrical about both axes. Also by this symmetry, there is a second focus S′$(ae, 0)$ and a second directrix $x = a/e$.

Further, since the equation can be rewritten $y^2/b^2 = x^2/a^2 - 1$ and the left-hand side is always positive, $x^2/a^2 - 1$ must be positive, hence there is no part of the curve for values of x between $+a$ and $-a$. On the other hand, since $x^2/a^2 = 1 + y^2/b^2$, y can have all values.

To summarise, the curve

$$\frac{x^2}{a^2} - \frac{y^2}{b^2} = 1$$

is a hyperbola of eccentricity $e(> 1)$ given by the equation

$$b^2 = a^2(e^2 - 1) \tag{20.23}$$

The foci are the points $(\pm ae, 0)$, the directrices the lines $x = \pm a/e$; $AA' = 2a$ is the transverse axis and O is the centre. The chord LS′L′ through S′ perpendicular to the major axis is the latus rectum.

Example 1 Find (i) the eccentricity, (ii) the co-ordinates of the foci, (iii) the equations of the directrices of the hyperbola $x^2/9 - y^2/16 = 1$.

(i) Comparing the equation with (20.22), we have $a^2 = 9$, $b^2 = 16$. Substituting in (20.23),

$$16 = 9(e^2 - 1)$$
$$25 = 9e^2$$
$$\tfrac{5}{3} = e$$

(ii) The co-ordinates of the foci are $(\pm ae, 0)$, that is $(\pm 5 \times \frac{5}{3}, 0)$ or $(\pm\frac{25}{3}, 0)$.

(iii) The equations of the directrices are $x = \pm a/e$, that is $x = \pm 3$.

Exercises 20i

Find (i) the eccentricities, (ii) the co-ordinates of the foci, (iii) the equation of the directrices and (iv) sketch the hyperbolae:

1 $x^2/4 - y^2/23 = 1$ **2** $y^2/9 - x^2/7 = 1$ **3** $x^2/4 - 4y^2/33 = 1$

4 $56x^2 - 25y^2 = 1400$ **5** $y^2/2 - x^2 = 1$ **6** $x^2 - y^2 = 25$

7 Find the length of the latus rectum of the hyperbola $x^2/9 - y^2/7 = 1$. Hence find the co-ordinates of the four points in which the latera recta meet the hyperbola. Verify that these co-ordinates satisfy the equation of the hyperbola.

8 The foci of a hyperbola are the points $(\pm 7, 0)$. Find the equation of the curve if $e = \frac{7}{6}$. If the eccentricity is unaltered but the foci are the points $(0, \pm 7)$, what is the equation?

9 The centre of a hyperbola is at the origin and its transverse axis lies along the x-axis. Find the equation of the hyperbola if it passes through the points $(6, \frac{11}{5})$ and $(-5, 0)$.

10 Referring to *Figure 20.10*, show that $PS = ex + a$ and that $PS' = ex - a$. Hence prove that the difference of the focal distances is constant and equal to the length of the transverse axis. [Hint: $PS = ePM$ and $PS' = ePM'$.] (See also Example 2 in Section 20.9 and Example 3 in Section 20.12.)

20.11 Properties of the hyperbola $x^2/a^2 - y^2/b^2 = 1$

Many of the results for the hyperbola can be obtained from the corresponding results for the ellipse by writing $-b^2$ in place of b^2.

(i) The equation of the tangent at the point (x_1, y_1) is

$$\frac{xx_1}{a^2} - \frac{yy_1}{b^2} = 1 \tag{20.24}$$

(ii) The equation of the normal at the point (x_1, y_1) is

$$\frac{y - y_1}{y_1/-b^2} = \frac{x - x_1}{x_1/a^2} \qquad (20.25)$$

(iii) The line $y = mx + c$ meets the hyperbola in real, or coincident points or not at all, according as c^2 is greater than, equal to or less than $a^2m^2 - b^2$.
(iv) The lines

$$y = mx \pm \sqrt{(a^2m^2 - b^2)} \text{ always touch the hyperbola} \qquad (20.26)$$

20.12 Parametric equations of the hyperbola $x^2/a^2 - y^2/b^2 = 1$

The most usual forms of the parametric equations are

$$x = a \sec\theta \qquad y = b \tan\theta \qquad (20.27)$$

(See also Exercises 20j, question 5.)

Example 1 Find the equations of the tangent and normal at any point on the hyperbola $x^2/a^2 - y^2/b^2 = 1$.

Any point on the hyperbola is $(a \sec\theta,\ b \tan\theta)$. Hence by (20.24), the equation of the tangent is

$$\frac{xa \sec\theta}{a^2} - \frac{yb \tan\theta}{b^2} = 1$$

that is

$$\frac{x}{a}\sec\theta - \frac{y}{b}\tan\theta = 1 \qquad (20.28)$$

The slope of this tangent is

$$\frac{b \sec\theta}{a \tan\theta} \quad \text{or} \quad \frac{b}{a \sin\theta}$$

Hence the equation of the normal is

$$(y - b\tan\theta) = \frac{-a\sin\theta}{\text{b}}(x - a\sec\theta)$$

which reduces to

$$ax\sin\theta + by = (a^2 + b^2)\tan\theta \qquad (20.29)$$

Example 2 P is any point on a hyperbola centre C. The normal at P meets the major axis at G and the ordinate at P meets the major axis at N. Prove that $\text{CG} = e^2\text{CN}$.

Let P be the point $(a\sec\theta,\ b\tan\theta)$, as shown in *Figure 20.11*. CN is the abscissa of P and therefore

$$\text{CN} = a\sec\theta \qquad \text{(i)}$$

From (20.29) the equation of PG is

$$ax\sin\theta + by = (a^2 + b^2)\tan\theta$$

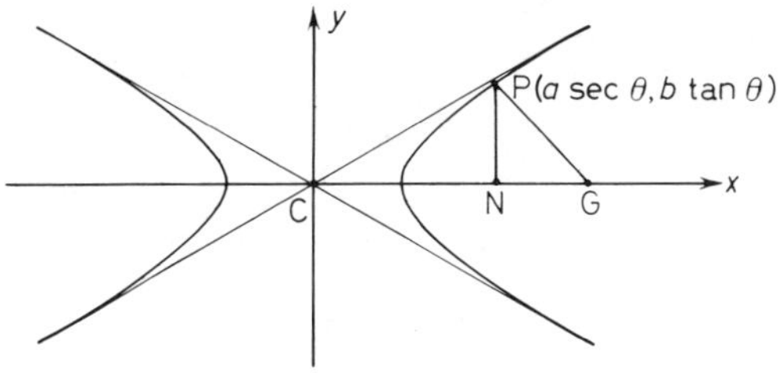

Figure 20.11

G lies on the x-axis ($y = 0$) and therefore its abscissa is given by

$$ax \sin\theta = (a^2 + b^2)\tan\theta$$

$$CG = \frac{a^2 + b^2}{a}\,\frac{\tan\theta}{\sin\theta}$$

$$= \frac{a^2 + b^2}{a}\sec\theta \qquad \text{(ii)}$$

From (i) and (ii)

$$\frac{CG}{CN} = \frac{a^2 + b^2}{a^2}$$

$$= \frac{a^2 + a^2(e^2 - 1)}{a^2} \qquad \text{(see (20.23))}$$

$$= e^2$$

Therefore

$$CG = e^2 CN$$

Example 3 If P is any point on a hyperbola whose foci are S and S′, prove that S′P − SP is constant.

Let P be the point $(a\sec\theta,\ b\tan\theta)$ on the hyperbola $x^2/a^2 - y^2/b^2 = 1$. The foci S′ and S are the points $(-ae, 0)$ and $(ae, 0)$ respectively. Thus

$$\begin{aligned} SP^2 &= (a\sec\theta - ae)^2 + b^2\tan^2\theta \\ &= a^2\sec^2\theta - 2a^2e\sec\theta + a^2e^2 + a^2(e^2 - 1)(\sec^2\theta - 1) \\ &= a^2\sec^2\theta - 2a^2e\sec\theta + a^2e^2 + a^2e^2\sec^2\theta - a^2\sec^2\theta - a^2e^2 + a^2 \\ &= a^2 - 2a^2e\sec\theta + a^2e^2\sec^2\theta \\ &= a^2(e\sec\theta - 1)^2 \end{aligned}$$

Therefore

$$SP = a(e\sec\theta - 1)$$

(Since $e > 1$ and $\sec\theta > 1$, this cannot be $1 - e\sec\theta$.) Similarly

$$S'P = a(e\sec\theta + 1)$$

Therefore

$$S'P - SP = 2a$$

Exercises 20j

1 Find the equations of the tangent and normal to the hyperbola. $9x^2 - 4y^2 = 36$ at the point $(4, 3\sqrt{3})$.

2 Show that the equation of the chord joining the points $(a\sec\theta, b\tan\theta)$ and $(a\sec\phi, b\tan\phi)$ on the hyperbola $x^2/a^2 - y^2/b^2 = 1$ is

$$\frac{x}{a}\cos\tfrac{1}{2}(\theta-\phi) - \frac{y}{b}\sin\tfrac{1}{2}(\theta+\phi) = \cos\tfrac{1}{2}(\theta+\phi)$$

Deduce the equation of the tangent at the point $(a\sec\theta, b\tan\theta)$.

3 Show that the two tangents to the hyperbola $x^2/4 - y^2 = 1$ which are parallel to the line $y = 2x - 3$ are a distance $2\sqrt{3}$ apart.

4 Find the condition for the line $lx + my = n$ to touch the hyperbola $x^2/a^2 - y^2/b^2 = 1$. By writing $+b^2$ in place of $-b^2$, deduce the condition for the same line to meet the ellipse $x^2/a^2 + y^2/b^2 = 1$.

5 Show that the point

$$\left[\frac{a}{2}\left(t+\frac{1}{t}\right), \frac{b}{2}\left(t-\frac{1}{t}\right)\right]$$

always lies on the hyperbola $x^2/a^2 - y^2/b^2 = 1$ for all values of t. Derive the equation of the tangent at this point.

6 P is any point on a hyperbola whose foci are S, S′. The tangent and normal at P meet the axis of the hyperbola at T and N respectively. Prove that PT, PN are the internal and external bisectors of the angle SPS′.

7 The pair of tangents from the point P to the hyperbola $x^2/a^2 - y^2/b^2 = 1$ are always at right angles. Show that the locus of P is the circle $x^2 + y^2 = a^2 - b^2$ (the director circle). [Hint: refer to Section 20.8, Example 2.]

8 Show that the eccentricities e_1 and e_2 of the hyperbolas $x^2/a^2 - y^2/b^2 = 1$ and $-x^2/a^2 + y^2/b^2 = 1$ satisfy the relation $1/e_1^2 + 1/e_2^2 = 1$.

9 The tangent and ordinate at the point P on the hyperbola $x^2/a^2 - y^2/b^2 = 1$ meet the x-axis at T and N respectively. If C is the centre of the hyperbola, show that $\text{CT}.\text{CN} = a^2$.

10 Find the equations and the points of contact of the tangents to the hyperbola $2x^2 - 3y^2 = 5$ which are parallel to $8x = 9y$.

20.13 Asymptotes of the hyperbola $x^2/a^2 - y^2/b^2 = 1$

The definition of an asymptote is that it is a straight line which meets a curve in two points at infinity, but which is not altogether at infinity.

The abscissae of the points of intersection of the line $y = mx + c$ and the hyperbola $x^2/a^2 - y^2/b^2 = 1$ are given by the equation

$$\frac{x^2}{a^2} - \frac{(mx+c)^2}{b^2} = 1$$

or, rearranged as a quadratic in $1/x$,

$$a^2(c^2+b^2)\frac{1}{x^2} + 2a^2mc\frac{1}{x} + (a^2m^2 - b^2) = 0 \qquad (20.30)$$

This equation has two zero roots if both $2a^2mc = 0$ and $a^2m^2 - b^2 = 0$.

That is, if $m = \pm b/a$ and $c = 0$, the line $y = mx + c$ meets the hyperbola in two points such that $1/x = 0$. If $1/x = 0$, x is infinite, and thus

$$y = +\frac{b}{a}x \quad \text{and} \quad y = -\frac{b}{a}x \tag{20.31}$$

both meet the curve in two points at infinity and are thus the asymptotes.

The lines both pass through the origin and are equally inclined to the x-axis at angles $\pm \arctan b/a$. Their combined equation is

$$\left(y - \frac{b}{a}x\right)\left(y + \frac{b}{a}x\right) = 0$$

that is

$$\frac{x^2}{a^2} - \frac{y^2}{b^2} = 0 \tag{20.32}$$

The lines are shown in *Figure 20.12* as LOL′ and KOK′.

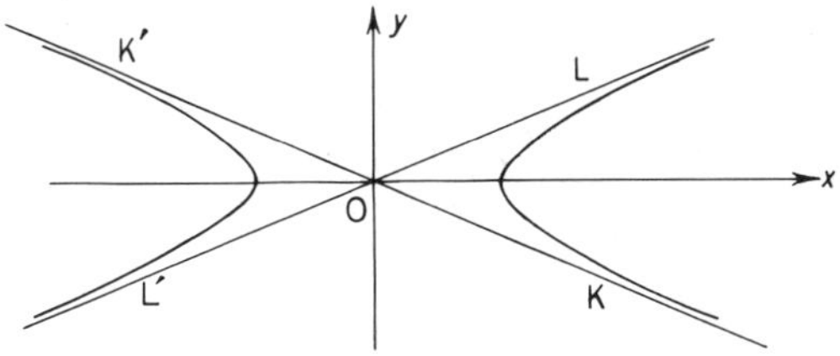

Figure 20.12

Example 1 Show that any straight line parallel to an asymptote will meet the curve in one point at infinity and one finite point.

Any line parallel to an asymptote has the equation

$$y = \pm\frac{b}{a}x + k \qquad (k \neq 0)$$

that is, its slope $m = \pm b/a$.

Hence from equation (20.30) the abscissae of the points of intersection of the line with the hyperbola are given by

$$a^2(k^2 + b^2)\frac{1}{x^2} + 2a^2\left(\pm\frac{b}{a}\right)k\frac{1}{x} + 0 = 0$$

The roots of this equation are

$$\frac{1}{x} = 0 \quad \text{and} \quad \frac{1}{x} = \mp\frac{2bk}{a(k^2 + b^2)}$$

that is, one value of x is infinite and, since $k \neq 0$, the other is finite.

Example 2 P is any point on the hyperbola $x^2/a^2 - y^2/b^2 = 1$ and the tangent at P meets the asymptotes in A and B. Show that P is the mid-point of AB.

Let P be the point $(a \sec\theta, b \tan\theta)$. The tangent at P is (see 20.28)

$$\frac{x}{a}\sec\theta - \frac{y}{b}\tan\theta = 1 \qquad \text{(i)}$$

The combined equation of the asymptotes is

$$\frac{x^2}{a^2} - \frac{y^2}{b^2} = 0 \qquad \text{(ii)}$$

From (i)

$$y = \frac{b}{\tan\theta}\left(\frac{x}{a}\sec\theta - 1\right)$$

Substituting (ii)

$$\frac{x^2}{a^2} - \cot^2\theta\left(\frac{x}{a}\sec\theta - 1\right)^2 = 0$$

$$\frac{x^2}{a^2}(1 - \operatorname{cosec}^2\theta) + \frac{2x}{a}\cot^2\theta\sec\theta - \cot^2\theta = 0$$

$$-\frac{x^2}{a^2}\cot^2\theta + \frac{2x}{a}\cot^2\theta\sec\theta - \cot^2\theta = 0$$

that is

$$\frac{x^2}{a^2} - \frac{2x}{a}\sec\theta + 1 = 0$$

which is a quadratic in x whose roots x_1, x_2 are the abscissae of the points A and B. Now

$$x_1 + x_2 = \frac{2\sec\theta}{a}\bigg/\frac{1}{a^2}$$
$$= 2a\sec\theta$$

that is

$$\tfrac{1}{2}(x_1 + x_2) = a\sec\theta$$

which is the abscissa of the point P.

Similarly, if x is eliminated from (i) and (ii), half the sum of the ordinates of A and B is equal to the ordinate of the point P. Hence P is the mid-point of AB.

20.14 The rectangular hyperbola

If the asymptotes of a hyperbola are at right angles, it is known as a rectangular hyperbola. The asymptotes will each be inclined at 45° to the x-axis. Hence from equations (20.31)

$$\frac{b}{a} = \pm 1$$

and the equation of the hyperbola can be written

$$\frac{x^2}{a^2} - \frac{y^2}{a^2} = 1$$

$$x^2 - y^2 = a^2 \tag{20.33}$$

Example 1 Show that the eccentricity of any rectangular hyperbola is $\sqrt{2}$.
Any rectangular hyperbola has the equation

$$\frac{x^2}{a^2} - \frac{y^2}{a^2} = 1$$

Hence from (20.23)

$$a^2 = a^2(e^2 - 1)$$

whence

$$e = \sqrt{2}$$

20.15 Parametric equations of $xy = c^2$

The equation $xy = c^2$ is always satisfied if

$$x = ct \qquad y = \frac{c}{t} \tag{20.34}$$

where t is a parameter. These are the parametric equations. $(ct, c/t)$ is any point on the curve as t varies.

20.16 The tangent and normal at the point $(ct, c/t)$ on the curve $xy = c^2$

If $xy = c^2$, then

$$y = \frac{c^2}{x}$$

and so

$$\frac{dy}{dx} = -\frac{c^2}{x^2}$$

Therefore, the gradient of the tangent at $(ct, c/t)$ is $-c^2/c^2t^2 = -1/t^2$. Hence the equation of the tangent is

$$\left(y - \frac{c}{t}\right) = -\frac{1}{t^2}(x - ct)$$

or

$$t^2y + x = 2ct \tag{20.35}$$

The equation of the normal is

$$\left(y - \frac{c}{t}\right) = t^2(x - ct)$$

or

$$ty - t^3x = c - ct^4 \tag{20.36}$$

Example 1 Show that the equation of the tangent at the point (x_1, y_1) to the curve $xy = c^2$ can be written $x_1y + xy_1 = 2c^2$. Verify that this agrees with equation (20.35).

$$xy = c^2$$

$$\frac{dy}{dx} = -\frac{c^2}{x^2}$$

Hence at the point (x_1, y_1), the slope of the tangent is $-c^2/x_1{}^2$ and its equation is

$$(y - y_1) = -\frac{c^2}{x_1{}^2}(x - x_1)$$

Since the point lies on the curve, $c^2 = x_1y_1$. Hence

$$y - y_1 = -\frac{y_1}{x_1}(x - x_1)$$

$$x_1y + y_1x = 2x_1y_1$$

that is

$$x_1y + y_1x = 2c^2$$

To verify that this agrees with equation (20.35), let $x_1 = ct$ and $y_1 = c/t$, then

$$cty + \frac{c}{t}x = 2c^2$$

$$ty + \frac{1}{t}x = 2c$$

or

$$t^2y + x = 2ct \quad \text{as in (20.35)}$$

Example 2 Find the co-ordinates of the vertices and the foci of the curve $xy = 18$, given that the equation of the rectangular hyperbola $x^2 - y^2 = a^2$ referred to its asymptotes as axes is $2xy = a^2$.

Comparing the given equation with $xy = \frac{1}{2}a^2$, it follows that $a = 6$. Referring to *Figure 20.13*, we have OA = OA′ = 6. Also since the hyperbola is rectangular $e = \sqrt{2}$. Hence OS = OS′ = $ae = 6\sqrt{2}$. Because SAOA′S′ is

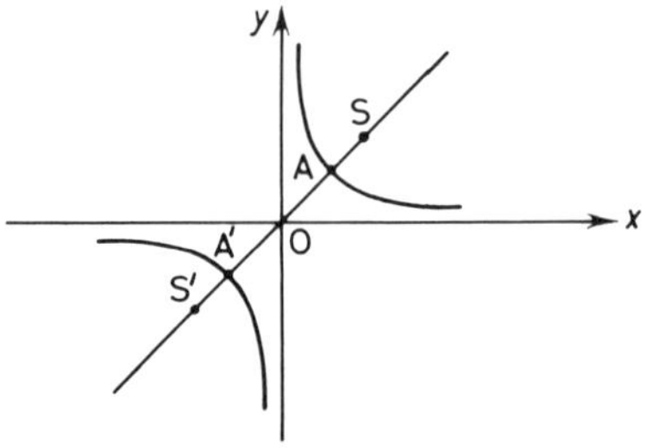

Figure 20.13

inclined at 45° to the axes, it follows that A, A′ are the points $(\pm 3\sqrt{2}, \pm 3\sqrt{2})$ and S, S′ are the points $(\pm 6, \pm 6)$.

Exercises 20k

1 Find the equations of the asymptotes and the co-ordinates of the vertices of the hyperbolae (i) $x^2/9 - y^2/4 = 1$ and (ii) $-x^2/9 + y^2/4 = 1$ and sketch the two curves on the same diagram.

2 For what values of m does the line $y = mx$ meet the hyperbola $x^2/a^2 - y^2/b^2 = 1$ in real and finite points?

3 Find the cartesian form of the equations of the following loci and sketch the curves:

(i) $x = 4t,\ y = \dfrac{4}{t}$ (ii) $x = t,\ y = -\dfrac{1}{t}$

(iii) $x = 1 + 3t,\ y = \dfrac{3}{t}$ (iv) $x = 2t - 1,\ y = -\dfrac{2}{t} + 1$

4 Find the equations of the tangents and normal at the point (4, 1) on the curve $xy = 4$.

5 Show that the equation of the line joining the points $P(t, 1/t)$ and $Q(u, 1/u)$ on the rectangular hyperbola $xy = 1$ is $x + tuy = t + u$. Deduce the equation of the tangent at P.

6 Find the equations of the tangents to the rectangular hyperbolae $x^2 - y^2 = 3$, $xy = 2$ at their points of intersection. Hence show that the curves cut at right angles.

7 Show that the normal to the hyperbola $xy = c^2$ at the point $(cp, c/p)$ cuts the hyperbola again at the point $(-c/p^3, -cp^3)$.

8 Show that the tangents to the rectangular hyperbola $x = ct$, $y = c/t$ at the points with parameters t_1 and t_2 meet at the point $P(x, y)$, where $x = 2c\dfrac{t_1 t_2}{t_1 + t_2}$, $y = \dfrac{2c}{t_1 + t_2}$. If $t_1 = 1/t_2$ find the locus of P.

9 Find the equation of the tangents to the hyperbola $x^2 - y^2 = 7$ which are parallel to $3y = 4x$ and find their points of contact. Find the area of the triangle which one of these tangents makes with the asymptotes.

10 Prove that the straight line $lx + my = n$ touches the rectangular hyperbola $xy = c^2$, if $n^2 = 4lmc^2$. Find the co-ordinates of the point of contact.

Exercises 20l

1 Find the locus of the mid-point of a straight line which moves so that it always cuts off a constant area k^2 from the corner of a square.

2 Show that the equation of the chord joining the points $P(cp, c/p)$, $Q(cq, c/q)$ on the curve $xy = c^2$ is $pqy + x = c(p + q)$. Hence or otherwise find the equation of the tangent at P. Find the co-ordinates of the point of intersection T of the tangents at P, Q. If p and q vary so that the chord PQ passes through the point $(a, 0)$, find the equation of the locus of T. [JMB]

3 Prove that the equation of the normal to the rectangular hyperbola $xy = c^2$ at the point $(ct, c/t)$ is $xt^3 - yt = c(t^4 - 1)$. Four normals to the curve from a point meet the curve at P, Q, R, S. Prove that the pairs of lines PQ, RS; PR, QS; PS, QR are such that the lines in each pair are perpendicular to each other. [SUJB]

4 Show that if the line $y = mx + c$ touches the hyperbola $x^2 - 3y^2 = 1$, then $3m^2 = 3c^2 + 1$. Obtain an equation for the gradients of the two tangents to the hyperbola from the point $P(x_0, y_0)$. Show that if these tangents are perpendicular then P lies on the circle $x^2 + y^2 = \frac{2}{3}$. [JMB]

5 Find the equation of the normal at the point (ct, ct^{-1}) on the rectangular hyperbola $xy = c^2$. The normal at the point P on $xy = c^2$ meets the hyperbola $x^2 - y^2 = a^2$ at Q and R. Prove that P is the mid-point of QR. Interpret this result geometrically when P is a point of intersection of the two curves. [JMB]

Exercises 20

1 Find the equation of the normal to the parabola $y^2 = 4ax$ at the point $(at^2, 2at)$. P, Q are two points on the parabola such that the chord PQ subtends a right angle at the vertex of the parabola. Find the locus of the point of intersection of the normals at P and Q.

2 A circle with centre at the point $(a, 0)$ and radius greater than a meets the parabola $x = at^2$, $y = 2at$ at the points P, Q. Prove that the tangents to the parabola at P and Q meet on the circle. [LU]

3 Show that the equation of the normal to the parabola $y^2 = 4ax$ at the point $P(at^2, 2at)$ is $y + tx = 2at + at^3$. If this normal meets the x-axis at Q, show that the mid-point M of PQ has the co-ordinates $(a + at^2, at)$. If P is a variable point on the parabola, find the cartesian equation of the locus of M. [AEB]

4 Show that the equation to the tangent to the hyperbola $x^2/a^2 - y^2/b^2 = 1$ at the point $P(a \sec\theta, b \tan\theta)$ is $\dfrac{x \sec\theta}{a} - \dfrac{y \tan\theta}{b} = 1$. Find also the equation of the normal. The ordinate at P meets an asymptote at Q. The tangent at P meets the same asymptote at R. The normal at P meets the x-axis at G. Prove that the angle RQG is a right angle. [JMB]

5 Show that the equation to the normal at the point $P(a \cos\theta, b \sin\theta)$ on the ellipse $x^2/a^2 + y^2/b^2 = 1$ is $ax/\cos\theta - by/\sin\theta = a^2 - b^2$. If the normal at P cuts the major and minor axes of the ellipse at G and H, show that as P moves on the ellipse the mid-point of GH describes another ellipse of the same eccentricity. [JMB]

6 The tangent to the parabola $y^2 = 4ax$ at the point $P(at^2, 2at)$ meets the x-axis at T. The straight line through P parallel to the axis of the parabola meets the directrix at Q. If S is the focus of the parabola, show that PQTS is a rhombus. If M is the mid-point of PT and N is the mid-point of PM, find the equation of the locus of (i) M, (ii) N. [AEB]

7 Prove that the equation of the normal to the rectangular hyperbola $xy = c^2$ at the point $P(ct, c/t)$ is $ty - t^3x = c(1 - t^4)$. The normal at P and the normal at the point $Q(c/t, ct)$, where $t > 1$, intersect at the point N. Show that OPNQ is a rhombus, where O is the origin. Hence, or otherwise, find the co-ordinates of N. If the tangents to the hyperbola at P and Q intersect at T, prove that the product of the lengths of OT and ON is independent of t. [JMB]

8 The tangents at the points P and Q on the parabola $y^2 = 4ax$ meet on the line $x = a$. Prove that the locus of the mid-point of PQ is $y^2 = 2a(x + a)$. [O]

9 Prove that the chord joining the points $P(ap^2, 2ap)$ and $Q(aq^2, 2ap)$ on the parabola $y^2 = 4ax$ has the equation $(p+q)y = 2x + 2apq$. A variable chord PQ of the parabola is such that the lines OP and OQ are perpendicular, where O is the origin. (i) Prove that the chord PQ cuts the axis of x at a fixed point, and give the x co-ordinate of this point. (ii) Find the equation of the locus of the mid-point of PQ. [C]

10 The point $P(a \sec t, b \tan t)$ on the hyperbola $x^2/a^2 - y^2/b^2 = 1$ is joined to the vertices $A(a, 0)$, $B(-a, 0)$. The lines AP, BP meet the asymptote $ay = bx$ at Q, R respectively. Prove that the x co-ordinate of Q is $\dfrac{a \cos \frac{1}{2}t}{\cos \frac{1}{2}t - \sin \frac{1}{2}t}$ and that the lengths of QR is independent of the value of t. [JMB]

11 Show that a circle meets the parabola $y^2 = 4ax$ in not more than four points. If three of these points coincide at $P(at^2, 2at)$ and the fourth is Q, prove that PQ and the tangent to the parabola at P make equal angles with the axis of the parabola. Show also that the centre of the circle lies on the curve $4(x-2a)^3 = 27ay^2$. [LU]

12 Show that the equation of the tangent at the point with parameter t to the curve $x = t^2$, $y = 2t$ is $ty = x + t^2$. Find the co-ordinates of the point T where the tangent at $P(9, 6)$ meets the tangent at $Q(1, -2)$. Find also the mid-points M and N of PT and QT respectively. Show that the line MN touches the curve and find the value of the parameter t at the point of contact. [LU]

13 Obtain the equation of the normal to the parabola $y^2 = 4ax$ at the point $P(at^2, 2at)$. The focal chord through P meets the parabola again at Q, and the normals at P and Q meet at R. Prove that R is the point

$$[a(t^2 + 1 + t^{-2}), \quad a(t - t^{-1})],$$

and find the equation of its locus as t varies. [SUJB]

14 The normal at a point $P(ct, c/t)$ of the rectangular hyperbola $xy = c^2$ meets the hyperbola again at Q. Prove that the parameter of Q is $-1/t^3$. If the normal at P meets the axes Ox and Oy at R and S, prove that the mid-point of RS is also the mid-point of PQ [O]

15 Prove that every point on the parabola $y^2 = 4ax$ can be expressed in the form $(a\mu^2, 2a\mu)$. A variable chord of the parabola has fixed length k. Prove that the locus of the mid-point of the chord has equation

$$(4ax - y^2)(y^2 + 4a^2) = k^2a^2$$ [SUJB]

16 Show that the equation of the normal to the parabola $y^2 = 4ax$ at the point $P(ap^2, 2ap)$ is $y + px = 2ap + ap^3$. Find the co-ordinates of R, the point of intersection of the normal at P and the normal at $Q(aq^2, 2aq)$. Given that the chord PQ passes through $S(a, 0)$, show that $pq = -1$ and find the equation of the locus of R. [AEB]

17 The points $P(ap^2, 2ap)$ and $Q(aq^2, 2aq)$ move on the parabola $y^2 = 4ax$, and $p + q = 2$. Show that the chord PQ makes a constant angle with the x-axis, and that the locus of the mid-point M of PQ is part of a line which is parallel to the x-axis. If also the point $R(ar^2, 2ar)$ moves so that $p - r = 2$, find in its simplest form the (x, y) equation of the locus of the mid-point N of PR. [JMB]

18 Show that the equation of the chord joining the points $P(ap^2, 2ap)$, $Q(aq^2, 2aq)$ of the parabola $y^2 = 4ax$ is $y(p+q) - 2x - 2apq = 0$. The variable chord PQ of the parabola $y^2 = 4ax$ passes through the fixed point (h, k). If the tangents to the parabola at P and Q meet at T, show that T lies on a fixed straight line. [JMB]

19 Show that the equation of the common tangent other than the y-axis, of the curves $y^2 = 4ax$ and $xy = 2a^2$ is $2y + x + 4a = 0$. This common tangent touches the curves at P and Q respectively. R is the point of intersection of the curves. Find the acute angle between PR and QR. [LU]

20 Prove that the line $y = mx + c$ touches the ellipse $x^2/a^2 + y^2/b^2 = 1$, if $c^2 = a^2m^2 + b^2$. Find in terms of m, a, b, the distance between the two tangents of slope m. If this distance is equal to the distance between the pair of tangents perpendicular to the first pair, show that it becomes $\sqrt{[2(a^2 + b^2)]}$. An ellipse in which the semi-axes a, b are in the ratio 3:2 touches the four sides of a square. Find the length of a side of the square in terms of a.

21 Find the equation of the normal to the parabola $y^2 = 4ax$ at the point $(at^2, 2at)$ and the co-ordinates of the point at which this normal cuts the x-axis. Show that the equation of the circle which touches this parabola at points $(at^2, 2at)$ and $(at^2, -2at)$ is

$$(x - 2a - at^2)^2 + y^2 = 4a^2(1 + t^2)$$

Find the values of t for which this circle passes through the point $(9a, 0)$. [LU]

22 Show that the tangent at $P(cp, c/p)$ to the rectangular hyperbola $xy = c^2$ has the equation $p^2y + x = 2cp$. The perpendicular from the origin to this tangent meets it at N, and meets the hyperbola again at Q and R. Prove that (i) the angle QPR is a right angle, (ii) as p varies, the point N lies on the curve whose equation is $(x^2 + y^2)^2 = 4c^2xy$. [C]

23 If the parabola $y^2 = 4ax (a > 0)$ cuts the hyperbola $xy = c^2$ at right angles, show that $c^4 = 32a^4$, and find the co-ordinate of the point of intersection in terms of a. [O]

24 Prove that the equation of the tangent at the point (x_1, y_1) on the ellipse $x^2/a^2 + y^2/b^2 = 1$ is $xx_1/a^2 + yy_1/b^2 = 1$. The tangent at the point $(2\cos\theta, \sqrt{3}\sin\theta)$ on the ellipse $x^2/4 + y^2/3 = 1$ passes through the point P(2, 1). Show that $\sqrt{3}\sin\theta + \sin\theta = \sqrt{3}$. Without using tables, calculator or slide rule, find all the solutions of this equation which are in the range $0° \leqslant \theta \leqslant 360°$. Hence obtain the co-ordinates of the points of contact, Q and R, of the tangents to the ellipse from P. Verify that the line through the origin and the point P passes through the mid-point of QR. [JMB]

25 A straight line has gradient m and passes through the point P(3, 4). Find the equation of this line in terms of m. Show that the line touches the parabola $y^2 = 4x$ if $3m^2 - 4m + 1 = 0$. If the tangents from P to the parabola touch the parabola at the points S and T, find the lengths of PS and PT. Show, without using tables or calculator, that $\tan \text{SPT} = -\frac{1}{2}$ and determine the area of triangle PST. [AEB]

21
Numerical methods

21.1 Graphical solution of equations

For any equation $f(x) = 0$ we have an associated function $y = f(x)$. The roots of the equation are those values of x which make y equal to zero. Thus the graph of the function $y = f(x)$ meets the x-axis at points whose x co-ordinates are the roots of the equation $f(x) = 0$.

Example 1 Find the approximate values of the roots of $2x^3 - 21x^2 + 60x - 35 = 0$.

Consider the function $y = 2x^3 - 21x^2 + 60x - 35$

$$\frac{dy}{dx} = 6x^2 - 42x + 60$$

$$= 6(x - 2)(x - 5)$$

$$\frac{d^2y}{dx^2} = 12x - 42$$

Thus there is a local maximum at the point (2, 17) and a local minimum at the point (5, −10). The values at the maximum and minimum are opposite

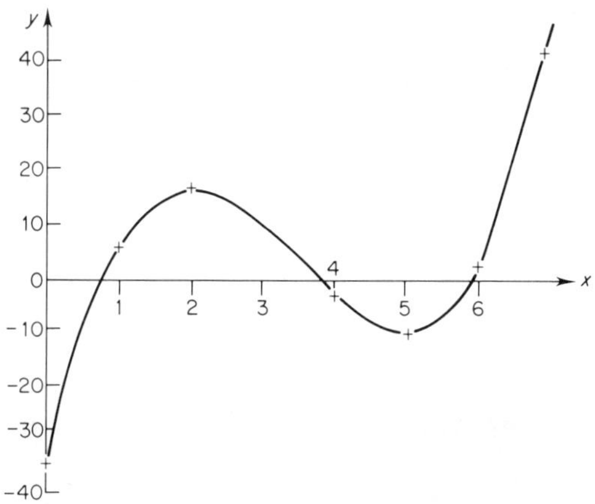

Figure 21.1

in sign and the graph must cross the x-axis at least once between $x = 2$ and $x = 5$. The graph of the function plotted for values of $x = 0(1)7$ (from $x = 0$ to $x = 7$ in steps of 1) is shown in *Figure 21.1*.

The graph cuts the x-axis at three points and the three roots of the equation $2x^3 - 21x^2 + 60x - 35 = 0$ are approximately 0·8, 3·8 and 5·9. Any of these values can be improved by drawing the relevant part of the graph on a much larger scale. Thus to improve on the value 0·8, we draw the graph of the function for values of $x = 0{\cdot}75(0{\cdot}1)0{\cdot}85$.

Expressing the equation $f(x) = 0$ in the form $g(x) = h(x)$ can be useful if the graphs of $y = g(x)$ and $y = h(x)$ are more easily drawn than that of $f(x)$. If $f(x) = h(x) - g(x)$, then $f(x) = 0$ when $h(x) = g(x)$. The graphs of $y = h(x)$ and $y = g(x)$ will intersect at a point with co-ordinates (x_1, y_1) such that

$$g(x_1) = y_1 = h(x_1)$$

therefore

$$h(x_1) - g(x_1) = 0$$
$$f(x_1) = 0$$

that is, $f(x_1) = 0$ and x_1 is a root of $f(x) = 0$.

Example 2 Find approximations to the positive roots of the equation $x^3 - 3x + 1 = 0$ by considering the graphs of $y = x^3 + 1$ and $y = 3x$ over the range of values of $x = 0(1)3$.

When $x^3 + 1 = 3x$, $x^3 - 3x + 1 = 0$. *Figure 21.2* shows the graphs of the functions $y = x^3 + 1$ and $y = 3x$. They intersect where $x = 0{\cdot}4$ approximately and where $x = 1{\cdot}5$ approximately.

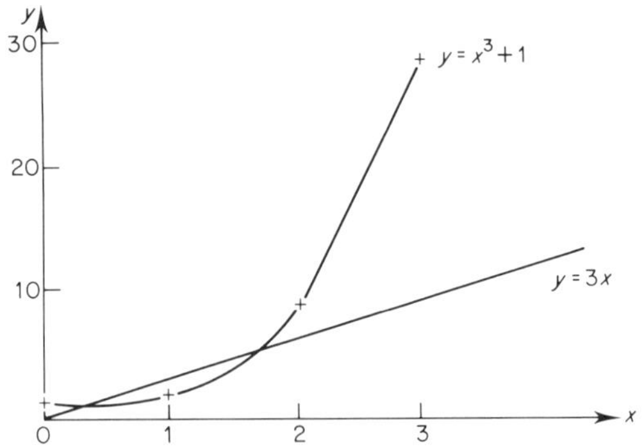

Figure 21.2

Better approximations can be obtained by drawing the graphs on a larger scale near these roots.

If the values of x, between which a single root is known to lie, are reasonably close together, it is possible to approximate to the root by *linear interpolation* without drawing a graph.

Let $P[a, f(a)]$, and $Q[b, f(b)]$ be points on the graph of $y = f(x)$ such that $f(a) > 0$ and $f(b) < 0$. Then OR (*Figure 21.3*) gives a better approximation to the root.

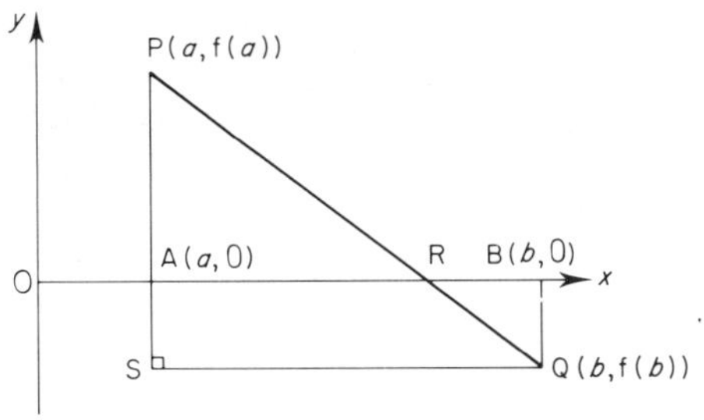

Figure 21.3

If PS is parallel to Oy and QS is parallel to Ox, then by similar triangles

$$\frac{\text{AR}}{\text{AB}} = \frac{\text{PA}}{\text{PS}} = \frac{|\text{f}(a)|}{|\text{f}(a)| + |\text{f}(b)|}$$

Therefore

$$x_{\text{R}} = \text{OR} = a + \text{AR} = a + \frac{|\text{f}(a)|}{|\text{f}(a)| + |\text{f}(b)|} \cdot (b - a) \qquad (21.1)$$

and this can be calculated without drawing the graph.

Example 3 Consider $\text{f}(x) = x^3 - 3x + 1 = 0$ (see Example 2).

$$\text{f}(0) = 1 > 0 \qquad \text{f}(1) = -1 < 0$$

Then by (21.1)

$$x_{\text{R}} = 0 + \frac{1}{1+1} = 0{\cdot}5$$

We can improve on this by a further application of (21.1):

$$\text{f}(0{\cdot}3) = 0{\cdot}127$$
$$\text{f}(0{\cdot}5) = -0{\cdot}375$$

Thus a better approximation is

$$0{\cdot}3 + \frac{0{\cdot}127}{0{\cdot}127 + 0{\cdot}375} \cdot (0{\cdot}2) \approx 0{\cdot}35$$

Exercises 21a

1 Show that the function $\text{f}(x) = x^4 + 6x - 10$ has a minimum value when $x = -\sqrt[3]{1{\cdot}5}$. Sketch the graph of the function and find an approximate root of the equation $x^4 + 6x - 10 = 0$. Improve your approximation by drawing another graph on a larger scale.

2 Given that $x = 5{\cdot}9$ is an approximate root of the equation $2x^3 - 21x^2 + 60x - 35 = 0$, find a better approximation.

3 For the equation $\text{f}(x) = 2x^3 - 21x^2 + 60x - 35 = 0$ show that $\text{f}(3) = 10$ and $\text{f}(4) = -3$. Hence show that $3{\cdot}77$ is an approximate root. By considering $\text{f}(3{\cdot}7)$ and $\text{f}(3{\cdot}8)$ show that $x = 3{\cdot}762$ is a better approximation.

4 On the same diagram draw graphs of the functions $y = \frac{1}{2}(\text{e}^x + \text{e}^{-x})$ and

$y = \sec x$ for values of $x = 4{\cdot}5(0{\cdot}1)5{\cdot}0$. Hence find an approximate value of a root of the equation $(e^x + e^{-x})\cos x - 2 = 0$.

5 On the same diagram draw graphs of the functions $y = e^x$ and $y = \operatorname{cosec} x$. Hence find the approximate value of the smallest positive root of $e^x \sin x - 1 = 0$.

6 Kepler's equation for finding the eccentric anomaly E given the mean anomaly M and the eccentricity e of a planetary orbit is $E - e \sin E = M$. Given $e = 0{\cdot}8$ and $M = 0{\cdot}5$ find an approximate value for E.

7 For the equation $x^3 - 3x + 1 = 0$ show that, if $f(x) = x^3 - 3x + 1$, then $f(-2) = -1, f(-1) = 3$. Hence find an approximate value for the negative root of the equation.

8 The equation $x^3 - 5x + 3 = 0$ has a root between 0 and 1. Verify that this is correct and find its approximate value.

9 The equation in question 8 may be written $x = \frac{1}{5}(x^3 + 3)$. Calculate the sequence of values $x_1, x_2, x_3, \ldots$ given by $x_1 = \frac{1}{5}(x_0{}^3 + 3)$, $x_2 = \frac{1}{5}(x_1{}^3 + 3)$, $x_3 = \frac{1}{5}(x_2{}^3 + 3)$, etc. starting with $x_0 = 0$. What do you notice?

10 On the same diagram draw the graphs of $y = x$ and $y = \cos x$. Show that there is a root between 0 and 1 of the equation $x - \cos x = 0$. Write this equation in the form $x = \cos x$ and starting with $x_0 = 0{\cdot}5$, calculate $x_1 = \cos x_0$, $x_2 = \cos x_1$, $x_3 = \cos x_2$, etc. What do you notice?

21.2 Newton's method

If we have an approximation to the root of an equation $f(x) = 0$ (which may have been obtained by the methods described in the previous section), then Newton's method (also known as the Newton–Raphson method) can be used to obtain a better approximation.

Figure 21.4 shows the graph of $y = f(x)$. The x co-ordinate of K is the root we require. Suppose x_0, the x co-ordinate of P, is an approximation to the root. Let PT be the tangent at P on the curve and T the point where the tangent meets the x-axis. Then, in general, OT will be a better approximation to the root.

$$\text{OT} = \text{OA} - \text{TA}$$

$$\frac{\text{PA}}{\text{TA}} = \tan \psi = f'(x_0)$$

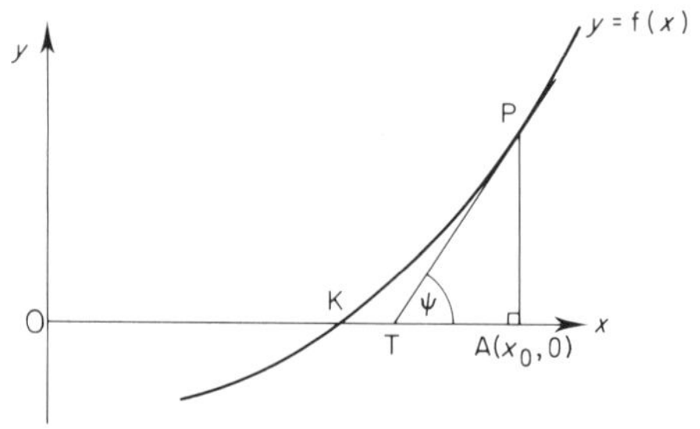

Figure 21.4

Therefore

$$TA = \frac{PA}{f'(x_0)} = \frac{f(x_0)}{f'(x_0)}$$

$$x_1 = x_0 - \frac{f(x_0)}{f'(x_0)}$$

We can similarly improve upon x_1:

$$x_2 = x_1 - \frac{f(x_1)}{f'(x_1)}$$

and in general

$$x_{r+1} = x_r - \frac{f(x_r)}{f'(x_r)} \qquad (21.2)$$

The re-calculations are continued until two successive approximations agree to the required accuracy.

Example 1 Find to four decimal places the root of the equation $x^3 - 3x + 1 = 0$ that lies between 0 and 1.

$$f(x) = x^3 - 3x + 1$$

$$f'(x) = 3x^2 - 3 = 3(x^2 - 1)$$

Take $x_0 = 0{\cdot}4$. The calculations that follow are done on a calculator.

$$f(0{\cdot}4) = -0{\cdot}136$$

$$f'(0{\cdot}4) = -2{\cdot}52$$

$$x_1 = 0{\cdot}4 - \frac{(-0{\cdot}136)}{(-2{\cdot}52)} = 0{\cdot}3460$$

$$f(0{\cdot}3460) = 0{\cdot}003\,422$$

$$f'(0{\cdot}3460) = -2{\cdot}641$$

$$x_2 = 0{\cdot}3460 + \frac{0{\cdot}003\,422}{2{\cdot}461} = 0{\cdot}3473$$

$$f(0{\cdot}3473) = -0{\cdot}000\,009\,615$$

$$f'(0{\cdot}3473) = -2{\cdot}638$$

$$x_3 = 0{\cdot}3473 - \frac{0{\cdot}000\,009\,615}{2{\cdot}638} = 0{\cdot}3473$$

Thus the required root is 0·3473.

The calculations can be tedious even when carried out with a calculator. The method is ideally suited to computation by a computer. A BASIC program written for a 'Pet' computer follows. The program is general in the sense that F = f(x) is evaluated by subroutine 1000 and D = f′(x) (D for derivative) is evaluated by subroutine 2000 for any value of x. These subroutines can be rewritten for different forms of f(x). The accuracy of the solution can be set by giving E a value sufficiently small.

```
READY.

10 PRINT "PROGRAM TO SOLVE F(X)=0"
20 PRINT "F(X) EVALUATED AT 1000"
30 PRINT "F'(X) EVALUATED AT 2000"
40 PRINT "ACCURACY REQUIRED":INPUT E
50 PRINT "INITIAL VALUE":INPUT Z
60 X=Z
70 GOSUB 1000
80 GOSUB 2000
90 Z=X-F/D
100 PRINT X,Z
110 IF ABS(Z-X)>E GOTO 60
120 PRINT "FINAL SOLUTION=";Z
130 END
1000 F=X*X*X-3*X+1
1010 RETURN
2000 D=3*(X*X-1)
2010 RETURN
READY.
```

The output for E = 0·000 01 and initial Z = 0·4 is shown below.

```
PROGRAM TO SOLVE F(X)=0
F(X) EVALUATED AT 1000
F'(X) EVALUATED AT 2000
·4              ·346031746
·346031746              ·347295726
·347295726              ·347296355
FINAL SOLUTION=·347296355
```

Newton's method will fail if the first approximation to the root is such that the value of $f(x_0)/f'(x_0)$ is not small enough (see *Figure 21.5*). The usual remedy is to improve the initial approximation to the root when the iterations will generally converge.

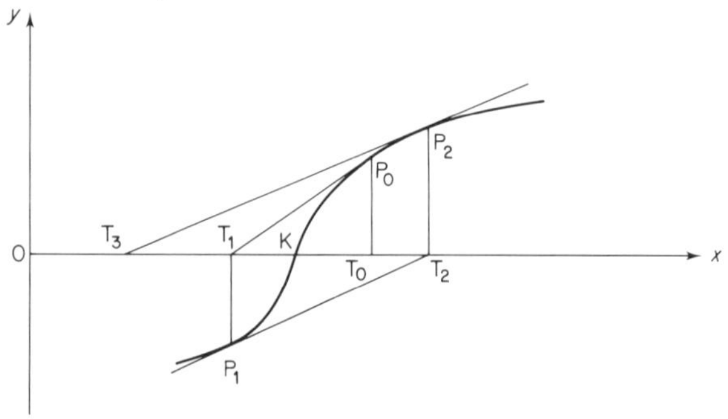

Figure 21.5

Exercises 21b

Use Newton's method to find to four decimal places the roots of the following equations. Use the initial value given.

1 $x^4 + 6x - 10 = 0$; $x_0 = 1{\cdot}25$
2 $2x^3 - 21x^2 + 60x - 35 = 0$; $x_0 = 0{\cdot}78$
3 $e^x \sin x - 1 = 0$; $x_0 = 0{\cdot}59$
4 $x - \cos x = 0$; $x_0 = 0{\cdot}8$
5 $e^{-x/2\pi} \sin x - 0{\cdot}5 = 0$; $x_0 = 0{\cdot}58$

21.3 Numerical integration: the trapezoidal rule

In Section 13.6 we showed that $\int_a^b f(x)\,dx$ is equal to the area A enclosed by the graph of $y = f(x)$, the x-axis and the ordinates at $x = a$ and $x = b$. Thus an approximate value for $\int_a^b f(x)\,dx$ can be obtained by finding an approximation to the value of A. One method is to use the trapezoidal rule.

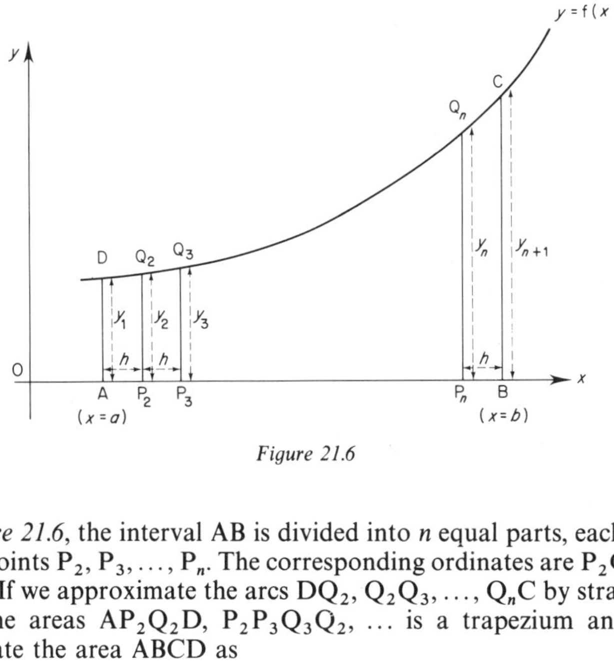

Figure 21.6

In *Figure 21.6*, the interval AB is divided into n equal parts, each of width h, by the points $P_2, P_3, \ldots, P_n$. The corresponding ordinates are P_2Q_2, P_3Q_3, $\ldots$, P_nQ_n. If we approximate the arcs DQ_2, Q_2Q_3, $\ldots$, Q_nC by straight lines, each of the areas AP_2Q_2D, $P_2P_3Q_3Q_2$, $\ldots$ is a trapezium and we can approximate the area ABCD as

$$\int_a^b f(x)\,dx \approx \tfrac{1}{2}h(AD + P_2Q_2) + \tfrac{1}{2}h(P_2Q_2 + P_3Q_3) + \ldots + \tfrac{1}{2}h(P_nQ_n + BC)$$
$$\approx \tfrac{1}{2}h(y_1 + 2y_2 + 2y_3 + \ldots + 2y_n + y_{n+1})$$

therefore

$$\int_a^b f(x)\,dx \approx \tfrac{1}{2}h(\text{first + last ordinate}) + h(\text{sum of all other ordinates}) \qquad (21.3)$$

It is not necessary to draw the graph of $y = f(x)$. The calculations can be set out as follows.

Example 1 Find by the trapezoidal rule an approximate value for $\int_0^1 \frac{dx}{1+x^2}$.

Use ordinates spaced at equal intervals of width $h = 0.1$.

x	0	0·1	0·2	0·3	0·4	0·5	0·6	0·7	0·8	0·9	1·0
$\frac{1}{1+x^2}$	1	0·9901	0·9615	0·9174	0·8621	0·8000	0·7353	0·6711	0·6098	0·5525	0·5000

Therefore

$$\int_0^1 \frac{dx}{1+x^2} \approx \frac{0{\cdot}1}{2}(1 + 0{\cdot}5) + 0{\cdot}1(0{\cdot}9901 + 0{\cdot}9615 + \ldots + 0{\cdot}5525)$$
$$= 0{\cdot}075 + 0{\cdot}1 \times 7{\cdot}0998$$
$$\approx 0{\cdot}7850.$$

By analytical methods

$$\int_0^1 \frac{dx}{1+x^2} = [\arctan x]_0^1 = \pi/4 = 0{\cdot}7854$$

Exercises 21c

Find the approximate value of the following definite integrals. Use the trapezoidal rule and ordinates spaced at equal intervals of width h as indicated.

1 $\int_0^{\pi/2} \cos x \, dx$, $h = \pi/12$

2 $\int_0^5 \log_e(1 + x) \, dx$, $h = 1$

3 $\int_0^2 e^{-x^2/2} \, dx$, $h = 0{\cdot}2$

4 $\int_0^{\pi/4} \tan x \, dx$, $h = \pi/24$

5 $\int_0^{\pi/2} \cos^2 \tfrac{1}{2}x \, dx$, $h = \pi/12$

21.4 Numerical integration: Simpson's rule

Simpson's rule is a better method of approximating to the value of a definite integral than the trapezoidal rule. In fact, it is exact when f(x) is a quadratic or a cubic function.

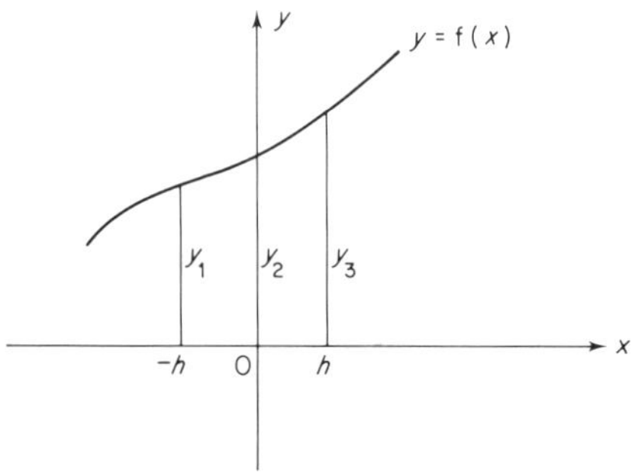

Figure 21.7

We first consider $\int_{-h}^{h} f(x)\,dx$ and obtain an approximation for this in terms of the ordinates at the end points of the interval and the ordinate at the mid-point of the interval. We denote these by y_1, y_3 and y_2 respectively (see *Figure 21.7*).

We can express the function f(x) in the interval $(-h, h)$ by its Maclaurin series (see Section 12.4):

$$f(x) = f(0) + xf'(0) + \frac{x^2}{2}f''(0) + \frac{x^3}{6}f'''(0) + \frac{x^4}{24}f''''(0) + \ldots$$

Therefore

$$\int_{-h}^{h} f(x)\,dx = \int_{-h}^{h}\left[f(0) + xf'(0) + \frac{x^2}{2}f''(0) + \frac{x^3}{6}f'''(0) + \frac{x^4}{24}f''''(0) + \ldots\right]dx$$

$$= \left[xf(0) + \frac{x^2}{2}f'(0) + \frac{x^3}{6}f''(0) + \frac{x^4}{24}f'''(0) + \frac{x^5}{120}f''''(0) + \ldots\right]_{-h}^{h}$$

$$= 2hf(0) + \frac{h^3}{3}f''(0) + \frac{h^5}{60}f''''(0) + \ldots$$

$$= \frac{h}{3}\left[6f(0) + h^2f''(0)\right] + \frac{h^5}{60}f''''(0) + \ldots \qquad \text{(A)}$$

Now

$$y_1 = f(-h) = f(0) - hf'(0) + \frac{h^2}{2}f''(0) - \frac{h^3}{6}f'''(0) + \frac{h^4}{24}f''''(0) - \ldots$$

$$y_2 = f(0)$$

$$y_3 = f(h) = f(0) + hf'(0) + \frac{h^2}{2}f''(0) + \frac{h^3}{6}f'''(0) + \frac{h^4}{24}f''''(0) + \ldots$$

therefore

$$y_1 + 4y_2 + y_3 = 6f(0) + h^2f''(0) + \frac{h^4}{12}f''''(0) + \ldots$$

and so

$$\frac{h}{3}[y_1 + 4y_2 + y_3] = \frac{h}{3}[6f(0) + h^2f''(0)] + \frac{h^5}{36}f''''(0) + \ldots \qquad \text{(B)}$$

From (A) and (B), ignoring terms in h^5 and above,

$$\int_{-h}^{h} f(x)\,dx \approx \frac{h}{3}(y_1 + 4y_2 + y_3)$$

Also from (A) and (B), the error in the approximation is of the order

$$\frac{h^5}{36}f''''(0) - \frac{h^5}{60}f''''(0) = \frac{h^5}{90}f''''(0)$$

Note that for a cubic function $ax^3 + bx^2 + cx + d$, $f''''(0)$ and higher derivatives are zero. Therefore, Simpson's rule is exact for cubic and quadratic functions.

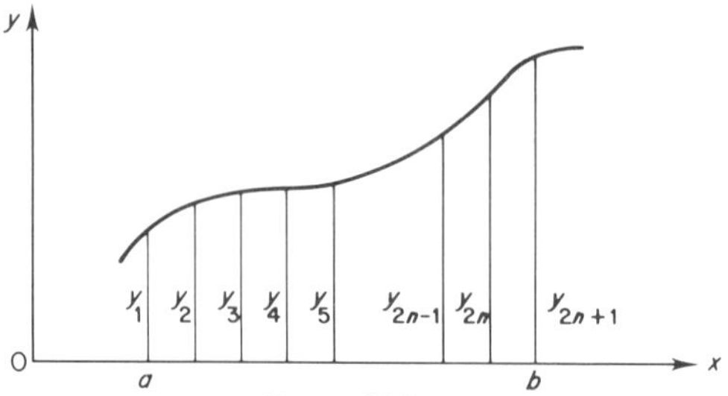

Figure 21.8

For $\int_a^b f(x)\,dx$, we divide the area into an *even* number of strips $2n$ and apply Simpson's rule to each pair of strips in turn (see *Figure 21.8*). Then

$$\int_a^b f(x)\,dx \approx \tfrac{1}{3}h(y_1 + 4y_2 + y_3 + y_3 + 4y_4 + y_5 + \ldots + y_{2n-1} + 4y_{2n} + y_{2n+1})$$

$$\approx \tfrac{1}{3}h[y_1 + y_{2n+1} + 4(y_2 + y_4 + y_6 + \ldots + y_{2n}) + 2(y_3 + y_5 + \ldots + y_{2n-1})]$$

Therefore

$$\int_a^b f(x)\,dx \approx \tfrac{1}{3}h(\text{first} + \text{last ordinate} + 4 \times \text{sum of even ordinates} + 2 \times \text{sum of remaining odd ordinates}) \qquad (21.4)$$

Example 1 Find the approximate value of $\int_0^1 \frac{dx}{1+x^2}$ using ten strips and Simpson's rule.

The range is 0 to 1 so that each strip is of width 0·1. Thus $h = 0{\cdot}1$. The work may be set out as follows.

x	*First and last ordinates*	*Even ordinates*	*Odd ordinates*
0·0	1		
0·1		0·9901	
0·2			0·9615
0·3		0·9174	
0·4			0·8621
0·5		0·8000	
0·6			0·7353
0·7		0·6711	
0·8			0·6098
0·9		0·5525	
1·0	0·5		
Totals	1·5	3·9311	3·1687
		4	2
		15·7244	6·3374

Therefore

$$\int_0^1 \frac{dx}{1+x^2} \approx \frac{0{\cdot}1}{3}(1{\cdot}5 + 15{\cdot}7244 + 6{\cdot}3374)$$

$$= 0{\cdot}7854$$

The result analytically is 0·7854, and by the trapezoidal rule (refer to Example 1 in Section 21.3) 0·7850.

Example 2 A river is 20 m wide. The depth of the river is found, in metres, at nine equidistant points on a line AB joining two points A, B on opposite banks of the river (AB is perpendicular to the direction of the flow) with the following results.

Distance from A in in metres	0·0	2·5	5·0	7·5	10·0	12·5	15·0	17·5	20·0
Depth in metres	0·6	0·8	2·4	3·6	2·5	1·6	1·2	0·8	0·4

At AB the water is flowing at 1·5 m/sec. Find the approximate number of cubic metres of water flowing down the river per second.

	First and last ordinates	*Even ordinates*	*Odd ordinates*
	0·6		
		0·8	
			2·4
		3·6	
			2·5
		1·6	
			1·2
		0·8	
	0·4		
Totals	1·0	6·8	6·1
		4	2
		27·2	12·2

$$\text{Cross-sectional area} \approx \frac{2{\cdot}5}{3}(1{\cdot}0 + 27{\cdot}2 + 12{\cdot}2)$$
$$\approx 33{\cdot}67\ \text{m}^2$$
$$\text{Volume of flow} \approx 33{\cdot}67 \times 1{\cdot}5$$
$$\approx 50{\cdot}5\ \text{m}^3/\text{sec}$$

A listing of a BASIC program to implement Simpson's rule follows. The program can be used interactively, the operator specifying the limits of integration and the number of strips (which must be even). The program listed considers $f(x) = 1/(1 + x^2)$ but subroutine 1000 can be rewritten to deal with any function.

```
READY.

10 PRINT "SIMPSON'S RULE FOR INTEGRATION"
20 PRINT "LOWER LIMIT":INPUT A
30 PRINT "UPPER LIMIT":INPUT B
40 PRINT "NO. OF STRIPS":INPUT N:M=INT(N/2)
50 IF M< >N/2 THEN PRINT "NO. OF STRIPS MUST BE
   EVEN":GOTO 40
60 H=(B-A)/N
```

```
70 X=A:GOSUB 1000:S=F
80 FOR I=2 TO N STEP 2
90 X=A+(I-1)*H:GOSUB 1000:S=S+4*F
100 X=A+I*H:GOSUB 1000:S=S+2*F
110 NEXT I
120 S=S-F
130 S=S*H/3
140 PRINT "INT=";S
150 END
1000F=1/(1+X*X)
1010 RETURN
READY.
```

Exercises 21d

1 Evaluate $\int_0^{\pi/2} \sin x \, dx$ by Simpson's rule. Use six strips ($h = \pi/12$). Compare your result with the exact value.

2 Given $\log_e 2 = 0{\cdot}6931$, find an approximate value for $\log_e 3$ by evaluating $\int_2^3 \frac{dx}{x}$ by Simpson's rule. Use ten strips ($h = 0{\cdot}1$). Compare your results with the value given by tables.

3 Find an approximation to the length L of an arc of an ellipse, where

$$L = \int_0^{\pi/2} \sqrt{(1 - 0{\cdot}9 \sin^2 \theta)} \, d\theta$$

Use four strips ($h = \pi/8$).

4 A particle moves along a fixed straight line AB. Its distance s metres from A at any time t seconds is given by

$$\frac{ds}{dt} = t\sqrt{(64 - t^3)}$$

Use Simpson's rule with eight strips to find approximately the distance travelled from $t = 0$ to $t = 2$.

5 The vertical depth of water a short distance behind a straight dam was measured at nine equidistant points on a line AB, with the following results.

Distance from A in metres	0	35	70	105	140	175	210	245	280
Depth in metres	0	53	87	99	105	100	68	36	0

AB is 280 m long and parallel to the dam face, which slopes uniformly into the water at an angle of 15° to the downward vertical. Calculate to the nearest 100 m^2 the wetted area of the dam's face.

21.5 Determination of laws

If two variables are related and we have measured pairs of values for the two variables, we can investigate the relationship between them by plotting the points on graph paper. A line graph is easily recognised by eye. In some cases where the connection is more complicated, we can sometimes find simple functions of the variables that are linearly related.

Example 1 The following table shows corresponding values of two variables F and C. Plot the values on a graph and hence estimate the relationship between the variables.

F	50	68	86	104	122	140
C	10	20	30	40	50	60

The plot of F against C is shown in *Figure 21.9*.

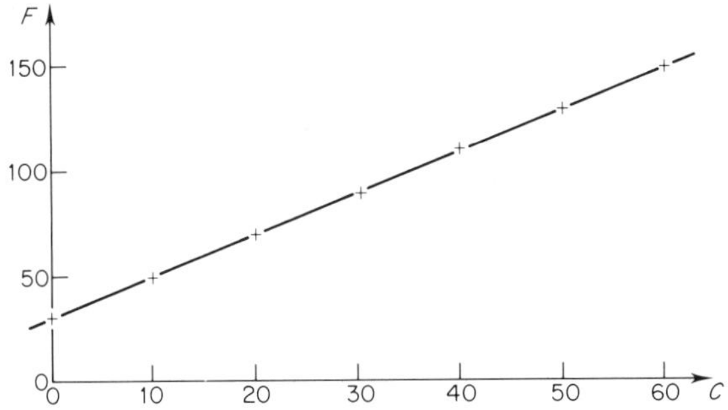

Figure 21.9

The graph is a straight line. Its intercept on the F-axis is 32 and its gradient is

$$\frac{140-50}{60-10}=\frac{9}{5}$$

Thus from (18.2), F and C are related by

$$F = 32 + \tfrac{9}{5}C$$

They are, in fact, the values of the same temperatures measured on the Fahrenheit and Celsius scales respectively.

Example 2 The variables x and y are related by $y = ax^n$, where a and n are constants. Determine a linear relationship between variables related to y and x and use the data given to estimate a and n.

y	3	4·2	5·2	6	6·7	7·3
x	1	2	3	4	5	6

$$y = ax^n$$

therefore

$$\begin{aligned}\log_e y &= \log_e a + \log_e x^n \quad \text{(see (1.27))}\\ &= \log_e a + n\log_e x \quad \text{(see (1.29))}\end{aligned}$$

Thus the relationship between $\log_e y$ and $\log_e x$ is linear. If we plot $\log_e y$

against $\log_e x$, the resulting straight line should intercept the $\log_e y$ axis at $\log_e a$ and have gradient n.

We have

$\log_e y$	1·099	1·44	1·649	1·791	1·902	1·988
$\log_e x$	0	0·693	1·099	1·386	1·609	1·791

These data are plotted in *Figure 21.10*.

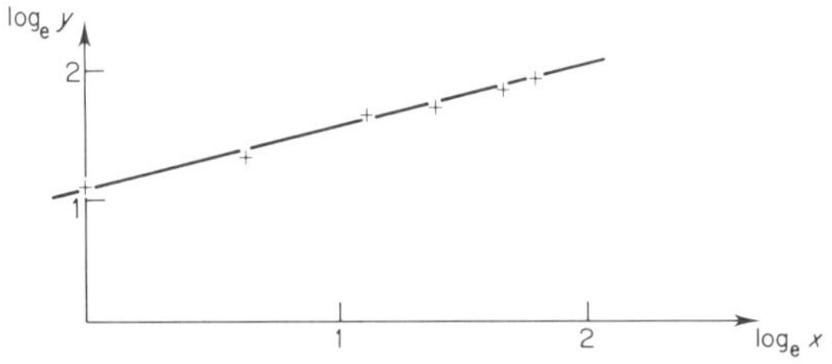

Figure 21.10

Thus $\log_e a = 1{\cdot}099$, whence $a = 3$. The gradient n is estimated to be

$$\frac{1{\cdot}988 - 1{\cdot}099}{1{\cdot}791 - 0} \approx 0{\cdot}5$$

Thus the relationship is of the form $y = 3\sqrt{x}$.

Exercises 21e

1 It is known that the relation between the variables x and y is of the form $y = ab^x$. Show that $\log_e y$ and x are linearly related.

2 The variables p and v are connected by the relation $p(v + a) = c$, where c and a are constants. What simple function of p is linearly related to v?

3 The variables x and y are connected by the relation $y = a + bx^2$. By plotting y against x^2 estimate a and b given the data below.

y	5·3	7·7	12·5	19·7	29·3
x	1	3	5	7	9

4 In an adiabatic expansion $pv^\gamma = c$ where p is the pressure and v is the volume of the gas. Show graphically that the following values obey this law and estimate the values of the constants γ and c.

p	79·8	30·3	17·8	11·5	6·5	3·2
v	1	2	3	4	6	10

5 The braking distance, b, of a vehicle travelling at speed v is given by $d = av^b$. Verify this and estimate the values of the constants a and b using the data below.

d metres	4	15	34	60	94
v km/h	10	20	30	40	50

Exercises 21

1 The height of the cap of a sphere of radius 5 cm is x cm. The volume of the cap is one-quarter the volume of the sphere. Show that $x^3-15x^2+125=0$. Find the value of x correct to three decimal places.

2 If in question 1 the volume of the cap is one-eighth of the volume of the sphere and the radius of the sphere is 4 cm, find an equation for x and find x correct to three decimal places.

3 Using Simpson's rule and eight strips, find the value of $\int_0^2 e^{-x^2/2}\,dx$ to four significant figures. [$e^{-x^2/2}$ is a function used in error theory and statistics.]

4 Sketch the curve $y=(x+1)(x-3)^2$. Hence sketch the curve $y^2=(x+1)\times(x-3)^2$. Use Simpson's rule with five ordinates to obtain an approximate value of the area in the first quadrant enclosed between the curve $y^2=(x+1)(x-3)^2$, the x-axis, the y-axis and the line $x=2$. Find also an approximation to the mean value of y in the range $0\leqslant x\leqslant 2$. [AEB]

5 Use Simpson's rule with five ordinates to find the mean value of $\sqrt{(\cos x)}$ with respect to x over the range $0\leqslant x\leqslant \pi/2$. Work as accurately as your tables allow. [AEB]

6 Verify that Simpson's rule gives the exact value of

$$\int_{-1}^{+1}(x^3+px^2+r)\,dx$$

7 Show that one solution of the equation $24x^3-149x^2-201x+476=0$ is between 1 and 2. Estimate its value by linear interpolation. Improve your estimate by finding two closer values between which the root lies and then use linear interpolation again.

8 The following pairs of values of x and y satisfy approximately a relation of the form $y=ax^n$, where a and n are integers. By plotting the graph of $\lg y$ against $\lg x$, find the values of the integers a and n. ($\lg N$ denotes $\log_{10} N$.)

x	0·7	0·9	1·1	1·3	1·5
y	1·37	2·92	5·32	8·80	13·50

Estimate the value of the integral $\int_{0.7}^{1.5} y\,dx$ (a) by Simpson's rule, using five ordinates and clearly indicating your method, (b) by using the relation $y=ax^n$ with the values found for a and n. [LU]

9 Use Simpson's rule with three ordinates (i.e. two strips of equal width) to evaluate $\int_{a-h}^{a+h}(x^3+1)\,dx$, where a and n are constants. By evaluating the integral directly, verify that the value obtained by Simpson's rule is exact. [JMB]

10 Sketch the curve $y=(1+x^2)^{1/2}$ for values of x between 0 and 1. Find the approximation to the integral $\int_0^1(1+x^2)^{1/2}\,dx$ given by the trapezium rule, using two trapezia, one with base from 0 to $\frac{1}{2}$, the other with base from $\frac{1}{2}$ to 1. By referring to your sketch determine whether your estimate is too large or too small, giving reasons. Evaluate the integral by Simpson's rule, using the same three ordinates. [O]

11 Use Simpson's rule to evaluate approximately $\int_0^\pi (1 - \cos x)\,dx$, by using seven ordinates (i.e. six strips of equal width). Evaluate the above integral also by direct integration. [JMB]

12 Tabulate the values of the expression $\sqrt{(1 + x^2)}$ at unit intervals from $x = 1$ to $x = 6$ inclusive, giving your answers to two decimal places. Use these values to find an estimate by the trapezoidal rule of the mean value of the expression in the range $1 \leqslant x \leqslant 6$. (All working must be shown.) [JMB]

13 The following values of x and y are believed to obey a law of the form $y - 3 = 20b^{x+a}$, where a and b are constants. Show graphically that they do obey this law and hence find approximate values of a and b.

x	0	2	5	7
y	8233	744	23	5

[AEB]

14 A smooth curve passes through the points given by the table below.

x	20	25	30	35	40	45	50
y	1·560	1·478	1·425	1·384	1·355	1·332	1·314

Using Simpson's rule with seven ordinates, find the area enclosed by the curve, the x-axis and the lines $x = 20$ and $x = 50$. It is believed that x and y are related by a law of the form $y = a + b/x$, where a and b are constants. Show graphically that this may be so and determine probable values of a and b. [AEB]

15 Evaluate $\int_0^{0\cdot4} x \ln(1 + x)\,dx$ using Simpson's rule with five ordinates and correcting your answer to two significant figures. Show clearly how your answer has been obtained. [LU]

16 Find the positive root of the equation $x = \sin 2x$ to two places of decimals, showing that your solution has this degree of accuracy. [O]

17 Show that the equation $x^3 - x - 2 = 0$ has a root between 1 and 2. Using Newton's approximation with starting point 1·5 (and showing all relevant working) determine, by means of two iterations, an approximation to this root, giving your answer to two decimal places. [JMB]

18 The table below gives the values of y obtained experimentally for the given values of x. Show graphically that, allowing for small errors of observation, y and x are related by the equation $y = k(1 + x)^n$. Find approximate values of k and n.

x	2	5	9	11	15
y	20·8	58·8	126·5	166·3	256·0

[AEB]

19 Use Simpson's rule with five ordinates to evaluate approximately $\int_0^{1/2} (1 - x^2)^{1/2}\,dx$. By using a suitable substitution, show that

$$\int_0^{1/2} (1 - x^2)^{1/2}\,dx = \frac{\pi}{12} + \frac{\sqrt{3}}{8}$$ [AEB]

20 Two ladders of lengths 4 m and 6 m are placed in an alleyway whose walls are vertical and whose roadway is horizontal. Each ladder rests with its edges in a vertical plane, its foot against one of the walls and its top resting on the other wall. They are found to cross at a height of 1·6 m above the roadway. Show that the height x metres of the top of the shorter ladder above the road satisfies the equation $5x^4-16x^3+100x^2-320x+256=0$. Hence find this height in metres correct to two decimal places.

21 On the same diagram draw the curves $y = -x\cot x$ and $x^2+y^2 = 9(y \geqslant 0)$. Show that they have a point of intersection near to $x = 2{\cdot}3$. Hence using Newton's formula, find a root of the equation $x\cot x + \sqrt{(9-x^2)} = 0$.

22 Given that $f(x) = \tan x - x$ show that $f(4{\cdot}4) < 0$ and $f(4{\cdot}6) > 0$. Use linear interpolation to find an approximation to the root of the equation $\tan x - x = 0$. Using Newton's formula, show that to four significant figures the root is 4·493.

23 Use Simpson's rule to evaluate $\int_0^3 \frac{dx}{x^2+9}$ and compare your result with the exact value.

24 Sketch the graphs of

$$y = e^{-x} \tag{1}$$

and

$$y = \ln(1+x) - 1 \qquad (x > -1) \tag{2}$$

and find the co-ordinates of the point of intersection of (2) with the x-axis. Use the Newton–Raphson process to find (to one decimal place) the x co-ordinate of the intersection of the curves (1) and (2).

25 Experiments are made upon two related quantities, x and y, and corresponding values are observed. It is known that the relation between x and y is of the form $y = ax^n$. Explain how the values of a and n may be obtained by plotting $\log y$ against $\log x$.

Water is discharged over a weir and it is found that for different heights, h, of the free surface of the water above the bottom of the weir, the discharge, Q, is given by the following table

h	4	6	8	10	12
Q	650	1740	3640	6360	9790

Show that these observations are consistent with a relation between Q and h of the form $Q = ah^n$ and give estimates, correct to one decimal place, for the values of a and n. [LU]

22
Vectors

22.1 Vectors and scalars

A *scalar* quantity has magnitude only and is not related to any definite direction in space. For example, time, volume, speed, mass and electric charge are scalar quantities. Scalars are completely specified by numbers which measure their magnitude in terms of some chosen unit, thus 3·3 min, 27 m^3, 9 km/h.

Other quantities exist which have both magnitude and direction, e.g. velocity, force, magnetic field intensity. Thus *9 km/h horizontally in a direction north-west* is speed in a given direction and is a *vector*. An example which does not have physical associations is the directed segment of a straight line $\overrightarrow{PQ}$. The length PQ is the magnitude of the vector. The direction is *parallel* to the line and in the sense from P to Q. We can represent all quantities which have both magnitude and direction by directed line segments.

A *vector* is a quantity which has both magnitude and direction and which can be compounded meaningfully* by the triangle or parallelogram rule described in Section 22.3.

In addition to representing a vector by $\overrightarrow{PQ}$, we shall also use bold faced italic type, such as $\boldsymbol{A}$, $\boldsymbol{a}$, to indicate a vector. The *modulus* of a vector $\boldsymbol{a}$ is the positive number which is a measure of the length of the directed segment and will be denoted by $|\boldsymbol{a}|$ or a.

It is important to remember that if $|\boldsymbol{a}| = |\boldsymbol{b}|$ it does not follow that the vectors are equal; they may have *different directions*.

Two special cases arise. If the modulus of a vector is zero (Q coincides with P), we refer to the zero vector $\mathbf{0}$ the direction of which is indeterminate. If the modulus of the vector is unity, it is referred to as a unit vector and will be denoted by $\hat{\boldsymbol{a}}$, $\hat{\boldsymbol{b}}$, ..., the circumflex indicating a unit vector.

It is important to realise that, in general (see Section 22.7), a vector is not fixed in space. In *Figure 22.1*, $\overrightarrow{AB}$, $\overrightarrow{CD}$, $\overrightarrow{EF}$ and $\overrightarrow{PQ}$ are all equal in length and parallel to one another and are four of the many possible representations of the same vector or, alternatively, they represent four equal vectors.

* Before a physical quantity is treated as a vector it should be proved theoretically, or shown experimentally, that it can be meaningfully compounded by the triangle rule. (Some directed quantities such as finite rotations do not give meaningful results when compounded in this way.) However, in this book we shall deal only with directed physical quantities which are vectors and their vector nature will be assumed without proof.

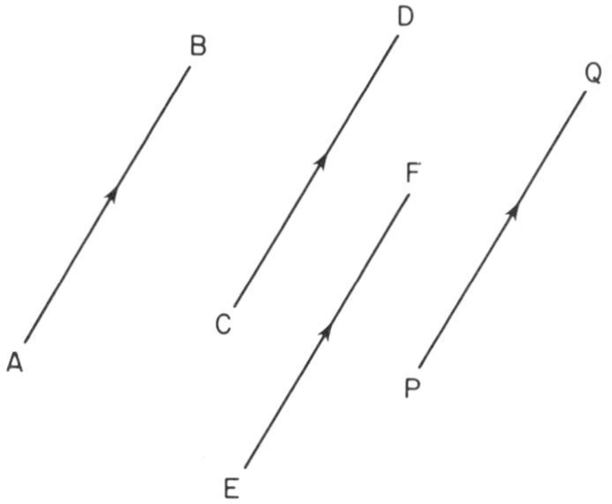

Figure 22.1

22.2 The angle between two vectors

Let the two vectors $\boldsymbol{a}$ and $\boldsymbol{b}$ be represented by $\overrightarrow{OP}$ and $\overrightarrow{OQ}$ (*Figure 22.2*).

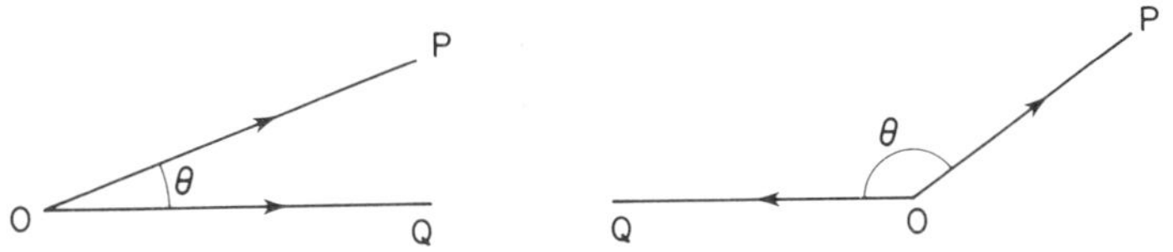

Figure 22.2

The angle between the two vectors is θ, $0 \leqslant \theta \leqslant \pi$, the angle between OP and OQ, where both vectors are drawn with the arrows pointing away from the point of intersection. When θ is equal to $\pi/2$, the vectors are said to be *perpendicular* and when θ is equal to 0 or π, they are said to be *parallel.*

Exercises 22a

1 Which of the following are vectors and which are scalars?
(a) energy (b) speed (c) electric potential
(d) weight (e) deceleration (f) velocity
(g) volume (h) temperature (j) force

2 Draw, on the same diagram, directed line segments to represent the following horizontal vector quantities:
(a) A force $\boldsymbol{F}_1$ of 10 N in a direction 45° E of N.
(b) A force $\boldsymbol{F}_2$ of 8 N in a direction 120° W of N.
(c) A displacement $\boldsymbol{d}$ of 10 m in a direction 150° E of N.
(d) A force $\boldsymbol{F}_3$ of 10 N in a direction 135° W of N.
(e) A velocity $\boldsymbol{v}_1$ of 10 m/s in a direction 45° E of N.
(f) An acceleration $\boldsymbol{f}$ of 8 m/s^2 in a direction due west.
(g) A velocity $\boldsymbol{v}_2$ of 10 m/s in a direction 135° E of N.
Are any of the forces equal and opposite? Can the same directed line segment represent two or more of these vectors?

3 In the previous question state the angle between the vectors: (a) $\boldsymbol{F}_1$ and $\boldsymbol{d}$;

(b) f and F_2; (c) F_2 and v_2. (d) Are any two of the vectors perpendicular to each other? (e) Are any two of the vectors parallel to each other?

4 Represent diagrammatically a horizontal acceleration f of 8 m/s^2 in a direction 20 degrees east of north. On the same diagram represent the following horizontal accelerations: (a) 24 m/s^2 in the direction of f; (b) 2 m/s^2 in the direction of f; (c) 16 m/s^2 in a direction at right angles to f (two cases).

5 A man walks 10 km north and then $7\frac{1}{2}$ km east. Represent these two displacements graphically on the same diagram and find the resulting displacement: (a) graphically, (b) analytically. [Note: this is an example of the triangular rule of vector addition.]

22.3 Addition and subtraction of vectors

Vectors have both magnitude and direction and it is difficult to say from first-hand reasoning how they should be combined. However, consider the special case of displacements.

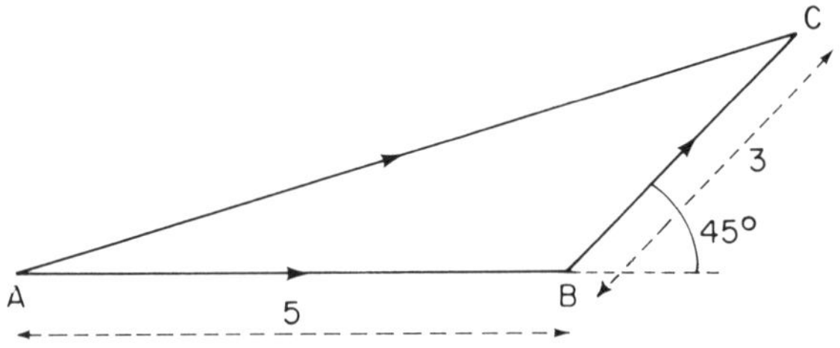

Figure 22.3

If from A, a man walks 5 km due east, this displacement can be represented completely by the directed segment $\overrightarrow{AB}$ of length 5 units (see *Figure 22.3*). If he now walks a further distance of 3 km NE, this further displacement can be represented by $\overrightarrow{BC}$ of length 3 units. It would seem reasonable to say that the final result of his journey is the same as if he had walked a distance represented by the length of AC in the direction from A to C. $\overrightarrow{AC}$ is called the *resultant* of $\overrightarrow{AB}$ and $\overrightarrow{BC}$. It is found in practice that vector quantities

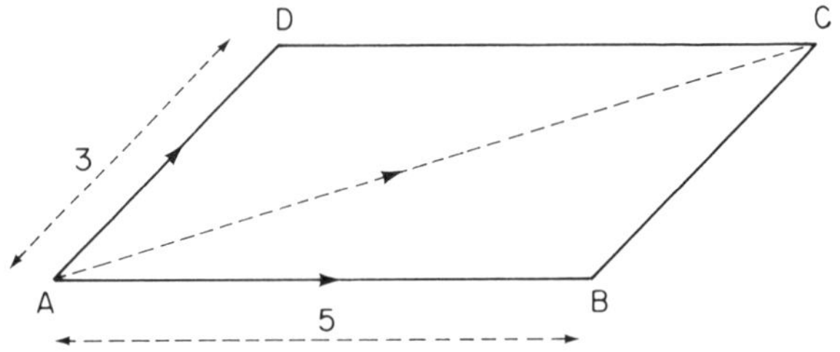

Figure 22.4

such as force, acceleration, velocity, etc., can be compounded in the same way, representing them by directed segments. The resultants thus obtained being meaningful, this is known as the *triangle rule.*

Complete the parallelogram ABCD (see *Figure 22.4*). Because the opposite sides are equal and parallel, it follows that $\overrightarrow{AD}$ also represents the second displacement of 3 km NE. Thus $\overrightarrow{AC}$ is also the resultant of $\overrightarrow{AB}$ and $\overrightarrow{AD}$. This is known as the *parallelogram rule.*

In the case of three vectors represented by $\overrightarrow{PQ}$, $\overrightarrow{QR}$, $\overrightarrow{RS}$ (see *Figure 22.5*), combining $\overrightarrow{PQ}$ and $\overrightarrow{QR}$ gives $\overrightarrow{PR}$, and a further application of the triangle rule to PRS gives the combination of $\overrightarrow{PR}$ and $\overrightarrow{RS}$ as $\overrightarrow{PS}$. This can be extended to any number of directed segments and so we have the *polygon rule*, that if a number of vectors are represented by $\overrightarrow{PQ}$, $\overrightarrow{QR}$, $\overrightarrow{RS}, \ldots, \overrightarrow{YZ}$, taken in order, then their resultant is the vector represented by $\overrightarrow{PZ}$.

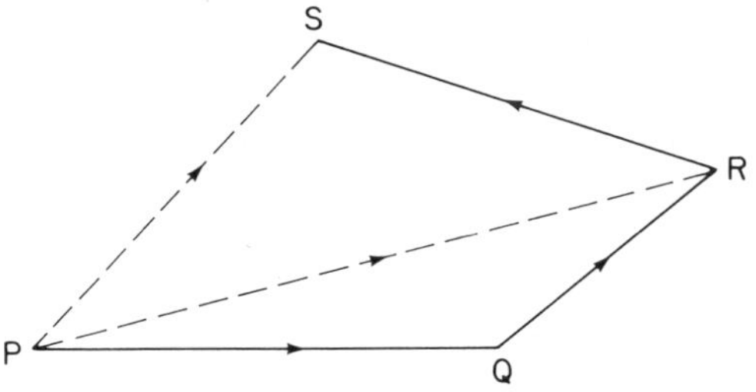

Figure 22.5

Example 1 A yacht is sailing with a velocity of 5 km/h due north. It is also being carried by a current of 2 km/h in a direction south-west. A man walks with a velocity of 3 km/h directly across the deck from starboard to port. Find the man's velocity relative to the shore.

The man is being carried by both the yacht and the current and he is also walking across the deck, thus the resultant of the three velocities is required (see *Figure 22.6*). An accurate scale drawing of the vector diagram gives $R = 5{\cdot}69$, $\theta = 50{\cdot}9°$.

Note that this example is a special case in which the three vectors can all be represented in the one plane. The more general case is dealt with later using components (see Section 22.5).

Exercises 22b

1 Find the magnitudes of two forces such that if they act at right-angles their resultants is $\sqrt{13}$ N, but if they act at an angle of 60° to each other, their resultant is $\sqrt{19}$ N.

2 Two vectors $\boldsymbol{P}$ and $\boldsymbol{Q}$ are at right-angles to each other. The magnitude of $\boldsymbol{P}$ is 5 units and the magnitude of their resultant is 13 units. Find the magnitude of $\boldsymbol{Q}$.

3 Find the vertical force and a force inclined at 30° to the horizontal which together have a resultant of $6\sqrt{3}$ N horizontally.

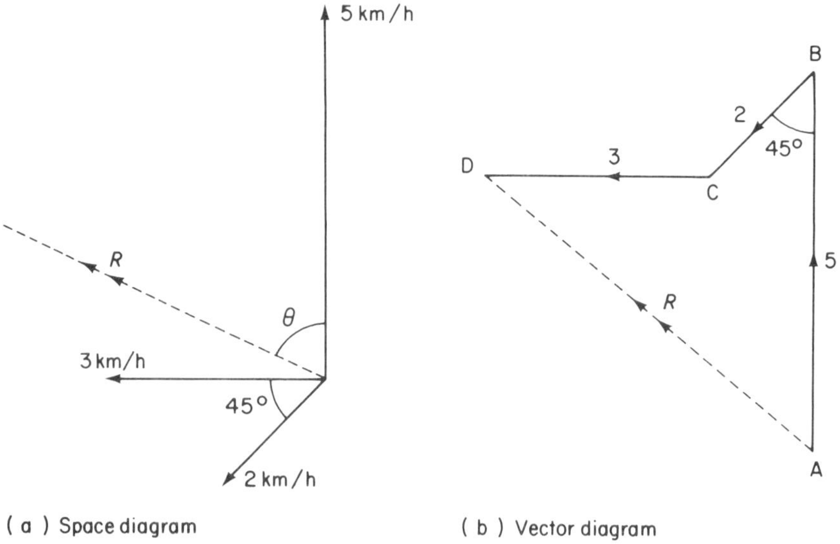

Figure 22.6

4 Three forces of magnitudes 140, 100 and 160 N act on a particle. If the particle is in equilibrium, find the angle between the 100 and 140 N forces.

5 Four horizontal forces have magnitudes and directions as follows: $3\sqrt{2}$ units, N 45° W; 8 units, N 150° E; 10 units, N 120° W; $5\sqrt{2}$ units, N 135° E. By drawing an accurate scale diagram, find the magnitude and direction of their resultant.

To indicate the method of compounding two or more vectors by the triangle or parallelogram rule, we shall use the + sign from algebra and define the addition of two vectors as follows:

If $\boldsymbol{a}$ is represented by the directed segment $\overrightarrow{PQ}$ and $\boldsymbol{b}$ by the directed segment $\overrightarrow{QR}$, then $\boldsymbol{a} + \boldsymbol{b}$ is defined as the vector which is represented by $\overrightarrow{PR}$ (see *Figure 22.7*).

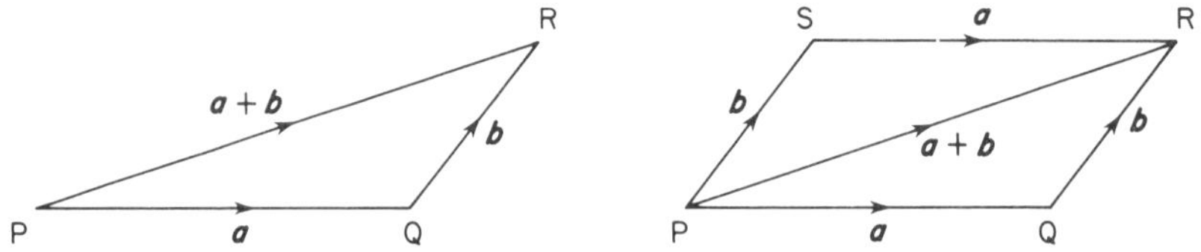

Figure 22.7

Complete the parallelogram PQRS. Then the opposite sides being equal and parallel, we have*

$$\overrightarrow{PQ} = \overrightarrow{SR} \quad \text{and} \quad \overrightarrow{PS} = \overrightarrow{QR}$$

* The equals sign is being used to indicate that PQ is both *equal and parallel* to SR. The arrows indicate the sense in which PQ, SR, ... are described. It follows that $\overrightarrow{PQ}$ and $\overrightarrow{SR}$ can be regarded as representations of the same vector.

and since both $\overrightarrow{PS} + \overrightarrow{SR}$ and $\overrightarrow{PQ} + \overrightarrow{QR}$ are equal to $\overrightarrow{PR}$, it follows that

$$\boldsymbol{b} + \boldsymbol{a} = \boldsymbol{a} + \boldsymbol{b} \tag{22.1}$$

That is, vectors satisfy the commutative law of addition.

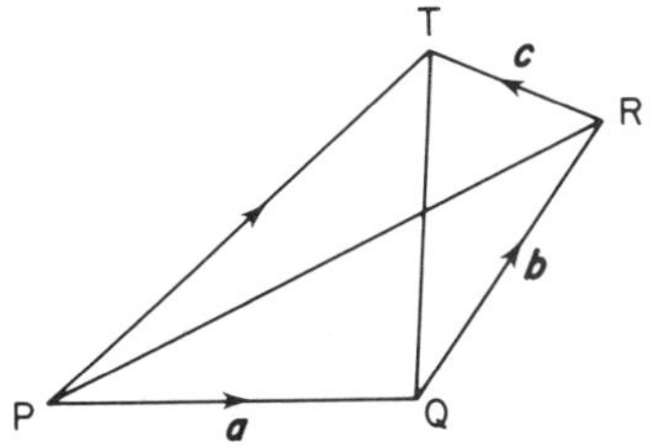

Figure 22.8

Now consider the addition of another vector, $\boldsymbol{c}$, represented by $\overrightarrow{RT}$ in *Figure 22.8*. From the law of addition

$$\overrightarrow{PR} = (\boldsymbol{a} + \boldsymbol{b}) \quad \text{and} \quad \overrightarrow{QT} = (\boldsymbol{b} + \boldsymbol{c})$$

Applying the law of addition again

$$(\boldsymbol{a} + \boldsymbol{b}) + \boldsymbol{c} = \overrightarrow{PR} + \overrightarrow{RT} = \overrightarrow{PT} = \overrightarrow{PQ} + \overrightarrow{QT} = \boldsymbol{a} + (\boldsymbol{b} + \boldsymbol{c}) \tag{22.2}$$

The argument can be extended to any number of vectors. Thus vectors satisfy the associative law of addition and we can write $\boldsymbol{a} + \boldsymbol{b} + \boldsymbol{c} + \ldots$ without brackets being needed to indicate the order of summation.

If the vector $\boldsymbol{b}$ is represented by $\overrightarrow{QR}$, we define $-\boldsymbol{b}$ to be the vector represented by $\overrightarrow{RQ}$, hence,

$$\boldsymbol{b} + (-\boldsymbol{b}) = \boldsymbol{0} \tag{22.3}$$

We now interpret $\boldsymbol{a} - \boldsymbol{b}$ as the sum of the vectors $\boldsymbol{a}$ and $-\boldsymbol{b}$, that is

$$\boldsymbol{a} - \boldsymbol{b} = \boldsymbol{a} + (-\boldsymbol{b}) \tag{22.4}$$

Referring to *Figure 22.9*,

$$\begin{aligned} \overrightarrow{SQ} &= \overrightarrow{SP} + \overrightarrow{PQ} = \overrightarrow{RQ} + \overrightarrow{PQ} \\ &= -\boldsymbol{b} + \boldsymbol{a} \\ &= \boldsymbol{a} - \boldsymbol{b} \end{aligned}$$

Note that the opposite diagonal of the parallelogram $\overrightarrow{PR}$ represents $\boldsymbol{a} + \boldsymbol{b}$.

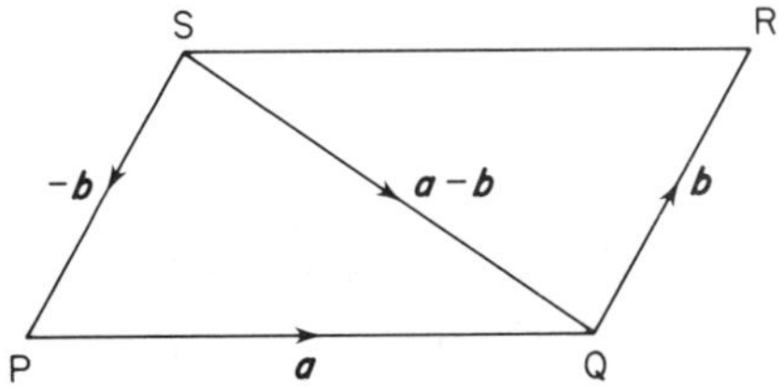

Figure 22.9

Example 2 Show that $|\boldsymbol{a}+\boldsymbol{b}| \leqslant |\boldsymbol{a}|+|\boldsymbol{b}|$. In what circumstances does the equality hold?

Referring to *Figure 22.9*, $|\boldsymbol{a}|$ is the number of units in the length of PQ, $|\boldsymbol{b}|$ is the number of units in the length of QR, $|\boldsymbol{a}+\boldsymbol{b}|$ is the number of units in the length of PR.

Since PQR is a triangle,

$$\mathrm{PR} \leqslant \mathrm{PQ}+\mathrm{QR}$$

or

$$|\boldsymbol{a}+\boldsymbol{b}| \leqslant |\boldsymbol{a}|+|\boldsymbol{b}|$$

This equality holds when PR = PQ + QR, that is, when P, Q, R (in that order) are collinear, i.e. $\boldsymbol{a}$ and $\boldsymbol{b}$ are parallel and in the same sense.

Example 3 The three co-terminal edges $\overrightarrow{\mathrm{AP}}$, $\overrightarrow{\mathrm{AD}}$, $\overrightarrow{\mathrm{AB}}$ of a cuboid ABCDPQRS represent respectively the three vectors $\boldsymbol{a}$, $\boldsymbol{b}$, $\boldsymbol{c}$. (Note that although the edges are equal in length they are in different directions and therefore represent different vectors.) Find in terms of $\boldsymbol{a}$, $\boldsymbol{b}$ and $\boldsymbol{c}$ the vectors represented by $\overrightarrow{\mathrm{AQ}}$, $\overrightarrow{\mathrm{AR}}$, $\overrightarrow{\mathrm{RP}}$ and $\overrightarrow{\mathrm{PC}}$ (see *Figure 22.10*).

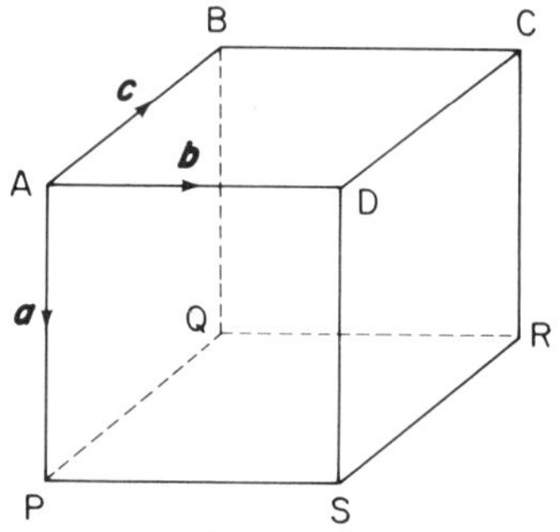

Figure 22.10

Since $\overrightarrow{\mathrm{AP}}$, $\overrightarrow{\mathrm{DS}}$, $\overrightarrow{\mathrm{CR}}$ and $\overrightarrow{\mathrm{BQ}}$ are all equal, in the same sense and parallel, they all represent the vector $\boldsymbol{a}$. Similarly, $\overrightarrow{\mathrm{AD}}$, $\overrightarrow{\mathrm{BC}}$, $\overrightarrow{\mathrm{PS}}$ and $\overrightarrow{\mathrm{QR}}$ all represent the vector $\boldsymbol{b}$ and $\overrightarrow{\mathrm{AB}}$, $\overrightarrow{\mathrm{DC}}$, $\overrightarrow{\mathrm{SR}}$ and $\overrightarrow{\mathrm{PQ}}$ all represent the vector $\boldsymbol{c}$. Now

$$\overrightarrow{\mathrm{AQ}} = \overrightarrow{\mathrm{AP}} + \overrightarrow{\mathrm{PQ}}$$

Therefore, $\overrightarrow{\mathrm{AQ}}$ represents the vector $\boldsymbol{a}+\boldsymbol{c}$.

$$\overrightarrow{\mathrm{AR}} = \overrightarrow{\mathrm{AP}} + \overrightarrow{\mathrm{PS}} + \overrightarrow{\mathrm{SR}}$$

Therefore, $\overrightarrow{\mathrm{AR}}$ represents the vector $\boldsymbol{a}+\boldsymbol{b}+\boldsymbol{c}$.

$$\begin{aligned}\overrightarrow{\mathrm{RP}} &= \overrightarrow{\mathrm{RS}} + \overrightarrow{\mathrm{SP}} \\ &= -\overrightarrow{\mathrm{SR}} - \overrightarrow{\mathrm{PS}}\end{aligned}$$

Therefore, $\overrightarrow{\mathrm{RP}}$ represents the vector $-\boldsymbol{c}-\boldsymbol{b}$.

$$\begin{aligned}\overrightarrow{\mathrm{PC}} &= \overrightarrow{\mathrm{PS}} + \overrightarrow{\mathrm{SD}} + \overrightarrow{\mathrm{DC}} \\ &= \overrightarrow{\mathrm{PS}} - \overrightarrow{\mathrm{DS}} + \overrightarrow{\mathrm{DC}}\end{aligned}$$

Therefore $\overrightarrow{\mathrm{PC}}$ represents the vector $\boldsymbol{b}-\boldsymbol{a}+\boldsymbol{c}$.

Exercises 22c

1 Sketch directed line segments to represent vectors $\boldsymbol{a}$, 3 units north, $\boldsymbol{b}$, 2 units north-east and $\boldsymbol{c}$, 4 units west. Now sketch directed line segments to represent the following vectors: (a) $\hat{\boldsymbol{c}}$, (b) $-\boldsymbol{c}$, (c) $\boldsymbol{a}+\boldsymbol{b}$, (d) $\boldsymbol{a}+\boldsymbol{c}$, (e) $\boldsymbol{a}-\boldsymbol{c}$, (f) $\boldsymbol{b}+\boldsymbol{b}$, (g) $\boldsymbol{a}+\boldsymbol{b}-\boldsymbol{c}$, (h) $\boldsymbol{a}+\boldsymbol{b}+\boldsymbol{c}$, (j) $\boldsymbol{a}-\boldsymbol{b}+\boldsymbol{c}$, (k) $\boldsymbol{a}-\boldsymbol{b}-\boldsymbol{c}$.

2 For the three vectors $\boldsymbol{a}$, $\boldsymbol{b}$, $\boldsymbol{c}$ defined in question 1, find the values of the following quantities: (a) $|\boldsymbol{b}|$, (b) $|\boldsymbol{a}+\boldsymbol{c}|$, (c) $|-\boldsymbol{c}|$, (d) $|\boldsymbol{a}+\boldsymbol{b}|$, (e) $|\boldsymbol{c}+\boldsymbol{b}|$, (f) $|\boldsymbol{c}-\boldsymbol{a}|$.

3 Show that $|\boldsymbol{a}-\boldsymbol{b}| \geqslant \big||\boldsymbol{a}|-|\boldsymbol{b}|\big|$. In what circumstances does the equality hold?

4 If $\boldsymbol{a}$ and $\boldsymbol{b}$ are two vectors such that $|\boldsymbol{a}|=|\boldsymbol{b}|=|\boldsymbol{a}+\boldsymbol{b}|$, find the angle between $\boldsymbol{a}$ and $\boldsymbol{b}$.

5 A pyramid has apex P and base ABCD. The four edges $\overrightarrow{\mathrm{PA}}$, $\overrightarrow{\mathrm{PB}}$, $\overrightarrow{\mathrm{PC}}$ and $\overrightarrow{\mathrm{PD}}$ represent respectively the vectors $\boldsymbol{a}$, $\boldsymbol{b}$, $\boldsymbol{c}$ and $\boldsymbol{d}$. Find in terms of some or all of $\boldsymbol{a}$, $\boldsymbol{b}$, $\boldsymbol{c}$, $\boldsymbol{d}$ the vectors represented by (a) $\overrightarrow{\mathrm{BC}}$, (b) $\overrightarrow{\mathrm{BA}}$, (c) $\overrightarrow{\mathrm{BD}}$, (d) $\overrightarrow{\mathrm{AC}}$.

6 In Example 3, find in terms of $\boldsymbol{a}$, $\boldsymbol{b}$ and $\boldsymbol{c}$ the vectors represented by (a) $\overrightarrow{\mathrm{PD}}$, (b) $\overrightarrow{\mathrm{SQ}}$, (c) $\overrightarrow{\mathrm{DQ}}$, (d) $\overrightarrow{\mathrm{CS}}$.

22.4 Multiplication of a vector by a scalar

From the law of addition of vectors it follows that $\boldsymbol{a}+\boldsymbol{a}$ is a vector in the same direction as $\boldsymbol{a}$ and whose magnitude is twice the magnitude of $\boldsymbol{a}$. We can denote it by $2\boldsymbol{a}$. By continued addition, $\boldsymbol{a}+\boldsymbol{a}+\boldsymbol{a}+\ldots$ to n terms is a vector in the direction of $\boldsymbol{a}$ whose magnitude is equal to n times the magnitude of $\boldsymbol{a}$. It can be denoted by $n\boldsymbol{a}$. We generalise this for any n, fractional or negative, as follows:

The product of a scalar λ and a vector $\boldsymbol{a}$ (written $\lambda\boldsymbol{a}$) is a vector, in the direction of $\boldsymbol{a}$ if λ is positive and in the opposite direction to $\boldsymbol{a}$ if λ is negative, of magnitude $|\lambda|.|\boldsymbol{a}|$.

Example 1 $\boldsymbol{F}$ is a force of 3 units in a direction N 50° E. What is meant by $-4\boldsymbol{F}$?

$-4\boldsymbol{F}$ is a force of magnitude $|-4|.|3| = 12$ units. The direction of $-4\boldsymbol{F}$ is opposite to the direction of F, that is, N 130° W. Therefore, $-4\boldsymbol{F}$ is a force of 12 units in a direction N 130° W.

It follows from the definition that if μ is also a scalar, $\mu(\lambda\boldsymbol{a})=(\mu\lambda)\boldsymbol{a}=\lambda(\mu\boldsymbol{a})$ and the associative law of multiplication holds. Also from the definition, $(\lambda+\mu)\boldsymbol{a}=\lambda\boldsymbol{a}+\mu\boldsymbol{a}$.

It is also true* that $\lambda(\boldsymbol{a}+\boldsymbol{b})=\lambda\boldsymbol{a}+\lambda\boldsymbol{b}$.

Division of a vector by a scalar m is defined as multiplication by $1/m$.

Note:

(i) If $\hat{\boldsymbol{a}}$ is a unit vector in the direction of $\boldsymbol{a}$, $\hat{\boldsymbol{a}}=\boldsymbol{a}/|\boldsymbol{a}|$.
(ii) If $\boldsymbol{b}$ is parallel to $\boldsymbol{a}$,

$$\frac{\boldsymbol{b}}{|\boldsymbol{b}|}=\frac{\boldsymbol{a}}{|\boldsymbol{a}|}$$

* See Mulholland and Phillips, *Applied mathematics for advanced level*, Butterworths.

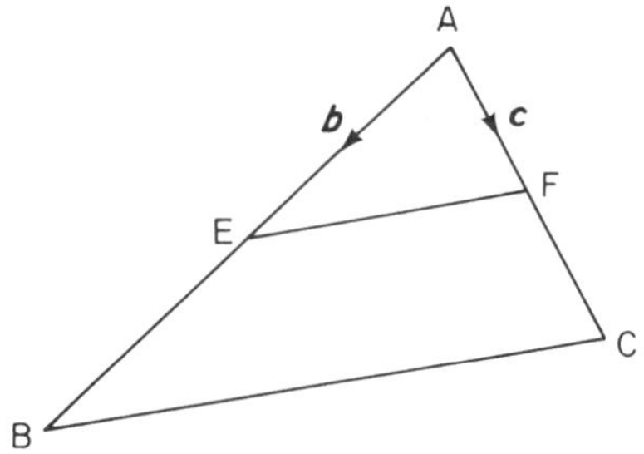

Figure 22.11

Example 2 In the triangle ABC the mid-points of the sides AB, AC are E and F respectively. Prove that EF is parallel to and equal to half the length of BC.

Let $\overrightarrow{AB}$ and $\overrightarrow{AC}$ represent the vectors $\boldsymbol{b}$ and $\boldsymbol{c}$ (see *Figure 22.11*).

$$\begin{aligned}\overrightarrow{BC} &= \overrightarrow{BA} + \overrightarrow{AC}\\ &= -\overrightarrow{AB} + \overrightarrow{AC}\end{aligned}$$

Therefore, $\overrightarrow{BC}$ represents the vector $\boldsymbol{c} - \boldsymbol{b}$ (i)

Since E and F are mid-points of AB and AC, $\overrightarrow{AE}$ and $\overrightarrow{AF}$ represent the vectors $\frac{1}{2}\boldsymbol{b}$ and $\frac{1}{2}\boldsymbol{c}$.

$$\begin{aligned}\overrightarrow{EF} &= \overrightarrow{EA} + \overrightarrow{AF}\\ &= -\overrightarrow{AE} + \overrightarrow{AF}\end{aligned}$$

Therefore, $\overrightarrow{EF}$ represents the vector $-\frac{1}{2}\boldsymbol{b} + \frac{1}{2}\boldsymbol{c}$. That is

$$\tfrac{1}{2}(\boldsymbol{c} - \boldsymbol{b}) \qquad \text{(ii)}$$

Now $\frac{1}{2}(\boldsymbol{c} - \boldsymbol{b})$ is a vector parallel and equal to half the magnitude of $(\boldsymbol{c} - \boldsymbol{b})$. Therefore, $\overrightarrow{EF}$ is parallel to, and equal to half the length of $\overrightarrow{BC}$.

Example 3 P is the apex of a rectangular pyramid whose base is ABCD. The edges $\overrightarrow{PA}$, $\overrightarrow{PB}$,$\overrightarrow{PC}$ represent respectively the vectors $\boldsymbol{a}$, $\boldsymbol{b}$, $\boldsymbol{c}$. Find in terms of $\boldsymbol{a}$, $\boldsymbol{b}$ and $\boldsymbol{c}$ the vectors represented by (i) $\overrightarrow{AD}$, (ii) $\overrightarrow{CD}$, (iii) $\overrightarrow{PD}$.

ABCD is a rectangle. Therefore (see *Figure 22.12*), $\overrightarrow{AD}$ is equal and parallel to $\overrightarrow{BC}$

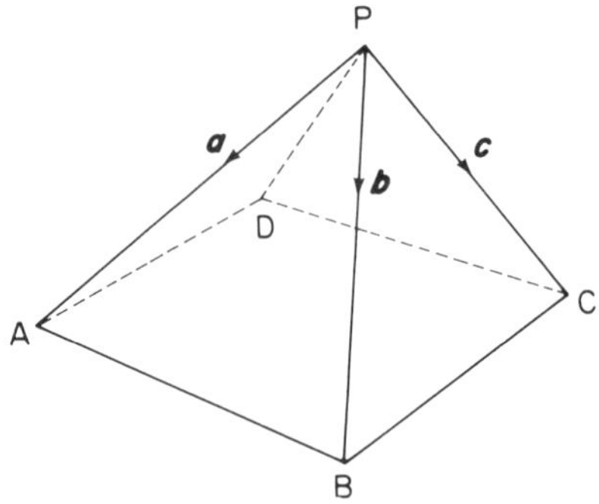

Figure 22.12

$$\overrightarrow{BC} = \overrightarrow{BP} + \overrightarrow{PC}$$
$$= -\overrightarrow{PB} + \overrightarrow{PC}$$

Therefore, $\overrightarrow{BC}$ and hence $\overrightarrow{AD}$ represents $\boldsymbol{c} - \boldsymbol{b}$. (i)

Similarly $\overrightarrow{CD}$ is equal and parallel to $\overrightarrow{BA}$ and

$$\overrightarrow{BA} = \overrightarrow{BP} + \overrightarrow{PA}$$
$$= -\overrightarrow{PB} + \overrightarrow{PA}$$

Therefore, $\overrightarrow{BA}$ and hence $\overrightarrow{CD}$ represents $\boldsymbol{a} - \boldsymbol{b}$ (ii)

$$\overrightarrow{PD} = \overrightarrow{PC} + \overrightarrow{CD}$$

Therefore, from (ii), $\overrightarrow{PD}$ represents $\boldsymbol{c} + \boldsymbol{a} - \boldsymbol{b}$.

Exercises 22d

1 ABCD is a plane quadrilateral. The mid-points of its sides are P, Q, R and S. Prove that PQRS is a plane parallelogram. Is the result also true if ABCD is a skew quadrilateral?

2 In the triangle ABC, D is the mid-point of the side BC. Prove that $\overrightarrow{AB} + \overrightarrow{AC} = 2\overrightarrow{AD}$.

3 In the parallelogram PQRS, A is the mid-point of PR and B is the mid-point of QS. Let $\overrightarrow{PQ} = \boldsymbol{a}$ and $\overrightarrow{PS} = \boldsymbol{b}$ and from the result of question 2 deduce that $\overrightarrow{PB} = \frac{1}{2}(\boldsymbol{a} + \boldsymbol{b})$. Find $\overrightarrow{PA}$ in terms of $\boldsymbol{a}$ and $\boldsymbol{b}$ and deduce that the diagonals of a parallelogram bisect each other.

4 In the tetrahedron ABCD, $\overrightarrow{AB}$, $\overrightarrow{AC}$ and $\overrightarrow{AD}$ represent vectors $\boldsymbol{a}$, $\boldsymbol{b}$ and $\boldsymbol{c}$ respectively. The points P, Q, R and S are the mid-points of the sides AB, BC, CD and DA respectively.

(i) Show that $\overrightarrow{PR}$ represents the vector $\frac{1}{2}(-\boldsymbol{a} + \boldsymbol{b} + \boldsymbol{c})$.

(ii) Find, in terms of $\boldsymbol{a}$, $\boldsymbol{b}$, $\boldsymbol{c}$, the vector represented by $\overrightarrow{SQ}$.

(iii) Deduce the vector represented by $\overrightarrow{TU}$ where T and U are the mid-points of AC and BD respectively.

5 Simplify the following expressions:

(i) $3(3\boldsymbol{a} - 2\boldsymbol{b} + \boldsymbol{c}) + 2(\boldsymbol{a} + 3\boldsymbol{b} - \boldsymbol{c})$

(ii) $5(\boldsymbol{a} - 3\boldsymbol{b} + 6\boldsymbol{c}) - 3(2\boldsymbol{a} - 2\boldsymbol{b} + 10\boldsymbol{c})$

(iii) $\frac{1}{2}(2\boldsymbol{a} + 6\boldsymbol{b} - \boldsymbol{c}) + 2(\boldsymbol{a} + 3\boldsymbol{b} + 2\boldsymbol{c})$

6 ABCD is a plane quadrilateral in which P, Q are the mid-points of AB, DC respectively. Show that $\overrightarrow{AD} + \overrightarrow{BC} = 2\overrightarrow{PQ}$. Is the result true if ABCD is a skew quadrilateral?

22.5 Resolution of a vector

We first define collinear and coplanar vectors.

Two or more vectors are said to be *coplanar* if they are all parallel to the same plane.

Two or more vectors are said to be *collinear* if they are all parallel to the same line.

Given a vector $\boldsymbol{r}$ and any three non-coplanar vectors, $\boldsymbol{a}$, $\boldsymbol{b}$ and $\boldsymbol{c}$, then $\boldsymbol{r}$ can be expressed as the sum of three vectors parallel to $\boldsymbol{a}$, $\boldsymbol{b}$ and $\boldsymbol{c}$.

Let $\hat{\boldsymbol{a}}$, $\hat{\boldsymbol{b}}$, $\hat{\boldsymbol{c}}$, be unit vectors parallel to $\boldsymbol{a}$, $\boldsymbol{b}$, $\boldsymbol{c}$, respectively. With any point O as origin construct a parallelepiped whose co-terminal edges OA, OB, OC are parallel to $\boldsymbol{a}$, $\boldsymbol{b}$, $\boldsymbol{c}$ respectively, and whose diagonal $\overrightarrow{\mathrm{OR}}$ represents $\boldsymbol{r}$ (see *Figure* 22.13).

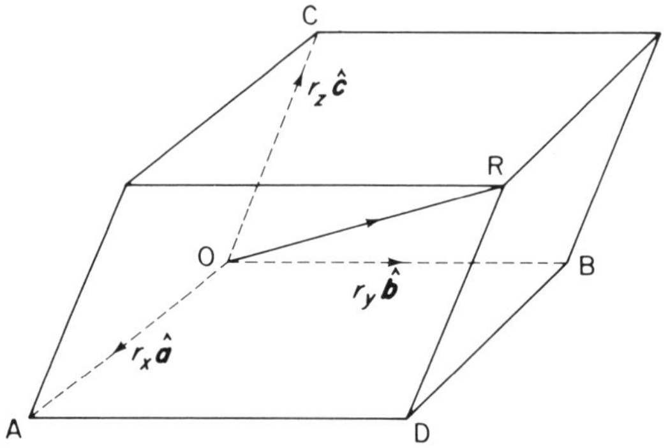

Figure 22.13

Then there exist real numbers r_x, r_y, r_z such that* $\overrightarrow{\mathrm{OA}} = r_x\hat{\boldsymbol{a}}$, $\overrightarrow{\mathrm{OB}} = r_y\hat{\boldsymbol{b}}$, $\overrightarrow{\mathrm{OC}} = r_z\hat{\boldsymbol{c}}$. ($r_x$ is positive or negative according as $\overrightarrow{\mathrm{OA}}$ is in the same or opposite direction to $\boldsymbol{a}$, similarly for r_y and r_z.)

$$\boldsymbol{r} = \overrightarrow{\mathrm{OR}} = \overrightarrow{\mathrm{OA}} + \overrightarrow{\mathrm{AD}} + \overrightarrow{\mathrm{DR}}$$

Opposite edges of a parallelepiped are equal and parallel. Thus

$$\overrightarrow{\mathrm{AD}} = \overrightarrow{\mathrm{OB}} \quad \text{and} \quad \overrightarrow{\mathrm{DR}} = \overrightarrow{\mathrm{OC}}$$

$$\boldsymbol{r} = \overrightarrow{\mathrm{OA}} + \overrightarrow{\mathrm{OB}} + \overrightarrow{\mathrm{OC}}$$

$$= r_x\hat{\boldsymbol{a}} + r_y\hat{\boldsymbol{b}} + r_z\hat{\boldsymbol{c}} \tag{22.5}$$

$r_x\hat{\boldsymbol{a}}$, $r_y\hat{\boldsymbol{b}}$, $r_z\hat{\boldsymbol{c}}$ are known as the component vectors and r_x, r_y, r_z as the components of $\boldsymbol{r}$ in the three given directions. Only one such parallelepiped can be constructed and the resolution is unique. Thus *equal vectors* have *equal components*. Conversely, if all three components of two vectors are equal, the vectors are equal.

COROLLARY In *two* dimensions two non-parallel vectors $\boldsymbol{a}$ and $\boldsymbol{b}$ define a plane and any vector $\boldsymbol{r}$ parallel to that plane can be expressed as the sum of two vectors parallel to $\boldsymbol{a}$ and $\boldsymbol{b}$, that is

$$\boldsymbol{r} = k\boldsymbol{a} + l\boldsymbol{b}$$

where k and l are scalars.

The most important case arises when the three directions defined by $\boldsymbol{a}$, $\boldsymbol{b}$, $\boldsymbol{c}$ are mutually perpendicular and form a rectangular frame of reference. The lines are labelled to form a right-handed set of axes, that is, rotation from Ox to Oy takes a right-handed corkscrew along Oz; similarly, a rotation from Oy to Oz takes a right-handed corkscrew along Ox; similarly for Oy (see *Figure 22.14*).

* The symbol = is now being used to mean *represents*.

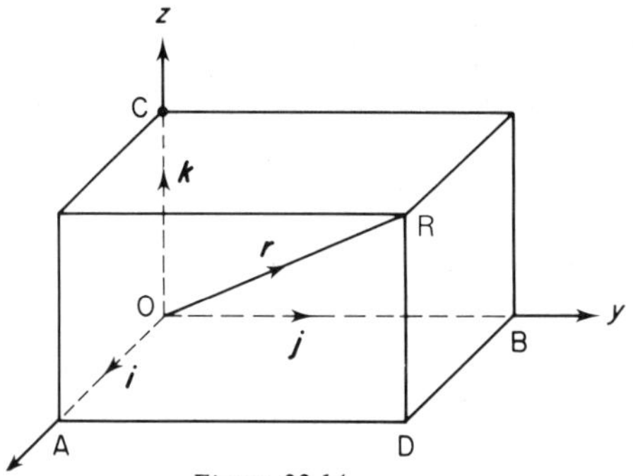

Figure 22.14

In this case, the unit vectors $\hat{a}$, $\hat{b}$ and $\hat{c}$ are known as $\boldsymbol{i}$, $\boldsymbol{j}$, and $\boldsymbol{k}$ (without circumflexes) and

$$\boldsymbol{r} = r_x\boldsymbol{i} + r_y\boldsymbol{j} + r_z\boldsymbol{k}$$

r_x, r_y and r_z are now referred to as the rectangular components, or simply, components if there is no fear of a misunderstanding.

Now

$$\begin{aligned} OR^2 &= OB^2 + BD^2 + DR^2 \quad \text{(see } \textit{Figure 22.14}\text{)} \\ &= OB^2 + BD^2 + DR^2 \\ &= OB^2 + OA^2 + OC^2 \\ r^2 &= r_x^2 + r_y^2 + r_z^2 \end{aligned} \tag{22.6}$$

From which we see that the modulus of a vector $\boldsymbol{r}$ with rectangular components r_x, r_y, r_z is

$$|\boldsymbol{r}| = \sqrt{(r_x^2 + r_y^2 + r_z^2)}$$

(Note the positive square root.)

It follows from the associative and distributive laws that

Rule I $\quad \boldsymbol{a} + \boldsymbol{b} = (a_x + b_x)\boldsymbol{i} + (a_y + b_y)\boldsymbol{j} + (a_z + b_z)\boldsymbol{k}$

Rule II $\quad m\boldsymbol{a} = ma_x\boldsymbol{i} + ma_y\boldsymbol{j} + ma_z\boldsymbol{k} \quad (m \text{ scalar})$

Example 1 If $\boldsymbol{a} = 3\boldsymbol{i} + 4\boldsymbol{j} - 12\boldsymbol{k}$, $\boldsymbol{b} = \boldsymbol{i} + 12\boldsymbol{k}$, $\boldsymbol{c} = \boldsymbol{i} - \boldsymbol{j} + \boldsymbol{k}$. Find $|\boldsymbol{a}|$, $|\boldsymbol{b}|$, $|\boldsymbol{c}|$, $|\boldsymbol{a} + \boldsymbol{b}|$ and $|\boldsymbol{a} + \boldsymbol{b} + \boldsymbol{c}|$.

From the equation $|\boldsymbol{r}| = \sqrt{(r_x^2 + r_y^2 + r_z^2)}$, we obtain

$$|\boldsymbol{a}| = \sqrt{[3^2 + 4^2 + (-12)^2]} = \sqrt{13}$$

$$|\boldsymbol{b}| = \sqrt{(1^2 + 0^2 + 12^2)} = \sqrt{145}$$

$$|\boldsymbol{c}| = \sqrt{[1^2 + (-1)^2 + 1^2]} = \sqrt{3}$$

$$|\boldsymbol{a} + \boldsymbol{b}| = |4\boldsymbol{i} + 4\boldsymbol{j}| = \sqrt{(4^2 + 4^2 + 0^2)} = 4\sqrt{2}$$

$$|\boldsymbol{a} + \boldsymbol{b} + \boldsymbol{c}| = |5\boldsymbol{i} + 3\boldsymbol{j} + \boldsymbol{k}| = \sqrt{(5^2 + 3^2 + 1^2)} = \sqrt{35}$$

Vectors and the vector algebra we are developing are eminently suitable for work in three-dimensions (and can be generalised to n-dimensions). In the case of two-dimensional applications, the r_z component is suppressed. Thus

$$|\boldsymbol{r}| = \sqrt{(r_x^2 + r_y^2)}$$

In general, $\boldsymbol{r} = a\boldsymbol{i} + b\boldsymbol{j} + c\boldsymbol{k}$ can be written as a column vector thus

$$\boldsymbol{r} = \begin{bmatrix} a \\ b \\ c \end{bmatrix}$$

Example 2 Given $\boldsymbol{r}_1 = \begin{bmatrix} 5 \\ -6 \\ -2 \end{bmatrix}$, $\boldsymbol{r}_2 = \begin{bmatrix} 7 \\ 3 \\ 1 \end{bmatrix}$, $\boldsymbol{r}_3 = \begin{bmatrix} 2 \\ 2 \\ -4 \end{bmatrix}$, find (i) $|\boldsymbol{r}_1|$, (ii) $|2\boldsymbol{r}_1 + \boldsymbol{r}_3|$, (iii) $|\boldsymbol{r}_1 - \boldsymbol{r}_2 - 3\boldsymbol{r}_3|$. (iv) Write in conventional form (using $\boldsymbol{i}, \boldsymbol{j}, \boldsymbol{k}$) the unit vector parallel to the vector $\boldsymbol{r}_2 - \boldsymbol{r}_3$.

(i)
$$\begin{aligned} \boldsymbol{r}_1 &= \begin{bmatrix} 5 \\ -6 \\ -2 \end{bmatrix} \\ &= 5\boldsymbol{i} - 6\boldsymbol{j} - 2\boldsymbol{k} \\ |\boldsymbol{r}_1| &= \sqrt{[5^2 + (-6)^2 + (-2)^2]} \\ &= \sqrt{65} \end{aligned}$$

(ii)
$$\begin{aligned} 2\boldsymbol{r}_1 + \boldsymbol{r}_3 &= 2\begin{bmatrix} 5 \\ -6 \\ -2 \end{bmatrix} + \begin{bmatrix} 2 \\ 2 \\ -4 \end{bmatrix} \\ &= 2(5\boldsymbol{i} - 6\boldsymbol{j} - 2\boldsymbol{k}) + (2\boldsymbol{i} + 2\boldsymbol{j} - 4\boldsymbol{k}) \\ &= 12\boldsymbol{i} - 10\boldsymbol{j} - 8\boldsymbol{k} \\ |2\boldsymbol{r}_1 + \boldsymbol{r}_3| &= \sqrt{[12^2 + (-10)^2 + (-8)^2]} \\ &= \sqrt{308} \end{aligned}$$

(iii)
$$\begin{aligned} \boldsymbol{r}_1 - \boldsymbol{r}_2 - 3\boldsymbol{r}_3 &= \begin{bmatrix} 5 \\ -6 \\ -2 \end{bmatrix} - \begin{bmatrix} 7 \\ 3 \\ 1 \end{bmatrix} - 3\begin{bmatrix} 2 \\ 2 \\ -4 \end{bmatrix} \\ &= (5\boldsymbol{i} - 6\boldsymbol{j} - 2\boldsymbol{k}) - (7\boldsymbol{i} + 3\boldsymbol{j} + \boldsymbol{k}) - 3(2\boldsymbol{i} + 2\boldsymbol{j} - 4\boldsymbol{k}) \\ &= -8\boldsymbol{i} - 15\boldsymbol{j} + 9\boldsymbol{k} \\ |\boldsymbol{r}_1 - \boldsymbol{r}_2 - 3\boldsymbol{r}_3| &= \sqrt{[(-8)^2 + (-15)^2 + 9^2]} \\ &= \sqrt{370} \end{aligned}$$

(iv)
$$\begin{aligned} \boldsymbol{r}_2 - \boldsymbol{r}_3 &= \begin{bmatrix} 7 \\ 3 \\ 1 \end{bmatrix} - \begin{bmatrix} 2 \\ 2 \\ -4 \end{bmatrix} \\ &= (7\boldsymbol{i} + 3\boldsymbol{j} + \boldsymbol{k}) - (2\boldsymbol{i} + 2\boldsymbol{j} - 4\boldsymbol{k}) \\ &= 5\boldsymbol{i} + \boldsymbol{j} + 5\boldsymbol{k} \\ |\boldsymbol{r}_2 - \boldsymbol{r}_3| &= \sqrt{(5^2 + 1^2 + 5^2)} \\ &= \sqrt{51} \end{aligned}$$

Unit vector in the direction of $(\boldsymbol{r}_2 - \boldsymbol{r}_3)$ is

$$\frac{(\boldsymbol{r}_2 - \boldsymbol{r}_3)}{|\boldsymbol{r}_2 - \boldsymbol{r}_3|} = \frac{5}{\sqrt{51}}\boldsymbol{i} + \frac{1}{\sqrt{51}}\boldsymbol{j} + \frac{5}{\sqrt{51}}\boldsymbol{k}$$

$$= (5\boldsymbol{i} + \boldsymbol{j} + 5\boldsymbol{k})/\sqrt{51}$$

After practice it will be found that the transformation from column vectors to $\boldsymbol{i}, \boldsymbol{j}, \boldsymbol{k}$ form can be omitted.

Example 3 Given that $\boldsymbol{a} = 2\boldsymbol{i} + 3\boldsymbol{j} - \boldsymbol{k}$, $\boldsymbol{b} = \boldsymbol{i} + 2\boldsymbol{k}$, $\boldsymbol{c} = \boldsymbol{i} + 2\boldsymbol{j}$, (i) show that there are no real values of the constants λ and μ such that $\boldsymbol{c} = \lambda\boldsymbol{a} + \mu\boldsymbol{b}$ (this proves that $\boldsymbol{a}$, $\boldsymbol{b}$, $\boldsymbol{c}$ are non-coplanar). (ii) Express $\boldsymbol{d} = -4\boldsymbol{j} + 8\boldsymbol{k}$ in terms of $\boldsymbol{a}$, $\boldsymbol{b}$ and $\boldsymbol{c}$.

(i) Let $\boldsymbol{c} = \lambda\boldsymbol{a} + \mu\boldsymbol{b}$, that is

$$(\boldsymbol{i} + 2\boldsymbol{j} + 0\boldsymbol{k}) = \lambda(2\boldsymbol{i} + 3\boldsymbol{j} - \boldsymbol{k}) + \mu(\boldsymbol{i} + 0\boldsymbol{j} + 2\boldsymbol{k})$$
$$= (2\lambda + \mu)\boldsymbol{i} + 3\lambda\boldsymbol{j} + (-\lambda + 2\mu)\boldsymbol{k}$$

Since we have assumed the vectors are equal, the components are equal. Therefore

$$1 = 2\lambda + \mu \qquad \text{(i)}$$
$$2 = 3\lambda \qquad \text{(ii)}$$
$$0 = -\lambda + 2\mu \qquad \text{(iii)}$$

From (i) and (ii), $\lambda = 3/2$, $\mu = -2$. Substituting these values in (iii),

$$-\lambda + 2\mu = -5\tfrac{1}{2} \neq 0$$

Therefore, no values of λ, μ exist for which $\boldsymbol{c} = \lambda\boldsymbol{a} + \mu\boldsymbol{b}$.

(ii) Let $\boldsymbol{d} = \lambda\boldsymbol{a} + \mu\boldsymbol{b} + \nu\boldsymbol{c}$, that is

$$(0\boldsymbol{i} - 4\boldsymbol{j} + 8\boldsymbol{k}) = \lambda(2\boldsymbol{i} + 3\boldsymbol{j} - \boldsymbol{k}) + \mu(\boldsymbol{i} + 0\boldsymbol{j} + 2\boldsymbol{k}) + \nu(\boldsymbol{i} + 2\boldsymbol{j} + 0\boldsymbol{k})$$
$$= (2\lambda + \mu + \nu)\boldsymbol{i} + (3\lambda + 2\nu)\boldsymbol{j} + (-\lambda + 2\mu)\boldsymbol{k}$$

Equating the components

$$2\lambda + \mu + \nu = 0$$
$$3\lambda + 2\nu = -4$$
$$-\lambda + 2\mu = 8$$

Hence

$$\lambda = -2 \qquad \mu = 3 \qquad \nu = 1$$

and

$$\boldsymbol{d} = -2\boldsymbol{a} + 3\boldsymbol{b} + \boldsymbol{c}$$

Exercises 22e

1 Given that $\boldsymbol{a} = \begin{pmatrix} 1 \\ 1 \\ -2 \end{pmatrix}$, $\boldsymbol{b} = \begin{pmatrix} 1 \\ 0 \\ 1 \end{pmatrix}$, $\boldsymbol{c} = \begin{pmatrix} 2 \\ -1 \\ 3 \end{pmatrix}$, find (i) $\boldsymbol{a} + \boldsymbol{b} - 3\boldsymbol{c}$, (ii) $|\boldsymbol{a} + \boldsymbol{b} + \boldsymbol{c}|$, (iii) $\boldsymbol{a} - 2\boldsymbol{b} + \boldsymbol{c}$, (iv) $|2\boldsymbol{a} + \boldsymbol{b} + 2\boldsymbol{c}|$.

2 Given that $\boldsymbol{a} = \begin{bmatrix} 2 \\ 0 \\ 1 \end{bmatrix}$, $\boldsymbol{b} = \begin{bmatrix} 1 \\ 3 \\ -1 \end{bmatrix}$, $\boldsymbol{c} = \begin{bmatrix} -5 \\ -1 \\ 1 \end{bmatrix}$, find (i) $\boldsymbol{a}+\boldsymbol{b}+3\boldsymbol{c}$, (ii) $|\boldsymbol{c}-\boldsymbol{a}|$, (iii) $2\boldsymbol{a}+\boldsymbol{b}+\boldsymbol{c}$, (iv) unit vector in the direction of $2\boldsymbol{a}+\boldsymbol{b}+\boldsymbol{c}$.

3 Show that the vectors $\boldsymbol{a} = \boldsymbol{i}+\boldsymbol{j}-\boldsymbol{k}$, $\boldsymbol{b} = 2\boldsymbol{i}+\boldsymbol{k}$, $\boldsymbol{c} = 3\boldsymbol{j}+2\boldsymbol{k}$ are non-coplanar (see Example 3) and express $\boldsymbol{d} = -\boldsymbol{i}+15\boldsymbol{j}+3\boldsymbol{k}$ in terms of $\boldsymbol{a}$, $\boldsymbol{b}$, $\boldsymbol{c}$.

4 Show that the vectors $\boldsymbol{a} = \begin{bmatrix} 5 \\ 2 \\ -2 \end{bmatrix}$, $\boldsymbol{b} = \begin{bmatrix} 2 \\ 0 \\ -1 \end{bmatrix}$, and $\boldsymbol{c} = \begin{bmatrix} 3 \\ 2 \\ 0 \end{bmatrix}$ are non-coplanar (see Example 3) and express the vector $\boldsymbol{d} = \begin{bmatrix} -2 \\ 2 \\ 1 \end{bmatrix}$ in terms of $\boldsymbol{a}$, $\boldsymbol{b}$, $\boldsymbol{c}$.

5 Show that the vectors $\boldsymbol{a} = \begin{bmatrix} 3 \\ -2 \\ 1 \end{bmatrix}$, $\boldsymbol{b} = \begin{bmatrix} 2 \\ -1 \\ 2 \end{bmatrix}$, $\boldsymbol{c} = \begin{bmatrix} 5 \\ -4 \\ -1 \end{bmatrix}$ are coplanar and hence express $\boldsymbol{c}$ in terms of $\boldsymbol{a}$ and $\boldsymbol{b}$.

22.6 Scalar (or dot) product

The justification for compounding two or more vectors by the triangle or parallelogram rule (see Section 22.3) was that this gave useful and practical results. The + and − signs from algebra were used to indicate such combinations. When considering how to 'multiply' two vectors, the criterion of the useful and practical is again applied. There are *two* 'vector products' which are mathematical models of practical results, the *scalar* (*dot*) *product* and the *vector* (*cross*) *product*. They are so named because the result of the first combination is a scalar and of the second a vector. In this book, we shall only consider the scalar product.

The *scalar product* of two vectors $\boldsymbol{a}$ and $\boldsymbol{b}$ is the scalar quantity $ab\cos\theta$, where a and b are the moduli of $\boldsymbol{a}$ and $\boldsymbol{b}$ and θ is the angle between them. The product is written as $\boldsymbol{a}.\boldsymbol{b}$ and is sometimes referred to as the dot product.

$$\boldsymbol{a}.\boldsymbol{b} = ab\cos\theta \qquad (22.7)$$

Note that the result is a scalar quantity.

One use of this definition is seen when we consider the work done by a force $\boldsymbol{F}$ whose point of application moves a distance r in a direction making an angle θ with the line of action of $\boldsymbol{F}$.

$$\begin{aligned} \text{Work done} &= \text{(magnitude of force in the direction of motion)} \\ &\quad \times \text{(distance moved)} \\ &= F\cos\theta \times r \\ &= Fr\cos\theta \end{aligned}$$

Using a vector notation because both force and displacement are vectors,

$$\boldsymbol{F}.\boldsymbol{r} = Fr\cos\theta = \text{work done}$$

From the definition,

$$\boldsymbol{b}.\boldsymbol{a} = ba\cos\theta = ab\cos\theta = \boldsymbol{a}.\boldsymbol{b} \tag{22.8}$$

that is, scalar multiplication is commutative.

We note two special cases:

(i) If $\boldsymbol{a}.\boldsymbol{b} = 0$, there are not just *two* possibilities, $\boldsymbol{a} = \boldsymbol{0}$ or $\boldsymbol{b} = \boldsymbol{0}$ but *three* because if $\theta = 90°$, $\cos\theta = 0$ and again $\boldsymbol{a}.\boldsymbol{b} = \boldsymbol{0}$ even though both $\boldsymbol{a}$ *and* $\boldsymbol{b}$ are not zero. Thus if $\boldsymbol{a}.\boldsymbol{b} = \boldsymbol{0}$ either $\boldsymbol{a} = \boldsymbol{0}$ *or* $\boldsymbol{b} = \boldsymbol{0}$ *or* $\theta = 90°$ (the vectors are perpendicular). Conversely, if the vectors are perpendicular, $\theta = 90°$, $\cos\theta = 0$ and $\boldsymbol{a}.\boldsymbol{b} = \boldsymbol{0}$.

(ii) If the vectors are collinear (see Section 22.5), $\cos\theta = \pm 1$ and $\boldsymbol{a}.\boldsymbol{b} = \pm ab$ according as they are in the same or opposite sense.

In the case of the unit vectors $\boldsymbol{i}, \boldsymbol{j}, \boldsymbol{k}$, since they are mutually perpendicular,

$$\left.\begin{aligned} \boldsymbol{i}.\boldsymbol{j} &= 0 = \boldsymbol{j}.\boldsymbol{i} \\ \boldsymbol{j}.\boldsymbol{k} &= 0 = \boldsymbol{k}.\boldsymbol{j} \\ \boldsymbol{k}.\boldsymbol{i} &= 0 = \boldsymbol{i}.\boldsymbol{k} \\ \text{and} \quad & \\ \boldsymbol{i}.\boldsymbol{i} = \boldsymbol{j}.\boldsymbol{j} &= \boldsymbol{k}.\boldsymbol{k} = 1 \end{aligned}\right\} \tag{22.9}$$

For any vector $\boldsymbol{a}$, the product $\boldsymbol{a}.\boldsymbol{a} = |\boldsymbol{a}|^2$, which is often written $\boldsymbol{a}^2$ or simply a^2.

Another interpretation of the scalar product is

$$\begin{aligned} \boldsymbol{a}.\boldsymbol{b} &= ab\cos\theta = a \times (b\cos\theta) \quad \text{or} \quad b \times (a\cos\theta) \\ &= a \times (\text{projected length of } \boldsymbol{b} \text{ on } \boldsymbol{a}) \\ &= a \times (\text{component of } \boldsymbol{b} \text{ in the direction of } \boldsymbol{a}) \end{aligned}$$

In the special case when either $\boldsymbol{a}$ or $\boldsymbol{b}$ is a unit vector, say $\hat{\boldsymbol{a}}$, then $|\hat{\boldsymbol{a}}| = 1$ and

$$\hat{\boldsymbol{a}}.\boldsymbol{b} = \text{projected length of } \boldsymbol{b} \text{ on } \boldsymbol{a}$$

Also, the component of a sum of vectors in any direction is equal to the sum of the components of the individual vectors in that direction. Hence

$$\begin{aligned} \boldsymbol{a}.(\boldsymbol{b}+\boldsymbol{c}) &= a \times (\text{component of } \boldsymbol{b}+\boldsymbol{c} \text{ in direction of } \boldsymbol{a}) \\ &= a \times (\text{component of } \boldsymbol{b}) + a \times (\text{component of } \boldsymbol{c}) \end{aligned}$$

Therefore

$$\boldsymbol{a}.(\boldsymbol{b}+\boldsymbol{c}) = \boldsymbol{a}.\boldsymbol{b} + \boldsymbol{a}.\boldsymbol{c}$$

Also, it is evident from the definitions of $\mathbf{a}.\boldsymbol{b}$ and $\lambda\boldsymbol{a}$ that

$$\lambda(\boldsymbol{a}.\boldsymbol{b}) = (\lambda\boldsymbol{a}).\boldsymbol{b} = \boldsymbol{a}.(\lambda\boldsymbol{b})$$

Thus scalar multiplication is distributive with respect to addition and commutative with respect to multiplication by a scalar, so that brackets can be removed and inserted as in ordinary algebra.

Let

$$\boldsymbol{a} = a_x\boldsymbol{i} + a_y\boldsymbol{j} + a_z\boldsymbol{k} \quad \text{and} \quad \boldsymbol{b} = b_x\boldsymbol{i} + b_y\boldsymbol{j} + b_z\boldsymbol{k}$$

Then

$$\begin{aligned}\boldsymbol{a}.\boldsymbol{b} &= (a_x\boldsymbol{i} + a_y\boldsymbol{j} + a_z\boldsymbol{k}).(b_x\boldsymbol{i} + b_y\boldsymbol{j} + b_z\boldsymbol{k})\\ &= a_xb_x\boldsymbol{i}.\boldsymbol{i} + a_xb_y\boldsymbol{i}.\boldsymbol{j} + a_xb_z\boldsymbol{i}.\boldsymbol{k} + a_yb_x\boldsymbol{j}.\boldsymbol{i} + a_yb_y\boldsymbol{j}.\boldsymbol{j} + a_yb_z\boldsymbol{j}.\boldsymbol{k}\\ &\quad + a_zb_x\boldsymbol{k}.\boldsymbol{i} + a_zb_y\boldsymbol{k}.\boldsymbol{j} + a_zb_z\boldsymbol{k}.\boldsymbol{k}\end{aligned}$$

Therefore, using the set of equations (22.9)

$$\boldsymbol{a}.\boldsymbol{b} = a_xb_x + a_yb_y + a_zb_z \tag{22.10}$$

Example 1 A force $\boldsymbol{F} = -2\boldsymbol{i} + 3\boldsymbol{j} + \boldsymbol{k}$ units has its point of application moved from A to B, where $\overrightarrow{\text{AB}} = 3\boldsymbol{i} + \boldsymbol{j} - 4\boldsymbol{k}$. Find the work done.

By the definition of a scalar product

$$\begin{aligned}\text{work done} &= \boldsymbol{F}.\overrightarrow{\text{AB}}\\ &= (-2\boldsymbol{i} + 3\boldsymbol{j} + \boldsymbol{k}).(3\boldsymbol{i} + \boldsymbol{j} - 4\boldsymbol{k})\\ &= (-2 \times 3) + (3 \times 1) + (1 \times -4) \quad \text{by (22.10)}\\ &= -7 \text{ units}\end{aligned}$$

The negative sign indicates that the work is done *on* the force.

Example 2 Find the resolved part of the vector $\boldsymbol{a} = 6\boldsymbol{i} - 3\boldsymbol{j} + 9\boldsymbol{k}$ in the direction of $\boldsymbol{b} = 2\boldsymbol{i} + 2\boldsymbol{j} - \boldsymbol{k}$.

Now

$$\begin{aligned}|\boldsymbol{b}| &= \sqrt{(2^2 + 2^2 + 1^2)}\\ &= 3\end{aligned}$$

Thus

$$\hat{\boldsymbol{b}} = \frac{\boldsymbol{b}}{|\boldsymbol{b}|} = (2\boldsymbol{i} + 2\boldsymbol{j} - \boldsymbol{k})/3$$

The resolved part of $\boldsymbol{a}$ in the direction of $\boldsymbol{b}$ is

$$\begin{aligned}\boldsymbol{a}.\hat{\boldsymbol{b}} &= (6\boldsymbol{i} - 3\boldsymbol{j} + 9\boldsymbol{k}).(2\boldsymbol{i} + 2\boldsymbol{j} - \boldsymbol{k})/3\\ &= -1\end{aligned}$$

Example 3 Show that the vectors $\boldsymbol{a} = \begin{pmatrix}2\\3\\1\end{pmatrix}$, $\boldsymbol{b} = \begin{pmatrix}5\\-2\\-4\end{pmatrix}$ and $\boldsymbol{c} = \begin{pmatrix}10\\-13\\19\end{pmatrix}$ are mutually perpendicular.

We note that $\boldsymbol{a} \neq 0$, $\boldsymbol{b} \neq 0$, $\boldsymbol{c} \neq 0$

$$\begin{aligned}\boldsymbol{a}.\boldsymbol{b} &= \begin{pmatrix}2\\3\\1\end{pmatrix} . \begin{pmatrix}5\\-2\\-4\end{pmatrix}\\ &= 2 \times 5 + 3 \times (-2) + 1 \times (-4)\\ &= 0\end{aligned}$$

Therefore, $\cos\theta = 0$, i.e. $\theta = 90°$, and $\boldsymbol{a}$ and $\boldsymbol{b}$ are perpendicular.

Similarly

$$\boldsymbol{a}.\boldsymbol{c} = \begin{pmatrix} 2 \\ 3 \\ 1 \end{pmatrix} . \begin{pmatrix} 10 \\ -13 \\ 19 \end{pmatrix}$$

$$= 2 \times 10 + 3 \times (-13) + 1 \times 19$$

$$= 0$$

Therefore, $\boldsymbol{a}$ and $\boldsymbol{c}$ are perpendicular.

$$\boldsymbol{b}.\boldsymbol{c} = \begin{pmatrix} 5 \\ -2 \\ -4 \end{pmatrix} . \begin{pmatrix} 10 \\ -13 \\ 19 \end{pmatrix}$$

$$= 5 \times 10 + (-2)(-13) + (-4)19$$

$$= 0$$

Therefore, $\boldsymbol{b}$ and $\boldsymbol{c}$ are perpendicular.

Exercises 22f

1 Given that $\overrightarrow{AB} = \boldsymbol{i} - 2\boldsymbol{j} + 3\boldsymbol{k}$; $\overrightarrow{AC} = 3\boldsymbol{i} - 4\boldsymbol{k}$, find (i) the length of the projection of $\overrightarrow{AB}$ on $\overrightarrow{AC}$ and (ii) the projection of $\overrightarrow{AC}$ on $\overrightarrow{AB}$.

2 Three points A, B, C are such that $\overrightarrow{AB} = \begin{pmatrix} 2 \\ -2 \\ 1 \end{pmatrix}$, $\overrightarrow{BC} = \begin{pmatrix} -4 \\ -8 \\ 1 \end{pmatrix}$. Find the angle ABC.

3 The vectors $\boldsymbol{a} = \begin{pmatrix} 2 \\ -2 \\ 1 \end{pmatrix}$, $\boldsymbol{b} = \begin{pmatrix} 1 \\ 3 \\ 4 \end{pmatrix}$, $\boldsymbol{c} = \begin{pmatrix} 11 \\ p \\ q \end{pmatrix}$ are mutually perpendicular. (i) Find the values of p and q. (ii) Write down, in terms of $\boldsymbol{i}$, $\boldsymbol{j}$, $\boldsymbol{k}$, unit vectors in the directions of $\boldsymbol{a}$, $\boldsymbol{b}$ and $\boldsymbol{c}$.

4 $\boldsymbol{a}$ and $\boldsymbol{b}$ are two non-zero vectors.
(i) If $|\boldsymbol{a}| = |\boldsymbol{b}|$, show that $(\boldsymbol{a} + \boldsymbol{b})$ and $(\boldsymbol{a} - \boldsymbol{b})$ are perpendicular.
(ii) If $\boldsymbol{a}$ and $\boldsymbol{b}$ are perpendicular, show that $|\boldsymbol{a} - \boldsymbol{b}| = |\boldsymbol{a} + \boldsymbol{b}|$.

5 If $\boldsymbol{a}.\boldsymbol{b} = \boldsymbol{a}.\boldsymbol{c}$ and $a \neq 0$, what can be said about the relation between $\boldsymbol{b}$ and $\boldsymbol{c}$?

6 Find the resolved part of the force $\begin{pmatrix} 3 \\ 2 \\ 4 \end{pmatrix}$ N in the direction of the vector $\begin{pmatrix} 3 \\ 4 \\ 12 \end{pmatrix}$

7 Given that $(\boldsymbol{a} - \boldsymbol{h}).(\boldsymbol{b} - \boldsymbol{c}) = 0$ and $(\boldsymbol{b} - \boldsymbol{h}).(\boldsymbol{c} - \boldsymbol{a}) = 0$, deduce that $(\boldsymbol{c} - \boldsymbol{h}).(\boldsymbol{a} - \boldsymbol{b}) = 0$.

8 Show that the vectors $\boldsymbol{a} = \begin{bmatrix} 6 \\ 2 \\ 3 \end{bmatrix}$ and $\boldsymbol{b} = \begin{bmatrix} 1 \\ 3 \\ -4 \end{bmatrix}$ are mutually perpendicular. A third vector $\boldsymbol{c} = \begin{bmatrix} 17 \\ p \\ q \end{bmatrix}$ is perpendicular to both $\boldsymbol{a}$ and $\boldsymbol{b}$, find the values of p and q.

9 If $\boldsymbol{a} = \begin{bmatrix} 1 \\ 1 \\ -2 \end{bmatrix}$, $\boldsymbol{b} = \begin{bmatrix} 1 \\ 0 \\ 1 \end{bmatrix}$, $\boldsymbol{c} = \begin{bmatrix} 2 \\ -1 \\ 3 \end{bmatrix}$, evaluate the following: (i) $\boldsymbol{a}.\boldsymbol{b}$, (ii) $2\boldsymbol{a}.(\boldsymbol{b} + 3\boldsymbol{c})$, (iii) $(5\boldsymbol{a} - 37\boldsymbol{b} + 15\boldsymbol{c}).\boldsymbol{c}$.

10 Given that $\boldsymbol{a} = 2\boldsymbol{i} + 3\boldsymbol{k}$, $\boldsymbol{b} = 5\boldsymbol{i} - \boldsymbol{j} + \boldsymbol{k}$, $\boldsymbol{c} = \boldsymbol{i} + \boldsymbol{j}$, evaluate (i) $(\boldsymbol{a}.\boldsymbol{b})\boldsymbol{c}$, (ii) $(\boldsymbol{a}.\boldsymbol{c})(\boldsymbol{b}.\boldsymbol{c})$, (iii) $(\boldsymbol{a} - \boldsymbol{c}).(\boldsymbol{b} - \boldsymbol{c})$, (iv) $(\boldsymbol{a}.\boldsymbol{c})\boldsymbol{b}$.

22.7 Position vectors

Generally, a vector has both size and direction but is not fixed in space. However, if its point of application is given then it is known as a *bound vector*. A force is a bound vector because its point of application must be known. A particular case of a bound vector arises when a point O is taken as origin and the displacement vector $\overrightarrow{OP}$ is used to locate a point P. In this case, the vector $\overrightarrow{OP}$ is known as the *position vector* of P with respect to O. We shall use $\boldsymbol{a}$, $\boldsymbol{b}$, ... for the position vectors of the points A, B, ...

Example 1 The point R divides the straight line joining the points A and B in the ratio $m:n$. If A and B have position vectors $\boldsymbol{a}$ and $\boldsymbol{b}$ with respect to an origin O, find the position vector of R with respect to O.

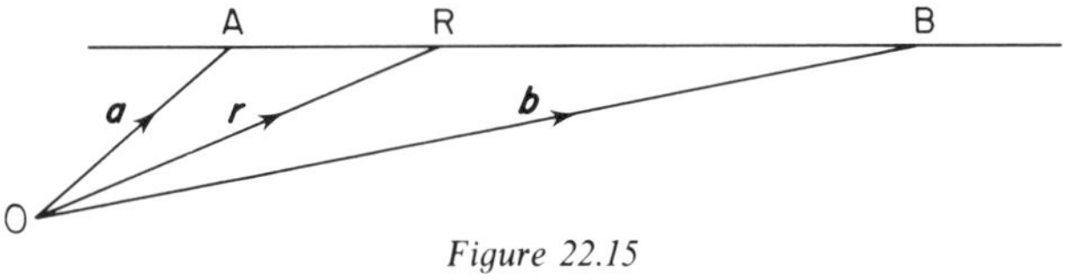

Figure 22.15

Referring to *Figure 22.15*, since ARB is a straight line, $\overrightarrow{AR}$ and $\overrightarrow{RB}$ are collinear. Also

$$\frac{AR}{RB} = \frac{m}{n}$$

or

$$nAR = mRB$$

Therefore

$$n\overrightarrow{AR} = m\overrightarrow{RB}$$

or

$$n(\boldsymbol{r} - \boldsymbol{a}) = m(\boldsymbol{b} - \boldsymbol{r})$$

Whence

$$r = \frac{n\boldsymbol{a} + m\boldsymbol{b}}{m + n} \quad (m + n \neq 0) \tag{22.11}$$

The result is true whether the ratio m/n is positive or negative. In the case of m/n negative, the point R is outside the segment AB.

Note that if R is the mid-point of AB, $m = n$ and $\boldsymbol{r} = (\boldsymbol{a} + \boldsymbol{b})/2$.

Example 2 Show that the medians AD, BE, CF of a triangle ABC have a common point of intersection, which divides each median in the ratio 2:1.

Let the position vectors of A, B, C be $\boldsymbol{a}$, $\boldsymbol{b}$ and $\boldsymbol{c}$, respectively. Since D, E, F are mid-points of the sides BC, CA, AB, respectively, their position vectors are D, $\frac{1}{2}(\boldsymbol{b} + \boldsymbol{c})$, E, $\frac{1}{2}(\boldsymbol{c} + \boldsymbol{a})$, F, $\frac{1}{2}(\boldsymbol{a} + \boldsymbol{b})$ (see *Figure 22.16*).

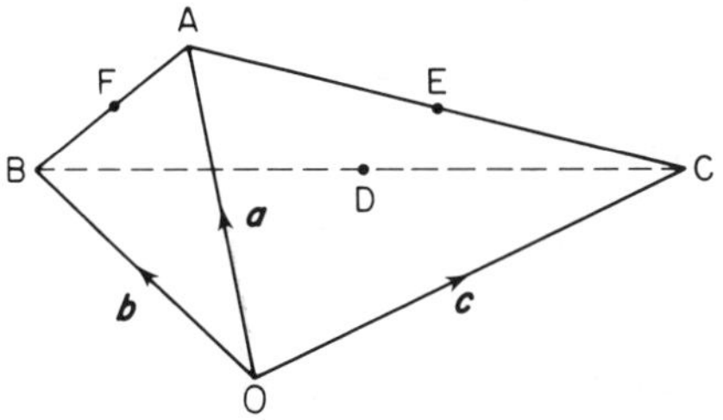

Figure 22.16

Let G be the point which divides AD in the ratio 2:1. Then, by the result of Example 1, the position vector of G is given by

$$\boldsymbol{g} = \frac{1 \times \boldsymbol{a} + 2 \times \frac{1}{2}(\boldsymbol{b} + \boldsymbol{c})}{1 + 2}$$
$$= (\boldsymbol{a} + \boldsymbol{b} + \boldsymbol{c})/3$$

and the symmetry of this result shows that G lies on BE and CF and divides them in the ratio 2:1.

If, in addition to the origin O, three mutually perpendicular unit vectors $\boldsymbol{i}$, $\boldsymbol{j}$, $\boldsymbol{k}$ specify a right-handed set of axes Ox, Oy, Oz (see Section 22.5 and *Figure*

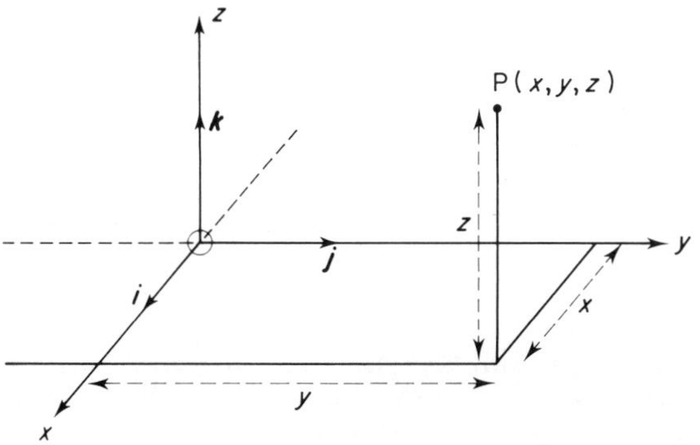

Figure 22.17

22.17), then the position of a point P can be specified by its perpendicular distances x, y, z from the three planes specified by $(\boldsymbol{j}, \boldsymbol{k})$, $(\boldsymbol{k}, \boldsymbol{i})$ and $(\boldsymbol{i}, \boldsymbol{j})$ (see *Figure 22.17*).

The position of P(x, y, z) is specified by the position vector $\overrightarrow{OP} = \boldsymbol{p} = x\boldsymbol{i} + y\boldsymbol{j} + z\boldsymbol{k}$.

Note: care must be taken not to confuse the co-ordinates (x, y, z) of a point P with the position vector $\overrightarrow{OP}$ written in the form of a column vector $\begin{pmatrix} x \\ y \\ z \end{pmatrix}$ even though the values of x, y and z are the same.

Example 3 Referred to O as origin, A and B are the points (2, −2, 1) and (6, −3, 2) respectively. Find the lengths of OA and OB and the area of the triangle AOB (leave the answer in surd form).

$$\boldsymbol{a} = \overrightarrow{OA} = 2\boldsymbol{i} - 2\boldsymbol{j} + \boldsymbol{k}$$

$$\boldsymbol{b} = \overrightarrow{OB} = 6\boldsymbol{i} - 3\boldsymbol{j} + 2\boldsymbol{k}$$

$$\overrightarrow{OA} = |\boldsymbol{a}| = \sqrt{[2^2 + (-2)^2 + 1^2]} = 3$$

$$\overrightarrow{OB} = |\boldsymbol{b}| = \sqrt{[6^2 + (-3)^2 + 2^2]} = 7$$

Let θ be the angle between OA and OB. Then

$$\begin{aligned} \text{OA} \times \text{OB} \times \cos\theta &= \boldsymbol{a}.\boldsymbol{b} \\ 3 \times 7 \times \cos\theta &= 2 \times 6 + (-2)(-3) + 1 \times 2 \\ 21 \cos\theta &= 20 \\ \cos\theta &= 20/21 \\ \sin\theta &= \sqrt{(1 - \cos^2\theta)} \\ &= \sqrt{1 - (20/21)^2} \\ &= \sqrt{41}/21 \\ \text{Area } \triangle\text{AOB} &= \tfrac{1}{2}\text{OA}.\text{OB}.\sin\theta \\ &= \tfrac{1}{2} \times 3 \times 7 \times \sqrt{41}/21 \\ &= \sqrt{41}/2 \end{aligned}$$

Exercises 22g

1 C is a point on the line AB such that AC/AB = t(a constant). The position vectors of A, B and C are respectively $\boldsymbol{a}$, $\boldsymbol{b}$ and $\boldsymbol{c}$. Show that $\boldsymbol{c} = (1 - t)\boldsymbol{a} + t\boldsymbol{b}$.

2 R is a point on the line AB such that AR : RB = 3 : q. The position vectors of A and B with respect to an origin O are respectively $\boldsymbol{a} = \boldsymbol{i} - \boldsymbol{j} + \boldsymbol{k}$, and $\boldsymbol{b} = 2\boldsymbol{i} + \boldsymbol{j} + 3\boldsymbol{k}$. Find the position vector of R in terms of q [refer to Example 1]. If the lines OA and OR are perpendicular, find the value of q.

3 In a triangle ABC the altitudes through the vertices A and B meet in H. The position vectors of A, B, C with respect to H are $\boldsymbol{a}$, $\boldsymbol{b}$, $\boldsymbol{c}$. Show that $\boldsymbol{a}.\boldsymbol{b} = \boldsymbol{b}.\boldsymbol{c} = \boldsymbol{c}.\boldsymbol{a}$ and deduce that the other altitude through C also passes through H.

4 The position vectors $\boldsymbol{a}$, $\boldsymbol{b}$, $\boldsymbol{c}$ of three points A, B, C are $\boldsymbol{a} = \begin{bmatrix} 4 \\ 0 \\ -1 \end{bmatrix}$,

$\boldsymbol{b} = \begin{bmatrix} 6 \\ 1 \\ -4 \end{bmatrix}$, $\boldsymbol{c} = \begin{bmatrix} 6 \\ 2 \\ 1 \end{bmatrix}$. Find the vectors $\overrightarrow{AB}$ and $\overrightarrow{AC}$ and deduce that AB and AC are at right angles.

5 The origin of co-ordinates is O. A and B are the points (2, 1, 2) and (2, 9, −6). Find the lengths of OA and OB and the area of triangle AOB leaving your answer in surd form [see Example 3].

22.8 The vector equation of a straight line

A straight line can be uniquely specified in a number of ways, for example, by (a) its direction and the position of a point on it, and (b) the position of two points on it.

The equation of the straight line is obtained by expressing the position vector $\boldsymbol{r}$ of a general point P on the locus in terms of the given conditions. To find the vector equation of a straight line through a given point A($\boldsymbol{a}$) and parallel to a given direction $\boldsymbol{b}$, proceed as follows.

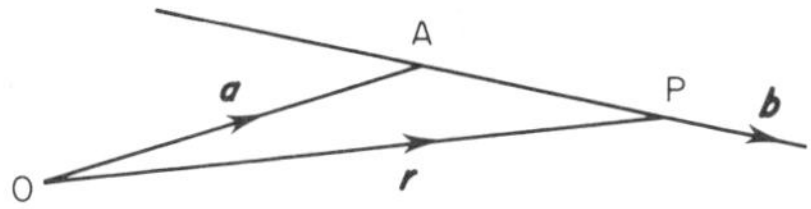

Figure 22.18

Let P($\boldsymbol{r}$) be any point on the line, then $\overrightarrow{AP}$ is parallel to $\boldsymbol{b}$ (see *Figure 22.18*) and is therefore equal to $t\boldsymbol{b}$, where t varies according to the position of P. Now

$$\boldsymbol{r} = \overrightarrow{OP} = \overrightarrow{OA} + \overrightarrow{AP}$$

therefore

$$\boldsymbol{r} = \boldsymbol{a} + t\boldsymbol{b} \tag{22.12}$$

which is the required equation.

For the particular case of a straight line through the origin, $\boldsymbol{a} = 0$ and we have

$$\boldsymbol{r} = t\boldsymbol{b} \tag{22.13}$$

(Note: t is *not* equal to AP unless $\boldsymbol{b}$ is a unit vector.)

Example 1 To find the vector equation of a straight line through two given points, A($\boldsymbol{a}$) and B($\boldsymbol{b}$).

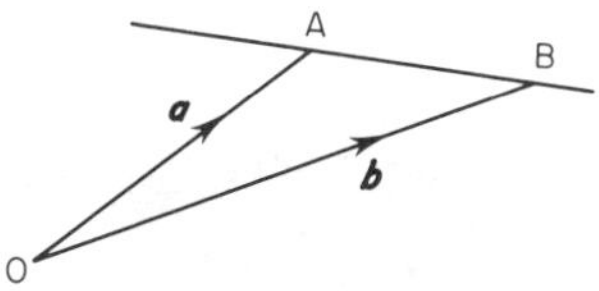

Figure 22.19

We note that $\overrightarrow{AB} = \boldsymbol{b} - \boldsymbol{a}$ (see *Figure 22.19*), so that we have a straight line through $A(\boldsymbol{a})$ parallel to $\boldsymbol{b} - \boldsymbol{a}$. Its equation is therefore

$$\boldsymbol{r} = \boldsymbol{a} + t(\boldsymbol{b} - \boldsymbol{a})$$

or

$$\boldsymbol{r} = (1 - t)\boldsymbol{a} + t\boldsymbol{b} \tag{22.14}$$

Example 2 The lines L_1 and L_2 have vector equations

$$\boldsymbol{r}_1 = (3\boldsymbol{i} + 2\boldsymbol{j} + 3\boldsymbol{k}) + t(\boldsymbol{i} + 2\boldsymbol{k})$$

and

$$\boldsymbol{r}_2 = (10\boldsymbol{i} - \boldsymbol{j} + 2\boldsymbol{k}) + s(3\boldsymbol{i} - \boldsymbol{j} + \boldsymbol{k})$$

where t and s are parameters. Show that L_1 and L_2 intersect and find the position vector of their point of intersection.

As the parameter t varies, $\boldsymbol{r}_1$ gives the position vectors of points on L_1. Similarly, as s varies, $\boldsymbol{r}_2$ gives the position vectors of points on L_2. If the lines L_1 and L_2 intersect then for some values of t and s, $\boldsymbol{r}_1$ and $\boldsymbol{r}_2$ are *identical*. Thus

$$\boldsymbol{r}_1 \equiv \boldsymbol{r}_2$$
$$(3\boldsymbol{i} + 2\boldsymbol{j} + 3\boldsymbol{k}) + t(\boldsymbol{i} + 2\boldsymbol{k}) \equiv (10\boldsymbol{i} - \boldsymbol{j} + 2\boldsymbol{k}) + s(3\boldsymbol{i} - \boldsymbol{j} + \boldsymbol{k})$$
$$(3 + t)\boldsymbol{i} + 2\boldsymbol{j} + (3 + 2t)\boldsymbol{k} \equiv (10 + 3s)\boldsymbol{i} + (-1 - s)\boldsymbol{j} + (2 + s)\boldsymbol{k}$$

For this to be true *all* the components of the vectors must be equal. Therefore, for the same values of s and t

$$3 + t = 10 + 3s \tag{i}$$
$$2 = -1 - s \tag{ii}$$
$$3 + 2t = 2 + s \tag{iii}$$

From (i) and (ii), $s = -3$, $t = -2$ and substituting these values in (iii) we have

$$3 + 2(-2) = -1 = 2 + (-3)$$

and the lines must meet. To find the position vector of the point of intersection we substitute $t = -2$ in $\boldsymbol{r}_1$,

$$\begin{aligned}\boldsymbol{r}_1 &= (3\boldsymbol{i} + 2\boldsymbol{j} + 3\boldsymbol{k}) - 2(\boldsymbol{i} + 2\boldsymbol{k}) \\ &= \boldsymbol{i} + 2\boldsymbol{j} - \boldsymbol{k}\end{aligned}$$

Check Substitute $s = -3$ in $\boldsymbol{r}_2$

$$\begin{aligned}\boldsymbol{r}_2 &= (10\boldsymbol{i} - \boldsymbol{j} + 2\boldsymbol{k}) - 3(3\boldsymbol{i} - \boldsymbol{j} + \boldsymbol{k}) \\ &= \boldsymbol{i} + 2\boldsymbol{j} - \boldsymbol{k} \quad \text{as before}\end{aligned}$$

Exercises 22h

1 Show that the lines

$$\boldsymbol{r} = (2\boldsymbol{i} + 2\boldsymbol{j} - 4\boldsymbol{k}) + t(\boldsymbol{i} + 3\boldsymbol{j} - 3\boldsymbol{k})$$

and

$$r = (i + j + k) + s(i + 2j - 4k)$$

intersect and find the position vector of their point of intersection.

2 If the lines $r = (i + 2j + k) + t(i - 2j + k)$ and $r = 2i + s(i + pj + 2k)$ intersect, find p and the position vector of their point of intersection.

3 The lines L_1 and L_2 are given by the equations

$$r = \begin{pmatrix} 2 \\ 3 \\ 1 \end{pmatrix} + t\begin{pmatrix} 2 \\ -1 \\ -1 \end{pmatrix} \quad \text{and} \quad r = \begin{pmatrix} 3 \\ 0 \\ 2 \end{pmatrix} + s\begin{pmatrix} 1 \\ -1 \\ 0 \end{pmatrix}$$

(i) Show that the lines do not intersect.
(ii) Calculate the acute angle between the directions L_1, L_2.
(iii) Verify that the vector $a = \begin{pmatrix} 1 \\ 1 \\ 1 \end{pmatrix}$ is perpendicular both to L_1 and to L_2.

4 A is the point (3, 2, 1) and B is the point (5, 4, 0). Find a vector equation for the line AB. Write down the co-ordinates of any point P on the line AB in terms of the parameter. Using the scalar product of $\overrightarrow{OP}$ and $\overrightarrow{AB}$ find the value of t when $\overrightarrow{OP}$ is perpendicular to $\overrightarrow{AB}$ and hence the co-ordinates of the foot of the perpendicular from O to the line AB.

5 Find the point of intersection of the line through the points (4, −2, 5) and (−2, 1, −1) and the line through the points (−1, −3, −3) and (8, 3, 15). Calculate the acute angle between the two lines.

22.9 The vector equation of a plane

A plane is completely specified when the length and direction of the perpendicular from the origin to the plane (the normal) are known.

Let O be the origin, $\overrightarrow{ON}$, the perpendicular from O to the plane, have length a and $\hat{n}$ be a unit vector in the direction ON (see *Figure 22.20*).

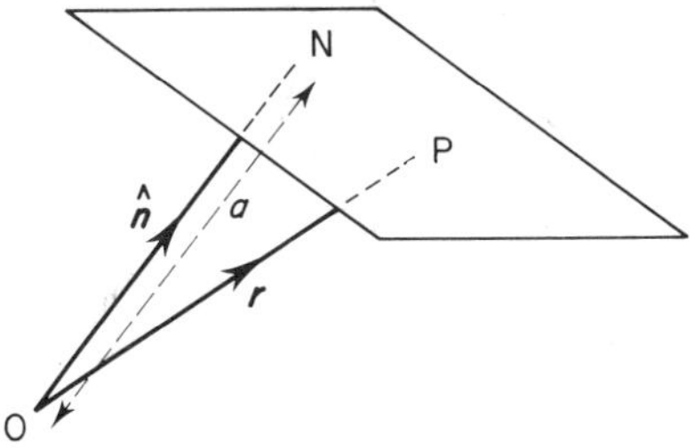

Figure 22.20

Then if r is the position vector of any point P on the plane, the resolved part of $\overrightarrow{OP}$ in the direction $\overrightarrow{ON}$ is always equal to a, that is

$$r \, . \, \hat{n} = a \tag{22.15}$$

which is the equation of the plane.

Note that $\hat{\boldsymbol{n}} = \boldsymbol{n}/|\boldsymbol{n}|$ and therefore

$$\begin{aligned} \boldsymbol{r}.\boldsymbol{n}/|\boldsymbol{n}| &= a \\ \boldsymbol{r}.\boldsymbol{n} &= a|\boldsymbol{n}| \\ \boldsymbol{r}.\boldsymbol{n} &= b \qquad (22.16) \end{aligned}$$

is another form of the equation where $b/|\boldsymbol{n}|$ is the perpendicular distance from the origin to the plane.

Example 1 O is the origin. A plane passes through the point A(3, 2, −6) and is perpendicular to OA. What is the equation of the plane in vector form?

$$\overrightarrow{OA} = 3\boldsymbol{i} + 2\boldsymbol{j} - 6\boldsymbol{k}$$

and because it is perpendicular (i.e. normal to the plane) we denote this by $\boldsymbol{n}$:

$$\begin{aligned} \boldsymbol{n} &= 3\boldsymbol{i} + 2\boldsymbol{j} - 6\boldsymbol{k} \\ \boldsymbol{n} &= \sqrt{[3^2 + 2^2 + (-6)^2]} = 7 \qquad \text{(i)} \\ \hat{\boldsymbol{n}} &= (3\boldsymbol{i} + 2\boldsymbol{j} - 6\boldsymbol{k})/7 \qquad \text{(ii)} \end{aligned}$$

From (i) and (ii) the required equation is

$$\boldsymbol{r}.(3\boldsymbol{i} + 2\boldsymbol{j} - 6\boldsymbol{k})/7 = 7$$

or

$$\boldsymbol{r}.(3\boldsymbol{i} + 2\boldsymbol{j} - 6\boldsymbol{k}) = 49$$

Example 2 O is the origin of co-ordinates. A plane P passes through the points A(5, 4, 0), B(2, 4, 3), C(6, 0, −3). Find the equation of the plane P.

$$\begin{aligned} \overrightarrow{AB} &= (2-5)\boldsymbol{i} + (4-4)\boldsymbol{j} + (3-0)\boldsymbol{k} \\ &= -3\boldsymbol{i} + 3\boldsymbol{k} \\ \overrightarrow{BC} &= (6-2)\boldsymbol{i} + (0-4)\boldsymbol{j} + (-3-3)\boldsymbol{k} \\ &= 4\boldsymbol{i} - 4\boldsymbol{j} - 6\boldsymbol{k} \end{aligned}$$

Let

$$\boldsymbol{n} = \overrightarrow{ON} = \boldsymbol{i} + \lambda\boldsymbol{j} + \mu\boldsymbol{k}$$

If ON is perpendicular to both AB and BC, i.e. perpendicular to the plane, then

$$\begin{aligned} \overrightarrow{ON}.\overrightarrow{AB} &= (\boldsymbol{i} + \lambda\boldsymbol{j} + \mu\boldsymbol{k}).(-3\boldsymbol{i} + 3\boldsymbol{k}) = 0 \\ -3 + 3\mu &= 0 \qquad \text{(i)} \end{aligned}$$

and

$$\begin{aligned} \overrightarrow{ON}.\overrightarrow{BC} &= (\boldsymbol{i} + \lambda\boldsymbol{j} + \mu\boldsymbol{k}).(4\boldsymbol{i} - 4\boldsymbol{j} - 6\boldsymbol{k}) = 0 \\ 4 - 4\lambda - 6\mu &= 0 \qquad \text{(ii)} \end{aligned}$$

From (i) and (ii), $\lambda = -\frac{1}{2}$, $\mu = 1$. Therefore, $\boldsymbol{n} = \boldsymbol{i} - \frac{1}{2}\boldsymbol{j} + \boldsymbol{k}$ is perpendicular to the plane.

$$\begin{aligned} |\boldsymbol{n}| &= \sqrt{[1^2 + (-\tfrac{1}{2})^2 + 1^2]} = \tfrac{3}{2}, \\ \hat{\boldsymbol{n}} &= \tfrac{2}{3}(\boldsymbol{i} - \tfrac{1}{2}\boldsymbol{j} + \boldsymbol{k}) \\ &= (2\boldsymbol{i} - \boldsymbol{j} + 2\boldsymbol{k})/3. \end{aligned}$$

Now $\vec{OA} = 5\boldsymbol{i} + 4\boldsymbol{j}$ and the projection of this in the direction of ON is

$$(5\boldsymbol{i} + 4\boldsymbol{j}) \cdot (2\boldsymbol{i} - \boldsymbol{j} + 2\boldsymbol{k})/3 = 2$$

which is the perpendicular distance from O to the plane. Therefore, the required equation $\boldsymbol{r} \cdot \hat{\boldsymbol{n}} = a$ becomes

$$\boldsymbol{r} \cdot (2\boldsymbol{i} - \boldsymbol{j} + 2\boldsymbol{k})/3 = 2$$

or

$$\boldsymbol{r} \cdot (2\boldsymbol{i} - \boldsymbol{j} + 2\boldsymbol{k}) = 6$$

Example 3 Given that P is the plane $\boldsymbol{r} \cdot \begin{pmatrix} 2 \\ 6 \\ 9 \end{pmatrix} = 33$, A is the point with co-ordinates (5, −1, 2) and O is the origin of co-ordinates, find

(i) The perpendicular distance from O to the plane.
(ii) Write down the equation of the line L which passes through A and is perpendicular to the plane.
(iii) Find the co-ordinates of the point B where the line L meets the plane P.
(iv) Find the position vector $\vec{OC} = \boldsymbol{c}$, where C is the reflection of A in the plane P.

(i) The equation of the plane is

$$\boldsymbol{r} \cdot \begin{pmatrix} 2 \\ 6 \\ 9 \end{pmatrix} = 33 \qquad \text{(i)}$$

so the normal to the plane is $\boldsymbol{n} = \begin{pmatrix} 2 \\ 6 \\ 9 \end{pmatrix}$

$$|\boldsymbol{n}| = \sqrt{(2^2 + 6^2 + 9^2)} = 11$$

Dividing both sides of (i) by 11

$$\boldsymbol{r} \cdot \frac{1}{11} \begin{pmatrix} 2 \\ 6 \\ 9 \end{pmatrix} = 33/11 = 3$$

$$\boldsymbol{r} \cdot \hat{\boldsymbol{n}} = 3$$

The perpendicular distance from O to the plane is 3.

(ii) The line L is perpendicular to the plane P and is therefore in the direction of its normal, vector $\begin{pmatrix} 2 \\ 6 \\ 9 \end{pmatrix}$. It also passes through the point A(5, −1, 2).

Hence $\vec{OA} = \begin{pmatrix} 5 \\ -1 \\ 2 \end{pmatrix}$. The equation of L is

$$\boldsymbol{r} = \begin{pmatrix} 5 \\ -1 \\ 2 \end{pmatrix} + t \begin{pmatrix} 2 \\ 6 \\ 9 \end{pmatrix}$$

(iii) The point B lies on both the plane P and the line L and hence its position vector $\boldsymbol{b}$ satisfies both equations, therefore

$$\boldsymbol{b}\cdot\begin{pmatrix}2\\6\\9\end{pmatrix} = 33 \qquad \text{(ii)}$$

and

$$\boldsymbol{b} = \begin{pmatrix}5\\-1\\2\end{pmatrix} + t\begin{pmatrix}2\\6\\9\end{pmatrix} \qquad \text{(iii)}$$

Taking the dot product of both sides of (iii) with the vector $\begin{pmatrix}2\\6\\9\end{pmatrix}$,

$$\boldsymbol{b}\cdot\begin{pmatrix}2\\6\\9\end{pmatrix} = \begin{pmatrix}5\\-1\\2\end{pmatrix}\cdot\begin{pmatrix}2\\6\\9\end{pmatrix} + t\begin{pmatrix}2\\6\\9\end{pmatrix}\cdot\begin{pmatrix}2\\6\\9\end{pmatrix}$$

From (ii), $\boldsymbol{b}\cdot\begin{pmatrix}2\\6\\9\end{pmatrix} = 33$ and evaluating the dot products we have

$$\begin{aligned}33 &= (10 - 6 + 18) + t(2^2 + 6^2 + 9^2)\\ 33 &= 22 + 121t\\ 1/11 &= t\end{aligned}$$

Therefore, the required position vector is

$$\boldsymbol{b} = \begin{pmatrix}5\\-1\\2\end{pmatrix} + \frac{1}{11}\begin{pmatrix}2\\6\\9\end{pmatrix}$$

$$\boldsymbol{b} = \begin{pmatrix}5\frac{2}{11}\\-\frac{5}{11}\\2\frac{9}{11}\end{pmatrix}$$

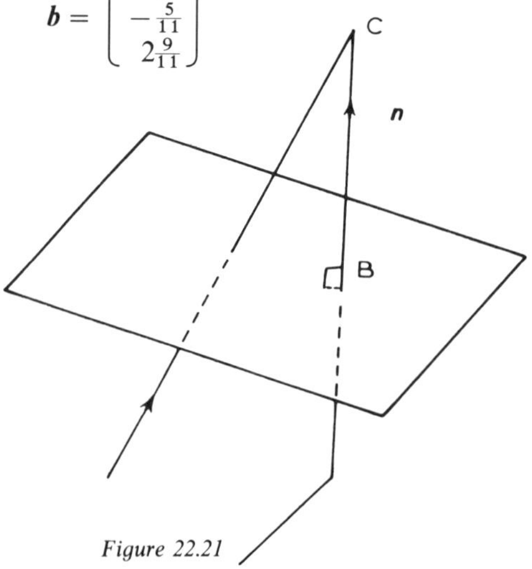

Figure 22.21

(iv) Refer to *Figure 22.21*. In the triangle OAC, B is the mid-point of AC. Therefore

$$\boldsymbol{a} + \boldsymbol{c} = 2\boldsymbol{b}$$
$$\boldsymbol{c} = 2\boldsymbol{b} - \boldsymbol{a}$$
$$= 2\begin{pmatrix} 5\frac{2}{11} \\ -\frac{5}{11} \\ 2\frac{9}{11} \end{pmatrix} - \begin{pmatrix} 5 \\ -1 \\ 2 \end{pmatrix} = \begin{pmatrix} 5\frac{4}{11} \\ \frac{1}{11} \\ 3\frac{7}{11} \end{pmatrix}$$

Exercises 22i

1 Find the acute angle between the normals to the planes P_1, $\boldsymbol{r}.\begin{pmatrix} 8 \\ 1 \\ -4 \end{pmatrix} = 36$ and P_2, $\boldsymbol{r}.\begin{pmatrix} -2 \\ -2 \\ 1 \end{pmatrix} = 27$. [This is also the angle between the planes.]

2 Explain why the two planes P_1, $\boldsymbol{r}.\begin{pmatrix} 2 \\ 2 \\ -1 \end{pmatrix} = 6$, and P_2, $\boldsymbol{r}.\begin{pmatrix} 2 \\ 2 \\ -1 \end{pmatrix} = 15$, are parallel. Find the perpendicular distance between them.

3 O is the origin. A plane passes through the point A(2, −6, 9) and is perpendicular to OA. Find its equation.

4 O is the origin. A plane passes through the points A(1, −9, −6), B(1, 3, 6) and C(7, 0, 6). Find the equation of the plane. Show that the point A with position vector $\boldsymbol{a} = \begin{pmatrix} 7 \\ -6 \\ 0 \end{pmatrix}$ lies on the plane.

5 Prove that the two planes L_1, $\boldsymbol{r}.\begin{pmatrix} 2 \\ 6 \\ -9 \end{pmatrix} = 55$, and L_2, $\boldsymbol{r}.\begin{pmatrix} 15 \\ 1 \\ 4 \end{pmatrix} = 44$, are perpendicular and find their perpendicular distances from the origin.

6 Two planes P_1, $\boldsymbol{r}.\begin{pmatrix} 2 \\ -3 \\ 5 \end{pmatrix} = 12$, and P_2, $\boldsymbol{r}.\begin{pmatrix} \frac{2}{3} \\ -1 \\ 1 \end{pmatrix} = 8/3$, intersect in the line L. Show that both the points A(4, 2, 2) and B(7, 4, 2) lie on L and hence find the equation of L.

7 Find the point of intersection A($\boldsymbol{a}$) of the line L, $\boldsymbol{r} = \begin{pmatrix} 3 \\ 0 \\ 1 \end{pmatrix} + t\begin{pmatrix} 1 \\ -2 \\ 1 \end{pmatrix}$, and the plane P, $\boldsymbol{r}.\begin{pmatrix} 3 \\ 2 \\ 2 \end{pmatrix} = 7$ [see Example 3].

8 Prove that the line L, $\boldsymbol{r} = (2\boldsymbol{i} + \boldsymbol{j}) + t(3\boldsymbol{i} + \boldsymbol{j} + 5\boldsymbol{k})$ lies wholly in the plane $\boldsymbol{r}.(\boldsymbol{i} + 2\boldsymbol{j} - \boldsymbol{k}) = 4$.

9 What is the condition that the line L, $\boldsymbol{r} = \boldsymbol{a} + t\boldsymbol{b}$, is parallel to the plane P, $\boldsymbol{r}.\boldsymbol{n} = k$? If the line L lies wholly in the plane show that $k = \boldsymbol{a}.\boldsymbol{n}$.

10 Prove that the line L, $\boldsymbol{r} = \begin{pmatrix} 1 \\ 2 \\ 3 \end{pmatrix} + t\begin{pmatrix} 2 \\ -1 \\ 1 \end{pmatrix}$, is parallel to both the planes P_1, $\boldsymbol{r}.\begin{pmatrix} 1 \\ 2 \\ 0 \end{pmatrix} = 5$, and P_2, $\boldsymbol{r}.\begin{pmatrix} 2 \\ 3 \\ -1 \end{pmatrix} = 5$. Also show that L is the line of intersection of the planes P_1 and P_2.

Exercises 22

1 Show that $|\boldsymbol{a}+\boldsymbol{b}+\boldsymbol{c}| \leqslant |\boldsymbol{a}| + |\boldsymbol{b}| + |\boldsymbol{c}|$, where $\boldsymbol{a}$, $\boldsymbol{b}$, $\boldsymbol{c}$ are any three vectors.

2 If $\boldsymbol{a}$ and $\boldsymbol{b}$ are two vectors such that $|\boldsymbol{a}| = |\boldsymbol{b}| = |\boldsymbol{a}-\boldsymbol{b}|$ find the angle between $\boldsymbol{a}$ and $\boldsymbol{b}$.

3 ABCDEF is a regular hexagon. If $\overrightarrow{AB}$, $\overrightarrow{BC}$ represents vectors $\boldsymbol{a}$, $\boldsymbol{b}$ respectively, find the vectors represented by $\overrightarrow{DE}$, $\overrightarrow{FE}$ and $\overrightarrow{DF}$.

4 Simplify the following expressions:
(i) $2(\boldsymbol{a}+2\boldsymbol{b}+\boldsymbol{c}) - 3(5\boldsymbol{a}-6\boldsymbol{c}) + 4(\boldsymbol{a}-\boldsymbol{b}-4\boldsymbol{c})$
(ii) $\frac{1}{2}(6\boldsymbol{a}+8\boldsymbol{b}) - (\boldsymbol{a}+\boldsymbol{b}+\boldsymbol{c}) + 3(\boldsymbol{a}-\boldsymbol{b}+\boldsymbol{c})$.

5 ABC is a triangle and the perpendiculars from A and B to the opposite sides meet at H. The position vectors of the points A, B, C, H with respect to an origin O are $\boldsymbol{a}$, $\boldsymbol{b}$, $\boldsymbol{c}$, $\boldsymbol{h}$ respectively. Prove that

$$(\boldsymbol{h}-\boldsymbol{a}).(\boldsymbol{c}-\boldsymbol{b}) = 0$$
$$(\boldsymbol{h}-\boldsymbol{b}).(\boldsymbol{a}-\boldsymbol{c}) = 0$$

and deduce from these that

$$(\boldsymbol{h}-\boldsymbol{c}).(\boldsymbol{b}-\boldsymbol{a}) = 0$$

What is the geometrical significance of this result?

6 Given the vectors $\boldsymbol{a} = \begin{pmatrix} 2 \\ 1 \\ 1 \end{pmatrix}$, $\boldsymbol{b} = \begin{pmatrix} 1 \\ -1 \\ 5 \end{pmatrix}$ and $\boldsymbol{c} = \begin{pmatrix} 3 \\ 4 \\ -2 \end{pmatrix}$, find a vector $\boldsymbol{p}$ which is perpendicular to both $\boldsymbol{a}$ and $\boldsymbol{b}$. Hence show that $\boldsymbol{c}$ is not coplanar with $\boldsymbol{a}$ and $\boldsymbol{b}$.

7 The position vectors of the four points A, B, C, D are $\overrightarrow{OA} = \begin{pmatrix} 1 \\ -9 \\ -6 \end{pmatrix}$, $\overrightarrow{OB} = \begin{pmatrix} 1 \\ 3 \\ 6 \end{pmatrix}$, $\overrightarrow{OC} = \begin{pmatrix} 7 \\ 0 \\ 6 \end{pmatrix}$, $\overrightarrow{OD} = \begin{pmatrix} 1 \\ 2 \\ -2 \end{pmatrix}$ respectively. Calculate the lengths of the projections of each of $\overrightarrow{OA}$, $\overrightarrow{OB}$, $\overrightarrow{OC}$ in the direction $\overrightarrow{OD}$. What two conclusions can be deduced from the results?

8 PABC is a tetrahedron and the displacements $\overrightarrow{PA}$, $\overrightarrow{PB}$, $\overrightarrow{PC}$ represent the vectors $\boldsymbol{a}$, $\boldsymbol{b}$, $\boldsymbol{c}$ respectively. Given that two pairs of opposite edges are perpendicular prove that the third pair of opposite edges are also perpendicular.

9 The equation of a plane P is $\boldsymbol{r}\,.\begin{pmatrix}4\\2\\1\end{pmatrix} = -20$. Write down a vector equation for the line L which passes through the point (5, −1, 4) and is at right-angles to the plane P. Using vector methods find the position vector $\boldsymbol{a}$ of the point A where the line L meets the plane P.

10 Given the plane P, $\boldsymbol{r}\,.\begin{pmatrix}1\\2\\-4\end{pmatrix} = 51$, and the point A whose position vector is $\boldsymbol{a} = \begin{pmatrix}2\\-1\\3\end{pmatrix}$, find the position vector $\boldsymbol{c}$ of C, the reflection of A in the plane P. [See Example 3 (iv).]

11 The equation of a line L is $\boldsymbol{r} = \begin{pmatrix}5\\1\\2\end{pmatrix} + t\begin{pmatrix}-1\\2\\-2\end{pmatrix}$. Show that the points B(4, 3, 0) and C(−1, 13, −10) lie on the line and that the length of BC is 15 units.

12 Given the points A(0, −1, 4), B(8, 7, 0) find a vector equation of AB in terms of a parameter t. O is the origin of co-ordinates. Find the co-ordinates of a point N on AB such that ON is perpendicular to AB. Deduce the co-ordinates of the reflection of O in AB.

13 A plane passes through the point with position vector $\boldsymbol{a}$ and is perpendicular to the direction of $\boldsymbol{n}$. If $\boldsymbol{r}$ is the position vector of a general point on the plane, write down the equation of the plane.

14 O is the origin, the position vector of a point A is $\boldsymbol{a}$ and $\hat{\boldsymbol{u}}$ is the unit vector along a line OL. The point B is the reflection of A in the line OL and C is the point where AB meets OL. Find the position vector of C with respect to O. By completing the parallelogram OALB, or otherwise, show that the position vector $\boldsymbol{b}$ of B is given by $\boldsymbol{b} = 2(\boldsymbol{a}\,.\,\hat{\boldsymbol{u}})\hat{\boldsymbol{u}} - \boldsymbol{a}$.

15 Given that the vector $\boldsymbol{a} = (1, u, v)$ is parallel to the two planes P_1, $\boldsymbol{r}\,.\begin{pmatrix}2\\3\\4\end{pmatrix} = 3$, P_2, $\boldsymbol{r}\,.\begin{pmatrix}1\\2\\3\end{pmatrix} = 2$, find the values of u and v. Verify that the point D with position vector $\boldsymbol{d} = \begin{pmatrix}1\\-1\\1\end{pmatrix}$ lies on both planes and hence write down an equation for L, their line of intersection. Show that a third plane P_3, $\boldsymbol{r}\,.\begin{pmatrix}5\\4\\3\end{pmatrix} = 4$, also passes through L.

16 The equations of two skew* lines are L_1, $\boldsymbol{r} = \begin{pmatrix}3\\2\\1\end{pmatrix} + t\begin{pmatrix}1\\5\\2\end{pmatrix}$,

* That is, no plane contains both lines.

L_2, $\boldsymbol{r} = \begin{pmatrix} -5 \\ -4 \\ 0 \end{pmatrix} + s\begin{pmatrix} 1 \\ 2 \\ 1 \end{pmatrix}$. The vector $\boldsymbol{p} = \begin{pmatrix} 1 \\ a \\ b \end{pmatrix}$ is perpendicular to both lines. Find the values of a and b. Find also the vector $\boldsymbol{a}$ represented by the line joining the point (3, 2, 1) on L_1 and (−5, −4, 0) on L_2. Calculate the projection of $\boldsymbol{a}$ in the direction $\boldsymbol{p}$. (This is the length of the common perpendicular to L_1 and L_2.)

17 Show that the three lines joining the mid-points of opposite edges of a tetrahedron ABCD meet at a point.

Answers

Chapter 1

Exercises 1a

1 $\frac{4}{3}$
2 $6\cdot653, -1\cdot653$
3 $-2, 1$
4 9
5 $1, -2, -3, -6$
6 $\pm 3, \pm 4$
7 ± 4
8 3
9 $-1, 2 \pm \sqrt{3}$
10 $\pm 1, 1 \pm \sqrt{2}$
11 $x = 1, y = 1; x = \frac{29}{11}, y = \frac{2}{11}$
12 $x = 1. y = -1; x = 5, y = -9$
13 $x = 2, y = 1; x = 11, y = 7$
14 $x = \pm 3, y = \mp 1$
15 $x = \pm 3, y = \mp 2; x = \pm 3/\sqrt{10}, y = \mp 11/\sqrt{10}$
16 $x = \pm 4, y = \mp 1; x = \pm 11/\sqrt{7}, y = \pm 4/\sqrt{7}$
17 $x = \pm 1, y = \pm 2; x = \pm 2, y = \pm 1$
18 $x = 3, y = 1; x = 1, y = 3;$

$$x = \frac{-5 \pm \sqrt{55}}{4}, y = \frac{-15}{2(-5 \pm \sqrt{55})}$$

19 $x = 5, y = 1; x = \frac{29}{3}, y = \frac{-25}{3}$
20 $x = \pm 1, y = \pm 4; x = \pm 4, y = \pm 1$

Exercises 1b

1 $(x-1)(y-1) = 1$
2 $(x-3)(y-2)^3 = 1$
3 $x^2y = y^2 - x^2$
5 $\dfrac{x^2}{a^2} + \dfrac{y^2}{b^2} = 1$
8 $a^2 - b^2 + 2 = 0$
9 $b^2 - a^2 = 4c$

Exercises 1c

1 $\frac{-11}{3} < x < \frac{8}{7}$
2 $2 < x < 3$
3 $2 < x < 3$
4 $-3 < x < \frac{9}{4}$
5 $2 < x < 3$
6 $-\frac{5}{3} \leqslant x \leqslant -1, x \geqslant \frac{3}{2}$
7 $\frac{2}{3} < x < 2$
8 $-1, -5$
9 $0, -2$
10 $x > 2$ or $x < -8$
11 $\frac{7}{4} < x < \frac{5}{2}, x \neq 2$

Exercises 1d

1 98 **2** 8 **3** 312
5 $m = -8, n = 8$
6 $(x-1)(x^2+4)$
7 $(x-1)(x-2)(x-3)$
8 $(x-2)(3x-4)(4x+3)$
9 $a = 3, b = -12$
10 $2x+5$

Exercises 1e

1 $\dfrac{2}{x-3}+\dfrac{4}{x+1}$
2 $x-1+\dfrac{3}{x+1}-\dfrac{2}{x-1}$
3 $\dfrac{5}{2(x-1)}+\dfrac{3}{2(x+5)}$
4 $\dfrac{1}{x-1}-\dfrac{2}{x+1}-\dfrac{4}{x+3}$
5 $\dfrac{3x+1}{x^2+4}-\dfrac{2}{x+1}$
6 $\dfrac{1}{2x+1}-\dfrac{1}{x^2+2x+3}$
7 $2x+\dfrac{1}{x-1}+\dfrac{x+1}{x^2+1}$
8 $\dfrac{1}{(x+1)^2}-\dfrac{1}{x+1}+\dfrac{2}{2x-3}$
9 $\dfrac{2}{x+1}+\dfrac{1}{(x+1)^2}-\dfrac{1}{x^2+1}$
10 $\dfrac{1}{x+1}-\dfrac{2}{(x+1)^2}+\dfrac{1}{(x+1)^3}$

Exercises 1f

1 243, $\frac{1}{216}$, 128, $\frac{1}{2}$
2 $\dfrac{3}{2}\dfrac{y^6}{x^5z^8}$, $\dfrac{b^{13/12}}{a^{5/12}c^{19/12}}$
4 4

Exercises 1g

1 3, -3, 4, -5, 3, $\frac{1}{3}$, -6, $-n$
3 1·386, 2·89

Exercises 1h

1 1·87 **2** 1·64 **3** 1·87, -5·11 **4** 1, 2
5 $\frac{1}{2}$, 2 **6** 0·431, 0·683 **7** $\frac{1}{2}$ **8** 2
9 $\sqrt{3}$, 9 **10** $x = 1{\cdot}92$, $y = 0{\cdot}66$

Exercises 1

1 $x = 3, y = 2; x = \frac{5}{3}, y = \frac{8}{3}$
2 $a = 2, b = 3, c = 2, d = 1$
3 $-1{\cdot}71$
4 $2+\sqrt{5}$
5 $x = \pm 4a, y = \pm a; x = \pm\frac{8}{3}a, y = \pm\frac{1}{3}a$
6 $\dfrac{3}{x-3}-\dfrac{3x+1}{x^2+4}$
8 1, -2, $\dfrac{-1\pm\sqrt{13}}{2}$
9 $\dfrac{-9}{3-y}-\dfrac{1}{1-y}+\dfrac{4}{(1-y)^2}$
10 $x > 6, -6 < x < -2$
11 $x > 1, -\frac{3}{2} < x < 0$
12 $\frac{1}{2}$, 1, 2
13 1, 5
15 27, $\sqrt[3]{3}$
16 $-4 < x < \frac{17}{8}, x > 3$

20 6

21 $\dfrac{-3 \pm \sqrt{5}}{2}, \dfrac{-5 \pm \sqrt{21}}{2}$

22 4, 6

23 $-0{\cdot}358$

24 $u = 6, v = 6; u = -\frac{3}{2}(1 \pm \sqrt{5}), v = -\frac{3}{2}(1 \mp \sqrt{5})$

25 3, 11

28 $y < -5$ or $y > \frac{1}{3}$

30 1

31 $x = -1, y = 2; x = -\frac{1}{2}, y = \frac{7}{4}$

32 $x = 1{\cdot}42, y = -0{\cdot}58$

33 $2 \leqslant x \leqslant 3, x > 4$

34 -2

36 $\frac{7}{3} < y < 4$

39 $-2, -\frac{1}{3}, \frac{3}{2}, 2$

40 $a = 3, b = -4, c = -12$

41 $\frac{17}{6} < x < \frac{13}{4}; x \neq 3$

42 $-2 < x < 2$

43 $y = 9, z = \frac{1}{3}$

44 $x = 7, y = 2$ or $x = -\frac{1}{2}, y = \frac{1}{2}$

45 13

46 $x = 9, y = -8; x = y = \frac{12}{7}$

47 $x = 3, y = 9; y = 9, x = 3$

48 $a = 6, n = -\frac{1}{2}$

49 $x = 2, y = -1; x = 3, y = 1$

50 $-4 < x < \frac{1}{2}, x > 5$

Chapter 2

Exercises 2a

1 (i) 2, 5, 8, 11, 14 (ii) $1, -\frac{1}{3}, \frac{1}{9}, -\frac{1}{27}, \frac{1}{81}$ (iii) 3, 8, 17, 32, 57
2 (i) r^3 (ii) $(-1)^{r+1}r^2$ (iii) $\frac{1}{2}(4)^{r-1}$
3 1, 2, 3, 5, 8, 13, 21
4 $a = 3, b = -4; -17$
5 (i) $r^3 - 1$ (ii) $2r$ (iii) $r^3 + 2r - 1$ (iv) 3^r (v) $-(-2)^r$ (vi) $3^r - (-2)^r$ (vii) $-(-6)^r$
6 $0, 2, 2, 0, -2, -2; 0$
7 $a = 3, b = -6, c = 3$; 90, 258
8 (i) 85 (ii) 363
9 $3r^2 - 3r - 1$

Exercises 2b

1 $-3, 1, 5$
2 $-1, 6, 13$
3 0, 5, 10, 15, 20, 25, 30, 35, 40, 45, 50, 55
4 206
5 10, 7
6 $-n^2$
7 (i) 124 (ii) $(5n^2 - 9n)/2$ (iii) $(n - 3n^2)/2$
8 1980
9 (i) $4n(3n+1)$ (ii) $\dfrac{n(n+1)(a+3b)}{2}$ (iii) $\frac{1}{2}(a-2b)n^2 + \frac{1}{2}(5a-2b)n$
10 (i) 4950 (ii) 1683 (iii) 4215

Exercises 2c

1 $1, -4, 16$

2 $-\frac{1}{3}, 1, -3$

3 10, 20, 40

4 -729

5 $\frac{729}{4}$

6 $\frac{3}{2}[1 - (\frac{2}{3})^8]$

7 $121(1+\sqrt{3})$ **8** 2, 9
10 7, 2

Exercises 2d

1 (i) $\dfrac{1}{1+x}$ (ii) $\frac{1}{4}$ (iii) $\frac{6}{11}$ **2** $\dfrac{x^2+4}{x^2-2x+4}$
3 $\frac{23}{99}$ **5** $x<-\frac{2}{3}$ or $x>4$

Exercises 2

1 5, 7, 9
2 6, 3
3 $\dfrac{1}{ar^{n-1}}\left(\dfrac{1-r^n}{1-r}\right)$
4 $n\log ar^{(n-1)/2}$
5 (i) $7p+7$ (ii) $(7p+7)(10p+11)$
6 (i) 247 500 (ii) 250 000 (iii) 1020
8 $a+(n-1)d$; $n+(n-1)(2^n-1)$
10 20, 79
11 $S=4$; $\frac{8}{3}, \frac{8}{9}, \frac{8}{27}$
12 9 or more
13 7, 360 m
14 332, 667
15 $r=-\frac{1}{2}$; $\frac{2}{3}a$
16 (i) 2 (ii) 4, 2·4 cm
17 $\dfrac{an(n+1)}{2}+\dfrac{5r(r^n-1)}{r-1}$; $a=8$, $r=2$, $S_{10}=10\,670$
18 $\frac{1}{10}$, 19·999 98, 9
20 $-\frac{1}{2}<x<1$
21 $\frac{1}{2}(n-m+1)(n+m)$ (i) $\dfrac{r^2-r+2}{2}$ (ii) $\frac{1}{2}(r^2+1)(r)$
23 (i) $x>0$ (ii) $x>0$ or $x<-\frac{1}{2}$
24 (i) $\dfrac{1-a^n}{1-a}$ (ii) $a^{n(n-1)/2}$ (iii) $\dfrac{n}{1-a}-\dfrac{(a-a^{n+1})}{(1-a)^2}$
28 1, $\dfrac{-2\pm\sqrt{2}}{2}$; (i) $\frac{8}{7}+\frac{2}{7}\sqrt{2}$ (ii) $-35+\frac{45}{2}\sqrt{2}$
30 $-2<x<2$; 6
31 $T_n=4n-1$; 3, 4; 23
32 $\frac{3}{16}$
33 (i) 1, $\frac{2}{3}$; 1, $\frac{4}{9}$ (ii) $\frac{53}{165}$
34 $-n-2n^2$
35 $-\frac{5}{2}, \frac{1}{2}$; 205

Chapter 3

Exercises 3a

1 (i) $1+2x+x^2$ (ii) $1+3x+3x^2+x^3$ (iii) $1+4x+6x^2+4x^3+x^4$
2 (i) $1+8x+28x^2+56x^3+70x^4+56x^5+28x^6+8x^7+x^8$
(ii) $1+9x+36x^2+84x^3+126x^4+126x^5+84x^6+36x^7+9x^8+x^9$

4 (i) 153 (ii) 55 (iii) 1287 (iv) 36
5 (i) $1 + 12x + 54x^2 + 108x^3 + 81x^4$
(ii) $1 - 5x + 10x^2 - 10x^3 + 5x^4 - x^5$
(iii) $1 - 14x + 84x^2 - 280x^3 + 560x^4 - 672x^5 + 448x^6 - 128x^7$
6 (i) $81 + 108x + 54x^2 + 12x^3 + x^4$
(ii) $64 - 192x + 240x^2 - 160x^3 + 60x^4 - 12x^5 + x^6$
7 (i) $8x^3 + 36x^2y + 54xy^2 + 27y^3$
(ii) $16x^4 - 160x^3y + 600x^2y^2 - 1000xy^3 + 625y^4$
8 $1\,140\,480x^3$
9 ± 2
10 22 680

Exercises 3b

1 $1 + 6x + 18x^2 + 32x^3 + 36x^4 + 24x^5 + 8x^6$
2 $1 - 4x + 14x^2 - 28x^3 + 49x^4 - 56x^5 + 56x^6 - 32x^7 + 16x^8$
3 80
4 $1 + 6x + 15x^2 + 20x^3 + 15x^4 + 6x^5 + x^6 \equiv (1 + x)^6$
5 3060 **6** $105x^{10}/32$ **7** 2160 **8** -360
9 $-\frac{1}{8}, -8\frac{9}{16}$
10 $x^7 + 14x^6y + 84x^5y^2 + 280x^4y^3 + 560x^3y^4 + 672x^2y^5 + 448xy^6 + 128y^7$; 1·149

Exercises 3c

1 $1 + x - x^2 + \frac{5}{3}x^3$; 1·0099 **2** 0·9355
4 (i) $1 - x + x^2 - x^3 + x^4 - x^5 + \ldots$
(ii) $1 - 2x + 3x^2 - 4x^3 + 5x^4 + \ldots$
(iii) $1 - 3x + \dfrac{3.4}{2}x^2 - \dfrac{4.5}{2}x^3 + \ldots$
5 (i) $1 - 3x + 9x^2 - 27x^3 + \ldots$
(ii) $\dfrac{1}{36} + \dfrac{x^3}{108} + \dfrac{x^6}{432} + \ldots$
(iii) $\dfrac{1}{27} - \dfrac{x}{27} + \dfrac{2x^2}{81} + \ldots$
6 $|x| < 3$ **7** $|x| < \left|\dfrac{a}{b}\right|$
9 $3 + 5x + 9x^2 + 17x^3 + 33x^4$

Exercises 3

1 $1 - 5x + 20x^2 - 50x^3 + 105x^4$
2 9·9499
3 $\frac{2}{3}x^3$ **4** 4·000 667
7 $\dfrac{5}{2 + x} - \dfrac{1}{1 - 3x} + \dfrac{2}{(1 - 3x)^2}$, $\dfrac{5}{2}\left(-\dfrac{1}{2}\right)^n + (2n + 1)3^n$

8 $2\left(1-\frac{x}{8}-\frac{x^2}{128}\right)$, $2+\frac{7x}{4}+\frac{175x^2}{64}$, $-\frac{1}{2}<x<\frac{1}{2}$

10 $\frac{5}{8}$ **15** 1·0198 **16** $\frac{21}{16}$

17 $1+3x+5x^2+3x^3$ **18** $1+2x-2x^2+4x^3$, 2·4495

19 3·317 **20** $p=-6$, $q=11$

26 $1+4x+\frac{15}{2}x^2+15x^3$ **27** $a=\pm\frac{8}{7}$, $b=\mp\frac{319}{343}$

28 $\frac{1}{x-3}+\frac{2}{x+1}$, $|x|<1$; $\frac{5}{3}-\frac{19x}{9}+\frac{53x^2}{27}$

29 $\frac{55}{53}$ **30** $a=-2$, $b=-1$; $\frac{73}{74}$

Chapter 4

Exercises 4a

1 (i) $\frac{7}{3}+0i$ (ii) $\pm 3+0i$
(iii) $0\pm i\sqrt{30}$ (iv) $-\frac{3}{2}\pm i\sqrt{31}/2$
(v) $0\pm 7i$ (vi) $-1\pm i\sqrt{7}$
(vii) $-2\pm 6i$ (viii) $\frac{1}{2}\pm i\sqrt{3}/2$

Exercises 4b

1 (i) 3 (ii) $18-i$
2 (i) $14-2i$ (ii) $12-16i$
3 $-3\pm 3i$
4 (i) $2i$ (ii) $-2+2i$ (iii) -4
6 (i) $\frac{-1-41i}{58}$ (ii) $\frac{-7-17i}{13}$
8 $\pm(7+3i)$, $\pm(6-i)$
9 (i) $x^5-10x^3+5x-i(5x^4-10x^2+1)$ (ii) $\frac{17-24i}{25}$
(iii) -1 (iv) $\frac{-(1+i)}{2}$ (v) $-\frac{1}{2}+i$ (vi) $\frac{11+2i}{125}$

Exercises 4c

2 $\sqrt{13}$, $\sqrt{17}$, $\sqrt{45}$, $\sqrt{2}$
3 (i) $2-3i$ (ii) $3+2i$ (iii) $-2+3i$
5 ± 24

Exercises 4d

1 (i) $3+4i$ (ii) $6-4i$ (iii) $3-7i$
2 (i) $4-2i$ (ii) $-6+3i$ (iii) $4+8i$
3 (i) $12+15i$ (ii) $-10+8i$
4 $2+23i$
6 $(3+2i)/13$

8 $y = 0$ **9** $x = 0$ **10** $x^2 + y^2 - 8x + 2y + 9 = 0$

13 $x = 0$ **14** $x^2 + y^2 = 16$ **15** $-i, \quad -\frac{1}{2} \pm i\frac{\sqrt{7}}{2}$

Exercises 4

1 (i) $1 - i$ (ii) $1 + 2i$ (iii) $-1 + 0i$
2 $\pm(3 + i)$
3 $(-7 \pm i\sqrt{31})/2; \quad -7, \; 20$
4 3 or 0
5 $\cot\frac{1}{2}\theta$

6 (i) $x^3 - 3xy^2, \; 3x^2y - y^3$ (ii) $\dfrac{x^2 - y^2}{(x^2 + y^2)^2}, \; \dfrac{-2xy}{(x^2 + y^2)^2}$

7 Length of a diagonal of parallelogram sides represented by z_1 and z_2
9 $2x^2 + 2y^2 - 5x + 2 = 0$
10 (i) 3 (ii) $a^3 + b^3$
11 $-2 \pm 4i$

13 (i) $2 - 11i$ (ii) $\dfrac{-7 - 9i}{10}$ (iii) $\dfrac{1}{2} - \dfrac{1}{2}i$ (iv) $\dfrac{53 - 9i}{10}$

14 $x = -5, \; y = -10$

17 $\frac{1}{2} + i\frac{\sqrt{3}}{2}$

18 (i) $(18 + i)/25$
(ii) (a) $4 - 7i$ (b) $7 + 4i$ (c) $-7 - 4i$
19 Circle has equation $x^2 + y^2 - 4x - 4 = 0$
21 $-i(\sqrt{2} + 1), \; i(\sqrt{2} - 1)$
25 $B(4 + i); \; C(5 + 2i)$ or $B(2 + 3i); \; C(3 + 4i)$

26 $-\dfrac{1}{2} \pm \dfrac{i\sqrt{3}}{2}; \; 1 \pm i\sqrt{3}, \dfrac{5}{2} \pm \dfrac{i\sqrt{3}}{2}; \; 28$

27 $x^2 + y^2 + 10x + 16 = 0, \; (-\frac{16}{5}, -\frac{12}{5})$
28 (i) $-1, \; -i; \; 3\pi/4$ (ii) $2, \; 2 - i; \; k = -10$

29 $a = \pm\dfrac{1}{\sqrt{2}}, \; b = \pm\dfrac{1}{\sqrt{2}}, \left(\dfrac{1}{\sqrt{2}} - 1\right) + \dfrac{1}{\sqrt{2}}i, \left(-\dfrac{1}{\sqrt{2}} - 1\right) - \dfrac{1}{\sqrt{2}}i$

30 (i) $x = \pm 2, \; y = \pm 1; \; \pm\dfrac{(1 + i)}{\sqrt{2}}$ (ii) $2, \; 1 - i$

Chapter 5

Exercises 5a

1 0, 3 **2** $-\frac{1}{3}$, 5 **3** $k \geqslant 2$ or $k \leqslant -10$

4 2 **9** 1, $-\frac{119}{125}$

Exercises 5b

4 $\dfrac{4ac-b^2}{4a}$ **5** 14 **6** 1 **10** $-20 \leqslant k \leqslant 5$

Exercises 5c

1 $\frac{61}{9}, \frac{55}{9}, \frac{3007}{81}$
2 (i) $x^2 - 19x + 25 = 0$ (ii) $25x^2 + 72x - 5 = 0$
3 (i) $q = 0$ (ii) $p = r$

5 (i) $\dfrac{\sqrt{(q^2-4rp)}}{p}$ (ii) $-q\dfrac{\sqrt{(q^2-4rp)}}{p^2}$

(iii) $\dfrac{\sqrt{(q^2-4pr)}}{p^3}(q^2-pr)$ (iv) $-q\dfrac{\sqrt{(q^2-4rp)}}{p^4}(q^2-2pr)$

7 $ac^2x^2 + b(b^2-3ac)x + a^2c = 0$
9 $ac(p+q)^2 = b^2pq$
10 $x^2 - 5x - 14 = 0$

Exercises 5

1 (i) $x^2 - 6x + 4 = 0$ (ii) $x^2 + 4x - 14 = 0$
(iii) $x^2 - 2ax + a^2 - 4b^2 = 0$
3 4 when $x = +3$ **4** $px^2 - 3(p+q)x + 7q = 0$
5 $-1, -\frac{1}{9}$
8 $7p^2$, $\sqrt{5}p$; $x^2 - 21\sqrt{5}p^2x - p^4 = 0$
9 0, -4
11 $k \leqslant -\frac{1}{2}$, $k \geqslant 3$
13 (i) $k < 1$ or $k > 9$ (ii) $k > 0$ **14** $a + b = 0$
16 $x^2 + x[p - \sqrt{(p^2-4q)}] - p\sqrt{(p^2-4q)} = 0$
18 $-\frac{21}{4}$ **19** 3
20 $(l_2n_1 - l_1n_2)^2 = (m_1n_2 - m_2n_1)(l_1m_2 - l_2m_1)$
21 0, 3, 8 **25** $-10 \leqslant k \leqslant 2$
26 (a) $a^2cx^2 + b(b^2-3ac)x + ac^2 = 0$
(b) $p \leqslant -1$; values are $\leqslant -2$ and $\leqslant 6$
27 (b) $k \leqslant 0$ or $\geqslant 3$ (c) $k \geqslant 3$
28 (i) $p = -4$, $q = 1$; $p = 3$, $q = -\frac{3}{4}$
(ii) $qx^2 + p(q+2)x + (q^2+4q+4) = 0$
29 (a) 0, k − 2 (b) $k = 7$, $-\frac{49}{4}$ (c) $-\frac{1}{2}$
30 $x^2 - 7x + 1 = 0$

Chapter 6

Exercises 6a

1 (i) $183°21'$ (ii) $-90°32'$ (iii) $36°$
2 (i) 4·102 (ii) 0·2455 (iii) $-2·2369$
3 (i) $129°$ (ii) $\frac{1}{3}$ radian
5 (i), (iii) and (iv); (ii) and (v)

Exercises 6b

1 $\frac{5}{13}$, $\frac{5}{12}$, $\frac{12}{5}$, $\frac{13}{12}$, $\frac{13}{5}$
2 $3/\sqrt{13}$, $2/\sqrt{13}$
3 (i) $a^2 \sin^2 \theta$ (ii) $\sin^5 \theta$
4 (i) $(\cos^2 \theta)/a^2$ (ii) $\sec \theta$
5 (i) $3 - 2c - 3c^2$ (ii) $\dfrac{1 - c^2 + 2c^3}{c^2}$ (iii) $\dfrac{2 - c^2}{\sqrt{(1 - c^2)}}$

Exercises 6c

1 (i) 0·3420 (ii) 0·7660 (iii) $-0·3640$ (iv) 0·3420
2 (i) $-\cos 10°$ (ii) $-\tan 50°$ (iii) $\cos 60°$ (iv) $-\sin 20°$
6 (i) -1 (ii) $1/\sqrt{2}$ (iii) $-\sqrt{3}$ (iv) $-\sqrt{3}/2$
7 (i) 0 (ii) -1 (iii) 0 (iv) -1 (v) 0 (vi) 1
8 $-\sqrt{\frac{2}{3}}$, $-\sqrt{\frac{1}{2}}$
9 $-\frac{3}{5}$, $-\frac{4}{5}$

Exercises 6d

3 They coincide **4** They coincide
5 (i) $\sqrt{3}/2$ (ii) 1 (iii) -1 (iv) $-\frac{1}{2}$
(v) $1/\sqrt{2}$ (vi) $\sqrt{3}$ (vii) -1 (viii) $-1/\sqrt{2}$

Exercises 6e

1 (i) $\frac{56}{65}$ (ii) $\frac{63}{65}$ (iii) $-\frac{56}{33}$, No
2 $\frac{1}{2}$
3 (i) $(\sqrt{3} - 1)/2\sqrt{2}$ (ii) $(\sqrt{3} + 1)/2\sqrt{2}$
6 (i) $\sin 10°$ (ii) $-\cos 70°$
7 (i) $\cos 10°$ (ii) $-\sin 50°$
8 (i) $\tan 70°$ (ii) $\tan 30°$
18 (i) $\sin A \cos B \cos C + \cos A \sin B \cos C + \cos A \cos B \sin C - \sin A \sin B \sin C$
(ii) $\cos A \cos B \cos C - \cos A \sin B \sin C - \sin A \cos B \sin C - \sin A \sin B \cos C$

Exercises 6f

1 $\frac{1}{2}$, -2
3 $\frac{24}{25}$, $3/\sqrt{10}$, $\frac{1}{3}$

15 (i) $\dfrac{(1+t)^2}{1+t^2}$ (ii) $\dfrac{2(1+t)}{1+t^2}$ (iii) $\dfrac{(1-t)^2}{1-t^2}$

Exercises 6h

1 (i) $\sqrt{2}\sin(\theta+45°)$ (ii) $\sqrt{2}\cos(\theta-45°)$
2 $5\cos(\theta+53°8')$
3 $\sqrt{13}\sin(\theta-56°19')$
4 $5\sin(2\theta+36°52')$
5 $\sqrt{10}$, $18°26'$

Exercises 6i

1 (i) $\pi/4$ (ii) $\pi/4$ (iii) $\pi/3$
2 (i) $-\pi/4$ (ii) $-\pi/6$ (iii) $\pi/6$
7 $\frac{1}{2}$

Exercises 6j

3 $30°39'$ **4** 131 m **5** $19{\cdot}5'$

Exercises 6

2 (i) 0·8988 (ii) −0·9336 (iii) 6·3138 (iv) 0·3256
3 (i) $-\frac{3}{5}$ (ii) $-\frac{4}{3}$ (iii) $-\frac{24}{25}$ (iv) $-\frac{7}{25}$

4 $\cos\theta=\sqrt{(1-s^2)}$, $\tan\theta=\dfrac{s}{\sqrt{(1-s^2)}}$, $\cot\theta=\dfrac{\sqrt{(1-s^2)}}{s}$,

$\sec\theta=\dfrac{1}{\sqrt{(1-s^2)}}$, $\operatorname{cosec}\theta=\dfrac{1}{s}$

14 $\dfrac{p^2+q^2-2q}{p^2+q^2+2q}$ **15** $(123°41', -\sqrt{52})$, $33°41'$, $213°41'$

16 $2-\sqrt{3}$, $(\sqrt{6}-\sqrt{2})/4$
20 2
23 $\dfrac{1-\sqrt{3}}{1+\sqrt{3}}$

24 (i) $\dfrac{3\tan A-\tan^3 A}{1-3\tan^2 A}$, $\dfrac{1-\tan^2 A}{2\tan A}$ (ii) $\dfrac{k\sqrt{(4-k^2)}}{k^2-2}$, $\dfrac{k^4-4k^2+2}{2}$

28 $\frac{1}{4}\pi$
29 $\pi-\arcsin 2x\sqrt{(1-x^2)}$, $\arccos(2x^2-1)$ for all x
32 $\frac{1}{6}$
33 $3+3\cos 2\theta+4\sin 2\theta\equiv 3+5\cos(2\theta-53°8')$
Max 8 when $\theta=26°34'$, min -2 when $\theta=116°34'$
38 $R=\sqrt{5}/2$, $\tan\alpha=\frac{1}{2}$; $1+\sqrt{5}/2$, $76°43'$
39 $\dfrac{a}{b}$, $\dfrac{2a-b+a^2b}{1+2ab-a^2}$

Chapter 7

Exercises 7a

1 $n180° + (-1)^n 56° 12'$; $56° 12'$, $123° 48'$
2 $n360° \pm 44° 34'$; $44° 34'$, $315° 26'$
3 $n180° + 64° 42'$; $64° 42'$ $244° 42'$
4 $n360° \pm 123° 54'$; $123° 54'$, $236° 6'$
5 $n180° - (-1)^n 28° 31'$; $208° 31'$, $331° 29'$
6 $n180° - 16° 42'$; $163° 18'$, $343° 18'$
7 $n60° + (-1)^n 10°$; $10°$, $50°$, $130°$, $170°$, $250°$, $290°$
8 $n30° - 8° 29'$; $21° 31'$, $51° 31'$, ..., $351° 31'$
9 $n1800° \pm 150°$; $150°$
10 $n180° + 15° \pm 27° 50'$; $42° 50'$, $222° 50'$, $167° 10'$, $347° 10'$
11 $n180° + (-1)^n 54° 24' - 18° 3'$; $36° 21'$, $107° 33'$
12 $\dfrac{n180°}{4 - 2(-1)^n}$; $0°$, $30°$, $180°$, $90°$, $360°$, $150°$, $210°$, $270°$, $330°$
13 $36n°$; $0°$, $36°$, $72°$, ..., $324°$, $360°$
14 $90n° + 45°$ or $180n° - 90°$; $90°$, $270°$, $45°$, $135°$, $225°$, $315°$
15 $\dfrac{n180° + (-1)^n 90°}{3 + 2(-1)^n}$; $18°$, $90°$, $162°$, $234°$, $306°$
16 $(2n + 1)18°$; $18°$, $54°$, $90°$, ..., $342°$
17 $90n°$ or $(2n + 1)30°$; $0°$, $90°$, $180°$, $270°$, $360°$, $30°$, $150°$, $210°$, $330°$
18 $90n°$ or $(2n + 1)22\tfrac{1}{2}°$; $0°$, $90°$, $180°$, $270°$, $360°$, $22\tfrac{1}{2}°$, $67\tfrac{1}{2}°$, ..., $337\tfrac{1}{2}°$
19 (i) $n360° \pm 120°$; $120°$, $240°$ (ii) $n180° + 60°$; $60°$, $240°$
20 $n360° - 120°$; $240°$

Exercises 7b

1 $n180° + 45°$ or $n180° + 26° 34'$; $45°$, $225°$, $26° 34'$, $206° 34'$
2 $n180° \pm 60°$; $60°$, $120°$, $240°$, $300°$
3 $n360° \pm 60°$; $60°$, $300°$
4 $n180° + 66° 2'$; $66° 2'$; $246° 2'$
5 $n180° + (-1)^n 14° 29'$; $14° 29'$, $165° 31'$
6 $n180° \pm 30°$; $30°$, $150°$, $210°$, $330°$
7 $n180° \pm (-1)^n 90°$; $90°$, $270°$
8 $n360° \pm 78° 28'$; $78° 28'$, $281° 32'$
9 $n180° \pm 40° 54'$; $40° 54'$, $139° 6'$, $220° 54'$, $319° 6'$
10 $n180° + 14° 2'$ or $n180° + 123° 41'$; $14° 2'$, $194° 2'$, $123° 41'$, $303° 41'$
11 $n180° + 45°$ or $n180° + 171° 52'$; $45°$, $225°$, $171° 52'$, $351° 52'$
12 $n90° + (-1)^n 9°$ or $n90° - (-1)^n 27°$; $9°$, $81°$, $189°$, $261°$, $117°$, $153°$, $297°$, $333°$
13 $n90°$ or $n180° \pm 60°$; $0°$, $90°$, $180°$, $270°$, $360°$, $60°$, $120°$, $240°$, $300°$
14 $n180°$ or $n360° \pm 80° 24'$; $0°$, $180°$, $360°$, $80° 24'$, $279° 36'$
15 $n60° - (-1)^n 30°$ or $n60° + (-1)^n 6° 29'$; $90°$, $210°$, $330°$, $6° 29'$; $53° 31'$, $126° 29'$, $173° 31'$, $246° 29'$, $293° 31'$.

Exercises 7c

1 $n360°$ or $n360° \pm 180°$; 0°, 180°, 360°
2 $n180°$ or $n180° + 108°26'$; 0°, 108°26', 180°, 288°26'
3 $n180° + 45°$ or $n180° - 18°26'$; 45°, 225°, 161°34', 341°34'
4 $n90° - 31°43'$ or $n90° + 35°47'$; 58°17', 148°17', 238°17', 328°17', 35°47', 125°47', 215°47', 305°47'
5 $n180° \pm 45°$ or $n120° \pm 20°$; 45°, 135°, 225°, 315°, 20°, 100°, 140°, 220°, 260°, 340°
6 $(2n+1)90°$ or $\dfrac{n180°}{2-3(-1)^n}$; 0°, 36°, 108°, 252°, 324°, 90°, 270°, 360°
7 $n360° \pm 180°$; 180°
8 $n90°$ or $n90° - (-1)^n 15°$; 0°, 90°, 180°, 270°, 360°, 105°, 165°, 285°, 345°
9 $n180°$, $n180° + (-1)^n 90°$, $n180° - (-1)^n 30°$; 0°, 90°, 180°, 360°, 210°, 330°
10 $n90° \pm 15°$; 15°, 75°, 105°, 165°, 195°, 255°, 285°, 345°

Exercises 7d

1 $n360° + 26°34' \pm 63°26'$; 90°, 323°8'
2 $n360° + 53°8'$; 53°8'
3 $n360° \pm 60° - 16°16'$; 43°44', 283°44'
4 $n720° \pm 112°38' + 112°38'$; 0°, 225°16'
5 $n180° \pm 33°27' - 5°40'$; 27°47', 207°47', 140°53', 320°53'
6 $n360° + 36°52' \pm 78°28'$; 115°20', 318°24'
7 $n120° - 15° \pm 45°$; 30°, 60°, 150°, 180°, 270°, 300°
8 $n180° - 7°54' \pm 27°22'$; 19°28', 144°44', 199°28', 324°44'
9 $n120° - 4°41' \pm 55°19'$; 50°38', 60°, 170°38', 180°, 290°38', 300°
10 $n360° \pm 135° + 45°$; 180°, 270°

Exercises 7

1 (i) 30°49', 59°11', 210°49', 239°11'
(ii) 13°53', 103°53', 193°53', 283°53', 76°7', 166°7', 256°7', 346°7'
(iii) 119°28', 299°28'
(iv) 0°, 36°, 108°, 180°, 252°, 324°, 360°
(v) $22\frac{1}{2}°$, $112\frac{1}{2}°$, $202\frac{1}{2}°$, $292\frac{1}{2}°$, 135°, 315°
2 (i) 36°52' (ii) 36°, 324°, 108°, 252°
3 45°, 225°, 171°52', 351°52'; $n\pi + 45°$, $n\pi + 171°52'$
4 0°, 180°, 360°, 210°, 330°
5 (i) 0°, 45°, 135°, 180° (ii) 35°16', 144°44' (iii) 90°
6 $n360° \pm 60°$
7 13°17' or 240°27'
8 (i) $n360°$ or $n120° + 30°$; 0°, 360°, 30°, 150°, 270°
(ii) $n360° + 71°34' - 18°26'$; 53°8'
9 $n360° \pm 60°$ or $n360° \pm 141°20'$
10 (i) 60°, 120°, 240°, 300° (ii) 53°48', 233°48', 159°54', 339°54' (iii) $172\frac{1}{2}°$
11 $360n° \pm 60°$
12 (i) 70°32', 120° (ii) 54°44', 125°16'

13 (i) 0°, 180°, 60°, 120°, 35° 16′, 144° 44′ (ii) 60°, 180°, 45°, 90°
14 $-\sin 3x \sin 2x$. 0°, 36°, 72°, 108°, 144°, 180°, $22\frac{1}{2}°$, $67\frac{1}{2}°$, $112\frac{1}{2}°$, $157\frac{1}{2}°$
15 60°, 120°, 240°, 300°, 45°, 135°, 225°, 315°
16 $(2n+1)\,30°$ or $n180° \pm 30°$
17 (i) $n360° - 36°\,52' \pm 113°\,35'$ (ii) $n360° + 73°\,44' \pm 78°\,28'$
18 $n180°$ or $n120°$
19 60°, 120°, 45°, 135°
20 $\cos A = k$, $\cos B = -\frac{1}{3}k$; $\cos A = -\frac{1}{3}k$, $\cos B = k$; $-1 \leqslant k \leqslant 1$
21 70° 32′, 289° 28′, 120°, 240°
22 (i) $n360° \pm 180°$, $n360° \pm 120°$ (ii) $n90° + 22\frac{1}{2}°$, $n180° - 45°$
(iii) $n45° - 2°\,30' + (-1)^n 3°\,23'$
23 −120°, −90°, −60°, 0°, 60°, 90°, 120°
24 $R = 13$, $\alpha = 67°\,23'$, 142° 54′, −8° 8′
25 $n180° - (-1)^n\,30°$
26 $n360° - 67°\,23' \pm 140°\,17'$
27 $\frac{4}{3}$ or -1
28 0°, 180°, 360°, 270°, 41° 50′, 138° 10′
29 45°, 135°, 225°, 90°, 210°, 330°, 315°
30 ± 2
31 19° 28′, 160° 32′, 194° 29′, 345° 31′
32 90°, 43° 10′
34 $n180°$, $n180° + 135°$
36 (a) 70° 32′, 120° (b) 45°
37 $65s^2 + 8s - 48$
40 $t_1 = \tan\frac{1}{2}\theta_1$, $t_2 = \tan\frac{2}{3}\theta_2$, $t_3 = \tan\frac{3}{2}\theta_3$, $t_4 = \tan\frac{4}{2}\theta_4$

$$\frac{t_1 + t_2 + t_3 + t_4 - (t_1t_2t_3 + t_2t_3t_4 + t_1t_3t_4 + t_1t_2t_4)}{1 - t_1t_2 - t_1t_3 - t_1t_4 - t_2t_3 - t_2t_4 - t_3t_4 + t_1t_2t_3t_4}$$

41 $\sqrt{5}\cos(x - 63°\,26')$; 110° 36′, 16° 16′
42 $-\sqrt{2} \leqslant a \leqslant \sqrt{2}$, 45°, 225°
43 $\sqrt{5}\sin(\theta + 26°\,34')$; 144° 52′, 342°
44 (i) $\cos x = -\frac{2}{3}$ or -1; $\tan^2 x = \frac{5}{4}$ or 0
(ii) $360n + 53.1°$, $360n - 36{\cdot}9°$
45 $\pi/2$, $3\pi/2$, $\pi/3$, $5\pi/3$, $2\pi/3$, $4\pi/3$
46 233° 8′, 328° 7′
47 18°
48 (i) 30°, 90°, 150° (ii) 120°
49 $\pi/6$, $\pi/3$, $2\pi/3$, $5\pi/6$
50 70°, 190°, 310°, 110°, 230°, 350°

Chapter 8

Exercises 8a

3 $B = 63°\,39'$, $C = 51°\,21'$, $a = 19{\cdot}20$ *or* $19{\cdot}21$ depending on the method of working
5 $c = 2$
7 W 26° 8′ N or 296° 8′
8 Correct to nearest minute $B = 70°\,54'$, $C = 65°\,13'$, $D = 43°\,54'$

10 (i) No possible triangle
(ii) $A = 44° 12'$, $B = 67° 54'$, $a = 4{\cdot}23$
(iii) $A = 51° 58'$, $B = 54° 40'$, $b = 674{\cdot}1$

Exercises 8

1 $A = 24° 34'$, $B = 79° 26'$, $b = 7{\cdot}09$
2 $a = 3\sqrt{5}$, $b = 2\sqrt{10}$
3 $\sin\theta = 0{\cdot}8127$, a $= 95{\cdot}7$
4 4·34 cm
5 $a = 116{\cdot}4$, $A = 34° 53'$, $C = 50° 55'$. $\Delta = 9172\ mm^2$
6 $b = 117$, $c = 56$
7 AC $= 5{\cdot}94$, $B = 115° 23'$, $D = 64° 37'$
8 Height $= \dfrac{[a^2 + (b-h)^2]\tan\theta}{a + (b-h)\tan\theta}$
9 AB $= 300\sqrt{(\operatorname{cosec}^2\alpha + \operatorname{cosec}^2\beta - 2\operatorname{cosec}\alpha\operatorname{cosec}\beta\cos\gamma)}$
10 $c = 9{\cdot}8$ cm
18 16°, 10°
22 AD $= 13\sqrt{3}/3$
24 60·1 m
25 $\dfrac{a^2\Delta}{16R^2}$
26 118 m
27 (i) 45° (ii) $\sqrt{\tfrac{8}{3}}a$ (iii) $\arccos(-\tfrac{1}{2})$
30 (i) 2646 m (ii) $A = 2529$ m, $B = 1732$ m (iii) 16° 46′
31 $\tfrac{1}{2}b^2\dfrac{\tan\alpha}{\sin\beta} + \tfrac{1}{2}b\sec\alpha(2l - b\tan\alpha\cot\beta)$
33 $d = \dfrac{6h\tan\alpha}{3 + \tan^2\alpha}$
36 45° 35′
37 (i) $\arcsin(1/\sqrt{3})$ (ii) $\arctan(1/\sqrt{2})$
38 $k \leqslant x \leqslant 2k$; $2\sqrt{3}k$
39 61°, 73·7°, 225·2°; 10·8 m
40 $3\sqrt{3}$ cm², $3\sqrt{7}$ cm²; 9

Chapter 9

Exercises 9a

1 $A = \tfrac{1}{2}x(100 - x)$
2 $V = \dfrac{1}{4\pi}h(20-h)^2$, $A = \pi r^2 + 2\pi r(20 - 2\pi r)$
3 $h = 10\sin x$
4 $h = \sqrt{(20x - 100)}$, $5 \leqslant x \leqslant 10$, $0 \leqslant h \leqslant 10$
5 $\theta = (20 - 2r)/r$

6 f(1) = −2, f(2) = 2, f(−1) = 2. All the real numbers in each case
7 $\phi(0) = 6$, $\phi(1) = 2$; $x = 2$ or 3
8 $G(0) = \frac{1}{4}$. All the real numbers. $0 < G(y) < \frac{1}{4}$
9 $F(0) = 1$, $F(\frac{1}{2}\pi) = -1$, $\theta = n\pi + \frac{1}{4}\pi$, $-\sqrt{2} \leqslant F(\theta) \leqslant \sqrt{2}$
10 All values except 1 and 2
11 $3 < x < 7$; $x \leqslant 3$ and $x \geqslant 7$
12 (i) $x = \frac{1}{2} + \frac{1}{2}y$ (ii) $x = \frac{1}{7}(y-3)$ (iii) $x = \arcsin y - \frac{1}{4}\pi$ (iv) $x = \sqrt[3]{y}$
13 (i) $y = x^3/(4+x)$ (ii) Two functions (a) $y = -x$, (b) $y = x - 1$ (iii) Not possible
14 (i) 1, $\frac{2}{3}$ (ii) y is not defined for $x = 1$ or 2
15 There are two functions (a) $y = \sqrt{(25 - x^2)}$, (b) $y = -\sqrt{(25 - x^2)}$ (i) (a) 4, (b) −4 (ii) (a) 3, (b) −3

Exercises 9b

1 50 m, 1250 m^2
2 $x = \frac{1}{2}$ or $x = 2$
3 0·86 radian
4 $3 + x_2 + 5$, 11
5 12 m/sec, 11 m/sec
6 $3 + x_2$, 4
7 10
8 7, 3·31, 3·0301, 3·000 030 0001; 3, $3x^2$
9 (i) −1 (ii) $-(1/x^2)$
10 10 m/sec, 20 m/sec

Exercises 9c

1 12
2 Yes, $-2x$, 2; 0, 2, No
3 Yes, $-2x$, $3x^2$; 0, 0, Yes
4 (i) $14x$ (ii) $4x^3 - 4x$ (iii) $2\cos 2x$ (iv) $-3\sin 3x$ (v) $\cos x - 2x$

Exercises 9

1 $A = r^2(1 + \sec\theta)\cot\frac{1}{2}\theta$
3 0, $\pi/4$. $-\pi/4$
4 1, 2; 0·001
5 (i) $y = \dfrac{3-4x}{1+x}$ (ii) $y = -x$
6 $y = 6/(3 + x^3)$; $\frac{3}{2}$
7 $y = x^2 - x - 2$; 4
8 ${x_1}^2 + x_1x_2 + {x_2}^2 + 3$, $4 + x + x^2$, 6
9 $5\frac{1}{10}$ m/sec, 5 m/sec
10 $x = 2$
11 2, −1, 2
12 u m/sec, a m/sec^2
13 $\sqrt{3}$
14 (i) $2x + 1$ (ii) $-4/x^5$
15 (i) $a\cos ax$ (ii) $-a\sin ax$
16 (i) $\sec^2 x$ (ii) $\sin 2x$
21 −2
23 $p = -3$, $q = 19$

Chapter 10

Exercises 10a

1 $35x^4 - 12x^3 + 2x$

2 $24x^2 - \cos x$

3 $-6\sin x - 16x - 8$

4 $7x^6 + 4x^3 + 3x^2$

5 $\cos x - \cos^2 x + \sin^2 x$

6 $6x + 4\cos x - 4x\sin x$

7 $8x^3 + 15x^2 + 12x + 4$

8 $16x(1 + \cos x)(1 + \sin x) + 8x^2(\cos x - \sin x)(\cos x + \sin x + 1)$

9 $\cos x - x\sin x + 6x$

10 $\sin x\,(3x^2 + 8x) + \cos x(4x^2 - 6x)$

11 $4x(x^2 + 1)$

12 $2(x^2 + 1)(5x^2 + 2x + 1)$

13 $6x(x^2 - 1)^2$

14 $3\sin x\cos x + 3x(\cos^2 x - \sin^2 x)$

15 $(9x^2 + 3)\sin x\cos x + 3x(x^2 + 1)(\cos^2 x - \sin^2 x)$

Exercises 10b

1 $\dfrac{x^2 + 2x}{(x + 1)^2}$

2 $\dfrac{\sin x - x\cos x}{(x + \sin x)^2}$

3 $\dfrac{2 + 2\cos x + x\sin x}{(1 + \cos x)^2}$

4 $-\operatorname{cosec} x\cot x$

5 $\dfrac{\sin x}{(1 + \cos x)^2}$

6 $\dfrac{-6}{(x + 1)^3}$

7 $\dfrac{x + \sin x(\sin x + \cos x)}{(\cos x + \sin x)^2}$

8 $\dfrac{x^4 + 6x^3 + 3x^2 + 6}{(x^2 + 3x + 2)^2}$

9 $\dfrac{\cos x(x^5 + x^4 - x^3 - x^2) - \sin x(x^4 + x^2 + 2x)}{(x + 1)^2(x^2 - 1)^2}$

10 $\dfrac{-2x^3 - 6x^2 + 6}{(x + 1)^2(x + 2)^2(x + 3)^2}$

11 $\dfrac{2\sin x\cos x + \sin^2 x}{\cos^2 x(\cos x + \sin x)^2}$

12 $\dfrac{4x^2 + 12x + 6}{(x^2 + 5x + 6)^2}$

13 1·08

14 $2/(2 + \sqrt{3})$

15 $\frac{1}{3}$, (0, 0); (−2, 6)

Exercises 10c

1 $\sec x\,(\sec x + \tan x)$

2 $\sec x\,(\sec^2 x + \tan^2 x)$

3 $-\cos x - \cot x\operatorname{cosec} x$

4 $\sec^2 x(1 + 2\tan x)$

5 $4\sec^2 x\tan x$

6 $(\sec x + \tan x)(1 - \sin x + \cos x + \tan x)$

7 $\dfrac{\sec x(\sin x\tan x + 2\sin x - \cos x)}{(\sin x + \cos x)^2}$

8 $\dfrac{\sec x\tan x}{(1 + \sec x)^2}$

9 $\dfrac{-2\sec^2 x}{(1 + \tan x)^2}$

10 (i) $v = (8t - 3)$ m/sec, 5 m/sec
$a = 8$ m/sec^2, 8 m/sec^2

(ii) $v = 2\pi(\cos 2\pi t - \sin 2\pi t)$ m/sec, 2π m/sec
$a = -4\pi^2(\cos 2\pi t + \sin 2\pi t)$ m/sec^2, $-4\pi^2$ m/sec^2

11 (i) $-2\cos 2x$ (ii) $\cos x/(1-\sin x)^2$
15 $\sin(x+\frac{1}{2}n\pi)$

Exercises 10d

1 $5(x-1)^4$
2 $10(2x-1)^4$
3 $5(4x-3)(2x^2-3x)^4$
4 $8(x+1)^7$
5 $3\sec 3x\tan 3x$
6 $5\sec^2 5x$
7 $\sin 4x+4x\cos 4x$
8 $2x\cos 3x-3x^2\sin 3x$
9 $3\sin^2 x\cos x$
10 $3x^2\cos x^3$
11 $6x\sec(3x^2+1)\tan(3x^2+1)$
12 $9\tan^2(3x-4)\sec^2(3x-4)$
13 $4x\sin(x^2+1)\cos(x^2+1)$
14 $(10x+7)(2x-1)^3$
15 $3\sin^2 x\cos x\tan 2x+2\sin^3 x\sec^2 2x$
16 $-4\sin^3 x\cos x$
17 $\dfrac{-3\cos^2 x}{(1+\sin x)^3}$
18 $\dfrac{-8x(1-x^2)}{(1+x^2)^3}$
19 $8\cos^7 x\sin x$
20 $\dfrac{4\sin 2x}{(1+\cos 2x)^2}$
21 $\dfrac{4\sin x\cos x}{(2+\sin^2 x)^2}$
22 $18\sec^3(\tan^2 3x)\tan(\tan^2 3x)$
$\tan 3x\sec^2 3x$
23 (i) $n\sin^{n-1}\theta\cos\theta$ (ii) $-m\cos^{m-1}\theta\sin\theta$
(iii) $\sin^{n-1}\theta\cos^{m-1}\theta(n\cos^2\theta-m\sin^2\theta)$

Exercises 10e

1 $\frac{3}{2}x^{1/2}$
2 $1/3\sqrt[3]{x^2}$
3 $\frac{7}{3}(\sqrt[3]{x})^4$
4 $6\left(x^3-\dfrac{3}{x^3}\right)\left(x^2+\dfrac{3}{x^4}\right)$
5 $\dfrac{-12x}{(2x^2-3)^4}$
6 $\dfrac{4x-1}{2\sqrt{(2x^2-x)}}$
7 $3x\sqrt{(x^2+1)}$
8 $\dfrac{-x}{(x^2+1)^{3/2}}$
9 $\dfrac{1-x-2x^2}{\sqrt{(1-x^2)}}$
10 $\dfrac{(1+x)^2(1+7x)}{2\sqrt{x}}$
11 $\dfrac{1}{2\sqrt{x}}\sec\sqrt{x}\tan\sqrt{x}$
12 $\frac{1}{2}\tan x\sqrt{(\sec x)}$
13 $\dfrac{\cos x}{2\sqrt{(1+\sin x)}}$
14 $\dfrac{1}{\sqrt{[(x+1)^3(x-1)]}}$
15 $-1/(1-\cos x)$

Exercises 10f

5 $\dfrac{2x}{\sqrt{(1-x^4)}}$
6 $\dfrac{-6}{\sqrt{(1-36x^2)}}$
7 $\dfrac{1}{1+(x+1)^2}$

8 $\dfrac{2}{x\sqrt{(x^4-1)}}$ **9** $\dfrac{2x}{1+x^4}$ **10** $\arcsin x + \dfrac{x}{\sqrt{(1-x^2)}}$

11 $\dfrac{-1}{(1+x)\sqrt{x}}$ **12** $\dfrac{1}{1+x^2}\left[\arctan\left(\dfrac{1+x}{1-x}\right) = \arctan x + \arctan 1\right]$

14 Because $2 \arctan x = \arctan\left(\dfrac{2x}{1-x^2}\right)$

Exercises 10g

1 $-\dfrac{y}{x} = -\dfrac{1}{x^2}$ **2** $\dfrac{1-2xy^2}{2x^2y-1}$

3 $-\sqrt{\left(\dfrac{y}{x}\right)}$ **4** $\dfrac{\sin x - 1}{1 - \sin y}$

5 $\dfrac{-y}{x+\cos y}$ **6** $-\dfrac{(1+y\cos xy)}{(1+x\cos xy)}$

7 -1 **8** $-\dfrac{(2x+3y)}{3x+2y}, \dfrac{10x^2+10y^2+30xy}{(3x+2y)^3}$

9 $-\frac{1}{3}, -\frac{20}{27}$ **10** $-\frac{8}{5}, -\frac{326}{125}$

Exercises 10h

1 $2/(3t)$ **2** $-\frac{2}{3}\cot\theta$

3 $-4\sin 2t$ **4** $\dfrac{2t\cos t - t^2 \sin t}{\sin t + t\cos t}$

5 $2t + t^2$ **6** $\dfrac{3t^2}{1-2t^3}$

7 $\dfrac{1}{2}\left(t-\dfrac{1}{t}\right), -\dfrac{(1+t^2)^3}{8t^3}$ **8** $-\frac{1}{2}t^3, \frac{3}{4}t^5$

9 $\dfrac{4t}{3t^2+1}$ **10** (i) $\frac{5}{4}$ sec (ii) 1 sec

Exercises 10

1 (i) $3x^3 - x$ (ii) $\dfrac{1}{4t\sqrt{t}}(9t^2 - t - 1)$

2 (i) $15x^4 + 4x^3 + 9x^2 + 2x$ (ii) $\sec\theta(1 + \tan\theta - \operatorname{cosec}\theta)$

3 (i) $-\dfrac{(1+x^2)}{(1-x^2)^2}$ (ii) $\dfrac{x^4+4x^2-1}{(x^2+1)^2}$

4 (i) $\dfrac{1}{2(1+\sqrt{x})^2\sqrt{x}}$ (ii) $\dfrac{1}{(1-\sqrt{x})^2\sqrt{x}}$

5 (i) $3\sin t\cos 3t + \cos t \sin 3t$ (ii) $\dfrac{2t\sqrt{(1-t^2)}\arcsin t + t^2}{\sqrt{(1-t^2)}}$

6 (i) $-\dfrac{(x^2+6x+16)}{(x^2-16)^2}$ (ii) $\dfrac{6(x^2-2)}{(x^2+3x+2)^2}$

7 (i) $-(5\cos t\sin 5t+\sin t\cos 5t)$ (ii) $\dfrac{(1+t^2)\arctan t+t}{1+t^2}$

8 (i) $4\sin^3 x\cos x$ (ii) $3\sec^3 x\tan x$

9 (i) $18\sec^3 6\theta\tan 6\theta$ (ii) $-10\theta\cot^4\theta^2\operatorname{cosec}^2\theta^2$

10 (i) $\sec\theta(\tan\theta+2\theta\tan^2\theta+\theta)$
(ii) $\sin x\cos 2x+x\cos x\cos 2x-2x\sin x\sin 2x$

11 (i) $-\dfrac{1}{x^2}\cos\dfrac{1}{x}$ (ii) $-\left(\cos\dfrac{1}{x}+x\sin\dfrac{1}{x}\right)\Big/x^3$

12 (i) $\dfrac{1-2\theta^3}{(1+\theta^3)^2}$ (ii) $\dfrac{3t^2}{\sqrt{(1-t^6)}}$

13 (i) $\dfrac{-1}{\sqrt{(1-\theta^2)}}$ (ii) $\dfrac{1}{2(2-\theta)(1-\theta)^{1/2}}$

14 (i) $\dfrac{\sec^2 x}{\sqrt{(1-\tan^2 x)}}$ (ii) $\dfrac{\cos x}{1+\sin^2 x}$

15 (i) $\dfrac{2}{\sqrt{(2-t^2)}}$ (ii) $\dfrac{-1}{1+2t+2t^2}$

16 (i) $\dfrac{2}{x^3}\sin\dfrac{1}{x^2}$ (ii) $\dfrac{2}{1+x^2}$

20 (i) $\dfrac{32x}{16-8x^2+17x^4}$ (ii) $\dfrac{1}{\sqrt{(5x-x^2-6)}}$

21 (i) $nx^{n-1}(\tan nx+x\sec^2 nx)$ (ii) $\dfrac{\frac{1}{2}\cos\frac{1}{2}x}{1+\sin^2\frac{1}{2}x}$

22 (i) $y^2+2xy\dfrac{dy}{dx}$ (ii) $\left(y-x\dfrac{dy}{dx}\right)\Big/y^2$

(iii) $\left(x\dfrac{dy}{dx}-y\right)\Big/x^2$ (iv) $\sin 2y\dfrac{dy}{dx}$

23 (i) $\dfrac{y-x^2}{y^2-x}$ (ii) $\dfrac{4xy^2-3x^2}{4y^3-4x^2y}$

24 (i) $\dfrac{3-x}{y-2}$ (ii) $\dfrac{1-x^2}{y^2-1}$

25 (i) $\dfrac{\sec^2 x}{\sqrt{(1-\tan^2 x)}}$ (ii) $\dfrac{ay-x^2}{y^2-ax}$

26 $\frac{5}{8}$

27 0, 0

28 2, 2

29 $-(20xy+432x^5y^5)/(6y^5-x)^3$

30 $-\dfrac{b}{a}\cot\theta$

31 $\dfrac{b}{a}\operatorname{cosec}\theta,\ -\dfrac{b}{a^2}\cot^3\theta$

32 $-\tan\theta$, $(\sec^4\theta\operatorname{cosec}\theta)/3a$

33 $\tan\frac{1}{2}t$

34 $\tan t$, $(\sec^3 t)/at$

35 $\frac{3}{4}$

39 $\dfrac{-(\theta\sin\theta+2\cos\theta)}{\theta^3},\ \dfrac{-(\theta^2\cos\theta-4\theta\sin\theta-6\cos\theta)}{\theta^4}$

41 $-\frac{1}{2}, \frac{3}{40}$

43 $2x\sin[2(x^2+1)]$

45 $-2\sin^3\theta\cos\theta$

Chapter 11

Exercises 11a

1 0·000 036 32 m/°C

2 4·77 cm/min

3 4π cm^2/sec, 10π cm^3/sec

4 500 m^3/min

5 10 cm/sec

6 0·265 m/min

7 20 m^2/min

8 $\dfrac{1}{9\pi}$ cm/sec

9 $-0{\cdot}03$ rad/sec

10 $\dfrac{1}{8\pi}$ cm

Exercises 11b

1 36 m/sec, 36 m/sec^2; $9t^2$ m/sec, $18t$ m/sec^2

2 8 m/sec

3 $3(t^2-1)$, after 1 sec

4 $a = 6t$ cm/sec^2. When $t=0$

5 $\frac{32}{27}$ m, 0 m; -4 m/sec^2, 4 m/sec^2

Exercises 11c

1 1·0006

2 0·7073, 0·7069; 0·7075, 0·7067; 0·7077, 0·7065

3 0·377 cm^2, 0·3779 cm^2

4 $4{\cdot}4\pi$ cm^3; $\dfrac{1}{21\pi}$ cm/sec

5 3% increase

6 Decrease of 40π cm^3

7 Decrease of 8π cm^3

9 0·4 cm^2

10 0·4%, 9960

Exercises 11d

1 $y-6x+11=0$, $6y+x-8=0$

2 $y+3x-3=0$, $3y-x+1=0$

3 $y = 2x$, $y + x - 1 = 0$, $y - 2x + 4 = 0$
4 $3y - 5x + 16 = 0$
5 $y + x - 2 = 0$, $y = x$
6 $y - x - a = 0$, $y + x - 3a = 0$
7 $y = 8$, $y = 4$
8 $27y - 135x - 40 = 0$, $y - 5x + 8 = 0$
9 $2y - x - 7 = 0$, $2y - x + 1 = 0$
10 $y = x$, $y - x - 4 = 0$
11 $x + y = a\sqrt{2}$
12 $y - x - a = 0$

Exercises 11e

1 Max of 0 when $x = -2$; min of -4 when $x = 0$
2 Max of $\frac{4}{27}$ when $x = \frac{1}{3}$; min of 0 when $x = 1$
3 Min of $-\frac{1}{2}$ when $x = -1$; max of $\frac{1}{2}$ when $x = 1$
4 Min of $\frac{1}{2}$, max of $\frac{3}{4}$, min of $\frac{-3}{2}$
5 $-\sqrt{(a^2 + b^2)}$
6 Height $2r/\sqrt{3}$, radius $r\sqrt{\frac{2}{3}}$
7 $A = r^2 \sin 2\theta(1 + \cos 2\theta)$
10 $\mathrm{XP} = \dfrac{av}{\sqrt{(u^2 - v^2)}}$

Exercises 11f

1 $(\frac{5}{8}, -\frac{251}{54})$
2 $(0, 2)$, $(\frac{2}{3}, \frac{38}{27})$
3 Min $(1, 3)$, max $(\frac{1}{3}, \frac{85}{27})$, point of inflexion $(\frac{2}{3}, \frac{83}{27})$

Exercises 11

1 $\dfrac{1}{4\pi}$ cm/sec
2 4·5 cm/sec away from lens
3 2·19 units/unit change in r
4 $10\,\mathrm{m}^3$/sec
5 $\dfrac{32}{15\pi}$ m/min
6 20, 16; 18 units
7 $\dfrac{\pi}{3}, \dfrac{2\pi}{3}; 0, \dfrac{\pi}{2}, \pi; \dfrac{\pi}{3} + \dfrac{\sqrt{3}}{2}$
9 Decreased by 1·02%
10 0·17 cm
11 Decreased by 0·42%
12 3, $y = 15x + 36$; $(-6, -54)$
13 $1 \pm \sqrt{2}$, $1 \pm \dfrac{2\sqrt{3}}{3}$
14 $y = \frac{4}{3}x - \frac{2}{3}$
15 (i) $y = 4x - 15$ (ii) $y = -\frac{1}{4}x + 2$
16 Min when $x = 0$, point of inflexion when $x = 1$
17 $a = \frac{1}{4}$, $b = \frac{3}{4}$, $c = -6$, $d = -\frac{10}{4}$; $(-4, \frac{35}{2})$
18 $-0{\cdot}29$
19 (i) Min $(\frac{1}{6}\pi, 3\sqrt{3})$, max $(\frac{5}{6}\pi, -3\sqrt{3})$ (ii) Min $(1, 0)$, max $(-1, \frac{8}{3})$

22 Max $(1, \frac{1}{2})$, min $(-1, -\frac{1}{2})$, points of inflexion $(0, 0)$ $\left(\sqrt{3}, \frac{\sqrt{3}}{4}\right)$, $\left(-\sqrt{3}, \frac{-\sqrt{3}}{4}\right)$
23 Min $(2, -\frac{1}{8})$, min $(-2, -\frac{1}{8})$
24 l
25 $\arcsin(\frac{1}{3})$
27 Max $(1, \frac{1}{5})$, min $(-1, 1)$
28 $-\frac{b}{3a}$
30 $y = -x + 2$, $y = -x + \frac{1}{2}\pi$
31 Min $(2, 0)$, $(2, 0)$, $(0, 4)$; min $(2, 4)$, does not cross x-axis, $(0, 8)$; min $(2, -4)$; $(0, 0)$, $(4, 0)$
33 $y = 2x$, $(1, 2)$
35 $\frac{1}{4}\pi - \frac{1}{2}A$
36 $614\,\text{cm}^3$
37 a
38 $(\frac{1}{6}\pi, -3\sqrt{3})$, $(\frac{5}{6}\pi, 3\sqrt{3})$
39 $(1, 0)$, $(-1, 6)$, $(3, 2)$; $(\frac{3}{2}, -\frac{1}{4})$; $x = 0$, $y = 3$
40 $(-1, -\frac{1}{2})$, $(1, \frac{1}{2})$
41 $(0, 3)$, $(\frac{4}{3}\pi, -\frac{3}{2})$, $(2\pi, -1)$
43 $64R^3/81$
44 $\frac{1}{3}\pi$, $4 + \pi$, 6
45 Maximum

Chapter 12

Exercises 12a

1 $1/x$
2 $-(2/x)$
3 $\cot x$
4 $a/(ax + b)$
5 $1/[2(x - 1)]$
6 $1/(\cos x \sin x)$
7 $\sec x$
8 $2\cot x$
9 $\log_e x$
10 $(1 - \log_e x)/x^2$
11 $-2/(1 - x^2)$
12 $\cos x \cot x - \sin x \log_e \sin x$
13 $\frac{-y(x + 1)}{x(y + 1)}$
15 $1/x$

Exercises 12b

1 $3e^{3x}$
2 $-2xe^{-x^2}$
3 $\cos x\, e^{\sin x}$
4 $-e^{-x}$
5 ae^{ax+b}
6 $2^x \log_e 2$
7 $2x \log_e 3 \times 3^{x^2}$
8 $e^x(\cos x - \sin x)$
9 $e^{-x^2}(1 - 2x^2)$
10 $\frac{2e^{2x}}{(1 + e^{2x})^2}$
11 $e^x\left(\log_e x + \frac{1}{x}\right)$
12 $2xe^{-x^2}(\cos x^2 - \sin x^2)$
15 4 or -1

Exercises 12c

1 $\sec^2 x\, e^{\tan x}$
2 $4x^3 e^{x^4}$
3 $x^{\sin x}\left(\dfrac{\sin x}{x} + \cos x \log_e x\right)$
4 $(\sin x)^x(x \cot x + \log_e \sin x)$
5 $(\log_e x)^x\left[\log_e(\log_e x) + \dfrac{1}{\log_e x}\right]$
6 $x^{x-1}\left(\dfrac{x-1}{x} + \log_e x\right)$
7 $e^x + x^x(1 + \log_e x)$
8 $2xe^{x^2} + x^{x^2+1}(1 + \log_e x^2)$
9 $\dfrac{-2}{3(x^2-1)}\left(\dfrac{x+1}{x-1}\right)^{1/3}$
10 $\dfrac{x}{x^4-1}\left(\dfrac{x^2-1}{x^2+1}\right)^{1/4}$

Exercises 12d

2 $-1 < x < 1$
3 0·5236, 0·4997, 0·5000, 0·5000 to 4 decimal figures
5 $\sin x \approx x - \frac{1}{6}x^3$, $\cos x \approx 1 - \frac{1}{2}x^2$
$\sin x \approx x - \frac{1}{6}x^3 + \frac{1}{120}x^5$, $\cos x \approx 1 - \frac{1}{2}x^2 + \frac{1}{24}x^4$
7 (i) $1 + nx + \dfrac{n(n-1)}{2!}x^2 + \ldots + nx^{n-1} + x^n$
(ii) $1 + nx + \dfrac{n(n-1)}{2!}x^2 + \dfrac{n(n-1)(n-2)}{3!}x^3 + \ldots$
9 0·5236, 0·5714, 0·5767 to 4 decimal figures
10 $a + bx + cx^2 + dx^3 + ex^4$

Exercises 12e

1 $\sqrt{e} \approx 1{\cdot}6487$, $\dfrac{1}{e} \approx 0{\cdot}3679$
2 $\log_e 1{\cdot}2 \approx 0{\cdot}1823$, $\log_e 0{\cdot}9 \approx -0{\cdot}1054$
3 (i) $1 + 2x + 2x^2 + \frac{4}{3}x^3 + \ldots$
(ii) $1 - 3x + \frac{9}{2}x^2 - \frac{9}{2}x^3 + \ldots$
(iii) $1 + x^2 + \frac{1}{2}x^4 + \frac{1}{6}x^6 + \ldots$
4 (i) $2x - 2x^2 + \frac{8}{3}x^3 - 4x^4 + \ldots$; $-\frac{1}{2} < x \leqslant \frac{1}{2}$
(ii) $-(3x + \frac{9}{2}x^2 + 9x^3 + \frac{81}{4}x^4 + \ldots)$; $-\frac{1}{3} \leqslant x < \frac{1}{3}$
(iii) $x^2 - \frac{1}{2}x^4 + \frac{1}{3}x^6 - \frac{1}{4}x^8 + \ldots$; $-1 < x < 1$
5 $-\frac{3}{4} < x \leqslant \frac{3}{4}$
6 $1 + \dfrac{1}{2!} + \dfrac{1}{4!} + \dfrac{1}{6!} + \ldots$
8 $-\frac{1}{4}\pi \leqslant x \leqslant \frac{1}{4}\pi$
9 0·6931

Exercises 12

1 (i) $3x - \frac{1}{x} + 6x\log_e x$ (ii) $e^x\left(\frac{1}{x} + \log_e 2x\right)$

(iii) $-2e^{-2x}(\cos 4x + 2\sin 4x)$

2 (i) $2e^{2t}/(1 + e^{2t})$ (ii) $\operatorname{cosec} t$ (iii) $-\operatorname{cosec} t$

4 $y/(x + y)$

5 $\frac{dy}{dx} = 5e^{3x}\cos(4x + \alpha)$; $\alpha = \arctan\frac{4}{3}$, $\cos\alpha = \frac{3}{5}$, $\sin\alpha = \frac{4}{5}$

$\frac{d^2y}{dx^2} = 25e^{3x}\cos(4x + 2\alpha)$

6 (i) $\frac{e^t(1 + e^{2t})}{(1 - e^{2t})^2}$ (ii) $\frac{1 - \log_e\theta}{(\theta + \log_e\theta)^2}$

7 (i) $e^{x\sin x}(x\cos x + \sin x)$ (ii) $\frac{1}{x}\cos(\log_e x)$

10 (i) $1/[2(x + 1)]$ (ii) $1/[2\sqrt{(x^2 - 1)}]$

13 $k = -7$

18 $1/2t^2$, $-1/2t^4$

19 $1/te^t$, $-(1 + t)/t^2e^{2t}$

20 Max of $1/e$ when $x = e$

22 $(1, 1/\sqrt{e})$, $(-1, 1/\sqrt{e})$

23 $y = 2x + 1$

24 $(2, 2/e^2)$

27 $(4, \frac{1}{2} + \log_e 4)$

28 $\pm 1/\sqrt{2}$

29 (i) $\frac{(\log_e x)^{\log_e x}}{x}[1 + \log_e(\log_e x)]$

(ii) $x^x(1 + \log_e x) + \sec^2 x\, e^{\tan x}$

30 $x^x(1 + \log_e x)\left(1 + \frac{x}{\alpha + 2}\right) + \frac{x^x}{\alpha + 2}$

31 $1/e$, $0 < k < 1/e$, $1{\cdot}43$

33 $0{\cdot}000\,334$

35 (i) $1 - 2x + 2x^2 - \frac{4}{3}x^3$ (ii) $1 - 2x - 2x^2 - 4x^3$, $0{\cdot}000\,40$

37 $x + \frac{x^2}{2} - \frac{2}{3}x^3 + \frac{x^4}{4} + \frac{x^5}{5} - \frac{x^6}{3} + \frac{x^7}{7} + \ldots + \frac{x^{3n-1}}{3n - 1} + x^{3n}\left(\frac{1}{3n} - \frac{1}{n}\right)$

$+ \frac{x^{3n+1}}{3n + 1} + \ldots$

38 (a) $-\frac{1}{2}\left(\sin^2\theta + \frac{\sin^4\theta}{2} + \frac{\sin^6\theta}{3} + \ldots + \frac{\sin^{2n}\theta}{n} + \ldots\right)$; $0 \leqslant x < \pi/2$

(b) $4 - x^2 + \frac{1}{3}x^3$; $2\log_e 2 - \frac{1}{4}x^2 + \frac{1}{12}x^3$; $\frac{(-1)^{n+1}}{n2^{n-1}}$

39 (a) $1 + x^2 + \frac{2}{3}x^3$
(b) $x - \frac{1}{2}x^2 + \frac{1}{3}x^3 - \frac{1}{4}x^4 + \ldots;$
$$2\left(\cos\theta + \frac{1}{3}\cos^3\theta + \frac{1}{5}\cos^5\theta + \ldots + \frac{1}{2n-1}\cos^{2n-1}\theta + \ldots\right)$$
40 0·6931, 1·0986, 1·3863, 1·6094, 1·7918
41 0·2231
42 $-\frac{2}{3}; \frac{2^n}{n!} - \frac{2^n}{(n-1)!} + \frac{2^{n-2}}{(n-2)!}$
44 Max for even r, min for odd r
46 $13e^{-2x}\cos 3x$
47 0
48 $-2x(1+x)e^{-2x}$, $(0, 1)$, $(-1, 0)$
49 $3x - \frac{5}{2}x^2 + 3x^3$, $6x + 6x^3$
50 $-1/e$

Chapter 13

Exercises 13a

Arbitrary constants in these exercises have been omitted.

1 $\frac{3}{8}x^{8/3}$, $\frac{x^{12}}{12}$, $3x^{+1/3}$, $\frac{3}{5}\sqrt[3]{x^5}$, $\log_e x$, $\frac{4}{9}\sqrt[4]{x^9}$, $\frac{x^{22}}{22}$

2 $-\frac{1}{2x^2}$, $-\frac{1}{x}$, $2\sqrt{x}$, $\log_e x$, $-\frac{1}{10x^{10}}$, $-\frac{3}{\sqrt[3]{x}}$, $\frac{5}{3}\sqrt[5]{x^3}$

3 $\arcsin\frac{1}{4}x$, $\arcsin x$, $\arcsin 2x$, $\arcsin 3x$, $\arcsin\frac{1}{6}x$
4 $\frac{1}{2}\arctan\frac{1}{2}x$, $\arctan x$, $\frac{1}{3}\arctan\frac{1}{3}x$, $3\arctan 3x$, $5\arctan 5x$
5 $\tan x$, $-\cot x$, $\frac{1}{2}x + \frac{1}{2}\sin x$, $\tan x - x$, $-\cot x - x$

6 $-\frac{5}{4x^4} - \frac{1}{3x^3} + \frac{1}{x^2}$, $\frac{\sqrt{x}(22 + 22x - 10x^5)}{11}$

7 $-\frac{a}{3x^3} - \frac{b}{5x^5} - \frac{c}{6x^6}$

$\frac{2}{n-4}x^{n-4} + \frac{3}{p-4}x^{p-4}$ $(n, p \neq 4)$

$-\frac{1}{(n-1)x^{n-1}} - \frac{3}{(n-2)x^{n-2}} - \frac{5}{(n-3)x^{n-3}}$ $(n \neq 1, 2, 3)$

8 $x - \frac{1}{2}x^4 + \frac{1}{7}x^7$, $x + 5x^2 + \frac{25}{3}x^3$, $2ab\log_e x - b^2/x + a^2x$
9 $-6\cos x$, $7\sin x$, $8\sin x + 6\cos x$
10 $y = x^3 - 3x^2 + 2x + 7$
11 $v = 5t - kt^2 + c$, $k = 4$
12 $x = -3\sin t$
13 $y = 3 - \cos x$
14 $y = \frac{1}{5}x^5 - \frac{1}{2}x^4 + \frac{1}{3}x^3$
15 $x = t^4 - 10t^3 + 16t^2$; 2 sec, 8 sec

Exercises 13b

1 (i) $\dfrac{(2x+3)^{11}}{22}$ (ii) $-\dfrac{(5-x)^{12}}{12}$ (iii) $\dfrac{2}{21}(7t+5)^{3/2}$
(iv) $\dfrac{2}{21}(3u-5)^{7/2}$

2 (i) $-\dfrac{1}{42(3x+1)^{14}}$ (ii) $\sqrt{(2x+1)}$ (iii) $\dfrac{2}{(1-x)^{1/2}}$
(iv) $-\frac{1}{2}(1-3y)^{2/3}$

3 (i) $-\frac{1}{3}\cos(3x+3)$ (ii) $\frac{1}{5}\sin(5u-1)$ (*iii*) $\cos(1-y)$
(iv) $\frac{1}{2}x-\frac{1}{4}\sin 2x$ (v) $\frac{1}{2}x+\frac{1}{4}\sin 2x$

4 (i) $-e^{2-x}$ (ii) $\frac{1}{5}e^{5(t+2)}$ (iii) $-\frac{1}{6}e^{1-6u}$

5 (i) $\frac{1}{2}\log_e(2x+1)$ (ii) $-\frac{1}{2}\log_e(1-2x)$ (iii) $\log_e(\sin x)$
(iv) $\frac{5}{2}\log_e(x^2+1)$ (v) $\log_e(x^2+x-1)$ (vi) $\log_e(2-e^{-t})$
(vii) $\frac{1}{2}\log_e(e^{x^2}+3)$ (viii) $\frac{1}{2}\log_e(\log_e u)$ [NB Another form of this is $\frac{1}{2}\log_e(\log_e u^2)$]
(ix) $\log_e(\log_e 3x)$ (x) $-\log_e(\cos x-\sin x)$

6 (i) $\frac{1}{2}\arctan\left(\dfrac{x-1}{2}\right)$ (ii) $\frac{1}{4}\arctan 4t$ (iii) $\arctan(x+\frac{1}{2})$

7 (i) $\frac{1}{4}\arcsin 4\theta$ (ii) $\frac{1}{2}\arcsin\left(\dfrac{2x+1}{4}\right)$ (iii) $\arcsin\left(\dfrac{u-1}{3}\right)$

8 (i) $\frac{1}{3}\log_e(x^3+1)$ (ii) $-\frac{2}{9}(2-3t)^{3/2}$ (iii) $\log_e(1+e^x)$

9 (i) $\frac{1}{2}\arctan 2u$ (ii) $\frac{1}{8}\log_e(1+4u^2)$ (iii) $\frac{1}{2}\arcsin 2u$

10 (i) $\log_e(1+\tan^2 x)$ (ii) $-\log_e(1-\sin^2 u)$

Exercises 13c

1 $y=1+x-\frac{2}{3}x^3$

2 $y=-\dfrac{2}{x}+1$

3 $v=92$ m/sec, 372 m

4 $v=u+\frac{1}{3}kt^3$ (k constant), $s=ut+\frac{1}{12}kt^4$

5 $3333\frac{1}{3}$ cm

Exercises 13d

1 (i) $\frac{1}{3}$ (ii) $16\frac{1}{4}$ (iii) 0

2 (i) $\frac{1}{3}$ (ii) $16\frac{1}{4}$ (iii) 0

3 (i) $4\frac{2}{3}$ (ii) $\frac{3}{4}$ (iii) $18\frac{3}{5}$

4 (i) $\pi/12$ (ii) $\pi/2$ (iii) $\pi/3\sqrt{3}$

5 (i) $-\frac{5}{72}$ (ii) $-\frac{5}{72}$ (iii) $-\frac{5}{72}$

6 (i) $12\frac{2}{3}$ (ii) $-12\frac{2}{3}$

7 (i) $-\frac{1}{3}$ (ii) $\frac{1}{5}$

8 (i) $\frac{1}{3}(e^9-e^6)$ (ii) $-\frac{1}{5}(e-e^6)$ (iii) $\frac{1}{2}(1-e^{-2})$

9 (i) $\log_e 3$ (ii) $\log_e\frac{5}{3}$ (iii) $\log_e 5$

10 (i) 0 (ii) $\pi/2$ (iii) $\pi/2$

11 $\frac{243}{4}$ square units

12 $\frac{8}{15}$ square units
13 2 square units
14 $\pi/2$ square units
15 $\frac{125}{6}$ square units

Exercises 13

1 (i) $\frac{1}{8}(2x-1)^4$ (ii) $\frac{1}{34}(2x-1)^{17}$
2 (i) $\frac{8}{7}x^7 - \frac{12}{5}x^5 + 2x^3 - x$ (ii) $\frac{8}{13}x^{13} - \frac{4}{3}x^9 + \frac{6}{5}x^5 - x$
3 (i) $\frac{2}{7}x^{7/2} - \frac{4}{5}x^{5/2} + \frac{2}{3}x^{3/2}$ (ii) $-\frac{1}{3}x^3 + \frac{1}{2}(a+b)x^2 - abx$
4 (i) $-6/x + \log_e x - \frac{1}{3}x^3$ (ii) $2x^{1/2} + \frac{4}{3}x^{3/2} + \frac{4}{5}x^{5/2}$
(iii) $\frac{15}{2}x^{2/3} + \frac{3}{5}x^{5/3} + \frac{9}{8}x^{8/3}$

5 (i) $\dfrac{3\sqrt{3}-5}{2}$ (ii) $\dfrac{4}{\sqrt{3}}$

6 $y = x^2 - x^3 + 1$
7 $y = x + 2x^2$
8 $y = C - k^2/x$
9 $y = 2x^2 + 1/2x - 1$
10 $1/\sqrt{2}$
11 $y = \frac{1}{2}x^2 - \log_e x + \frac{3}{2}$
12 $\frac{1}{2}\pi + 1$
13 343/6
14 $2\frac{2}{3}$
16 $v = 4s$
17 $y = 9 + 6x - 3x^2$, 32
18 1 sec, $21\frac{1}{3}$ m
20 $\frac{2}{3}$, $\frac{4}{3}$
21 $\log_e 5$

22 $\dfrac{1}{3(2-x)} + \dfrac{1}{3(1+x)}$, $\frac{1}{3}\log_e 4$

23 Vertical asymptote $x = 6$; max at $x = 4$, min at $x = 8$; $\frac{15}{2} - 8\log_e 2$
24 Max at $x = \frac{3}{5}$, min at $x = 3$; 243/10

Chapter 14

Exercises 14a

Constants of integration in these exercises have been omitted.

1 $\frac{1}{2}x^2 + x + \log_e(x-1)$
2 $\frac{1}{3}x^3 + \frac{1}{2}x^2 + x + \log_e(x-1)$
3 $x + \log_e(x-1)$
4 $-\frac{1}{3}t - \frac{1}{9}\log_e(1-3t)$
5 $-\frac{1}{6}t^2 - \frac{1}{9}t - \frac{1}{27}\log_e(1-3t)$
6 $x - \log_e(1-x)$
7 $\frac{7}{8} - \log_e 2$
8 $\frac{1}{4}t - \frac{1}{16}\log_e(1+4t)$
9 $-\frac{1}{4}\log_e 5$
10 $\dfrac{1}{b}x - \dfrac{a}{b^2}\log(a+bx)$

Exercises 14b

1 $5\arctan(x+1)$
2 $\frac{7}{2}\arctan\frac{1}{2}(x-3)$
3 $\pi/4$
4 $5x + \arctan x$
5 $\frac{1}{18}\arctan\frac{1}{6}(3x-1)$
6 $\pi/2$
7 $2 - 2\arctan 2$
8 $\frac{1}{3}x^3 - 9x + 27\arctan\frac{1}{3}x$
9 $x + \dfrac{2}{\sqrt{3}}\arctan\left(\dfrac{2x+1}{\sqrt{3}}\right)$
10 $\frac{27}{16} + \pi/8$

Exercises 14c

1 $\frac{1}{2}\log_e(x^2+16)+\frac{7}{4}\arctan\frac{1}{4}x$
2 $\frac{3}{2}\log_e(x^2+36)-\frac{5}{6}\arctan\frac{1}{6}x$
3 $\log_e(x^2+2x+10)+\frac{1}{3}\arctan\frac{1}{3}(x+1)$
4 $\frac{3}{2}\log_e(x^2-6x+10)+14\arctan(x-3)$
5 $-\frac{3}{2}\log_e(x^2-8x+25)-\frac{11}{3}\arctan\frac{1}{3}(x-4)$
6 $\dfrac{1}{2}\log_e(x^2-x+1)+\dfrac{1}{\sqrt{3}}\arctan\left(\dfrac{2x-1}{\sqrt{3}}\right)$
7 $\frac{1}{2}x^2+6x+13\log_e(x^2-6x+10)+18\arctan(x-3)$
8 $\dfrac{5}{6}\log_e(3x^2-12x+13)+\dfrac{11}{\sqrt{3}}\arctan[(\sqrt{3})(x-2)]$
9 $x-\log_e(x^2+2x+5)-\frac{3}{2}\arctan\frac{1}{2}(x+1)$
10 $\frac{1}{2}x^2+x-\frac{7}{4}\log_e(2x^2+2x+5)-\frac{1}{2}\arctan\frac{1}{3}(2x+1)$

Exercises 14d

1 $-\log_e(x+2)+2\log_e(x+3)$
2 $\log_e(2x-1)-\log_e(3x-1)$
3 $5\log_e x-6\log_e(1-x)$
4 $2x+\log_e(x-2)+\log_e(2x+1)$
5 $\frac{1}{2}x^2+3x+2\log_e x-\frac{7}{2}\log_e(2x+1)$
6 $\log_e(2x+3)-\log_e(3x-1)-\dfrac{17}{3(3x-1)}$
7 $\log_e(x-1)+3\log_e(x+2)-2/(x-1)$
8 $\log_e(x-1)+\frac{1}{2}\log_e(2x+1)-1/(2x+1)$
9 $2x-\frac{7}{3}\log_e(3x+1)+5\log_e x+6/x$
10 $\frac{1}{2}\log_e(x^2+4)-\frac{1}{2}\log_e(x^2+8)$
11 $\dfrac{1}{8}\arctan\dfrac{x}{2}-\dfrac{1}{8\sqrt{2}}\arctan\dfrac{x}{2\sqrt{2}}$
12 $\frac{3}{2}\log_e(2x+3)+\frac{1}{4}\log_e(2x^2-1)$
13 $3\log_e(x+1)-\log_e(x^2-x+1)$
14 $\frac{1}{3}\arctan x-\frac{1}{6}\arctan\frac{1}{2}x$
15 $\log_e(x-1)-\frac{1}{2}\log_e(x^2+9)-\frac{1}{3}\arctan\frac{1}{3}x$
16 $\log_e(x+1)-\frac{1}{2}\log_e(x^2+4)+\frac{1}{2}\arctan\frac{1}{2}x$
17 $\dfrac{1}{(p-q)}\log_e\left(\dfrac{x-p}{x-q}\right)$
18 $\dfrac{x^3}{3}+9x-\dfrac{27}{2}\log_e\left(\dfrac{x+3}{x-3}\right)$
19 $\dfrac{a}{(a-b)(a-c)}\log_e(x-a)+\dfrac{b}{(b-a)(b-c)}\log_e(x-b)$
$+\dfrac{c}{(c-a)(c-b)}\log_e(x-c)$

20 $\dfrac{1}{2(a^2-b^2)}\log_e\left(\dfrac{x^2+b^2}{x^2+a^2}\right)$

Exercises 14e

1 $-\frac{1}{2}e^{-x^2}$

2 $-\frac{1}{3}(a^2-x^2)^{3/2}$

3 $-\sqrt{(9-x^2)}$

4 $\frac{1}{4}(\log_e x)^4$

5 $-\log_e(1-\sin x)$

6 $\frac{1}{2}\log_e(\cos^2 x+2\sin^2 x)$

7 $-\dfrac{1}{30(x^5+6)^6}$

8 $\frac{2}{5}(x+1)^{5/2}-\frac{2}{3}(x+1)^{3/2}$

9 $\frac{1}{2}(\log_e x)^2$

10 $\log_e(\log_e x)$

11 $\frac{1}{6}\tan^6 x$

12 $-e^{1/x}$

13 $\dfrac{1}{3}\log_e\left(\dfrac{1+x^3}{2+x^3}\right)$

14 $\frac{1}{45}(x^3-2)^{15}$

15 $\frac{1}{2}(\arcsin x)^2$

16 $\frac{1}{4}$

17 $\log_e 2$

18 $\dfrac{1}{n+1}(\log_e 2)^{n+1}$

19 $\dfrac{1}{6}\log_e\left(\dfrac{3e^2+2}{5}\right)$

20 $\frac{1}{2}\log_e(e+1)(e^2+1)$

21 $\pi/24$

22 $\dfrac{1}{2}\log_e\dfrac{1+e^2}{(1+e)^2}$

23 $5\pi^2/288$

24 0

25 $\cos 4-\cos 2$

Exercises 14f

1 $\frac{1}{3}(x^2-a)^{3/2}$

2 $\frac{1}{3}(x^2+4)^{3/2}$

3 $-\dfrac{1}{x}\sqrt{(16-x^2)}-\arcsin\frac{1}{4}x$

4 $-\sqrt{(1-x^2)}$

5 $-\arccos x-\sqrt{(1-x^2)}$

6 $\frac{2}{5}(x+1)^{5/2}-\frac{2}{3}(x+1)^{3/2}$

7 $\frac{2}{7}(x-1)^{7/2}+\frac{4}{5}(x-1)^{5/2}+\frac{2}{3}(x-1)^{3/2}$

8 $\frac{1}{3}$

9 $\frac{1}{3}(5\sqrt{5}-8)$

10 $\frac{1}{8}\pi-\frac{1}{4}$

Exercises 14g

1 $-\frac{1}{18}\cos 9x-\frac{1}{10}\cos 5x$

2 $-\frac{1}{22}\cos 11x+\frac{1}{10}\cos 5x$

3 $\frac{1}{22}\sin 11x+\frac{1}{2}\sin x$

4 $\frac{1}{4}\sin 2x-\frac{1}{24}\sin 12x$

5 $\frac{1}{5}\sin^5 x-\frac{1}{7}\sin^7 x$

6 $-\frac{1}{7}\cos^7 x+\frac{2}{9}\cos^9 x-\frac{1}{11}\cos^{11} x$

7 $\frac{2}{3}(\sin x)^{3/2}$

8 $1/\cos x+2\cos x-\frac{1}{3}\cos^3 x$

9 0

10 $+\frac{1}{162}$

Exercises 14h

1 $-(x+1)e^{-x}$

2 $-\frac{1}{2}$

3 $\frac{1}{2}\theta^2 \sin 2\theta + \frac{1}{2}\theta \cos 2\theta - \frac{1}{4}\sin 2\theta$

4 $e^x(x^3 - 3x^2 + 6x - 6)$

5 $-x^3 \cos x + 3x^2 \sin x + 6x \cos x - 6 \sin x$

6 $\frac{32}{5}\log_e 2 - \frac{31}{25}$

7 $\frac{1}{2}x^2 \arcsin x - \frac{1}{4}\arcsin x + \frac{1}{4}x\sqrt{(1-x^2)}$

8 $\frac{1}{6}x^6 \log_e 3x - \frac{1}{36}x^6$

9 $\pi\left(\frac{1}{\sqrt{3}} - \frac{1}{4}\right) - \frac{1}{2}\log_e 2$

10 $-\frac{\theta \cos m\theta}{m} + \frac{\sin m\theta}{m^2}$

11 $\frac{1}{4}\pi - \frac{1}{2}\log_e 2$

12 $x(\log_e x)^2 - 2x \log_e x + 2x$

13 $(x + \sin x)\tan \frac{1}{2}x + \cos x$ or $x \tan\frac{1}{2}x$

14 $-x \operatorname{cosec} x + \log_e(\operatorname{cosec} x - \cot x)$

15 $\frac{1}{2}\sec x \tan x + \frac{1}{2}\log_e(\sec x + \tan x)$

Exercises 14i

1 $\frac{1}{2}e^x(\sin x + \cos x)$

2 $-\frac{1}{13}e^{-2x}(3\cos 3x + 2\sin 3x)$

3 $\frac{1}{101}e^{5x}(2\sin\frac{1}{2}x + 20\cos\frac{1}{2}x)$

4 $\frac{1}{2}x\sqrt{(16+x^2)} + 8\log_e[x + \sqrt{(16+x^2)}]$

5 $-\frac{1}{2}\operatorname{cosec} x \cot x - \frac{1}{2}\log_e(\operatorname{cosec} x - \cot x)$

Exercises 14

2 (i) $\frac{1}{7}(2x+1)^{7/2}$ (ii) $-\frac{1}{3}\log_e(1-3x)$
(iii) $\tan x + \frac{1}{3}\tan^3 x$ (iv) $\frac{1}{3}\sec^3 x - \sec x$
(v) $\arcsin \frac{1}{2}(x+1)$ (vi) $\arctan(x+3)$
(vii) $\log_e(3 + \sin x)$ (viii) $3\log_e(1 + e^{2x})$
(ix) $\log_e(\arcsin x)$ (x) $-\frac{1}{14}\cos 7x + \frac{1}{6}\cos 3x$
(xi) $-2\sqrt{\cos x}$ (xii) $\frac{1}{4}\sin^4 x - \frac{1}{6}\sin^6 x + C_1$ or $-\frac{1}{4}\cos^4 x + \frac{1}{6}\cos^6 x + C_2$

3 (i) $3\log_e(x-2) - 2\log_e(x-1)$
(ii) $x + 2\log_e\left(\frac{x+1}{x}\right)$
(iii) $-\frac{1}{5}\frac{1}{(x+1)} + \frac{2}{25}\log_e(x+1) - \frac{1}{25}\log_e(x^2+4) - \frac{3}{50}\arctan\frac{1}{2}x$
(iv) $-\frac{1}{8}\log_e 9$
(v) $\log_e(x-1) - \frac{1}{2}\log_e x - \frac{1}{4}\log_e(x^2+4) - \frac{1}{2}\arctan\frac{1}{2}x$
(vi) $-\frac{1}{10}\log_e(x+1) + \frac{1}{20}\log_e(x^2+9) + \frac{3}{10}\arctan\frac{1}{3}x$
(vii) $3\log_e(2+x) + \log_e(1-x) - 2/(1-x)$

4 $y = 6 - 2/x$

6 (i) 1 (ii) $\frac{1}{4}\pi - \frac{1}{2}\log_e 2$ (iii) $\frac{1}{12}\pi - \frac{1}{6} + \frac{1}{6}\log_e 2$ (iv) $\frac{1}{32}\pi^2$

7 (i) $\frac{1}{12}\log_e\frac{3}{8} + \frac{1}{3\sqrt{2}}\arctan\frac{1}{\sqrt{2}}$ (ii) $\log_e 4$ (iii) $\frac{8}{15}$ (iv) $\pi^2 - 4$

8 (a) (i) $\frac{1}{4}$ (ii) $\frac{1}{9}(1-4e^{-3})$
(b) $\frac{1}{2}(1-\log_e 2)$
9 (a) $\frac{1}{4}\log_e 5$ (b) $\sqrt{2\pi}-4$
10 (i) (a) $48+\log_e 2$ (b) $\dfrac{3\sqrt{2}}{10}$
(ii) $\frac{1}{2}[x^2-\log_e(x^2+1)]$
11 0·287, 0·605, 1·07
12 (i) $\frac{2}{5}(6-x)^{5/2}-4(6-x)^{3/2}$
(ii) $x+\frac{1}{2}\log_e(x^2+4)-\frac{3}{2}\arctan\frac{1}{2}x$
(iii) $x-\sin x\cos x-\frac{1}{3}\sin^3 x$
13 0
14 π
15 $\dfrac{1}{a-b}\log_e\left(\dfrac{x-a}{b-x}\right)$, $\arcsin\sqrt{\left(\dfrac{x-a}{b-a}\right)}$
16 $\dfrac{1}{2\sqrt{2}}\arctan\dfrac{u}{\sqrt{2}}-\dfrac{1}{4\sqrt{2}}\log_e\left(\dfrac{v-\sqrt{2}}{v+\sqrt{2}}\right)$
17 $n=0$; $x\log_e x-x$
$n=1$; $\frac{1}{2}(\log_e x)^2$
$n\neq 1$ or 0; $\dfrac{1}{(1-n)}\dfrac{1}{x^{n-1}}\left[\log_e x-\dfrac{1}{(1-n)}\right]$
19 (i) $\dfrac{\pi^2}{32}$ (ii) $\dfrac{1}{n+1}[1+(-1)^n]$
21 $-\frac{1}{13}(3e^{\pi}+2)$
24 $\arcsin\{\sqrt{x}-\sqrt{[x(1-x)]}\}$
25 (i) $\frac{1}{2}(5-2x)^{3/2}-\frac{23}{2}(5-2x)^{1/2}$ (ii) $x+9\log_e(x-2)-6\log_e(x-1)$
26 (i) $\dfrac{1}{5}$ (ii) $\dfrac{8}{21}$ (iii) $\dfrac{2}{9\sqrt{3}}$
28 $\frac{1}{2}[\arcsin x+x\sqrt{(1-x^2)}]$
29 (a) $\dfrac{x^4}{4(1+x^2)^2}+C_1$ (b) $-\dfrac{1}{2(x^2+1)}+\dfrac{1}{4}\dfrac{1}{(x^2+1)}+C_2$
31 (a) $\frac{1}{4}\pi+\frac{1}{2}$ (b) $\arctan(e^x)$
32 (ii) $x\tan x+\log_e(\cos x)$
33 (a) $\frac{2}{3}$ (b) $2/e$ (c) $\frac{4}{15}$
34 (a) $a=\frac{12}{13}$, $b=\frac{5}{13}$; $\frac{12}{13}x+\frac{5}{13}\log_e(3\sin x+2\cos x)$ (b) $\frac{22}{3}$
35 $\alpha=\frac{3}{2}$
36 (i) $\dfrac{2-\sqrt{3}}{2}$ (ii) $\dfrac{e-2}{e}$
37 $A=2$, $B=1$, $C=1$
38 (i) $\frac{1}{2}e^{2x}-2e^x+x$ (ii) $\frac{2}{3}(1+\sin\theta)^{3/2}$
39 $1-\frac{1}{4}\pi$

Chapter 15

Exercises 15a

1 $\frac{1}{2}\log_e 2$
2 $8\frac{2}{3}+\log_e 3$
3 $\frac{1}{6}$
4 OCA = 12, OBA = 4
5 $13\frac{1}{2}$
6 $3\frac{26}{27}$
7 $\frac{1}{3}$
8 (i) $\frac{1}{4}$, $\frac{1}{4}$ (ii) $11\frac{1}{4}$, $\frac{7}{12}$
9 (i) $\pi/\sqrt{3}$ (ii) $\sqrt{6}\pi/6$ (iii) $2\pi/\sqrt{5}$
10 $\frac{1}{3}$

Exercises 15b

1 2666·66 N/cm^2
2 (i) $2/\pi$, $1/\sqrt{2}$ (ii) 0, $1/\sqrt{2}$ (iii) 0, 1 (iv) 0, $I/\sqrt{2}$
5 665

Exercises 15c

1 $55\frac{13}{15}\pi$
3 $\frac{1}{4}\pi^2$
4 $\pi/2$
5 $2\pi ah^2$
6 πa^2h
7 πa^2b, πab^2
9 $\pi/30$
10 $\frac{2}{9}\pi(8e^9+1)$

Exercises 15d

1 (a) $\bar{x}=\frac{3}{5}$, $\bar{y}=0$ (b) $\bar{x}=\pi/2$, $\bar{y}=\pi/8$ (c) $\bar{x}=\frac{3}{7}$, $\bar{y}=0$
2 (a) $\bar{x}=\frac{2}{3}$, $\bar{y}=0$ (b) $\bar{x}=\pi/2$, $\bar{y}=0$ (c) $\bar{x}=\frac{2}{5}$, $\bar{y}=0$
3 (a) One third of the way up a median
(b) With the straight edges as axes $\bar{x}=\bar{y}=4a/3\pi$
(c) One quarter of the way up the axis of symmetry
(d) On the axis of symmetry distance

$$\frac{h}{4}\left(\frac{3a^2+2ab+b^2}{a^2+ab+b^2}\right)$$

from the smaller end
5 $\bar{x}=\bar{y}=256a/315\pi$

Exercises 15e

3 $a^2\pi^3/384$
4 $\frac{3}{8}\pi a^2+a^2$
5 $\frac{1}{4}(e^{3\pi}-1)$

Exercises 15

1 $39\frac{1}{105}$
2 $\frac{8}{3}$, $(\frac{3}{8}\pi, 0)$
3 (0, 0), (2, 2), $\frac{5}{8}$, $2\frac{6}{7}\pi$
6 $\frac{8}{3}$
7 $1/\sqrt{2}$
8 $\frac{1}{3}\pi$
9 $\frac{3}{2}\pi+1$, $\frac{9}{4}\pi^2+2\pi$
10 $\log_e\frac{9}{4}$, $\frac{2}{3}\pi$
11 $\bar{x}=2\frac{1}{2}$, $\bar{y}=1\frac{11}{70}$
12 $\frac{2}{3}a^2$, $(\frac{9}{20}a, \frac{9}{10}a)$

13 (i)$\frac{1}{2}$ (ii) $\frac{1}{8}\pi^2$ (iii) $(\frac{1}{4}\pi - \frac{1}{2}, \frac{1}{8}\pi)$

14 $(\frac{9}{5}a, \frac{18}{5}a)$

15 $\frac{8}{15}\pi a^3$

16 $(1, 6)$, $(2, 3)$, area $= (4\frac{2}{3} - 6\log_e 2)$

18 $\bar{x} = 1{\cdot}35$, $\bar{y} = 2{\cdot}70$

19 $t = 0$, $\bar{x} = 1{\cdot}6$

20 $a = \frac{1}{2}$, $b = -1$

22 $c^2 = a^2 + b^2 + ab$
(i) $\frac{1}{2}b^2\theta$ (ii) $\frac{1}{2}(a^2 + ab)\theta$

23 (i) 12π (ii) 16

24 4π

25 $P = (a - b)/c^2$, $Q = 0$, $R = b$; $2\pi c(3a^2 + 4ab + 8b^2)/15$

26 $y = \dfrac{1}{x-3} - \dfrac{1}{1+x} + \dfrac{2}{(x-3)^2}$, $\dfrac{1}{2}\left(\dfrac{4}{3} + \log_e \dfrac{15}{7}\right)$, $-\dfrac{10}{9} + \dfrac{28}{27}x - \dfrac{26}{27}x^2$

27 $\left(\dfrac{1}{e-1}, \dfrac{e+1}{4}\right)$, $\left(\dfrac{e^2+1}{2(e^2-1)}, 0\right)$

28 (i) $\frac{4}{3}$ (ii) $\frac{32}{15}$

30 $\frac{4}{3}\pi a^2 b$

Chapter 16

Exercises 16a

1 (i) $y = -\cos x + C$ (ii) $x = \frac{1}{3}t^3 - 2t + C$
(iii) $x = Ce^{3t}$ (iv) $y = \frac{5}{2}t^2 + At + B$

2 (i) $y = x\dfrac{dy}{dx}$ (ii) $\dfrac{dy}{dx} = 2y$

(iii) $x^2\dfrac{d^2y}{dx^2} = 2y$ (iv) $\dfrac{d^2y}{dx^2} - \dfrac{dy}{dx} - 6y = 0$

4 $x = 10e^{-kt}$

5 $11{\cdot}2$ km/sec

6 Terminal velocity $= 980/k$

7 $y = Ae^{-x}$, $y = -\frac{1}{2}x^2 + B$

8 $i = \dfrac{E}{R}(1 - e^{-Rt/L})$, $i \to \dfrac{E}{R}$ as $t \to \infty$

9 $V = \dfrac{abV_1}{(b-a)}\left(\dfrac{1}{r} - \dfrac{1}{b}\right)$

10 $y = e^{-2}e^{2x}$

Exercises 16b

1 (i) $y(C - \arctan x) = 1$ (ii) $y = \dfrac{2}{1 + Ae^{x^2}}$

(iii) $\log_e\left(\dfrac{1+y}{1-y}\right) + \dfrac{2}{1+x} = C$ (iv) $-e^{-y} = e^x + C$

2 $s = \frac{1}{30}(\arctan \frac{1}{3}v - \arctan \frac{1}{3}u)$

3 $r = Ae^{-\theta/k}$

4 $y = \dfrac{5 - x}{1 + 5x}$

5 $y^2 = x^2 + 2x + A$

7 $y = \dfrac{3}{3 - 2x}$

9 $x^2 + y^2 = C$

10 $H = \dfrac{45e^{t/15} - 36}{5e^{t/15} + 4}$

Exercises 16c

1 $x = 5e^{-5t} + 7e^{5t}$
2 $x = 3 \sin 2t + 4 \cos 2t$
3 (i) 6 cm/sec (ii) 10 cm/sec (iii) 5 cm
4 (i) $v = 0$ (ii) $v = 12$ cm/sec (iii) $x = 0$
5 $x = 5e^{6t} + 2$

Exercises 16

1 (a) $\sin x \cos y = C$
(b) $\log_e y = \frac{1}{3}x^3 - x + C$
(c) $ye^x = Cx$

2 $\left(\dfrac{dy}{dx}\right)^3 + x\left(\dfrac{dy}{dx}\right) = y$

3 $y = \frac{1}{2}x^2 + A$, $y = Ce^x$

4 $(1 - y^2)(1 + x^2)^2 = C$

5 $\log_e[(9 + y^2)/9] = 2 \arctan x$

9 $y = Ce^{-kt}$, 120 days, 30 g

10 $xy = C$

12 $y = (2 + \sin^3 x)/(1 - \sin^3 x)$

13 $x\dfrac{dz}{dx} = \dfrac{9 - z^2}{1 + z}$

15 $x = \dfrac{60 + 40e^{2t/5}}{3 + e^{2t/5}}$, $v = 2$, $t = \frac{5}{2}\log_e 3$

17 $y = \dfrac{x}{2}\left(\dfrac{e^{x-1} - 1}{e^{x-1} + 1}\right)$

18 $\dfrac{(y - x)^2}{xy^2} = C$

20 $y = A \cos x + B \sin x - x^2$
24 $\log_e(1 - \sin y) = (1 - \sin x)$
25 (a) $\log_e(x + y + 1) = y - \frac{1}{2}x^2 + C$
(b) $(1 + y^2) = \frac{1}{5}(1 + x^2)$
26 $kv = g(2e^{-kt} - 1)$, $k^2x = g(2 - 2e^{-kt} - kt)$
27 (i) $v = 8$ (ii) $x = 4$ (iii) $v = 1$
29 $a = \frac{1}{5}\log_e 2$
30 $x = \dfrac{7V^2}{P}\left[C - v + \frac{1}{2}V \log_e\left(\dfrac{V + v}{V - v}\right)\right]$

31 $\frac{1}{2}\log_e(e^2 + \log_e 4)$

32 $2(1 + x^2)(1 - y^2) = 3$

33 (a) $x = 2 + \frac{1}{12}\log_e\left[\frac{5(3y-2)}{3y+2}\right]$ (b) $\frac{dp}{dt} = kp$

34 $y = (x + 2)e^{-x}$, $x + y - 2 = 0$

35 $y = \arccos[A(1 - e^x)]$

Chapter 17

Exercises 17a

3 A$(2, -\pi/4)$, B$(5, -\pi/9)$, C$(3, 0)$, D$(2, -\pi/4)$, E$(6, \pi/9)$, F$(4, -\pi/4)$, G$(6, 5\pi/6)$

4 $(0, 3)$, $(2, -2\surd 3)$, $(-5, 0)$, $(-\surd 3, -1)$, $(0, -3)$

5 $(2\surd 2, \pi/4)$, $(5, -126° 52')$, $(5, \pi/2)$, $(13, 157° 23')$, $(3, 0)$, $(3\surd 5, -26° 34')$

6 A and F, B and E, C and H, D and J

7 A$(2, \pi/4)$, B$(2, 7\pi/12)$, C$(2, 11\pi/12)$, D$(2, -3\pi/4)$, E$(2, -5\pi/12)$, F$(2, -\pi/12)$

8 P must lie on the positive portion of the x-axis

9 (a) On the whole line through the origin making an angle $\pi/4$ with Ox

(b) On the whole line through the origin making an angle $3\pi/4$ with Ox

10 B$(7, 8)$, D$(-5, -4)$

Exercises 17b

1 AB = BC = CD = DA = 5; AC = BD = $5\surd 2$

2 A$(3, -4)$, B$(-3, 4)$, AB = 10

3 AB = CD = $3\surd 13/5$, AD = BC = $\surd 2$, AC$(\surd 137/5) \neq$ BD$(\surd 197/5)$, hence ABCD is a parallelogram

4 AB = $2\surd 13$, BC = $\surd 13$, AC = $\surd 65$, hence AC^2 = AB^2 + BC^2, $\angle$ABC = 90° and AB = 2BC

5 $X = 12$ or -6

6 All distances equal to $2\surd(a^2 + b^2)$

7 AB = $3\surd 2$, AC = $\surd 2$, CB = $2\surd 2$

10 P$(5, 3)$

Exercises 17c

1 Internal $(-3, -4)$, external $(-9, -13)$

2 $(8, 4)$

3 P$(2, 6)$, Q$(17, 36)$

4 (a) Externally in ratio 1:2 (b) internal bisector (c) externally in ratio 2:1

5 CP:PD = 2:3, P$(5, 5)$

6 A$'(-5\frac{1}{2}, -1\frac{1}{2})$, G$(-2, 1)$

7 $P_2\left(\frac{2p + 4a}{6}, \frac{2q + 4b}{6}\right)$, $P_5\left(\frac{5p + a}{6}, \frac{5q + b}{6}\right)$

8 P(0, 12). The internal and external bisectors are at right angles and since P(0, 12) lies on Oy then the internal bisector is Ox

9 $A'\left(\frac{x_2+x_3}{2}, \frac{y_2+y_3}{2}\right)$, $G\left(\frac{x_1+x_2+x_3}{3}, \frac{y_1+y_2+y_3}{3}\right)$

10 3:1

Exercises 17d

1 13 square units
2 (i) 1 square unit (ii) $2\frac{1}{2}$ square units (iii) 0 (iv) $3\frac{1}{2}$ square units
4 $AB^2 + BC^2 = 52 + 13 = 65 = AC^2$; $\triangle$ABC is 13 square units
7 P$(1, 1\frac{1}{2})$. Area $\triangle$BPD is zero hence P lies on line BD. $\triangle PAB = \triangle PCD = 3\frac{1}{2}$ square units
8 Area ABCD = 32 square units
9 $\triangle ABD = \triangle ACD = 12$ square units. The quadrilateral is re-entrant

Exercises 17e

1 $(x-3)^2 + (y+4)^2 = 49$
2 $3x^2 + 3y^2 - 44x - 46y + 239 = 0$
3 $x^2 + y^2 - 9x - 6y + 26 = 0$
4 $r = 8 \sin \theta$
5 $(x-2)^2 + (y-8)^2 = 169$
6 $xy = c^2$, $x^2/a^2 + y^2/b^2 = 1$
7 $y^2 - 4x - 6y + 13 = 0$
8 $3y^2 - x^2 = 0$
9 $r \sin \theta = 6$
10 $4x^2 + 3y^2 - 48 = 0$

Exercises 17f

1 $(-3, 5)$ **2** (5, 13), (12, 6)
3 $(1, -2)$, $(-1, 0)$ **4** $(10, \pi/6)$, $(10, -\pi/6)$
5 Common point is $(-4, 3)$ **6** Two coincident points $(1. -1)$
7 The common points are co-incident (7, 7) or the distance between centres equals the difference between the radii
8 $(3\sqrt{3}, \pi/6)$. Note that the origin does not satisfy both equations simultaneously
9 (7, 9), $(-1, 5)$
10 (4, 1)

Exercises 17g

1 $3x - 2y + 9 = 0$
2 $5x + y - 22 = 0$
3 $2y - 5x - 10 = 0$
4 $x + 5y = 0$
5 $2x - 3y + 4 = 0$
6 $y^2 = 4ax$
7 $\frac{x^2}{4} + \frac{y^2}{16} = 1$
8 $x^2 + y^2 = 25$

9 $\frac{x^2}{4} - \frac{y^2}{16} = 1$

10 $x^2 + y^2 = 25$

Exercises 17h

1 $r = 3$
2 $r = 6 \sin \theta$
3 $r^2 \cos \theta (\cos \theta + \sin \theta) = 1$
4 $r^2 \sin 2\theta = 2c^2$
5 $r^2 \cos^2 2\theta = c^2$
6 $y = x (x \geqslant 0)$
7 $y = \sqrt{3}x$
8 $x^2 + y^2 = 9$
9 $y = 5$
10 $y^2 = 9 - 6x$
11 $x^2 - y^2 = a^2$
18 $(\sqrt{2}, \angle \pi/4)$
19 $(4, \angle \pi/3)$, $(4, \angle -2\pi/3)$
20 $(a, \angle 2\pi/3)$, $(a, \angle -2\pi/3)$

Exercises 17

1 $(0, -6)$, $(-1, 1)$, $(\sqrt{3}, 1)$, $(0, 8)$, $(-1, -1)$, $(-5, 0)$
2 $(5\sqrt{2}, -\pi/4)$, $(2, -5\pi/6)$, $(3\sqrt{2}, 3\pi/4)$, $(2, \pi/3)$, $(2, \pi/2)$, $(3, 0)$
3 A and F, C and H, G and J
4 AB $= \sqrt{17}$, BC $= \sqrt{40}$, CD $= 5$, DA $= 6$, area 16 square units
6 $(3, -2)$
7 A(5, 0), B(3, $-\pi/2$), C(5, $-\pi/2$), D(3, 0), E(5, 0), F(1, $3\pi/4$), G(3, $\pi/4$)
8 $(6, 8)$, $(-18, -40)$
9 $(3:1)$, $(3:-1)$
10 $4x^2 + 3y^2 - 16x + 24y + 52 = 0$
11 $a = -2$ or $7\frac{1}{2}$
12 $(8, 5)$
13 Sides 15, 20, 25, area 150 square units
14 (i) 3:2 (ii) 5:-2 (iii) 3:-5
15 Centroid $(\frac{2}{3}, \frac{17}{3})$, circumcentre (3, 4)
16 36 square units
17 $x^2 + y^2 = 27$
18 2 units
19 $b = -4$ and $a = -5$, AB $= \frac{1}{6}\sqrt{5}$
20 $r = 6 \sin \theta$
21 $a = 1$
24 $x^2 - 2x - 2y + 2 = 0$
25 (i) $(7, -4)$ (ii) $(7, 4)$ (iii) $(-7, -4)$
26 $y^2 = 64x$, $(x - 4)(6 - y) = 1$
28 Mid-point of AC $(-1, -1)$, D is point $(9, -10)$
29 $xy =$ constant
30 $r(\cos \theta - \sin \theta) = 5$, $x - y = 5$

Chapter 18

Exercises 18a

1 (i) $y = 0{\cdot}36x + 5$
(ii) $y = 1{\cdot}73x - 4$
(iii) $y = 1{\cdot}19x$
(iv) $y = 2{\cdot}75x$
2 (i) $y = 5x$
(ii) $3x + 2y = 0$
(iii) $y = \frac{3}{4}x - 3$
(iv) $y = -2$

3 (i) Slope $\frac{4}{3}$, intercept -2
(ii) Slope -1, intercept $\frac{3}{2}$
(iii) Slope $\frac{5}{3}$, intercept 0
(iv) Slope 0, intercept 6

4 Slope is $\dfrac{y_2 - y_1}{x_2 - x_1}$

Exercises 18b

1 (i) $7y - 2x - 11 = 0$ (ii) $y + 3x - 16 = 0$
(iii) $y + x - 3 = 0$ (iv) $2y + 3x - 11 = 0$
(v) $2x + 5y + 10 = 0$ (vi) $y + x - 6\sqrt{2} = 0$

2 $\dfrac{x}{-5} + \dfrac{y}{3} = 1,\ \dfrac{5}{\sqrt{34}}y - \dfrac{3}{\sqrt{34}}x = \dfrac{15}{\sqrt{34}}$

3 $m = -\dfrac{5}{2},\ \dfrac{2}{\sqrt{29}}y + \dfrac{5}{\sqrt{29}}x = \dfrac{31}{\sqrt{29}},\ x/\frac{31}{5} + y/\frac{31}{2} = 1$

4 $x \cos \alpha + y \sin \alpha = a$

5 $\theta = \alpha + 90^\circ\ (0 < \alpha \leqslant \pi/2)$
$\theta = 90^\circ - \alpha(-\pi \leqslant \alpha \leqslant 0)$

Exercises 18c

1 $(-1, -1)$

2 (a) $(2, 3)$ (b) $(2\frac{1}{2}, -3)$
(c) $\left(\dfrac{ab}{a+b}, \dfrac{ab}{a+b}\right)$ (d) $(2, 8)$

3 (i) and (iii) are coincident

4 $k = 5, k = 3$, lines (ii) and (iii) are parallel; $k = 6$, lines (i) and (ii) are parallel

5 $7x + 4y = 0$

9 $k = 5$, the point $(3, 4)$

10 $5x + 7y - 35 < 0$
$4x - 11y - 28 < 0$
$14x + 3y + 68 > 0$

Exercises 18d

1 (a) $8^\circ 8'$ (b) $75^\circ 32'$ (c) 90° (d) 90°

2 $2x - 3y + 8 = 0$

3 $x + y - 5 = 0$

4 75°

5 AB $y - 2x + 3 = 0, \sqrt{20}$
BC $2y + x - 9 = 0, \sqrt{20}$
CA $y + 3x - 22 = 0, \sqrt{40}$

Exercises 18e

1 (a) $6\frac{1}{5}$ (b) $-\frac{4}{5}$ (c) 3 (d) $\frac{10}{13}$

2 $4\frac{16}{25}$ opposite side to the origin; $-2\frac{1}{5}$ same side as origin

3 No, it is one of the excentres

4 $a = 1$
5 $(2, 4)$

Exercises 18f

1 (a) $y + x = 0$ (b) $x + 4y + 3 = 0$ (c) $2x + 3y + 1 = 0$
2 $2x - 6y + 17 = 0$, $6x + 2y + 11 = 0$
4 $2y + 3x = 0$, $y - 2x = 0$
5 (a) $14x - 21y + 95 = 0$ (b) $7x + 14y - 33 = 0$

Exercises 18

1 (a) PQ is $y + 4x - 4 = 0$
QR is $7y + 2x + 24 = 0$
PR is $3y - x + 1 = 0$
(b) $2y - 7x + 7 = 0$ (c) $y + 3x - 2 = 0$
(d) $(\frac{11}{13}, -\frac{7}{13})$ (e) 13 square units
2 $2y = 4x - 13$
4 (a) $4x + 3y - 39 = 0$ (b) $3x - 4y + 2 = 0$
5 $2y - x - 9 = 0$, $y + 2x + 2 = 0$
6 $\dfrac{x}{4} + \dfrac{y}{-5} = 1$, $\dfrac{5}{\sqrt{41}}x - \dfrac{4}{\sqrt{41}}y = \dfrac{20}{\sqrt{41}}$
8 $k = -4$, three distinct lines through the point $(-2, -1)$
$k = 2$, three coincident lines
$k = -2$, two of the lines are parallel
9 7
11 $x_1 - \dfrac{a}{a^2 + b^2}(ax_1 + by_1 + c)$, $y_1 - \dfrac{b}{a^2 + b^2}(ax_1 + by_1 + c)$
12 $2x + y = 2$
13 $\frac{1}{3}(BC^2 + CA^2 + AB^2)$
14 $m_1m_2m_3m_4 = -1$
15 $7y - 3x + 102 = 0$
17 Outside
18 $B\left(\dfrac{-2k}{3}, -\dfrac{k}{6}\right)$, $C\left(\dfrac{k}{2}, k\right)$, locus $x + 5y = 0$
19 $3x - y = 3$, $x + 3y = 11$
20 $\dfrac{x - h}{a} = \dfrac{y - k}{b}$
22 $15x + 8y = 85$
24 $k = 3$ or $k = -\frac{5}{3}$
25 $(2, 2)$
26 $3x + 2y = 13$
27 $x \cos 2\alpha + y \sin 2\alpha = 2p \cos \alpha$
28 $y = 2x - x^2$
29 (a) -3 and $\frac{1}{3}$ (b) $(5, 1)$, $(5, 3)$
30 $a_1b_2c_3 - a_1b_3c_2 + a_2b_3c_1 - a_2b_1c_3 + a_3b_1c_2 - a_3b_2c_1 = 0$

Chapter 19

Exercises 19a

1 (i) $x^2 + y^2 - 6x - 14y + 33 = 0$
(ii) $x^2 + y^2 + 6x + 14y + 22 = 0$
(iii) $x^2 + y^2 - 10x = 0$
(iv) $x^2 + y^2 + 6y - 7 = 0$

2 (i) Centre $(-1, -3)$, radius 2
(ii) Centre $(-\frac{3}{2}, -\frac{2}{3})$, radius $\sqrt{21}/6$
(iii) Centre $(\frac{5}{2}, 0)$, radius $\frac{5}{2}$
(iv) Centre $(0, \frac{7}{2})$, radius 2
(v) Centre $(a, -b)$, radius $\sqrt{(a^2 - b^2)}$
(vi) Centre $(-a, a)$, radius $\sqrt{2}a$

3 $x^2 + y^2 - 14x + 12y - 15 = 0$

4 $x^2 + y^2 - 3x - y - 2 = 0$

5 $x^2 + y^2 - 6x + 4y + 11 = 0$

8 $x^2 + y^2 - 26x - 26y + 169 = 0$

9 $y + 4x - 13 = 0$

10 $x^2 + y^2 - 5x - y + 4 = 0$

Exercises 19b

1 $x^2 + y^2 - 10y = 0$, $r = 5$

2 $x^2 + y^2 - 11x - 7y + 30 = 0$, $r = \frac{5}{2}\sqrt{2}$

3 $x^2 + y^2 - 5x - y + 4 = 0$, $r = \frac{1}{2}\sqrt{10}$

4 $2x^2 + 2y^2 + x - 11y - 1 = 0$, $r = \frac{1}{4}\sqrt{130}$

5 $x^2 + y^2 + 3x - 3y - 38 = 0$, $r = \frac{1}{2}\sqrt{170}$

Exercises 19c

1 $3x + 4y - 45 = 0$

2 $7x - 9y + 59 = 0$

3 $11x + 7y + 91 = 0$

4 $15y - 8x - 56 = 0$

5 $x + y - 14 = 0$ is the equation of the common tangent

Exercises 19d

1 (i) $t = 3\sqrt{5}$ (ii) $t = \sqrt{43}$

2 A is inside, B outside, C is on the circle

3 $k = 80$

4 $(1, 2)$, $(-1, 4)$

5 $15y - 8x = \pm 289$

6 $5y - 3x = 0$

8 (i) $(10, 2)$ (ii) 2 units

9 (i) 2, $(1, 3)$ (ii) $(6, 3)$, $(-2, -1)$; $x^2 + y^2 - 23x + 36y - 15 = 0$

10 $x^2 + y^2 - 10x - 6y + 9 = 0$, $15y + 8x = 0$

Exercises 19

1 $x^2 + y^2 - 10x - 10y + 25 = 0$; $(8, 9)$ or
$x^2 + y^2 - 60x - 60y + 900 = 0$; $(12, 6)$

3 $4x - 3y + 6 = 0$; $4x - 3y - 14 = 0$

4 Centre $(3, -1)$ and radius 1; centre $(-1, -5)$ and radius 5
5 $c = 12$ or -38
6 $(a\cos\theta, a\sin\theta)$
7 $y = mx + 18$, $-\frac{11}{21}$
8 $(2, 3)$, 1, PA $= 2$, $\tan\theta = 2$
9 $(-4, -5)$, $\sqrt{45}$; $(1, 0)$, $\sqrt{5}$; $\sqrt{50}$; $3\sqrt{2}$
11 $(5, 8)$, $(5, -2)$
12 $x^2 + y^2 - 4x - 5y + 4 = 0$, $9y + 40x = 0$
13 $x^2 + y^2 + 4x - 4 = 0$, centre $(-2, 0)$, radius $2\sqrt{2}$, $x^2 + y^2 - 28x - 4 = 0$
15 $x^2 + y^2 - 2x - 6y = 0$, $x^2 + y^2 - 12x - 16y + 80 = 0$, $x + y - 8 = 0$
16 $3x^2 + 3y^2 - 12x - 6y + 11 = 0$
17 $3x - 4y - 9 = 0$; $(-1, -3)$, $(\frac{23}{25} - \frac{39}{25})$; $x^2 + y^2 + \frac{1}{2}x + 4y + 2\frac{1}{2} = 0$
18 $2\frac{1}{2}$ units
19 2 units
21 (i) $x^2 + y^2 - 7x + 19y + 70 = 0$ (ii) $1:5$; $\frac{1}{5}$
23 $p = \frac{1}{5}$, $q = \frac{3}{5}$, $(2, 1)$, $\sqrt{5}$
25 $x^2 + y^2 = 1$
26 (i) $(-\frac{17}{5}, \frac{11}{5})$ (ii) $4x + 3y + 7 = 0$ (iii) $12\frac{1}{24}$ square units
27 $(2, 1)$, 1; $y = 0$ and $3y - 4x = 0$; $(2\frac{10}{13}, 2\frac{11}{13})$
28 $(-1, 0)$, $(5, 0)$, $\pm\frac{4}{3}$, 0, $-\frac{8}{9}$
29 $(g_1 - g_2)^2 + (f_1 - f_2)^2 = [\sqrt{(g_1^2 + f_1^2 - c_1)} \pm \sqrt{(g_2^2 + f_2^2 - c_2)}]^2$
30 $x^2 + y^2 - 4x - 6y - 12 = 0$, $3x + 4y - 43 = 0$, $(-1, -1)$

Chapter 20

Exercises 20a

1 (i) $(2, 0)$, $x = -2$
(ii) $(-6, 0)$, $x = 6$
(iii) $(0, -\frac{1}{4})$, $y = \frac{1}{4}$
(iv) $(0, 3)$, $y = -3$
(v) $(-\frac{2}{3}, 0)$, $x = \frac{2}{3}$
2 $(2, 0)$
3 Relative to axes horizontally and vertically through the lowest point of the rod; $x^2 = 360y$
4 (i) $y = x + 1$, $y = -x + 3$
(ii) $x = y - 3$, $y = -x - 9$
5 $(27, -18)$
6 (i) $y^2 = 10x + 2y + 4$
(ii) $x^2 - 2xy + y^2 + 8x + 8y - 16 = 0$
(iii) $16x^2 - 24xy + 9y^2 + 118x + 174y + 316 = 0$
8 $(24, -24)$

Exercises 20c

2 1 unit
3 Tangent $y - tx + 3t^2 = 0$, normal $ty + x = 6t + 3t^3$
4 $y^2 = 4ax + 9a^2$

Exercises 20d

1 $a = 5,\ b = \frac{1}{2},\ y = 5x$
3 $\lambda = 0$ or $\frac{3}{19}$, $x - y + 1 = 0,\ -25x + 10y - 4 = 0$
4 $y^2 = 2a(x - a)$. Vertex $(a, 0)$, focus $(3a/2, 0)$

Exercises 20e

1 $e = \frac{3}{5}$, $(\pm 6, 0)$, $x = \pm\frac{50}{3}$, area $= 80\pi$
2 $e = \frac{3}{5}$, $(0, \pm 6)$, $y = \pm\frac{50}{3}$, area $= 80\pi$
3 $e = \frac{4}{5}$, $(\pm 2, 0)$, $x = \pm\frac{25}{8}$, area $= 15\pi/4$
4 $e = 1/\sqrt{3}$, $(\pm\sqrt{2}, 0)$, $x = \pm 3\sqrt{2}$, area $= 2\sqrt{6}\pi$
5 $e = 1/\sqrt{2}$, $(0, \pm 1)$, $y = \pm 2$, area $= \sqrt{2}\pi$
6 $e = \frac{3}{5}$, foci $(+4, -2), (-2, -2)$, area 20π. Directrices $x = \frac{28}{3}$ and $x = -\frac{22}{3}$
7 $\frac{288}{13}$, $(\pm 5, \pm\frac{144}{13})$
8 $8x^2 + 3y^2 = 35$, $e = \sqrt{\frac{5}{8}}$
9 Semi-major axis 5, semi-minor axis 3; $\dfrac{x^2}{9} + \dfrac{y^2}{25} = 1$
10 $a = 3$, $b = \sqrt{5}$; $x = \frac{1}{2}$, $x = 9\frac{1}{2}$; $(5, 2)$; $\dfrac{(x-5)^2}{9} + \dfrac{(y-2)^2}{5} = 1$

Exercises 20f

1 $x + y = 5,\ x - y + 1 = 0$
2 $2x + 5y = 12,\ 5x - 2y = 1$
3 $by - ax = 2ab,\ ay + bx = a^2 - b^2$
4 $y = 2x \pm \sqrt{14}$
5 $y = 2x \pm 6$
6 $y = 3x \pm 7$
10 $\dfrac{x^2}{49} + \dfrac{y^2}{33} = 1$

Exercises 20g

1 $\dfrac{x}{a}\cos\theta + \dfrac{y}{b}\sin\theta = 1$

Exercises 20h

1 $2x + 3y = 12,\ 2y - 3x = \pm\sqrt{\frac{97}{2}}$
2 ab
3 $\dfrac{a^2}{4x^2} + \dfrac{b^2}{4y^2} = 1$
6 $e = \frac{7}{17}$
8 $x_2 \approx ae^2 - \frac{1}{2}ae^2\theta^2$
10 Centre $\left[0, \dfrac{1}{2}\left(\dfrac{b}{\sin\theta} - \dfrac{a^2 - b^2}{b}\sin\theta\right)\right]$
radius $\dfrac{1}{2}\left(\dfrac{b}{\sin\theta} + \dfrac{a^2 - b^2}{b}\sin\theta\right)$

Exercises 20i

1 $e = \frac{3}{2}\sqrt{3}$, $(\pm 3\sqrt{3}, 0)$, $x = \pm\frac{4}{9}\sqrt{3}$
2 $e = \frac{4}{3}$, $(0, \pm 4)$, $y = \pm\frac{9}{4}$
3 $e = \frac{7}{4}$, $(\pm\frac{7}{2}, 0)$, $x = \pm\frac{8}{7}$
4 $e = \frac{9}{5}$, $(\pm 9, 0)$, $x = \pm\frac{25}{9}$
5 $e = \sqrt{\frac{3}{2}}$, $(0, \pm\sqrt{3})$, $y = \pm\frac{2}{3}\sqrt{3}$
6 $e = \sqrt{2}$, $(\pm 5\sqrt{2}, 0)$, $x = \pm\frac{5}{2}\sqrt{2}$
7 $\frac{14}{3}$, $(\pm 4, \pm\frac{7}{3})$
8 $\dfrac{x^2}{36} - \dfrac{y^2}{13} = 1$, $\dfrac{y^2}{36} - \dfrac{x^2}{13} = 1$
9 $\dfrac{x^2}{25} - \dfrac{y^2}{11} = 1$

Exercises 20j

1 $\sqrt{3}x - y = \sqrt{3}$, $x + \sqrt{3}y = 13$
4 $a^2l^2 - b^2m^2 = n^2$, $a^2l^2 + b^2m^2 = n^2$
5 $bx(t^2 + 1) - ay(t^2 - 1) = 2abt$
10 $8x - 9y = \pm 5$, $(\pm 4, \pm 3)$

Exercises 20k

1 (i) $y = \pm\frac{2}{3}x$; $(0, \pm 2)$
(ii) $y = \pm\frac{2}{3}x$; $(\pm 2, 0)$
2 $-b/a < m < b/a$
3 (i) $xy = 16$ (ii) $xy = -1$
(iii) $(x - 1)y = 9$ (iv) $(x + 1)(y - 1) = -4$
4 $4y + x = 8$, $y - 4x + 15 = 0$
5 $x + t^2y = 2t$
6 $2x - y = 3$, $2y + x = 4$ and $2x - y = -3$, $2y + x = -4$
8 $y = x$
9 $4x - 3y = \pm 7$, $(4, 3)$ and $(-4, -3)$. 7 square units
10 $\left(\dfrac{n}{2l}, \dfrac{2c^2l}{n}\right)$

Exercises 20l

1 $2xy = k^2$ (rectangular hyperbola)
2 $y = 2c^2/a$
4 $m^2(x_0^2 - 1) - 2x_0y_0m + y_0^2 + \frac{1}{3} = 0$
5 When P is a point of intersection of the two curves, the normal to $xy = c^2$ is a tangent to $x^2 - y^2 = a^2$. Hence the curves cut at right angles.

Exercises 20

1 $y^2 = 16a(x + 2a)$
3 $y^2 = a(x - a)$
6 (i) $x = 0$ (ii) $2y^2 = 9ax$
7 $\left(\dfrac{c(1 + t^2)}{t}, \dfrac{c(1 + t^2)}{t}\right)$

9 (i) $4a$ (ii) $y^2 = 2a(x - 4a)$
12 T$(-3, 2)$, M$(3, 4)$, N$(-1, 0)$, 1
13 $y^2 = a(x - 3a)$
16 $[2a + a(p^2 + pq + q^2), -apq(p + q)]$, $y^2 = a(x - 3a)$
17 $y^2 = 4a(x - a)$
18 $ky - 2ax = 2ah$
19 arctan 3
20 $\dfrac{2\sqrt{(a^2m^2 + b^2)}}{\sqrt{(1 + m^2)}}, \dfrac{a}{3}\sqrt{26}$
21 $y + tx = 2at + at^3$; $(2a + at^2, 0)$; $\pm\sqrt{3}$, $\pm\sqrt{15}$
23 $(2a, 2\sqrt{2a})$
24 $0°, 60°, 360°$; $(2, 0)$, $(1, \frac{3}{2})$
25 $y = mx + 4 - 3m$; $\sqrt{8}$, $\sqrt{40}$; 4

Chapter 21

Exercises 21a

1 1·25
2 5·96
4 4·73
5 0·59
6 1·26
7 −1·75
8 0·66
9 0·6, 0·643, 0·653, 0·656, 0·657 to 3 decimal places
10 0·5, 0·878, 0·639, 0·8027, 0·6947, 0·768, 0·719, 0·7523, 0·7300, 0·745, 0·735, 0·742, 0·737, 0·740, 0·738, 0·739

Exercises 21b

1 1·2542
2 0·7809
3 0·5885
4 0·7391
5 0·5804

Exercises 21c

1 0·994
2 5·683
3 1·195
4 0·348
5 1·283

Exercises 21d

1 1·000, 1
2 1·0986, 1·0986
3 1·106
4 15·59
5 20 200 m^2

Exercises 21e

2 $1/p$
3 $y = 5 + 0{\cdot}3x^2$
4 $\gamma = 1{\cdot}4$, $c = 80$
5 $d = 0{\cdot}0375v^2$

Exercises 21

1 3·264
2 $x^3 - 12x^2 + 32 = 0$, 1·769
3 1·196
4 7·40, 2·68
5 0·75
6 $2r + \frac{2}{3}p$
7 1·33
8 $a = 4{\cdot}0$, $n = 3$; (a) 4·83 (b) 4·82
9 $2a^3h + 2ah^3 + 2h$
10 1·163, too large; 1·147

11 3·1415, π.
Agreement to 7 decimal places
12 3·67
13 $a = -5$, $b = 0{\cdot}3$
14 42·02; $a = 1{\cdot}15$, $b = 8{\cdot}2$
15 0·187
16 0·95
17 1·52
18 $k = 4$, $n = 1{\cdot}5$
19 0·478
20 2·34
21 2·2789
22 4·454
23 0·2618, $\frac{1}{12}\pi$
24 1·718 28, 2·1 (2·08)
25 $a = 20$, $n = 2{\cdot}5$

Chapter 22

Exercises 22a

1 Scalars: (a), (b), (c), (g), (h). Vectors: (d), (e), (f), (j).
2 (a) and (d) are equal and opposite; (a) and (e) represent two or more vectors.
3 (a) 105° (b) 30° (c) 105°
(d) $\boldsymbol{F}_1$ and $\boldsymbol{v}_2$; $\boldsymbol{v}_1$ and $\boldsymbol{v}_2$
$\boldsymbol{F}_2$ and $\boldsymbol{d}$; $\boldsymbol{F}_3$ and $\boldsymbol{v}_2$
(e) $\boldsymbol{F}_1$ and $\boldsymbol{F}_3$; $\boldsymbol{F}_3$ and $\boldsymbol{v}_1$
5 12·5 km at an angle arctan $(\frac{3}{4})$ E of N

Exercises 22b

1 3 N, 2 N
2 $|\boldsymbol{Q}| = 12$
3 6 N, 12 N
4 $\arccos(-\frac{1}{7}) = 98°\,13'$
5 14·2 units, N 169° 11′ W

Exercises 22c

2 (a) 2 (b) 5 (c) 4 (d) 4·64 (e) 2·95 (f) 5
3 $\boldsymbol{a}$ and $\boldsymbol{b}$ are parallel and in the same sense
4 120°
5 (a) $\boldsymbol{c} - \boldsymbol{b}$ (b) $\boldsymbol{a} - \boldsymbol{b}$ (c) $\boldsymbol{d} - \boldsymbol{b}$ (d) $\boldsymbol{c} - \boldsymbol{a}$
6 (a) $\boldsymbol{b} - \boldsymbol{a}$ (b) $\boldsymbol{c} - \boldsymbol{b}$ (c) $\boldsymbol{a} + \boldsymbol{c} - \boldsymbol{b}$ (d) $\boldsymbol{a} - \boldsymbol{c}$

Exercises 22d

1 Yes
3 $\overrightarrow{PA} = \frac{1}{2}(\boldsymbol{a} + \boldsymbol{b})$
4 (ii) $\frac{1}{2}(\boldsymbol{a} + \boldsymbol{b} - \boldsymbol{c})$ (iii) $\frac{1}{2}(\boldsymbol{a} - \boldsymbol{b} + \boldsymbol{c})$
5 (i) $11\boldsymbol{a} + \boldsymbol{c}$ (ii) $-\boldsymbol{a} - 9\boldsymbol{b}$ (iii) $3\boldsymbol{a} + 9\boldsymbol{b} + 3\frac{1}{2}\boldsymbol{c}$
6 Yes

Exercises 22e

1 (i) $\begin{bmatrix} -4 \\ 4 \\ -10 \end{bmatrix}$ (ii) $\sqrt{20}$ (iii) $\begin{bmatrix} 1 \\ 0 \\ -1 \end{bmatrix}$ (iv) $\sqrt{58}$

2 (i) $\begin{bmatrix} -12 \\ 0 \\ 3 \end{bmatrix}$ (ii) $5\sqrt{2}$ (iii) $\begin{bmatrix} 0 \\ 2 \\ 2 \end{bmatrix}$ (iv) $\dfrac{1}{\sqrt{2}}\begin{bmatrix} 0 \\ 1 \\ 1 \end{bmatrix}$

3 $d = 3a - 2b + 4c$
4 $d = \frac{3}{2}a - 4b - \frac{1}{2}c$
5 $c = 3a - 2b$

Exercises 22f

1 (i) $-\frac{9}{5}$ (ii) $-(9/\sqrt{14})$ **2** $\arccos(-\frac{1}{3})$
3 $p = 7, q = 8$; $\hat{a} = (2i - 2j + k)/3$; $\hat{b} = (i + 3j + 4k)/\sqrt{26}$;
$c = (11i + 7j - 8k)/3\sqrt{26}$
5 Either $b = c$ or a is perpendicular to $b - c$
6 5 N **8** $p = -27, q = -16$
9 (i) -1 (ii) -32 (iii) 0
10 (i) $13i + 13j$ (ii) 8 (iii) 9 (iv) $10i - 2j + 2k$

Exercises 22g

2 $r = [(q + 6)i + (3 - q)j + (q + 9)k]/(q + 3)$; $q = -4$

4 $\vec{AB} = \begin{pmatrix} 2 \\ 1 \\ -3 \end{pmatrix}, \quad \vec{AC} = \begin{pmatrix} 2 \\ 2 \\ 2 \end{pmatrix}$

5 OA = 3, OB = 11; area = $4\sqrt{17}$

Exercises 22h

1 $3i + 5j - 7k$
2 $p = -2$; $4(i - j + k)$
3 30°

4 $r = \begin{pmatrix} 3 \\ 2 \\ 1 \end{pmatrix} + t\begin{pmatrix} 2 \\ 2 \\ -1 \end{pmatrix}$ or $r = \begin{pmatrix} 5 \\ 4 \\ 0 \end{pmatrix} + s\begin{pmatrix} 2 \\ 2 \\ -1 \end{pmatrix}$

$(3 + 2t, 2 + 2t, 1 - t)$ or $(5 + 2s, 4 + 2s, -s)$; $t = -1$ or $s = -2$;
(1, 0, 2)
5 (2, −1, 3), arccos (16/21)

Exercises 22i

1 $\arccos(\frac{-22}{27})$
2 Because their normals are in the same direction; 3 units

3 $r \cdot \begin{pmatrix} 2 \\ -6 \\ 9 \end{pmatrix} = 121$

4 $r \cdot \begin{pmatrix} 1 \\ 2 \\ -2 \end{pmatrix} = -5$

5 5, $2\sqrt{2}$

6 $r = \begin{bmatrix} 4 \\ 2 \\ 3 \end{bmatrix} + t\begin{bmatrix} 3 \\ 2 \\ 0 \end{bmatrix}$ or $r = \begin{bmatrix} 7 \\ 4 \\ 2 \end{bmatrix} + s\begin{bmatrix} 3 \\ 2 \\ 0 \end{bmatrix}$

7 $\boldsymbol{a} = \begin{bmatrix} -1 \\ 8 \\ 3 \end{bmatrix}$ **9** $\boldsymbol{b}.\boldsymbol{n} = 0$

Exercises 22

2 60°
3 $-\boldsymbol{a}, \boldsymbol{b}, -(\boldsymbol{a}+\boldsymbol{b})$
4 (i) $-9\boldsymbol{a}$ (ii) $5\boldsymbol{a}+2\boldsymbol{c}$
5 The altitudes of a triangle are concurrent

6 $\boldsymbol{p} = \begin{bmatrix} 2 \\ -3 \\ -1 \end{bmatrix}$

7 All three are $-\frac{5}{3}$. The four points A, B, C, D are coplanar and OD is normal to the plane.

9 $\boldsymbol{a} = \begin{bmatrix} -3 \\ -5 \\ 2 \end{bmatrix}$ **10** $= \begin{bmatrix} 8 \\ 11 \\ -21 \end{bmatrix}$

12 $\boldsymbol{r} = \begin{bmatrix} 0 \\ -1 \\ 4 \end{bmatrix} + t\begin{bmatrix} 8 \\ 8 \\ -4 \end{bmatrix}$; N$(\frac{4}{3}, \frac{1}{3}, \frac{10}{3})$; $(\frac{8}{3}, \frac{2}{3}, \frac{20}{3})$

13 $(\boldsymbol{r}-\boldsymbol{a}).\boldsymbol{n} = 0$
14 $(\boldsymbol{a}.\hat{\boldsymbol{u}})\hat{\boldsymbol{u}}$

15 $u = -2, v = 1$; L is $\boldsymbol{r} = \begin{bmatrix} 1 \\ -1 \\ 1 \end{bmatrix} + t\begin{bmatrix} 1 \\ -2 \\ 1 \end{bmatrix}$

16 $a = 1, b = -3$; $\boldsymbol{a} = \begin{bmatrix} 8 \\ 6 \\ 1 \end{bmatrix}$; $\sqrt{11}$

Index